ESSDERC 2022 - IEEE 52nd European Solid-State Device Research Conference (ESSDERC 2022)

Milan, Italy
19-22 September 2022

IEEE Catalog Number: CFP22543-POD
ISBN: 978-1-6654-8498-5

**Copyright © 2022 by the Institute of Electrical and Electronics Engineers, Inc.
All Rights Reserved**

Copyright and Reprint Permissions: Abstracting is permitted with credit to the source. Libraries are permitted to photocopy beyond the limit of U.S. copyright law for private use of patrons those articles in this volume that carry a code at the bottom of the first page, provided the per-copy fee indicated in the code is paid through Copyright Clearance Center, 222 Rosewood Drive, Danvers, MA 01923.

For other copying, reprint or republication permission, write to IEEE Copyrights Manager, IEEE Service Center, 445 Hoes Lane, Piscataway, NJ 08854. All rights reserved.

*** This is a print representation of what appears in the IEEE Digital Library. Some format issues inherent in the e-media version may also appear in this print version.

IEEE Catalog Number: CFP22543-POD
ISBN (Print-On-Demand):   978-1-6654-8498-5
ISBN (Online):   978-1-6654-8497-8

Additional Copies of This Publication Are Available From:

Curran Associates, Inc
57 Morehouse Lane
Red Hook, NY 12571 USA
Phone: (845) 758-0400
Fax: (845) 758-2633
E-mail: curran@proceedings.com
Web: www.proceedings.com

ESSDERC 2022 TABLE OF CONTENTS

Joint Keynote 1

Date: Tuesday, September 20, 2022
Time: 9:20 - 10:00
Room: Aula Magna
Chair(s):   Andrea Baschirotto, University of Milano-Bicocca

The Future of Short Reach Interconnect .... N/A

Davide Tonietto

Huawei Technologies Co., Ltd., Canada

Joint Keynote 2

Date: Tuesday, September 20, 2022
Time: 10:00 - 10:40
Room: Aula Magna
Chair(s):   Davide Giacomini, Infineon Technologies

**Semiconductors Take the Driver's Seat - Challenges and Opportunities for the Car
of the Future** .... N/A

Tim Gutheit

Infineon Technologies AG, Germany

Joint Keynote 3

Date: Wednesday, September 21, 2022
Time: 8:30 - 9:10
Room: Aula Magna
Chair(s):   Renato Lombardi, Huawei

Integrated Circuits as Key Enabler for Today's Smart MEMS Sensors .... N/A

Dirk Droste[1], Horst Symanzik[1], Timo Gießelmann[1], Markus Ulm[1], Ivano Galdi[2], Riccardo Campagna[2]

[1]Bosch Sensortec GmbH, Germany; [2]Robert Bosch S.p.A., Italy

Joint Keynote 4

Date: Wednesday, September 21, 2022
Time: 11:10 - 11:50
Room: Aula Magna
Chair(s):   Davide Pandini, STMicroelectronics

**The Next "Automation Age": How Semiconductor Technologies Are Changing Industrial
Systems and Applications** .... N/A

Domenico Arrigo, Claudio Adragna, Vincenzo Marano, Rachela Pozzi, Fulvio Pulicelli, Francesco Pulvirenti

STMicroelectronics, Italy

ESSDERC Keynote 1

Date: Tuesday, September 20, 2022
Time: 16:40 - 17:20
Room: Aula Magna
Chair(s):

Quantum Computing Technology and Roadmap 25

Heike Riel

IBM Research Zürich, Switzerland

ESSDERC Keynote 2

Date: Wednesday, September 21, 2022
Time: 14:20 - 15:00
Room: Aula Magna
Chair(s): Piero Malcovati, University of Pavia

SiC Power Device Mass Commercialization 31

Victor Veliadis

North Carolina State University, United States

ESSDERC Keynote 3

Date: Thursday, September 22, 2022
Time: 8:30 - 9:10
Room: Plenary Auditorium
Chair(s): Francesco Maria Puglisi, University of Modena and Reggio Emilia

On the Scaling Potential of Transistors with Low Dimensional Materials

Iuliana Radu, Sheng-Kai Su, Terry Hung, Greg Pitner, Albert Cheng, Edward Chen, Han Wang, Jin Cai, Philip Wong

Taiwan Semiconductor Manufacturing Company, United States; Taiwan Semiconductor Manufacturing Company, Taiwan

Devices & Circuits for Quantum Computing

Date: Tuesday, September 20, 2022
Time: 11:10 - 12:50
Room: Aula 1B
Chair(s): Domenico Zito, Aarhus University
Veeresh Deshpande, Helmholtz Zentrum Berlin

Coupling Control in the Few-Electron Regime of Quantum Dot Arrays Using 2-Metal Gate Levels in CMOS Technology .... N/A

Bruna Cardoso Paz[1], Victor El-Homsy[1], David Niegemann[1], Bernhard Klemt[1], Emmanuel Chanrion[1], Vivien Thiney[2], Baptiste Jadot[2], Pierre-André Mortemousque[2], Benoit Bertrand[2], Thomas Bédécarrats[2], Heimanu Niebojewski[2], François Perruchot[2], Silvano De Franceschi[3], Maud Vinet[2], Matias Urdampilleta[1], Tristan Meunier[1]

[1]Institut Néel, France; [2]CEA-Leti, France; [3]CEA-IRIG, France

Multi-Gate FD-SOI Single Electron Transistor for Hybrid SET-MOSFET Quantum Computing .... N/A

Fabio Bersano[1], Franco De Palma[1], Fabian Oppliger[1], Floris Braakman[2], Ionut Radu[3], Pasquale Scarlino[1], Martino Poggio[2], Adrian Mihai Ionescu[1]

[1]École Polytechnique Fédérale de Lausanne, Switzerland; [2]University of Basel, Switzerland; [3]Soitec, France

Cryogenic Comparator Characterization and Modeling for a Cryo-CMOS 7b 1-GSa/s SAR ADC .... N/A

Gerd Kiene, Aishwarya Gunaputi Sreenivasulu, Ramon W.J. Overwater, Masoud Babaie, Fabio Sebastiano

Delft University of Technology, The Netherlands

A Cryogenic SRAM Based Arbitrary Waveform Generator in 14 nm for Spin Qubit Control .... N/A

Mridula Prathapan[1], Peter Mueller[1], Christian Menolfi[1,2], Matthias Brändli[1], Marcel Kossel[1], Pier Andrea Francese[1], David Heim[1], Maria Vittoria Oropallo[1], Andrea Ruffino[1], Cezar Zota[1], Thomas Morf[1]

[1]IBM Research Zürich, Switzerland; [2]Cisco Systems Inc.

A 28nm 6.5-8.1GHz 1.16mW/qubit Cryo-CMOS System-on-Chip for Superconducting Qubit Readout .... N/A

Steven Van Winckel[1], Alican Caglar[1,2], Benjamin Gys[1], Steven Brebels[1], Anton Potocnik[1], Bertrand Parvais[1], Piet Wambacq[1], Jan Craninckx[1]

[1]imec, Belgium; [2]Vrije Universiteit Brussel, Belgium

Devices & Circuits for Emerging Technologies

Date: Thursday, September 22, 2022
Time: 11:20 - 12:40
Room: Aula 1F
Chair(s):   Claudio Bruschini, EPFL
            Lotte Geck, Forschungszentrum Jülich

Non-Volatile Ternary Content Addressable Memory Based on Phase Change Nanoelectromechanical (NEM) Relay .... N/A

Mohammad Ayaz Masud, Luis Hurtado, Gianluca Piazza

Carnegie Mellon University, United States

A 24V Thin-Film Ultrasonic Driver for Haptic Feedback in Metal-Oxide Thin-Film Technology Using Hybrid DLL Locking Architecture .... N/A

Jonas Pelgrims[1], Kris Myny[1,2], Wim Dehaene[2]

[1]Katholieke Universiteit Leuven, Belgium; [2]imec, Belgium

A Monolithic SPAD-Based Random Number Generator for Cryptographic Application .... N/A

Nicola Massari[1], Alessandro Tontini[1], Luca Parmesan[1], Matteo Perenzoni[1,2], Miloš Grujić[3], Ingrid Verbauwhede[3], Thomas Strohm[4], Dayo Oshinubi[4], Ingo Herrmann[4], Andreas Brenneis[4]

[1]Fondazione Bruno Kessler, Italy; [2]Sony Semiconductor Solutions Corporation, Italy; [3]ITEC, Belgium; [4]Robert Bosch GmbH, Germany

Quantum-Correlated Photon-Pair Source with Integrated Feedback Control in 45 nm CMOS .... N/A

Danielius Kramnik[1], Imbert Wang[2], J.M. Fargas Cabanillas[2], Anirudh Ramesh[3], Sidney Buchbinder[1], Panagiotis Zarkos[1], Christos Adamopoulos[1], Prem Kumar[3], M.A. Popović[2], Vladimir Stojanović[1]

[1]University of California-Berkeley, United States; [2]Boston University, United States; [3]Northwestern University, United States

Digital ML Processors

Date: Tuesday, September 20, 2022
Time: 11:10 - 12:50
Room: Aula 1E
Chair(s): Marian Verhelst, KU Leuven
Andreas Burg, EPFL

A 4.6-8.3 TOPS/W 1.2-4.9 Tops CNN-Based Computational Imaging Processor with Overlapped Stripe Inference Achieving 4K Ultra-HD 30fps .... N/A

Yu-Chun Ding, Kai-Pin Lin, Chi-Wen Weng, Li-Wei Wang, Huan-Ching Wang, Chun-Yeh Lin, Yong-Tai Chen, Chao-Tsung Huang

National Tsing Hua University, Taiwan

A 40nm 5.6TOPS/W 239GOPS/mm² Self-Attention Processor with Sign Random Projection-Based Approximation .... N/A

Seong Hoon Seo, Soosung Kim, Sung Jun Jung, Sangwoo Kwon, Hyunseung Lee, Jae W. Lee

Seoul National University, Korea

A 28nm 8-Bit Floating-Point Tensor Core Based CNN Training Processor with Dynamic Activation/Weight Sparsification .... N/A

Shreyas Kolala Venkataramanaiah[1], Jian Meng[1], Han-Sok Suh[1], Injune Yeo[1], Jyotishman Saikia[1], Sai Kiran Cherupally[1], Yichi Zhang[2], Zhiru Zhang[2], Jae-Sun Seo[1]

[1]Arizona State University, United States; [2]Cornell University, United States

A Differentiable Neural Computer for Logic Reasoning with Scalable Near-Memory Computing and Sparsity Based Enhancement .... N/A

Yuhao Ju[1], Shiyu Guo[1], Zixuan Liu[1], Tianyu Jia[2], Jie Gu[1]

[1]Northwestern University, United States; [2]Peking University, China

SIF-NPU: A 28nm 3.48 TOPS/W 0.25 TOPS/mm² CNN Accelerator with Spatially Independent Fusion for Real-Time UHD Super-Resolution .... N/A

Sumin Lee[1], Ki-Beom Lee[1], Sunghwan Joo[2], Hong Keun Ahn[1], Junghyup Lee[1], Dohyung Kim[1], Bumsub Ham[1], Seong-Ook Jung[1]

[1]Yonsei University, Korea; [2]Samsung Electronics Co., Ltd., Korea

Resistive Memory Based In-memory Compute

Date: Tuesday, September 20, 2022
Time: 11:10 - 12:50
Room: Aula 1F
Chair(s): Manuel Le Gallo, IBM
          Cristiano Calligaro, RedCat Devices

A 40nm RRAM Compute-in-Memory Macro with Parallelism-Preserving ECC for Iso-Accuracy Voltage Scaling .... N/A

Wantong Li, James Read, Hongwu Jiang, Shimeng Yu

Georgia Institute of Technology, United States

In-Memory Realization of In-Situ Few-Shot Continual Learning with a Dynamically Evolving Explicit Memory .... N/A

Geethan Karunaratne[1,4], Michael Hersche[1,4], Jovin Langenegger[1,4], Giovanni Cherubini[1], Manuel Le Gallo[1], Urs Egger[1], Kevin Brew[2], Sam Choi[2], Injo Ok[2], Claire Silvestre[2], Ning Li[2], Nicole Saulnier[2], Victor Chan[2], Ishtiaq Ahsan[2], Vijay Narayanan[3], Luca Benini[4], Abu Sebastian[1], Abbas Rahimi[1]

[1]IBM Research Zürich, Switzerland; [2]IBM Research Albany, United States; [3]IBM T. J. Watson Research Center, United States; [4]ETH Zürich, Switzerland

An Embedded PCM Peripheral Unit Adding Analog MAC In-Memory Computing Feature Addressing Non-Linearity and Time Drift Compensation .... N/A

Alessio Antolini[1], Andrea Lico[1], Eleonora Franchi Scarselli[1], Antonio Gnudi[1], Luca Perilli[1], Mattia Luigi Torres[2], Marcella Carissimi[2], Marco Pasotti[2], Roberto Antonio Canegallo[2]

[1]University of Bologna, Italy; [2]STMicroelectronics, Italy

A Highly Integrated Crosspoint Array Using Self-Rectifying FTJ for Dual-Mode Operations: CAM and PUF .... N/A

Sehee Lim[1], Youngin Goh[2], Young Kyu Lee[1], Dong Han Ko[1], Junghyeon Hwang[2], Minki Kim[2], Yeongseok Jeong[2], Hunbeom Shin[2], Sanghun Jeon[2], Seong-Ook Jung[1]

[1]Yonsei University, Korea; [2]Korea Advanced Institute of Science and Technology, Korea

Physical Implementation of a Tunable Memristor-Based Chua's Circuit .... N/A

Manuel Escudero[1], Sabina Spiga[1], Mauro Di Marco[2], Mauro Forti[2], Giacomo Innocenti[3], Alberto Tesi[3], Fernando Corinto[4], Stefano Brivio[1]

[1]CNR-IMM, Italy; [2]University of Siena, Italy; [3]University of Florence, Italy; [4]Politecnico di Torino, Italy

Synaptic Devices

Date: Tuesday, September 20, 2022
Time: 11:10 - 12:50
Room: Aula 1D
Chair(s): Stefan Slesazeck, NaMLab
Alessandro Spinelli, Politecnico di Milano

Ferroelectric Schottky Barrier MOSFET as Analog Synapses for Neuromorphic Computing ... N/A

Fengben Xi[1,2], Andreas Grenmyr[1,2], Jiayuan Zhang[1,2], Yi Han[1,2], Jin-Hee Bae[1], Detlev Grützmacher[1,2], Qing-Tai Zhao[1]

[1]Peter Grünberg Institute, Germany; [2]RWTH Aachen University, Germany

Interplay Between Charge Trapping and Polarization Switching in MFDM Stacks Evidenced by Frequency-Dependent Measurements .... N/A

Justine Barbot, Jean Coignus, Nicolas Vaxelaire, Catherine Carabasse, Olivier Glorieux, Messaoud Bedjaoui, François Aussenac, François Andrieu, Francois Triozon, Laurent Grenouillet

CEA-Leti, Université Grenoble Alpes, France

Frequency Modulation of Conductance Level in PCM Device for Neuromorphic Applications .... N/A

Ahmed Trabelsi, Carlo Cagli, Tifenn Hirtzlin, Olga Cueto, Marie-Claire Cyrille, Elisa Vianello, Valentina Meli, Veronique Sousa, Guillaume Bourgeois, François Andrieu

CEA-Leti, France

Thermal Switching of TiO_2-Based RRAM for Parameter Extraction and Neuromorphic Engineering .... N/A

Alessandro Milozzi[1], Daniel Reiser[2,3], Andreas Drost[2], Thomas Neuner[2,3], Marc Tornow[2,3], Daniele Ielmini[1]

[1]Politecnico di Milano, Italy; [2]Fraunhofer EMFT Research Institution for Microsystems and Solid State Technologies, Germany; [3]Technical University of Munich, Germany

Improvement of FTJ On-Current by Work Function Engineering for Massive Parallel Neuromorphic Computing .... N/A

Suzanne Lancaster[1], Quang T. Duong[1], Erika Covi[1], Thomas Mikolajick[1,2], Stefan Slesazeck[1]

[1]NaMLab gGmbH, Germany; [2]Technische Universität Dresden, Germany

SRAM Based In-memory Compute

Date: Thursday, September 22, 2022
Time: 14:20 - 16:00
Room: Aula 1D
Chair(s): Adam Teman, Bar-Ilan University
Antoine Frappé, IEMN CNRS

A 4-Bit Mixed-Signal MAC Macro with One-Shot ADC Conversion .... N/A

Xiangxing Yang[1], Nan Sun[2]

[1]The University of Texas at Austin, United States; [2]Tsinghua University, China

A 28nm 1.644TFLOPS/W Floating-Point Computation SRAM Macro with Variable Precision for Deep Neural Network Inference and Training .... N/A

Sangsu Jeong, Jeongwoo Park, Dongsuk Jeon

Seoul National University, Korea

All-Digital Time-Domain Compute-in-Memory Engine for Binary Neural Networks with 1.05 POPS/W Energy Efficiency .... N/A

Jie Lou, Christian Lanius, Florian Freye, Tim Stadtmann, Tobias Gemmeke

RWTH Aachen University, Germany

A 1.23-GHz 16-Kb Programmable and Generic Processing-in-SRAM Accelerator in 65nm ...... N/A

Amitesh Sridharan[1], Shaahin Angizi[2], Sai Kiran Cherupally[1], Fan Zhang[1], Jae-Sun Seo[1], Deliang Fan[1]

[1]Arizona State University, United States; [2]New Jersey Institute of Technology, United States

A 1-to-4b 16.8-POPS/W 473-TOPS/mm² 6T-Based In-Memory Computing SRAM in 22nm FD-SOI with Multi-Bit Analog Batch-Normalization .... N/A

Adrian Kneip, Martin Lefebvre, Julien Verecken, David Bol

Université Catholique de Louvain, Belgium

Optoelectronic Devices & Negative Capacitance FET

Date: Tuesday, September 20, 2022
Time: 15:10 - 16:10
Room: Aula 1D
Chair(s):   Sara Pellegrini, STMicroelectronics
          Clara Moldovan, EPFL

Fully Integrated Si:HfO$_2$ Negative Capacitance 2D-2D WSe$_2$/SnSe$_2$ Subthermionic Tunnel FETs .... N/A

Sadegh Kamaei, Ali Saeidi, Xia Liu, Carlotta Gastaldi, Clara Moldovan, Jürgen Brugger, Adrian M. Ionescu

École Polytechnique Fédérale de Lausanne, Switzerland

Electroluminescence of Si$_x$Ge$_{1-x-y}$Sn$_y$ / Ge$_{1-y}$Sn$_y$ Pin-Diodes Grown on a GeSn Buffer .... N/A

Lukas Seidel[1], Sören Schäfer[1], Michael Oehme[1], Dan Buca[2], Giovanni Capellini[3,4], Jörg Schulze[5], Daniel Schwarz[1]

[1]University of Stuttgart, Germany; [2]Peter Grünberg Institute, Germany; [3]IHP - Leibniz Institute for High Performance Microelectronics, Germany; [4]Roma Tre University, Italy; [5]University of Erlangen-Nuremberg, Germany

GeSn-on-Si Avalanche Photodiodes for Short-Wave Infrared Detection .... N/A

Maurice Wanitzek[1], Michael Oehme[1], Christian Spieth[1], Daniel Schwarz[1], Lukas Seidel[1], Jörg Schulze[2]

[1]University of Stuttgart, Germany; [2]University of Erlangen-Nuremberg, Germany

Sensor Interfaces

Date: Wednesday, September 21, 2022
Time: 9:20 - 10:40
Room: Aula 1E
Chair(s): Marco Grassi, University of Pavia
Georgi Radulov, TU Eindhoven

A Drift-Compensated Magnetic Spectrometer for Point-of-Care Wash-Free Immunoassays Using a Concurrent Dual-Frequency Oscillator .... N/A

Jui-Hung Sun[1], Bill Ling[2], Md. Abdullah-Al Kaiser[1], Constantine Sideris[1]

[1]University of Southern California, United States; [2]California Institute of Technology, United States

A Full Current-Mode Timing Circuit with Dark Noise Suppression for the CERN CMS Experiment .... N/A

Edgar Albuquerque[1], Ricardo Bugalho[1], Luis Oliveira[3], Tahereh Niknejad[1,2], Jose Silva[1,2], Alessio Boletti[2], João Varela[1,2]

[1]PETsys Electronics, Portugal; [2]Laboratório de Instrumentação e Física Experimental de Partículas, Portugal; [3]CTS-UNINOVA, Portugal

A 2.74pJ/Conversion 0.0018mm² Temperature Sensor with On-Chip Gain and Offset Correction .... N/A

Yuting Shen, Mariska van der Struijk, Kevin Pelzers, Hanyue Li, Eugenio Cantatore, Pieter Harpe

Eindhoven University of Technology, The Netherlands

An Integrated Optical Transceiver Circuit for Power Delivery and Bi-Directional Data Communication in a Medical Catheter Device .... N/A

Alexander Frank[1], Jens Anders[1], Joachim Burghartz[1], Bart Kootte[2], Jean Schleipen[2], Peter Jutte[2]

[1]IMS CHIPS, Germany; [2]OSYPKA AG, Germany; [3]Philips Research Eindhoven, The Netherlands

Image Sensors

Date: Wednesday, September 21, 2022
Time: 12:00 - 13:00
Room: Aula 1E
Chair(s): Robert Henderson, University of Edinburgh
Daniele Perenzoni, Sony

A Reconfigurable 224×272-Pixel Single-Photon Image Sensor for Photon Timestamping, Counting and Binary Imaging at 30.0-μm Pitch in 110nm CIS Technology .... N/A

Leonardo Gasparini[1], Manuel Moreno García[1], Majid Zarghami[1], André Stefanov[2], Bruno Eckmann[2], Matteo Perenzoni[1,3]

[1]Fondazione Bruno Kessler, Italy; [2]University of Bern, Switzerland; [3]Sony Seminconductor Solutions, Italy

Statistical Measurements and Monte-Carlo Simulations of DCR in SPADs .... N/A

Mathieu Sicre[1,2,3], Megan Agnew[4], Christel Buj[1,3], Caroline Coutier[3], Dominique Golanski[1], Rémi Helleboid[1], Bastien Mamdy[1], Isobel Nicholson[4], Sara Pellegrini[4], Denis Rideau[1], David Roy[1], Francis Calmon[2]

[1]STMicroelectronics, France; [2]INSA Lyon, France; [3]CEA-Leti, France; [4]STMicroelectronics, United Kingdom

A Scalable 64×64 Pixels Monolithic HV-CMOS Sensor for Hadron Therapy with 1ns Time Stamping Capability and In-Pixel ADC .... N/A

Nicola Massari[1], Alessio D'Andragora[1], Matteo Perenzoni[2], Andrej Selijak[3], Carlos Chavez Barajas[4], Alan Taylor[4], Jon Taylor[4], Gianluigi Casse[4], John Pettingell[5], Ignacio Di Biase[5]

[1]Fondazione Bruno Kessler, Italy; [2]Sony Seminconductor Solutions Corporation, Italy; [3]Jozef Stefan Institute, Slovenia; [4]University of Liverpool, United Kingdom; [5]Rutherford Cancer Centres, United Kingdom

Biomedical Sensors

Date: Thursday, September 22, 2022
Time: 11:20 - 12:40
Room: Aula 1E
Chair(s): Mirjana Banjevic, Sensirion
Matthias Kuhl, University of Freiburg

A 32-ch Neuromodulator with Redundant Voltage Monitors Avoiding Blocking Capacitors ..... N/A

Stefan Reich[1], Markus Sporer[1], Joachim Becker[1], Stefan B. Rieger[2], Martin Schüttler[2], Maurits Ortmanns[1]

[1]University of Ulm, Germany; [2]CorTec GmbH, Germany

Fully Implantable 192×256 SPAD Sensor with Global-Shutter and Micro-LEDs for Bidirectional Subdural Optical Brain-Computer Interfaces .... N/A

Eric H. Pollmann, Yatin Gilhotra, Heyu Yin, Kenneth L. Shepard

Columbia University, United States

A 1.8μW 5.5mm³ ADC-Less Neural Implant SoC Utilizing 13.2pJ/Sample Time-Domain Bi-Phasic Quasi-Static Brain Communication with Direct Analog to Time Conversion .... N/A

Baibhab Chatterjee, K. Gaurav Kumar, Shulan Xiao, Gourab Barik, Krishna Jayant, Shreyas Sen

Purdue University, United States

A 50 Mb/s Full HBC TRX with Adaptive DFE and Variable-Interval 3x Oversampling CDR in 28nm CMOS Technology for a 75 Cm Body Channel Moving at 0.75 Cycle/Sec .... N/A

Jaehyun Ko[1], Iksu Jang[1], Chanho Kim[1], Jihoon Park[1], Changjae Moon[1], Sooeun Lee[2], Byungsub Kim[1]

[1]Pohang University of Science and Technology, Korea; [2]Samsung Electronics Co., Ltd., Korea

Medical Imaging

Date: Thursday, September 22, 2022
Time: 11:20 - 12:40
Room: Aula 1D
Chair(s): Angel Rodriguez-Vasquez, Universidad de Sevilla
Michiel Pertijs, TU Delft

An Implantable Power Extraction Circuit with Integrated PMUTs for Wireless Power Delivery .... N/A

Oi-Ying Wong[1,2], Dries Tabruyn[1,2], Veronique Rochus[1], Nick Van Helleputte[1]

[1]imec, Belgium; [2]Katholieke Universiteit Leuven, Belgium

A PMUT Transceiver Front-End with 100-V TX Driver and Low-Noise Voltage Amplifier in BCD-SOI Technology .... N/A

Lara Novaresi[2], Piero Malcovati[1], Andrea Mazzanti[1], Edoardo Bonizzoni[1], Marco Terenzi[2], Stefano Ottaviani[2], Davide Ugo Ghisu[2], Fabio Quaglia[2], Alessandro Stuart Savoia[3]

[1]University of Pavia, Italy; [2]STMicroelectronics, Italy; [3]Roma Tre University, Italy

A Portable CMOS-Based MRI System with 67×67×83 μm³ Image Resolution .... N/A

Daniel Krüger[1,2], Aoyang Zhang[1], Henry Hinton[1], Victor M. Arnal[1], Yi-Qiao Song[1,4], Yiqiao Tang[4], Ka-Meng Lei[3], Jens Anders[2], Donhee Ham[1]

[1]Harvard University, United States; [2]University of Stuttgart, Germany; [3]University of Macau, China; [4]Schlumberger-Doll Research, United States

A Sub-0.01° Phase Resolution 6.8-mW fNIRS Readout Circuit Employing a Mixer-First Frequency-Domain Architecture .... N/A

Cheng Chen[1], Zhouchen Ma[1], Yaxin Liu[1], Zhenhong Liu[1], Linfeng Zhou[1], Yan Wu[1], Liang Qi[1], Yongfu Li[1], Mohamad Sawan[2], Guoxing Wang[1], Jian Zhao[1]

[1]Shanghai Jiao Tong University, China; [2]Westlake University, China

Memory Devices

Date: Tuesday, September 20, 2022
Time: 14:30 - 16:10
Room: Aula 1E
Chair(s): Andrea Redaelli, STMicroelectronics
           Innocenzo Tortorelli, Micron Technologies

Multilayer Structure in SeAsGeSi-Based OTS for High Thermal Stability and Reliability Enhancement 225

Camille Laguna[1,2], Mathieu Bernard[1], Julien Garrione[1], Niccolo Castellani[1], Valentina Meli[1], Simon Martin[1], François Aussenac[1], Denis Rouchon[1], Névine Rochat[1], Emmanuel Nolot[1], Guillaume Bourgeois[1], Marie-Claire Cyrille[1], Liviu Militaru[2], Abdelkader Souifi[3], François Andrieu[1], Gabriele Navarro[1]

[1]CEA-Leti, Université Grenoble Alpes, France; [2]University Lyon 1, France; [3]Université de Lyon, France

Characterization of Reset State Through Energy Activation Study in Ge-GST Based ePCM 229

Matteo Baldo[1], Lorenzo Turconi[2], Alessandro Motta[1], Elisa Petroni[1], Luca Laurin[1], Daniele Ielmini[2], Andrea Redaelli[1]

[1]STMicroelectronics, Italy; [2]Politecnico di Milano, Italy

Enhanced Thermal Confinement in Phase-Change Memory Targeting Current Reduction 233

Clement De Camaret[1,2], Guillaume Bourgeois[2], Olga Cueto[2], Valentina Meli[2], Simon Martin[2], Dominique Despois[2], Virginie Beugin[2], Niccolo Castellani[2], Marie-Claire Cyrille[2], François Andrieu[2], Julien Arcamone[2], Yannick Le-Friec[1], Gabriele Navarro[2]

[1]STMicroelectronics, France; [2]CEA-Leti, France

TiTe/Ge$_2$Sb$_2$Te$_5$ Bi-Layer-Based Phase-Change Memory Targeting Storage Class Memory 237

Giusy Lama, Mathieu Bernard, Julien Garrione, Nicolas Bernier, Niccolo Castellani, Guillaume Bourgeois, Marie-Claire Cyrille, François Andrieu, Gabriele Navarro

CEA-Leti, France

Design Exploration of IGZO Diode Based VCMA Array Design for Storage Class Memory Applications 241

Mohit Gupta, Manu Perumkunnil, Andrea Fantini, Saeideh Alinezhad Chamazcoti, Woojin Kim, Marie Garcia Bardon, Gouri Sankar Kar, Arnaud Furnémont

imec, Belgium

Heterogenous Integration

Date: Tuesday, September 20, 2022
Time: 14:30 - 16:10
Room: Aula 1F
Chair(s): Erik Lind, Lund University
Michael Waltl, TU Wien

Device Optimization for 200V GaN-on-SOI Platform for Monolithicly Integrated Power Circuits 245

Olga Syshchyk[1], Thibault Cosnier[1], Zheng-Hong Huang[2], Deepthi Cingu[1], Dirk Wellekens[1], Anurag Vohra[1], Karen Geens[1], Pavan Vudumula[3], Urmimala Chatterjee[1], Stefaan Decoutere[1], Tian-Li Wu[2], Benoit Bakeroot[4]

[1]imec, Belgium; [2]National Yang Ming Chiao Tung University,Taiwan; [3]Katholieke Universiteit Leuven, Belgium; [4]IMEC and Ghent University, Belgium

Effect of Post Annealing on the Electrical Characteristics and Deep Level Defects of Ga_2O_3/SiC Heterojunction Diodes 249

Dong-Wook Byun, Min-Yeong Kim, Soo-Young Moon, Myeong-Cheol Shin, Michael A. Schweitz, Sang-Mo Koo
Kwangwoon University, Korea

BEoL Integrated Hafnium Zirconium Oxide Varactors for Tunable mmWave Applications 253

Sukhrob Abdulazhanov[1], Dang Khoa Huynh[1], Quang Huy Le[1], David Lehninger[1], Thomas Kämpfe[1], Gerald Gerlach[2]

[1]Fraunhofer Institute for Photonic Microsystems, Germany; [2]Technische Universität Dresden, Germany

In-Depth Electrical Characterization of Deca-Nanometer InGaAs MOSFET Down to Cryogenic Temperatures for Low-Power Quantum Applications 257

Francesco Serra Di Santa Maria[1], Christoforos Theodorou[1], Francis Balestra[1], Gerard Ghibaudo[1], Eunjung Cha[2], Cezar B. Zota[2]

[1]Université Grenoble Alpes, IMEP-LAHC, France; [2]IBM Research Zürich, Switzerland

III-V HBTs on 300 mm Si Substrates Using Merged Nano-Ridges and its Application in the Study of Impact of Defects on DC and RF Performance 261

Abhitosh Vais[1], Sachin Yadav[1], Y. Mols[1], Bjorn Vermeersch[1], Komal Vondkar Kodandarama[1], Marina Baryshnikova[1], Geert Mannaert[1], Reynald Alcotte[1], Guillaume Boccardi[1], Piet Wambacq[1], Bertrand Parvais[1,2], Robert Langer[1], Bernardette Kunert[1], Nadine Collaert[1]

[1]imec, Belgium; [2]Vrije Universiteit Brussel, Belgium

Advanced CMOS Modeling

Date: Tuesday, September 20, 2022
Time: 17:30 - 18:30
Room: Aula 1D
Chair(s): Benjamin Iniguez, University Rovira i Virgili
Viktor Sverdlov, TU Wien

Co-Integration Process Compatible Input/Output (I/O) Device Options for GAA Nanosheet Technology 265

Gautam Gaddemane[1], Krishna K. Bhuwalka[2], Philippe Matagne[1], Gerhard Rzepa[3], Maarten Van de Put[1], Sybren Santermans[1], Oskar Baumgartner[3], Hao Wu[2], Geert Hellings[1]

[1]imec, Belgium; [2]Huawei Technologies Co., Ltd., Belgium; [3]Global TCAD Solutions GmbH, Austria

Cryogenic RF Characterization and Simple Modeling of a 22 nm FDSOI Technology 269

Hung-Chi Han[1], Farzan Jazaeri[1], Antonio D'Amico[1], Zhixing Zhao[2], Steffen Lehmann[2], Claudia Kretzschmar[2], Edoardo Charbon[1], Christian Enz[1]

[1]École Polytechnique Fédérale de Lausanne, Switzerland; [2]GlobalFoundries, Germany

A Novel Approach to Modeling Insulator Wave-Function Penetration and Interface Roughness Scattering in MOSFETs 273

Zlatan Stanojević, Lee-Chi Hung, Chen-Ming Tsai, Markus Karner

Global TCAD Solutions GmbH, Austria

Sensors Device Modeling

Date: Tuesday, September 20, 2022
Time: 9:20 - 10:40
Room: Aula 1F
Chair(s): Denis Rideau, STMicroelectronics
Zlatan Stanojevic, Global TCAD Solutions

On the Convergence of the Recurrence Solution of Mcintyre's Local and Non-Local Avalanche Triggering Probability Equations for SPAD Compact Models 277

Dorian Saint-Pierre[1], Raphaël Clerc[1], Remi Helleboid[2], Denis Rideau[2]

[1]Institut d'Optique Graduate School, France; [2]STMicroelectronics, France

A Self-Sustaining Single Photon Avalanche Diode Model 281

Sven Rink[1], Vincent Quenette[1], Jean-Robert Manouvrier[1], André Juge[1], Gilles Gouget[1], Denis Rideau[1], Raul-Andres Bianchi[1], Dominique Golanski[1], Bastien Mamdy[1], Jean-Baptiste Kammerer[2], Wilfried Uhring[2], Christophe Lallement[2], Sara Pellegrini[3], Megan Agnew[3], Bruce Rae[3]

[1]STMicroelectronics, France; [2]Université de Strasbourg, France; [3]STMicroelectronics, Scotland

Thermal Sensing Performances of Thin-Film Lateral PiN Diodes at 80 K and 300 K 285

Adrien Fournol, Jérémy Blond, Abdelkader Aliane, Hacile Kaya, Jérôme Meilhan, Laurent Dussopt

Université Grenoble Alpes, France

MEMS Optical Microphone Based on Light Phase Modulation 289

Niccolò de Milleri[1], Güclü Onaran[2], Andreas Wiesbauer[1], Andrea Baschirotto[3]

[1]Infineon Technologies AG, Austria; [2]Infineon Technologies AG, Germany; [3]University of Milano-Bicocca, Italy

Reliability & Characterization

Date: Tuesday, September 20, 2022
Time: 9:20 - 10:40
Room: Aula 1D
Chair(s): Gianluca Fiori, University of Pisa
Claire Fenouillet, CEA

Filament Localization and Characterization in HfO₂ ReRAM Cells Using Laser Stimulation 293

Franco Stellari[1], Ernest Y. Wu[2], Martin M. Frank[1], Leonidas E. Ocola[2], Takashi Ando[1], Peilin Song[1]

[1]*IBM T.J. Watson Research Center, United States;* [2]*IBM Research, United States*

Role of Conductive-Metal-Oxide to HfOₓ Interfacial Layer on the Switching Properties of Bilayer TaOₓ/HfOₓ ReRAM 297

Tommaso Stecconi[1], Youri Popoff[2], Roberto Guido[1], Donato Falcone[1], Mattia Halter[2], Marilyne Sousa[1], Folkert Horst[1], Antonio La Porta[1], Bert Jan Offrein[1], Valeria Bragaglia[1]

[1]*IBM Research Zürich, Switzerland;* [2]*ETH Zürich, Switzerland*

Impact of Gold Interconnect Microstructure on Electromigration Failure Time Statistics 301

Hajdin Ceric, Roberto Lacerda de Orio, Siegfried Selberherr

Technische Universität Wien, Austria

Influence of Metal on Schottky Barrier Inhomogeneity in Ga₂O₃ Schottky Barrier Diodes 304

Min-Yeong Kim, Geon-Hee Lee, Hee-Jae Lee, Dong-Wook Byun, Michael A. Schweitz, Sang-Mo Koo

Kwangwoon University, Korea

Compact Modeling

Date: Wednesday, September 21, 2022
Time: 12:00 - 13:00
Room: Aula 1F
Chair(s): Jean-Michel Sallese, EPFL
Wladek Grabinski, MOS-AK

Compact Modeling of Phase Change Memory with Parameter Extractions 308

Feilong Ding[1], Xi Li[2], Yihan Chen[3], Zhitang Song[2], Runsheng Wang[1], Clarissa Cyrilla Prawoto[4], Mansun Chan[4], Lining Zhang[1], Ru Huang[1]

[1]*Peking University, China;* [2]*Shanghai Institute of Microsystem and Information Technology, CAS, China;*
[3]*The Chinese University of Hong Kong, China;* [4]*The Hong Kong University of Science and Technology, China*

A Novel Approach to Measure and Model Plasma Noise in Avalanche Diodes 312

Elmar Gondro, Joost Willemen, Peter Bauer

Infineon Technologies AG, Germany

Self-Consistent Automated Parameter Extraction of RRAM Physics-Based Compact Model 316

Tommaso Zanotti, Paolo Pavan, Francesco Maria Puglisi

University of Modena and Reggio Emilia, Italy

RF Power Device Optimization & Characterization

Date: Wednesday, September 21, 2022
Time: 15:20 - 16:40
Room: Aula 1E
Chair(s): Christoforos Theodorou, INP Grenoble
          Frederic Allibert, SOITEC, France

InP DHBT Test Structure Optimization Towards 110 GHz Characterization 320

Nil Davy[1], Marina Deng[2], Virginie Nodjiadjim[1], Chhandak Mukherjee[2], Muriel Riet[1], Colin Mismer[1], Jérémie Renaudier[3], Cristell Maneux[2]

[1]*III-V Lab, France; [2]University of Bordeaux, France; [3]Nokia Bell Labs, France*

Trap Behavior of Metamorphic HEMTs with Pulsed IV and 1/f Noise Measurement 324

Ki-Yong Shin[1], Ju-Won Shin[1], Surajit Chakraborty[1], Walid Amir[1], Chan-Soo Shin[2], Tae-Woo Kim[1]

[1]*University of Ulsan, Korea; [2]Korea Advanced Nano Fab Center, Korea*

Analysis and Optimization of an Analog MOSFET with a Slit Well at Channel Center Towards Higher Output Resistance 328

Hiroki Fujii, Jaehyun Yoo, Dawon Jeong, Seongsik Min, Myoungsoo Kim, Uihui Kwon, Dae Sin Kim

Samsung Electronics Co., Ltd., Korea

Highly Robust and Reliable Power Amplifiers in 22FDX and 45RFSOI Technologies 332

Alice Bossuet[1], Alexis Divay[1], Baudouin Martineau[1], Cedric Dehos[1], Benjamin Blampey[1], Yvan Morandini[2]

[1]*CEA-Leti, France; [2]Soitec, France*

Memory Modeling

Date: Wednesday, September 21, 2022
Time: 15:20 - 16:40
Room: Aula 1F
Chair(s): Sadayuki Yoshitomi, Toshiba
Jens Trommer, Namlab

Joint Modeling of Multi-Domain Ferroelectric and Distributed Channel Towards Unveiling the Asymmetric Abrupt DC Current Jump in Ferroelectric FET 336

Simon Thomann[1], Kai Ni[2], Hussam Amrouch[1]

[1]*University of Stuttgart, Germany;* [2]*Rochester Institute of Technology, United States*

Polarization Switching and AC Small–Signal Capacitance in Ferroelectric Tunnel Junctions 340

Mattia Segatto[1], Marco Massarotto[1], Suzanne Lancaster[2], Q.T. Duong[2], Antonio Affanni[1], Riccardo Fontanini[1], Francesco Driussi[1], Daniel Lizzit[1], Thomas Mikolajick[2,3], Stefan Slesazeck[2], David Esseni[1]

[1]*University of Udine, Italy;* [2]*NaMLab gGmbH, Germany;* [3]*Technische Universität Dresden, Germany*

Multi-Level Operation of FeFETs Memristors: The Crucial Role of Three Dimensional Effects 344

Daniel Lizzit, Thomas Bernardi, David Esseni

University of Udine, Italy

Spin Torques in ULTRA-Scaled MRAM Devices 348

Simone Fiorentini[1], Mario Bendra[1], Johannes Ender[1], Roberto Lacerda de Orio[1], Wolfgang Goes[2], Siegfried Selberherr[1], Viktor Sverdlov[1]

[1]*Technische Universität Wien, Austria;* [2]*Silvaco Europe Ltd., United Kingdom*

Advanced Devices & VLSI Processing

Date: Wednesday, September 21, 2022
Time: 9:20 - 10:50
Room: Aula 1F
Chair(s): Jerome Dubois, NXP
Gunnar Malm, KTH

The Environmental Footprint of IC Production: Meta-Analysis and Historical Trends 352

Thibault Pirson, Thibault Delhaye, Alex Pip, Grégoire Le Brun, Jean-Pierre Raskin, David Bol

Université Catholique de Louvain, Belgium

Experimental Fabrication of an ESF3 Floating Gate Flash Cell in an FD-SOI Process 356

Nicki Mika[1], Thomas Melde[1], Stefan Dünkel[1], Michael Otto[1], Fancois Weisbuch[1], Peter Krottenthaler[1], Thomas Mikolajick[2]

[1]GlobalFoundries, Germany; [2]Technische Universität Dresden, Germany

A Novel Energy-Efficient Salicide-Enhanced Tunnel Device Technology Based on 300 mm Foundry Platform Towards AIoT Applications .... N/A

Kaifeng Wang[1], Qianqian Huang[1,3,4], Yongqin Wu[2], Ye Ren[2], Renjie Wei[1], Zhixuan Wang[1], Libo Yang[1], Fangxing Zhang[1], Kexing Geng[1], Yiqing Li[1], Mengxuan Yang[1], Jin Luo[1], Ying Liu[1], Kai Zheng[2], Jin Kang[2], Le Ye[1], Lining Zhang[1], Weihai Bu[2], Ru Huang[1,3,4]

[1]Peking University, China; [2]Semiconductor Technology Innovation Center, China; [3]Chinese Institute for Brain Research, China; [4]Beijing Advanced Innovation Center for Integrated Circuits, China

GeSn Vertical Gate-All-Around Nanowire n-Type MOSFETs 364

Yannik Junk[1], Marvin Frauenrath[2], Yi Han[1], Omar Concepción Diaz[1], Jin-Hee Bae[1], Jean-Michel Hartmann[2], Detlev Grützmacher[1], Dan Buca[1], Qing-Tai Zhao[1]

[1]Peter Grünberg Institute, Germany; [2]CEA-Leti, France

Defects & Traps Modeling

Date: Wednesday, September 21, 2022
Time: 9:20 - 10:50
Room: Aula 1D
Chair(s): Cristell Maneux, University of Bordeaux
Valeria Vadalà, University of Milano-Bicocca

Defects Motion as the Key Source of Random Telegraph Noise Instability in Hafnium Oxide ... 368

Sara Vecchi, Paolo Pavan, Francesco Maria Puglisi

University of Modena and Reggio Emilia, Italy

A Novel Temperature Estimation Technique Exploiting Carrier Emission from Buffer Traps ... 372

Marcello Cioni, Nicolò Zagni, Alessandro Chini

University of Modena and Reggio Emilia, Italy

Metastability of Negatively Charged Hydroxyl-E' Centers and Their Potential Role in Positive Bias Temperature Instabilities ... 376

Christoph Wilhelmer, Dominic Waldhoer, Markus Jech, Al-Moatasem Bellah El-Sayed, Lukas Cvitkovich, Michael Waltl, Tibor Grasser

Technische Universität Wien, Austria

Silicon-Impurity Defects in Calcium Fluoride: A First Principles Study ... 380

Dominic Waldhoer[1], Bibhas Manna[1], Al-Moatasem Bellah El-Sayed[1], Theresia Knobloch[1], Yury Illarionov[1,2], Tibor Grasser[1]

[1]Technische Universität Wien, Austria; [2]Ioffe Institute, Russia

GaN HEMT Technology

Date: Wednesday, September 21, 2022
Time: 14:20 - 16:00
Room: Aula 1F
Chair(s): Susanna Reggiani, University of Bologna
Mikael Östling, KTH Royal Insti of Technology

Impact of Channel Thickness Scaling on the Performance of GaN-on-Si RF HEMTs on Highly C-Doped GaN Buffer 384

Alireza Alian[1], Raul Rodriguez[1], Sachin Yadav[1], Uthayasankaran Peralagu[1], Arturo Sibaja Hernandez[1], Vamsi Putcha[1], Ming Zhao[1], Rana Elkashlan[1,2], Bjorn Vermeersch[1], Hao Yu[1], Erik Bury[1], Ahmad Khaled[1], Nadine Collaert[1], Bertrand Parvais[1,2]

[1]imec, Belgium; [2]Vrije Universiteit Brussel, Belgium

Characterization of GaN-Based HEMTs Down to 4.2 K for Cryogenic Applications .... N/A

Bolun Zeng, Haochen Zhang, Chao Luo, Zikun Xiang, Yuanke Zhang, Mingjie Wen, Qiwen Xue, Sirui Hu, Yue Sun, Lei Yang, Haiding Sun, Guoping Guo

University of Science and Technology of China, China

A Novel Approach to Analyze the Reliability of GaN Power HEMTs Operating in a DC-DC Buck Converter 392

Giuseppe Capasso, Mauro Zanuccoli, Andrea Natale Tallarico, Claudio Fiegna

University of Bologna, Italy

A Novel Depletion Mode p-GaN Island HEMT and its Use in a Monolithically Integrated Start-Up Circuit 396

Loizos Efthymiou, Martin Arnold, Giorgia Longobardi, Florin Udrea

Cambridge GaN Devices Ltd., United Kingdom

Novel Normally-Off AlGaN/GaN-on-Si MIS-HEMT Exploiting Nanoholes Gate Structure 400

Mamta Pradhan, Matthias Moser, Mohammed Alomari, Joachim N. Burghartz, Ingmar Kallfass

University of Stuttgart, Germany

ESSDERC

IEEE 52nd European Solid-State Device Research Conference

September 19-22, 2022
Milan, Italy

IEEE

Electron Devices Society

ESSDERC 2022 - IEEE 52nd European Solid-State Device Research Conference (ESSDERC)

On Line PROCEEDINGS

ESSDERC

IEEE 52nd European Solid-State Device Research Conference

Co - Organizer

Financial Sponsorship

under the Patronage of

SPONSORS

DIAMOND SPONSORS

SPONSORS

GOLD SPONSORS

SONY

SYNOPSYS®

SILVER SPONSORS

am OSRAM

ANALOG
DEVICES

INVENTVM

KIOXIA

knowles

Melexis
INSPIRED ENGINEERING

power
integrations

RedCat Devices

ESSDERC · 2022

SPONSORS

SILVER SPONSORS

SΛMSUNG

⬡TDK

BRONZE SPONSORS

cādence®

SPRINGER NATURE

xfab

Gap in pagination due to unavailable papers.

Pages 1-24

Quantum Computing Technology and Roadmap

Heike Riel
IBM Research
Rüschlikon, Switzerland
hei@zurich.ibm.com

Abstract—**The progress of Quantum Computing has significantly accelerated in recent years due to increased resources for research and development across the world. The focus has changed from just building quantum devices to establishing a quantum computing system including software and applications. An overview of the recent advancements of superconducting quantum computing systems is provided including the achievements made to scale processors to 127-qubit and scale up to 4000 qubits in 2025. Three key metrics are defined, scale, quality, and speed to holistically measure performance. The computational capabilities of today's quantum chips can be extended by adding classical resources using circuit knitting techniques to reach quantum advantage.**

Keywords—quantum computing, superconducting Josephson junctions, roadmap for quantum computing

I. INTRODUCTION

In the last 70 years computing has been advanced in unimaginable ways and heroic engineering and scaling efforts have propelled the development of successive generations of technology for the acquisition, processing, and storage of digital information. Supercomputers have evolved and reached a performance of more than 10^{18} floating point operations per second (ExaFLOPs) and have played a pivotal role in pushing the frontiers of science. Despite these computational advances, there are still many relevant mathematical problems that are intractable to classical super-computers but could be solved by Quantum Computers [1-3].

Quantum Computers are a completely new computing paradigm and are based on the laws of quantum physics. The information is encoded in quantum bits (qubits) as basic unit of quantum information analogous to digital bits in classical computers. A qubit is a two-level quantum mechanical system that is in a coherent superposition of the ground state |0> and the exited state |1>, enlarging the computational state space. Ideally the power of quantum computing scales exponentially with 2^N, with N the number of qubits. In addition, quantum entanglement creates strong correlations between qubit states leading to increased information in the combined system. In short, quantum computing utilizes the quantum physical phenomena of superposition, entanglement, and interference accessing unique resources which cannot be mimicked by classical computers to solve computationally hard problems.

We are developing circuit-based programmable quantum systems towards universal quantum computers. Hereby the system is initialized, and the state evolves by a sequence of unitary gate operations to a final state that is measured. A universal quantum gate set comprising one-qubit and two-qubit gates is used where any Hamilton operator can be implemented. Quantum circuits are the basic unit of quantum computation analogous to logic circuits for classical computing. Error correction is needed to build fault-tolerant quantum processors and benefit of the proven super-polynomial speed-up for certain problems, e.g., Shor's algorithm [1].

Significant progress has been achieved in the development of quantum computing in recent years. Whereas a few years back the field was dominated by exploratory physics experiments, it moved into engineering quantum systems. The central goal of our efforts is to build the entire quantum computing stack from ground up, scaling and optimizing the hardware (HW), the computing architecture, creating the software (SW) stack and exploring applications of quantum computers to achieve the desired quantum advantage.

II. SUPERCONDUCTING QUANTUM BITS

Our quantum processors utilize superconducting qubits built on silicon that can be designed to exhibit "atom-like" energy spectra with desired properties (see Fig. 1). This type of qubit is based on a superconducting Josephson junction (JJ), comprising two superconducting electrodes with a thin insulator sandwiched in between and an area of about 100nm x 100nm [4-6]. This so-called transmon qubit is shunted with a large capacitance C (Fig. 1a) to decrease charge noise sensitivity [4] and thus increase the coherence times. The JJ acts as a non-linear inductor resulting in an anharmonic energy spectrum with non-equidistant levels important for selective qubit control. This allows the ground and first excited energy level to be selected as |0> and |1> state for the computational basis and selectively control them with microwave pulses in the range of about 5 GHz corresponding to the energy separation E_{01} (Fig. 1c).

The measurement, control and coupling of the qubits is performed utilizing microwave resonators with techniques of circuit quantum electro-dynamics [7]. Superconducting qubits

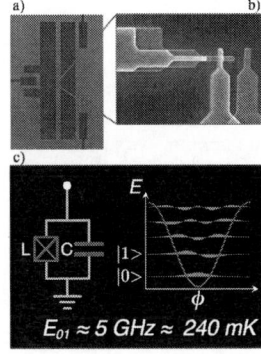

Fig. 1.   Superconducting Qubit based on a JJ. b) Scanning electron micrograph (top view) of a JJ with a junction area of 100nm x 100nm. c) Qubit circuit with JJ as non-linear inductor L shunted with capacitance C, yields an anharmonic energy spectrum. E_{01} is the energy separation between two lowest energy levels.

978-1-6654-8498-5/22 $31.00 © 2022 IEEE

offer a rich parameter space of possible qubit properties and operation regimes, with predictable performance in terms of transition frequencies, anharmonicity, and complexity. Due to the energy scales ($E_{01} \approx 5$ GHz ≈ 240 mK) the operation of super-conducting qubits requires cryogenic temperature of about 10-20 mK. At this temperature the junction operates nearly free of dissipation.

The qubit states are controlled by applying sequences of microwave pulses at around 5 GHz for tens of nanoseconds. Consequently, very fast gate operation times can be achieved which is a significant advantage of superconducting qubits. The number of gate operations run within the coherence time and thus the depth of a quantum circuit is limited by the ratio of coherence and gate time. Over the last 20 years continuous increase in coherence times from nano- to milliseconds has been achieved and further optimization is ongoing (see Sec. IV). Another benefit of superconducting qubits is that they can be built on silicon substrates using standard fabrication technologies enabling the scalability to high qubit numbers. So far, we have not yet encountered any fundamental physical principles that would prohibit further improvements and performance gain. Which makes superconducting qubits today the preferred technology for quantum computing.

III. IBM QUANTUM SYSTEMS

Similarly, to a classical computer the IBM Quantum Systems consist of multiple components that work together to serve as an advanced cloud-based quantum computing platform. The first fully integrated quantum computing system we demonstrated is called IBM Quantum System One as shown schematically in Fig. 2. It includes various technologies that are co-developed, such as the quantum processor unit (QPU); electronic components and wiring; the cryogenic platform; control HW and input/output (I/O) as well as the tight integration with classical computing.

The quantum HW is designed for stability, and auto calibrated to give repeatable and predictable performance from high-quality qubits. System engineering efforts have gone into the cryogenic system to achieve stable and robust operation and to well isolate the quantum environment. This includes engineering of thermalization, coaxial cabling for I/O signal integrity, microwave signal isolators, and cryogenic amplifiers. High precision electronics in compact form factors has been developed to control large numbers of qubits within very demanding parameters. Quantum firmware is designed to manage the system health and enable system upgrades without downtime for users. In addition, quantum resources are tightly coupled with classical computation to provide secure cloud access and hybrid execution of quantum algorithms.

With the technical advances at every level of the system, quantum computers have transitioned from the laboratory environment to data centers. More than 6 years ago we made the first 5-qubit processor accessible worldwide via the cloud. Today more than 24 quantum systems with quantum processors up to 127 qubits have been deployed and are available through the cloud. The number of registered users has reached more than 400'000 and is rising. More than 1.4 trillion quantum circuits have been run in total, and over 4 billion run on a typical day and continuously increasing. In parallel to all the system improvements, an open-source quantum SW development kit Qiskit [8] has been established to run quantum programs and simplify using quantum computers for further research and applications.

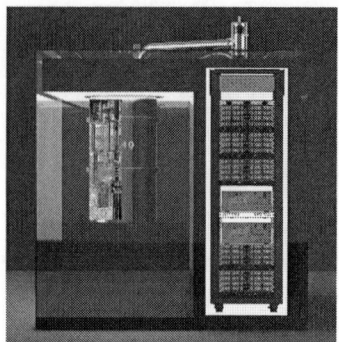

Fig. 2.   The IBM Quantum System 1 is shown schematically. It comprises the cryogenic platform, components and wiring, the qubit processor as well as room temperature control electronics.

IV. QUANTUM COMPUTING PERFORMANCE

Measuring and benchmarking the performance of quantum computing is a key area of research and development. Getting the metric right is critical for comparing different technologies, for determining the right goals and establishing a roadmap to further advance quantum computing. In that regard it is necessary to create holistic benchmarks that capture all the components that will translate to increased performance on real world applications. The performance of quantum computers requires three crucial attributes: Scale, Quality, and Speed. Progress on all three areas has been already achieved and is needed to demonstrate quantum advantage which is the point when quantum computers are either faster, cheaper, or more accurate than classical computers at a relevant task.

A. Scale

Scale is measured by numbers of qubits which indicates the amount of information that can be encoded in the quantum system and thus the size of the computable problems. Scaling quantum HW is a key objective that requires developing technologies to increase the number of qubits while maintaining quantum coherence on the processor.

Several technical challenges have been solved the last couple of years and enabled the scaling from 5 qubits in 2016, to the 27-qubit processor Falcon in 2019, to the 65-qubit chip Hummingbird in 2020 and to the 127-qubit chip named Eagle in 2021. Starting with the 27-qubit chip we implemented an optimized qubit layout called heavy-hex as shown in Fig. 3(b) [9]. Various qubit topologies have been tested and the heavy-hex lattice results in an optimized tradeoff between qubit connectivity, qubit frequency collisions and spectator errors [10]. Zero-frequency collisions are reduced compared to other topologies [11], providing more flexibility in choosing the qubit frequency thus enabling the realization of the 27- and 65-qubit processor. The heavy-hex layout proves beneficial also for the implementation of error correction codes [12].

A prerequisite for scaling to very high qubit numbers is a high yield in device fabrication with the required specifications; hitting the target operation frequency is crucial. The superconducting qubit frequency is determined by the JJ which is very sensitive to the thin tunnel junction. Fabrication fluctuations of the superconductor-oxide-superconductor tunnel junction lead to variations of the qubit frequency in the range of 600 MHz. This large distribution results in frequency collisions that prohibit scaling. Therefore, a laser annealing technique was developed to tune the frequency after

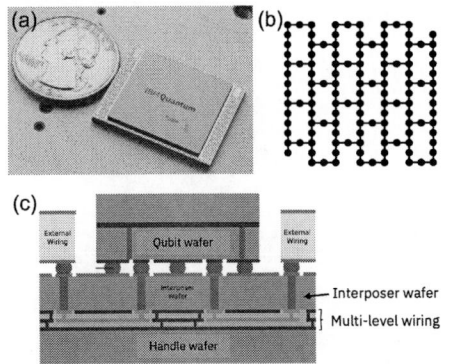

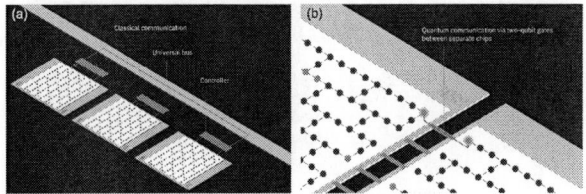

Fig. 4. (a) Three 133-qubit chips are connected via a classical link. (b) Quantum communication via two-qubit gates between separate chips.

Fig. 3. (a) 127-Qubit processor Eagle. (b) Qubit layout of the 127-qubit chip. In the heavy-hex topology each unit cell of the lattice consists of a hexagonal placement of qubits with an additional qubit on each edge. (c) Packaging scheme of Eagle with handle-wafer, interposer wafer and qubit wafer. Superconducting multi-level wiring for control and readout signals is formed on backside of interposer.

fabrication [11] to a desired frequency with a significantly reduced spread of only ca. 130 MHz. A 10x increase in tolerance and a boost in device yield is achieved that is essential for scaling up the number of qubits in a QPU.

Another important technical challenge for large qubit processors is the signal readout, as many components in the cryostat are required. For the 65-qubit chip that was demonstrated in 2020, this problem was overcome by implementing readout multiplexing. Eight readout cavities are measured through one amplifier line multiplexed through one Purcell filter. This technique required only nine amplifiers for the 65-qubit chip and is important for larger processors.

Scaling above 100 qubits requires a completely different approach in packaging due to the I/O challenge which increases with numbers of qubits. The requirement to read out and control all qubits cannot be accomplished by in plane wiring. Applying microwave activated gate operations requires the ability to deliver microwave signals with high-fidelity, and with low crosstalk. Therefore, for the 127-qubit processor a new packaging scheme was developed as shown in Fig. 3. The package comprises a qubit wafer bump-bonded to an interposer, and the handle wafer. In this processor generation superconducting multi-layer wiring (MLW) is used. This additional layer is well-isolated from the quantum device and is used to route control and readout signals and enables signal delivery into large chips. The MLW level consists of three metal layers, patterned, and with planarized dielectric between each level, and it includes through substrate vias (TSVs) connecting the metal levels. These vias act like a Faraday cage, preventing capacitive crosstalk between circuit elements. The created transmission lines are fully fenced from each other and isolated from the quantum device. Thus, MLW and TSVs provide a natural shielding against crosstalk which is an important source of errors for superconducting QPUs. Measurements have shown that the crosstalk in the 127-qubit processor is much reduced compared to the 27-qubit chip due to the improved packaging technology [13].

The next step in scaling is the demonstration of the 433-qubit chip named Osprey planned for end of 2022. The challenge here is the number of signal lines and the space they require in the cryostat. Traditional microwave wires take too much space, therefore cryo-flex cables [14] are developed to achieve a 10x increase in density. Breaking the 1000-qubit barrier is a major milestone targeted in 2023 by demonstrating the 1127-qubit processor named Condor. This requires major

technical breakthroughs, e.g., addressing miniaturization and optimization of various system components, electronics signal delivery, improving device quality including two-qubit gate errors of 10^{-4}. With this work we test the limits of single-chip QPUs and their control that will be integrated into the next generation IBM Quantum System Two.

Condor is an important step for scaling, but we also need to develop a path towards quantum systems with hundreds of thousands, and millions of high-quality qubits in the future. A giant chip is not the best solution, instead a modular approach can provide a successful future path. Therefore, we develop ways to link chips together into a modular system capable of scaling without physics limitations. In a first approach QPUs will be linked by classical communication links. This will be demonstrated by introducing Heron, a 133-qubit processor with control HW that allows for real-time classical communication between separate processors as shown in Fig. 4a. This approach allows parallelization of computing facilitating new algorithmic approaches like "knitting techniques" described in section VI. In the second approach the size of the quantum chip is extended by creating multi-chip processors enabled by quantum communication via two-qubit gates between separate chips (see Fig 4b). Following this approach, a 408-qubit processor called Crossbill will be made in 2024 from three chips connected by chip-to-chip couplers that allow for a continuous realization of the heavy-hex lattices across multiple chips. Due to the quantum communication the combination of the three smaller chips creates one large quantum processor. The third approach supports quantum parallelization by quantum communication between processors. A 462-qubit processor called Flamingo with a built-in quantum communication link will be fabricated and at least three Flamingo processors will be linked into a 1,386-qubit system in 2024. This longer distance link is expected to result in slower and lower-fidelity gates across processors but will be considered in HW-SW architecture optimization. In the next step all approaches are combined to demonstrate a 4,158-qubit system called Kookaburra in 2025. It comprises three 1,386 qubit multi-chip processors with quantum communication links. The combination of these technologies - classical parallelization, multi-chip quantum processors, and quantum parallelization - will enable the scaling of quantum processors in the future. The described path to large-scale quantum systems is summarized in the Quantum Development Roadmap in Fig. 8.

B. Quality

Quality is a quite complex parameter and can be measured by the quantum volume (QV). This single HW-agnostic parameter measures the largest random circuit with two-qubit gates of equal width (number of qubits involved) and depth (the number of discrete time steps during which the circuit can run gates before the qubits decohere) that can be successfully implemented [15]. The QV quantifies the largest quantum computational space a QPU can utilize, and thus higher QV

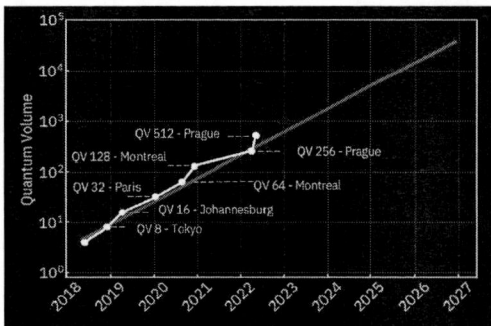

Fig. 5. Increase of the Quantum Volume (QV) of IBM Quantum processors (named Tokyo, Johannesburg, Paris, Montreal and Prague). The QV was doubled each year over the last five years reaching a value of 512 on the Falcon r10 QPU called Prague.

directly equate to higher performance. This metric considers all relevant HW parameters, comprising the performance parameters (coherence, calibration errors, crosstalk, spectator errors, gate and measurement fidelity, initialization fidelity) as well as design parameters such as connectivity and gate set. It also includes the SW behind the circuit optimization, e.g., compilers which can optimize circuits to minimize the effect of decoherence. The QV is architecture-independent and can be applied to any system that can run quantum circuits.

QV is a holistic metric because it cannot be increased by just improving one aspect of the system, but rather requires all parts of the system to be improved in a synergistic manner. In general, low operation errors, high connectivity and large calibrated gate sets are needed to achieve large QV. Over the last five years we doubled the QV each year as indicated in Fig. 5 reaching a QV of 512 on the Falcon r10 QPU called Prague. This increase was enabled by faster and higher-fidelity gates with the bulk of the two-qubit gates approaching 99.9% gate fidelity. Incorporating improvements of the compiler, the calibration of two-qubit gates as well as the noise handling and readout enabled by microwave pulse tuning were required to achieve the QV improvements [16].

An important parameter for QV is the coherence time which depends on a variety of factors like qubit design, microwave engineering, shielding, thermalization and filtering as well as material quality and semiconductor fabrication processes. Important progress has been achieved recently increasing the T1 coherence time by 3x achieving 300 µs in the 27-qubit core processors revision 8 (R8) compared to Falcon R5 (Fig. 6). In test devices T1 times in the millisecond range are achieved.

Another crucial parameter for QV is the error rate in particular two-qubit gate errors are a bottleneck today. Advances in gate fidelities immediately translate to measurable improvements in computation. We have demonstrated continued improvements in CNOT error rates as shown in Fig. 7. The colors indicate various processors differing in design (20-qubit Penguin, 27-qubit Falcon, 65-qubit Hummingbird) and revision. The heavy-hex topology resulted in a reduction of the frequency collisions and spectator errors and thus a reduced error rate [17]. With the recent revision of the 27-qubit Falcon processor R10 we achieved a CNOT error rate of 8×10^{-4} and two-qubit gate fidelity of 99.9%.

C. Speed

The third key metric for measuring quantum computing performance is speed. A metric called circuit layer operations per second (CLOPS) was proposed which measures the number of primitive circuits that the quantum system can execute per unit of time [18]. This metric is system-agnostic and captures the full dependencies across HW and SW of circuit executions. To ensure a meaningful comparison between different systems the metric comprises a variety of parameters to cover all system aspects. This includes data transfer, run-time compilation, latencies and initialization of control electronics, gate and measurement times, reset time of qubits, delays between circuits, processing results as well as parameterized updates. The benchmark can be used to finding the bottlenecks for improving speed. Subdividing the CLOPS into three main parts facilitates further analysis: 1. circuit execution which is the time spent running the circuit on the device, 2. Circuit delay which is the delay time between each shot of each circuit on the device, 3. run-time compilation and data transfer which describes the time spent on preparing the circuits to run on the device and data transfers. An analysis of QPUs with same QV but different number of qubits shows that the circuit delay is 10x larger than gate time. Run-time compilation and data transfer are 100x larger limiting the overall utilization. Hence, the system architecture comprising classical and quantum computing is crucial.

Today the CLOPS achieved are 1.4k measured on a 5-qubit chip with QV=32 and 2.4k on a Falcon R5 [13], targeting 10k CLOPS end of year 2022. Improving speed can be achieved by changes in the SW stack, reducing initialization effort as well as the amount of data needed to be loaded. In addition, the time to compile the circuits into the instructions needed for the control electronics must be reduced and a new high-performance compiler must be developed to further improve CLOPS.

Physical qubit architectures may effect the repetition time, gate times, reset times, and measurement times and can vary significantly across technologies. For example, the repetition rate and gate rate of superconducting qubits can be orders of

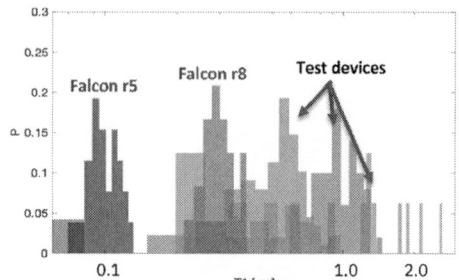

Fig. 6. Distribution of measured coherence times T1 in 27-qubit Falcon R5 and R8 processors and test devices.

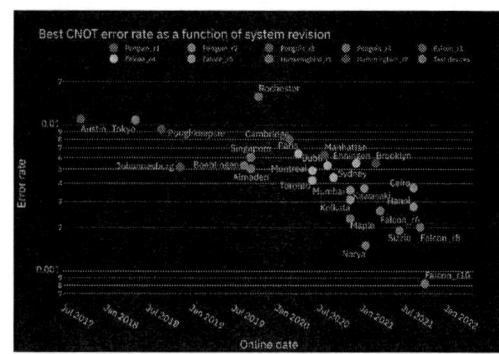

Fig. 7. Best CNOT error rate as function of system revision of IBM Quantum processors. The colors indicate different quantum processor designs and revisions.

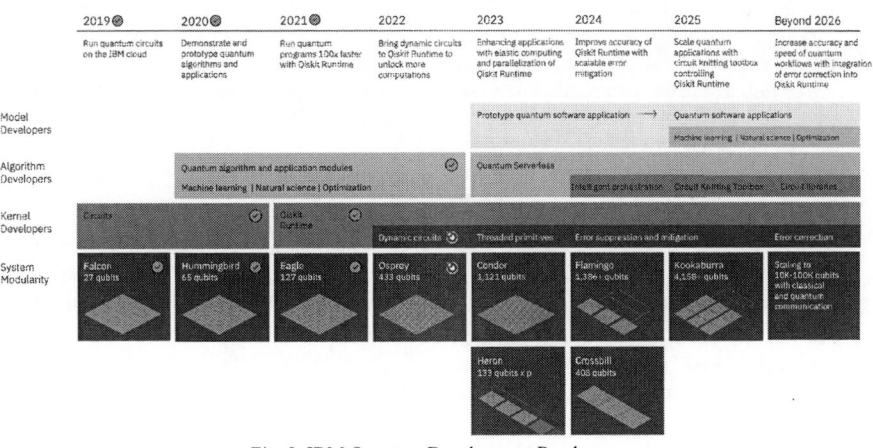

Fig. 8: IBM Quantum Development Roadmap.

magnitude faster than the ones of ion trap qubits [19, 20] which significantly impacts the CLOPS. The speed also depends heavily on the runtime architecture and runtime compilation. Therefore, Qiskit Runtime has been introduced enabling a tight integration between classical and quantum computing.

V. IBM QUANTUM DEVELOPMENT ROADMAP

Quantum computing is the beginning of a new era, and it requires an entire ecosystem to be built up. Creating a technical roadmap to identify and target key demonstrations facilitates solving the biggest challenges and achieving the technological progress required. Therefore, we have created the IBM Quantum Technology Roadmap laying out the path to increasingly larger and improved quantum processors leading to the million-plus qubit devices of the future as shown in Fig. 8. The roadmap summarizes HW and SW goals and describes the modular HW scaling approach for the next few years, as discussed in section IV. It lays out the development of SW tools and services deployed via the cloud, on top of the evolving quantum HW. The SW stack comprises different levels of abstraction to serve various types of developers starting from kernel developers who are HW-aware, to algorithm developers and finally model developers, each one making use of the tools created one level below.

Crucial to building these quantum-centric supercomputers is the interaction of the quantum HW with the classical computer as quantum applications always require classical processing. To optimize this interaction Qiskit Runtime was introduced; a runtime environment of co-located classical and quantum systems that support containerized execution of quantum circuits to increase speed and quality. A significant speed increase was achieved for simulating molecules with the Variational Quantum Eigensolver (VQE) algorithm which requires numerous iterations between quantum and classical resources during calculation. Qiskit Runtime supports this iterative efficient running of different circuits and updating the future circuits based on the measurement of the previous one. A combination of improvements at the algorithmic level, the system SW, the control, the gate and measurement fidelities together with the use of Quiskit Runtime enabled a significant speedup of 120x for VQE simulations. Qiskit runtime will be expanded with threaded primitives that support, e.g., the use

of parallelized quantum processors, and of error suppression and mitigation tools crucial for Quantum Advantage.

Another speed up of computation will be achieved by dynamic circuits. These are circuits in which future states depend on outcomes of measurements that happen during the circuit. They involve the evolution of the quantum state as well as periodic mid-circuit measurements of a subset of qubits and concurrent processing of the resulting classical information within timescales shorter than the execution times of the circuits [19]. It has been demonstrated that by resetting the state multiple times an increased fidelity of zero-state preparation can be achieved. Dynamic circuits extend what the HW can do by reducing circuit depth, by allowing for alternative models of constructing circuits, and by enabling parity checks of the fundamental operations at the heart of quantum error correction. Dynamic circuits will be at the core for future quantum circuit libraries, algorithms, and applications.

To establish the SW stack, a first suite of quantum algorithms and application modules has been developed comprising Optimization, Natural Sciences and Machine Learning. At the next stage a serverless programming model will be introduced to enable flexible quantum-classical resource combinations without requiring developers to be HW and infrastructure experts. Further functionalities will be developed and integrated to enable easy use and scaling of quantum computing services through the cloud. The roadmap shown in Fig. 8 presents a clear path to develop quantum-centric supercomputers, where quantum resources will be woven together with classical computing into a compute fabric comprising the HW and SW as well as application workflows.

VI. TOWARDS QUANTUM ADVANTAGE

The big interest in quantum computers is the potential to solve specific mathematical problems that are intractable to classical machines. Many important mathematical problems in science and business belong to this class and are currently investigated: this includes (a) simulation of quantum systems, (b) algebraic problems like factoring, differential equations, and (c) quantum search problems like Quantum Monte Carlo sampling, optimization, and graph problems. Theoretical evidence is given that quantum algorithms drastically outperform classical ones for some problems when using fault tolerant quantum computers. Although quantum computers have seen tremendous improvements in their scale, quality, and speed in recent years they are still noisy and cannot yet run error-corrected computations. This noise accumulates over the execution of a quantum algorithm and thus limits the execution to restrictively small quantum circuits. Today the big challenge is achieving quantum advantage. A current research focus is to find the right problems that are solvable with quantum circuits and are known to be difficult to simulate classically. To accelerate the path towards quantum advantage by enhancing the accuracy of computation error mitigation techniques have proven to be beneficial [21-24]. Quantum error mitigation provides a collection of tools and methods that allow to evaluate accurate expectation values from noisy, shallow depth quantum circuits, even before the introduction of fault tolerance. These techniques may achieve practical runtime reductions for complex problems and thus achieve quantum advantage.

Another valuable approach to increase the complexity of problems that can be calculated on today's available HW is a smart combination of quantum and classical resources called circuit knitting. Three different techniques, embedding, entanglement forging, and circuit cutting are investigated showing promising results. Circuit cutting expands the reach of quantum computers with partitioning and post-processing techniques that augment small quantum systems with CPUs and GPUs [25, 26]. Hereby a large quantum circuit is cut into smaller subcircuits at less entangled connections. The subsystems are independently executed by quantum processors with less quality and size requirements. This approach can be implemented with the classically coupled QPUs and offers a practical strategy for hybrid quantum-classical advantage in quantum computing applications.

Entanglement forging is a method to simulate a given quantum system using only half as many qubits on a quantum computer. It was used to successfully simulate the ground state energy of a water molecule representing 10 spin-orbitals on just five qubits achieving highly accurate results on the order of 1-10 milli Hartree [27]. The idea is to divide the problem into two weakly entangled halves that can be calculated separately on a quantum computer and then use classical resources to calculate the entanglement between them, thus reconstructing the larger problem by classical processing. The problems which fit best the technique have few correlations between the two halves of the original quantum system. Examples are molecular systems with limited entanglement between the spin-up and -down orbitals.

The third method is an embedding scheme that enables the partitioning of electronic structure calculations into an active subsystem treated with a high-level quantum algorithm and an environment described at the Hartree-Fock or density functional theory level [28]. Thus, the quantum calculations can be restricted to a critical subset of molecular orbitals that can fit on current quantum computers, while the remaining electrons provide the embedding potential computed using a classical algorithm. This approach is very interesting for chemical processes as there the electronic structure predictions depend only on a small set of frontier orbitals. Significant improvements in accuracy have been shown on test systems like molecular nitrogen, molecular oxygen, water, and oxirane C_2H_4O with the latter demonstrating the applicability of the proposed quantum embedding scheme also

to complex organic molecules. The possibility of partitioning the solution of the electronic structure problem into an active component treated by the quantum system and an inert environment component solved at the HF or DFT theory will enable the use of quantum computers in the solution of important problems in physics, chemistry, biology, and medicine [28].

VII. SUMMARY

Significant progress has been made on all levels of the full stack of quantum computing systems and a roadmap to scale the HW and develop the SW to provide performance and ease of use has been established. Superconducting qubits are a preferred technology for scalable quantum HW delivering high quality circuits and computational speed, with opportunities for continued improvements. The three key metric scale, quality and speed, provide a HW-agnostic and holistic performance measure of the system that is a prerequisite for further improvements. A tight integration between quantum and classical computing resources is essential for best performance. The techniques of quantum error mitigation and circuit knitting together with the continuous improvements in performance of quantum computing create a continuous path towards achieving quantum advantage for valuable problems.

ACKNOWLEDGMENT

The Paper summarizes work across the global team of IBM Quantum and IBM Research. The author gratefully acknowledges the contributions of all team members.

REFERENCES

[1] P. W. Shor, SIAM J. Comput., vol. 26, p. 1484, 1997.

[2] L. K. Grover, Phys. Rev. Lett., vol. 79, p.325, 1997.

[3] M. H. Yung, et al. Adv. in Chem. Phys. vol. 154, S. Kais, Ed., Wiley, 2014.

[4] M. H. Devoret, R. J. Schoelkopf, Science, vol. 339, p. 1169, 2013.

[5] A. D. Corcoles, et al. Nature Communications, vol. 6, p. 6797, 2015.

[6] P. Krantz, et al., Applied Physics Reviews , vol. 6, p. 021318, 2019.

[7] A. Blais, et al. Rev. Mod. Phys. vol. 93, p. 25005, 2021.

[8] https://qiskit.org

[9] C. Chamberland, et al. Phys. Rev. X, vol. 10, p. 011022, 2020.

[10] M. Takita, et al., Phys. Rev. Lett. vol. 119, p. 180501, 2017.

[11] J.B. Hertzberg et al., npj Quantum Information, vol. 7, p. 129, 2021.

[12] E. Chen, et al., Phys. Rev. Lett. vol. 128, p. 110504, 2022.

[13] https://research.ibm.com/blog/eagle-quantum-processor-performance

[14] Harris et al., Rev. of Scientific Instruments, vol. 83, p. 086105, 2012.

[15] A. W. Cross et al., Phys. Rev. A vol. 100, p. 032328, 2019.

[16] P. Jucevic et al., Quantum Sci. Technol. vol. 6, p. 025020, 2021.

[17] N. Sundaresan, et al. arXiv:2007.02925, 2020.

[18] A. Wack, et al., arXiv:2110.14108v2, 2021.

[19] A.D.Corcoles et al. Phys. Rev. Lett. vol. 127, p. 100501 2021.

[20] J. M. Pino, et al. Nature vol. 592, p. 209, 2021.

[21] K. Temme, et al. Phys. Rev. Lett. vol. 119, p. 180509, 2017.

[22] Y. Li, S.C. Benjamin, Phys. Rev. X, vol. 7, p. 021050, 2017.

[23] A. Kandala et al. Nature vol. 567, p. 491, 2019.

[24] Y. Kim et al. arXiv:2108.09197v1, 2021.

[25] S. Bravyi, et al. PRX vol. 6, p. 021043, 2016.

[26] T. Peng, et al. PRL vol. 125, p. 150504, 2020.

[27] A. Eddins et al. PRX Quantum vol. 3, p. 010309, 2022.

[28] Rossmanek et al. J. Chem. Phys. vol. 154, p. 114105, 2021.

SiC Power Device Mass Commercialization

V. Veliadis
PowerAmerica/North Carolina State University, 930 Main Campus Drive, Suite 200
Raleigh NC, USA
jvveliad@ncsu.edu

Abstract—In an increasingly electrified technology driven world, power electronics is central to the entire clean energy manufacturing economy. Silicon (Si) power devices have dominated power electronics due to their low cost volume production, excellent starting material quality, ease of fabrication, and proven reliability. Although Si power devices continue to improve, they are approaching their operational limits primarily due to their relatively low bandgap, critical electric field, and thermal conductivity that result in high conduction and switching losses, and poor high temperature performance. Silicon Carbide's (SiC) compelling efficiency and system benefits have led to significant development efforts over the last two decades and today planar and trench MOSFETs, and JFETs are commercially available from several vendors as discrete components or in high power modules in the 650 V to 3.3 kV voltage range. High impact application opportunities, where SiC devices are displacing their incumbent Si counterparts, have emerged and include automotive and rail power electronics with reduced losses and reduced cooling requirements; novel data center topologies with reduced cooling loads and higher efficiencies; variable frequency drives for efficient high power electric motors at reduced overall system cost; more efficient, flexible, and reliable grid applications with reduced system footprint; and "more electric aerospace" with weight, volume, and cooling system reductions contributing to energy savings. In particular, SiC insertion in electric vehicles brings major competitive advantages and is a volume application opportunity that can spur manufacturing economies of scale and lower system costs. As SiC continues to grow, the industry is lifting the last barriers to mass commercialization that include higher than Si device cost, relative lack of wafer planarity, the presence of basal plane dislocations, reliability and ruggedness concerns, and the need for a workforce skilled in SiC power technology to keep up with the rising demand. It should be noted that in many applications, insertion of SiC reduces overall system cost compared to Si even though SiC devices can cost 2-3 more than their Si counterparts. This is due to the passive component and cooling system simplifications enabled by the efficient high frequency SiC operation. In this paper, we will review key aspects of SiC technology and discuss overcoming barriers to mass commercialization.

Keywords—SiC, substrate, epitaxy, fabrication, etch, heated implantation, Ohmic contact, silicide, SiC fab infrastructure, SiC foundry, fabless, basal plane dislocations, bipolar degradation, 200 mm

I. SiC Wafers

Today, the SiC wafer represents 50-65% of the overall SiC device cost [1], a consequence of its unique complex fabrication specifics. Conventional SiC substrates are primarily grown by the seeded sublimation technique at temperatures of ~2500 °C, which creates process control challenges. Crystal expansion is limited requiring the use of large high-material quality seeds, and the sublimation growth rates can be relatively low in the order of 0.5-2 mm/h. Dislocations propagate through the boule and are present in the device wafers. Furthermore, SiC material's hardness, which is comparable to that of diamond, makes sawing and polishing SiC substrates slow and costly relative to Si.

The epitaxial layers, where SiC devices are fabricated, are grown by chemical vapor deposition (CVD) in horizontal or planetary reactors at 1500-1650 °C. Pressure typically ranges from 30 to 90 Torr and growth rates can be as high as 46 mm/hr. Epitaxial growth is done on 4-degree off-cut substrates to maintain the polytype stability of the substrate. Epitaxy goals are to restrict "performance-degrading" defect propagation from the substrate into the epitaxy and ensure that any "performance-degrading" defects propagating from the substrate propagate as benign defects into the epitaxy. As defects in SiC wafers limit large-area device yields, and numerous devices are paralleled in modules to increase current output, tight epitaxial doping and thickness uniformities are highly desirable, particularly as wafer size increases. Fig. 1 shows a JFET pinch-off voltage map (a parameter that depends on doping) of an 84.3% overall yield wafer [2]. The yield is based on the screening parameter ranges that appear in the top right table. The pinch-off voltage screening ranges appear in the bottom right table. "Failed" JFETs are shown in gray and are primarily located at the wafer edges. The 84.3% yield includes the contribution of devices located at the wafer edges. The large uniformly situated white squares contain experimental devices and characterization structures. Each color-coded wafer map area corresponds to JFETs with pinch-off voltages within the range shown in the bottom right table. The "bull's eye" wafer map structure (concentric circular regions of different colors) correlates well with the measured radial doping variation in the epitaxy of this "older generation" 3-inch 4H-SiC wafer. The radial doping variation can render "outer" devices out of specification and lower yields.

978-1-6654-8498-5/22 $31.00 © 2022 IEEE

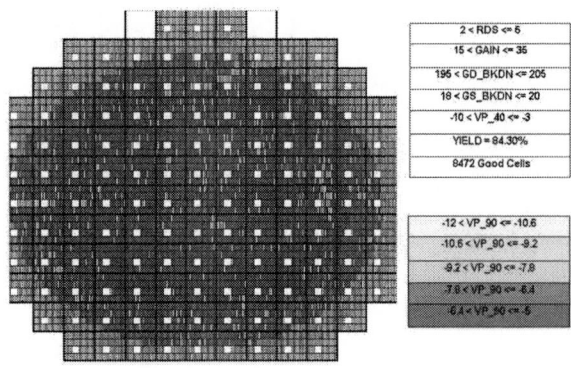

2 < RDS <= 5	
15 < GAIN <= 35	
195 < GD_BKDN <= 205	
18 < GS_BKDN <= 20	
-10 < VP_40 <= -3	
YIELD = 84.30%	
8472 Good Cells	

-12 < VP_90 <= -10.6
-10.6 < VP_90 <= -9.2
-9.2 < VP_90 <= -7.8
-7.8 < VP_90 <= -6.4
-6.4 < VP_90 <= -5

Fig. 1. JFET pinch-off voltage wafer map." The "bull's eye" structure (concentric circular regions of different colors) correlates well with the measured radial doping variation in the epitaxy of this "older generation" 3-inch 4H-SiC wafer. The pinch-off voltage screening parameter ranges appear in the bottom right table.

Overall, SiC wafer fabrication (substrate and epitaxy) is more complex and slower than that of silicon. The result is more expensive wafers, and ultimately higher device costs. A key part of the vertical integration occurring in today's SiC industry is securing internal substrate and epitaxy wafer capabilities to eliminate purchasing profit margins. In addition, opportunities for disruptive SiC substrate formation, boule slicing, sawing/polishing, etc., have a high return and are being sought by several companies [3].

Presently, the majority of SiC device production is on 150 mm wafers. 200 mm SiC wafers were demonstrated in 2015, and a seven-year or so period historically passes before they become commercially available. It is highly desirable that defect density and the cost of material per cm2 are the same or lower for 200 mm vs. 150 mm wafers. In addition, wafer planarity should not be worse in the 200 mm wafers. Due to the large fab overheads, and assuming 200 mm tools are in place, the cost of processing a wafer is to a first approximation unrelated to its size. So processing a 200 mm wafer will produce about 1.8 times more devices than a 150 mm wafer at the same processing cost. Of course, a 200 mm wafer will be more expensive than a 150mm wafer, and that needs to be factored into the overall cost equation. Many silicon fabs are starting to also process SiC wafers. Given the plethora of 200 mm mature Si fabs with fully depreciated tools, there are many large 200 mm Si companies waiting on the sidelines to enter SiC production when 200 mm wafers become commercially available. These are companies that have established 200 mm silicon wafer production and do not want to retool to fabricate at the 150 mm SiC wafer size that is currently commercially available. Therefore, when 200 mm wafers become commercially available, several 200 mm fabs are expected to produce SiC devices. To illustrate the economic benefits of moving to larger wafer size, let us assume starting material cost parity per cm² between 150 mm and 200 mm wafers. Further assuming a $1500 cost for a fully processed 150 mm SiC wafer, and 60% of that coming from the starting wafer material (40% fabrication cost), a back of the envelope calculation points to a 17% device cost reduction when switching to 200 mm. The same fabrication cost scenario, with now 50% of the overall cost representing the starting wafer material, allows for a 22% device cost reduction when switching to 200 mm wafers. These calculations do not include the additional processing cost reductions that will come with streamlined mass production in large 200 mm volume fabs.

II. SiC DEVICE FABRICATION

For mass SiC commercialization, high yielding fabrication processes are required. Numerous well-established processes from silicon technology have been successfully transferred to SiC. However, SiC material properties necessitate optimization of specific processes including wafer thinning, etching, heated implantation and anneal, and low resistivity Ohmic contact formation [2]. SiC is inert against chemical solvents and only dry etching is practical. Furthermore, the hardness of SiC results in low photoresist selectivity and a "hard" mask, usually composed of metal or dielectrics, is required for SiC photolithographic patterning and etch. 0.7 μm deep SiC trenches, etched using a "silicon" reactive-ion etch (RIE) tool and a Cr/Al mask, are shown in Fig. 2. Cr assists with adhesion of metal layers to the underlying SiC surface. The RIE was fluorine based for higher mask/SiC selectivity. RIE settings were optimized to eliminate micro-masking and achieve vertical etched sidewall profile formation.

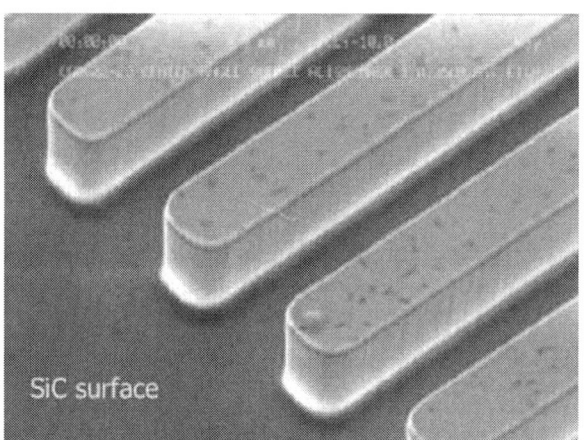

Fig. 2. Reactive-ion etched 0.7 μm deep vertical SiC mesas using a Cr/Al mask. Cr assists with adhesion of metal layers to the underlying SiC surface. The RIE was fluorine based.

Conventional thermal diffusion is not realistic in doping SiC due to its high melting point and the low diffusion constant of dopants within SiC. Heated ion-implantation is typically performed for doping densities of 1016-1020 cm-3 (the higher doping densities facilitating ohmic contact formation), and room temperature implantation can work well for low implant doses (~1015 cm-3). Nitrogen/phosphorus and aluminum are the preferred impurities for n-type and p-type SiC doping, respectively. The as-implanted depth profiles are retained after the anneal for Al, P, and N as expected from their low diffusion constants. The lack of diffusion makes it easy to form shallow junctions and difficult to form deep ones. After ion implantation, a 1600-1800 °C anneal is performed for lattice damage recovery, and high dopant electrical activation. A protective cap layer covering the SiC wafer protects its surface from degradation due to Si desorption and migration of surface atoms, Fig. 3.

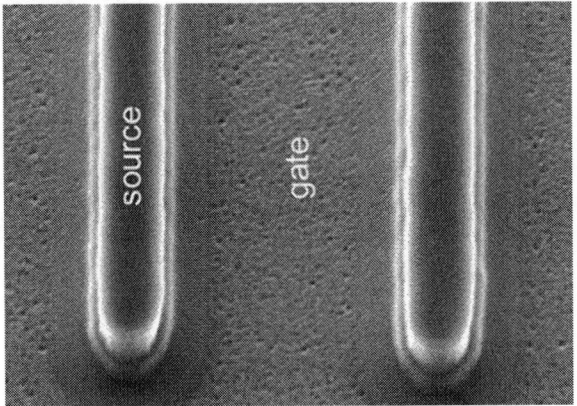

Fig. 3. Scanning electron microscopy image of a "post p+ ion-implantation" 1650 °C annealed SiC wafer in the presence of a carbon protective cap layer. Excellent surface morphology and high device yield are attained.

The high value of the SiC/metal barrier results in rectifying metal contacts and post metal deposition anneal is required for ohmic contact formation. Typically, a 50-100 nm Ni layer is blanket deposited and patterned on the wafer for the simultaneous ohmic contact formation on n-type and p-type doped regions, Fig. 4. Depending on the specifics of the fabrication process, isolating the source from the gate areas with dielectrics can facilitate high yields in the subsequent high temperature processing.

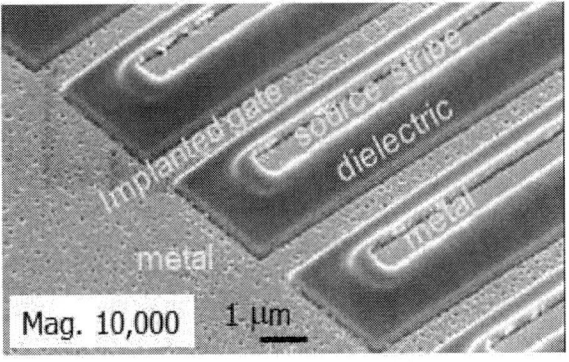

Fig. 4. Scanning electron microscopy image of a patterned Ni layer on a SiC wafer surface. Dielectric isolates the metalized p+ implanted gate areas (pitted surface) from the n-doped source stripes.

High temperature annealing of the Ni patterned wafer creates Ni-silicide for low resistivity ohmic contact formation. Rapid thermal annealing (RTA) at 950 °C, using standard silicon fabrication equipment, was used to create Ni-silicide with no metal strings, Fig. 5. The dielectric isolates the source from the gate areas eliminating shorting during the high temperature silicide process.

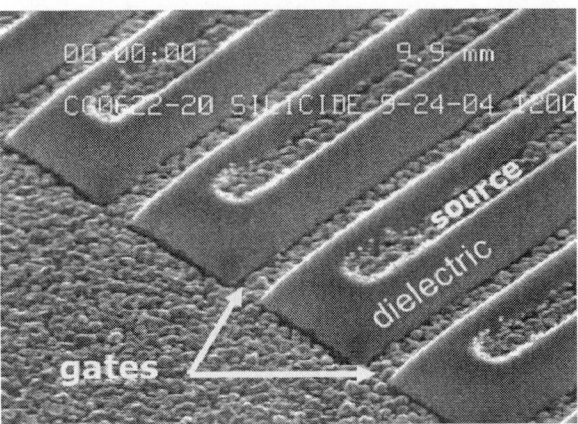

Fig. 5. Scanning electron microscopy image of the Ni patterned SiC wafer of Fig. 4, after a 950 °C rapid-thermal-annealing event. Ni silicide is formed with no shorting of the p-gates to the n-source regions.

Unlike Si wafers, SiC wafers are transparent. This complicates the use of "silicon" tools for CD-SEM and metrology measurements, as the focal plane is determined with the use of an optical microscope. SiC-specific wavelength metrology/inspection tools are now available from multiple vendors. Another issue is the relative lack of flatness of SiC wafers, compared to those of Si, which can complicate photolithography. In addition, the high-temperature SiC processing can further degrade wafer flatness, occasionally rendering wafers unusable. This is particularly problematic with the thick epitaxy wafers used in +3.3 kV device fabrication. Efforts are underway to produce flatter starting SiC wafers, and to minimize flatness degradation during fabrication. Lastly, the poor SiC/SiO2 interface quality reduces inversion layer mobility. Thus, passivation techniques including annealing in nitrides are utilized to improve the SiC/SiO2 interface quality similar to the case of silicon [4].

III. DEFECTS, RELIABILITY, AND RUGGEDNESS

The majority of "killer" defects have been virtually eliminated in modern SiC wafers. Basal-Plane-Dislocations (BPDs) are the major remaining defect degrading device performance [5]. BPDs can propagate from the wafer substrate through the thickness of the epitaxial layers where devices are fabricated. BPDs can also be generated during the high-temperature ion-implantation fabrication process. In commercial wafers, more than 95% of substrate BPDs propagate as relatively "benign" threading edge dislocations in epitaxial layers grown off axis by CVD [6].

When bipolar current flows through a SiC device, electron–hole pair recombination at BPDs in the drift layer provides the energy to activate dislocation glides that give rise to stacking faults and degradation. To investigate the impact of BPDs on the electrical characteristics of ion-implanted SiC transistors 17 JFETs with 100-μm drift epitaxial layers (10 kV rated) were stressed at a fixed gate-drain DC bipolar current density of 100 A/cm2 for 5 hours. Representative curves are presented in Fig. 6 [7].

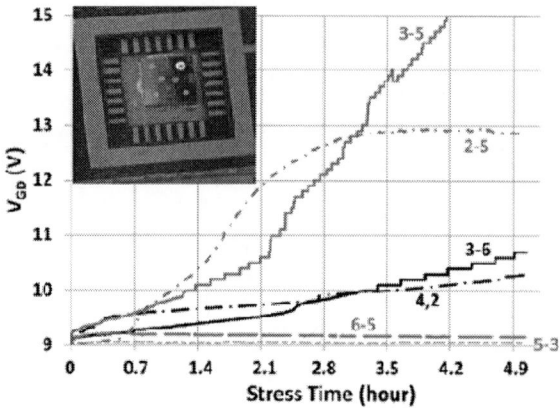

Fig. 6. Representative 100 μm drift layer SiC JFET forward gate–drain voltages as a function of time for a fixed DC 100-A/cm2 gate–drain bipolar stress. The compliance is set at VGD = 15 V. A biased JFET, with its gate–drain diode's bipolar current giving rise to blue/violet electroluminescence, is shown in the inset.

At 100 A/cm2, the gate-drain p-n junction is turned on, as evidenced by the emission of blue/violet electroluminescence at the edges of the JFET (inset photograph of Fig. 6), and bipolar current flows. Of the 17 JFETs stressed, six exhibit no forward gate-drain voltage degradation. This confirms that with optimized process flows and implantation recipes, BPDs are not generated during fabrication. Nine JFETs exhibit intermediate voltage degradation, and two exhibit severe voltage degradation. Bipolar current in the presence of BPDs leads to forward gate-drain p-n junction and ON-state conduction degradations as shown in Fig. 7. Interestingly, transistor BPD-related electrical degradations can be fully reversed by annealing at 350 °C, while non-degraded characteristics remain unaffected by this annealing.

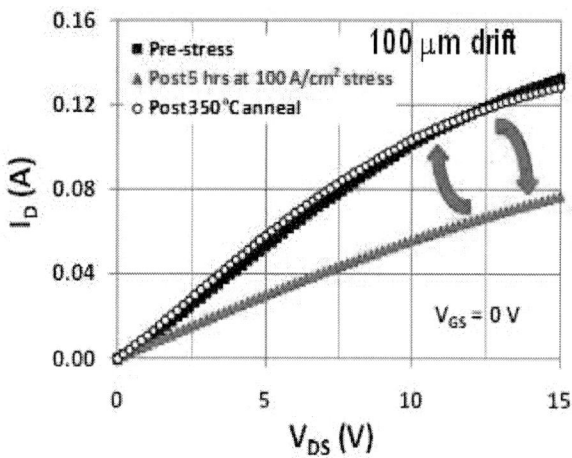

Fig. 7. BPD-related JFET ON-state conduction degradation and full recovery by annealing. The black squares, red triangles, and open circles represent the ON-state conduction characteristics prior to bipolar stress, after 5 hours of 100-A/cm2 bipolar stress, and after a 350 °C anneal, respectively.

Threshold voltage instability is the main remaining reliability concern in SiC MOSFETs, which are the dominant transistors in SiC-based power electronics applications. It is primarily due to oxide traps at the SiC/gate-oxide interface. A positive shift in the SiC transistor's threshold voltage has the deleterious effect of increasing conduction losses, while a negative shift is undesirable as it can spontaneously turn the device on. Valuable SiC reliability data has been accumulated over years of field operation and is driving device optimization. SiC devices can be made more rugged by leveraging design trade-offs. This, combined with intelligent and fast gate drives, can provide adequate circuit protection [8].

IV. U.S. SiC FAB INFRASTRUCTURE

Device manufactures have developed IP for several SiC processing steps and compete on both design and processing. Although SiC is not fully CMOS-compatible, the SiC industry has leveraged Si technology processes and infrastructure by making the relatively small financial investments required to adapt existing fabs. Today, SiC manufacturing has matured and its fab infrastructure now mirrors that of Si. Integrated SiC device manufactures coexist with foundries and fabless companies, and design houses provide know-how and IP that can be leveraged to accelerate entry to market, Fig. 8.

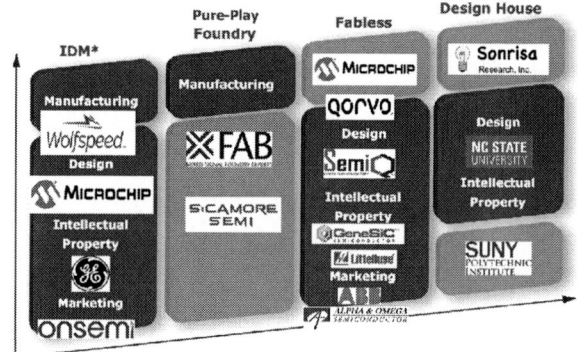

Fig. 8. The U.S. SiC fab infrastructure mirrors that of Si. It consists of Integrated Device Manufacturers, Foundries, Fabless companies, and Design houses.

SiC device fabrication in volume fabs alongside Si has emerged as a cost-reduction model exploiting silicon manufacturing economies of scale. Through re-purposing older fully depreciated 150 mm (and 200 mm in the near future) Si foundries, SiC power devices can be manufactured with the relatively small investments necessary to support the unique SiC processing steps. Minimizing fabrication cost by exploiting the mature Si volume production assumes the fab is loaded close to capacity with standard Si and SiC processes running on the same lines. In addition, aggregating the demand for SiC substrates and epilayers in volume fabs contributes to lower material costs. Lower fabrication costs in a fully depreciated Si+SiC "capacity loaded" fab, coupled with decreased material costs can lead to significant price reductions for SiC devices. This approach offers a new opportunity for outdated Si foundries, which have not kept up with the channel length reductions of the last two decades, to continue manufacturing legacy Si parts while ramping up SiC fabrication that requires relatively modest ~0.3 micron design rules [9].

V. SiC Commercialization Initiatives

Helping to exploit the potential for energy savings and technological innovation are several initiatives promoting the adoption of wide bandgap power devices. In the US, a number of government programs played a major role in supporting early work to develop advanced crystal growth of SiC, wafer fabrication, and device processing technologies. As far back as the late 1980s, organizations such as the Air Force Research Laboratory, the Army Research Laboratory, the Office of Naval Research, the Missile Defense Agency, and the Defense Advanced Research Projects Agency provided hundreds of millions of dollars to fund what has been decades of work at universities, industry, and government laboratories. Efforts initially focused on developing critical enabling technologies such as high-quality substrates and epitaxy; and unit process steps such as ion-implantation, implant activation, and gate oxidation. Success ensured a domestic source for the US Department of Defense Wide Bandgap systems. More recently, the Advanced Manufacturing Office of the US Department of Energy and North Carolina State University have formed PowerAmerica, a member-driven consortium of industry, universities, and national labs accelerating the commercialization of energy efficient silicon carbide and gallium nitride power semiconductor technologies. The PowerAmerica membership network spans the wide-bandgap technology ecosystem, from materials to device developers and fabs to module manufacturers to end users, as well as universities that educate and supply the future workforce, Fig. 9.

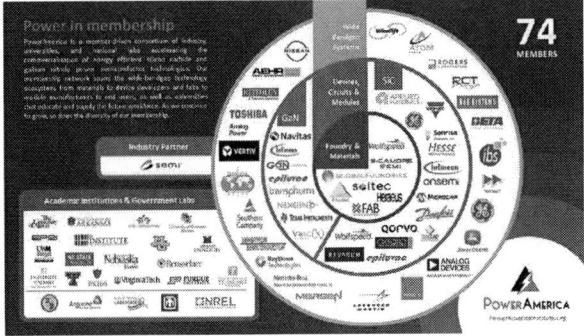

Fig. 9. The PowerAmerica membership network spans the wide-bandgap technology ecosystem.

Working with its members, PowerAmerica invested US$145M in 196 industry/university collaborative power WBG projects in 2016-2020 focusing on device design and process integration, semiconductor fab development, modules, reliability, and circuit integration addressing all major applications including automotive and rail traction, on board chargers, aerospace, photovoltaic, flexible alternative current

transmission systems, high voltage DC systems, microgrids, energy storage, motor drives, UPS, and data centers. Execution of these seed projects demonstrated the WBG competitive system advantages and catalyzed further industry investments in these technologies that are creating high-tech manufacturing jobs and energy savings; both crucial to the economy and national security. A timeline snapshot of the project topic areas is shown in Fig. 10.

PowerAmerica $145M investment	2016	2017	2018	2019	2020	Total
Industry Projects	14	14	16	15	14	**73**
Number of Academia Projects (top row in Education & Workforce, bottom	6	9	4	9	8	36
row in power applications)	12	16	15	15	18	**76**
National Lab Projects	2	2	2	2	3	**11**
Yearly projects	34	41	37	41	43	**196**
Industry Projects 2016-2020						**73**
Academic Projects 2016-2020						**112**

67 Low-Voltage SiC Device/Fab Projects

28 Module/Reliability Projects

24 Low-Voltage GaN Applications Projects

40 Low-Voltage SiC Applications Projects

21 Medium -Voltage SiC Device/Fab Projects

20 Education Projects

4 additional projects in 202¹

Fig. 10. Timeline snapshot of the 196 PowerAmerica project topics performed in 2016-2020.

Through these hands-on projects, PowerAmerica trained 410 full-time university students in applied WBG technology. Furthermore, PowerAmerica educational activities engaged over 4100 attendees in tutorials, short courses, and webinars evangelizing the merits and competitive advantages of SiC and GaN power technologies.

One of the strengths of PowerAmerica is that it has the reach and depth to connect companies and practitioners across the wide bandgap supply chain. It provides its members with unparalleled opportunities to effortlessly make connections, create partnerships, advance technology innovation, grow their business, and build their brand. Members have complimentary access to online members-only business and technical content – including market research and presentations – and this provides powerful context and data for making sound technical and business decisions. By participating in regularly held member-only meetings, individual companies are able to deliver a collective, amplified voice on issues that affect the wide bandgap industry. This influences its direction and shapes its growth. The work of PowerAmerica also includes the execution of member-initiated, pre-competitive projects. These are selected by members and financed with membership funds. Working on projects of common interest and sharing generated IP is a cost-effective way to spur technological innovation and overcome barriers limiting industry growth. Finally, there are

978-1-6654-8498-5/22 $31.00 © 2022 IEEE

PowerAmerica education and workforce benefits for the members. They include access to industry tailored short courses at reduced cost, specialized tutorials, opportunities for internships and talent recruitment, and interaction with experts across the wide bandgap supply chain.

VI. SUMMARY

The compelling efficiency and system benefits of SiC are leading to wide adoption with insertion in electric vehicles being the volume application that enables manufacturing economies of scale and lower system costs. Industry is removing the last barriers to mass SiC commercialization that include higher than Si device cost, the presence of basal plane dislocations, and reliability and ruggedness concerns. The wafer represents a disproportionately high percentage of the overall SiC device cost, compared to that of Si, and technological improvements and transition to volume 200 mm wafer production will help lower device costs. SiC manufacturing has matured, the non-CMOS compatible processes are fully developed, and today's SiC fab infrastructure mirrors that of Si. Basal plane dislocations are the major remaining yield lowering defect to overcome. Gate-oxide interface quality improvements will further enhance performance and alleviate any lingering reliability concerns.

Through the PowerAmerica WBG commercialization initiative, the US has built a robust and resilient power SiC manufacturing supply chain that is continuing to grow.

REFERENCES

[1] https://www.exa-watt.com/power-electronics

[2] V. Veliadis, Silicon Carbide Junction Field Effect Transistors (SiC JFETs), in: Wiley Encyclopedia of Electrical and Electronics Engineering, Online (2014) pp. 1-37.

[3] https://www.soitec.com/en/press-releases/soitec-announces-joint-development-program-with-applied-materials-on-next-generation-silicon-carbide-substrates

[4] S. Wolf, R. N. Tauber, Silicon Processing for the VLSI Era, vol.1, Lattice Press, California, 1986, pp. 222-223.

[5] E. Van Brunt, A. Burk, D. J. Lichtenwalner, R. Leonard, S. Sabri, D. A. Gajewski, A. Mackenzie, B. Hull, S. Allen, and J. W. Palmour, Mat. Sci. Forum 924 (2018) 137-140.

[6] T. Kimoto, Jpn. J. Appl. Phys. 54 (2015) 040103-1-27.

[7] V. Veliadis, H. Hearne, E. Stewart, M. Snook, W. Chang, J. Caldwell, H. Ha, N. El-Hinnawy, P. Borodulin, R. Howell, D. Urciuoli, and C. Scozzie, IEEE Elec. Dev. Lett. 33, (2012) 952-954.

[8] A. Kumar, S. Parashar, S. Sabri, E. Van Brunt, S. Bhattacharya, and V. Veliadis, IEEE 30th Inter. Sym. on Power Semiconductor Devices and ICs (ISPSD), (2018) 423-426.

[9] V. Veliadis, Compound Semiconductor Magazine 25, vol. 36, pp. 36-42, 2019.

Gap in pagination due to unavailable papers.

Pages 37-224

Multilayer Structure in SeAsGeSi-based OTS for High Thermal Stability and Reliability Enhancement

C. Laguna[1,2], M. Bernard[1], J. Garrione[1], N. Castellani[1], V. Meli[1], S. Martin[1], F. Aussenac[1], D. Rouchon[1], N. Rochat[1], E. Nolot[1], G. Bourgeois[1], M. C. Cyrille[1], L. Militaru[2], A. Souifi[3], F. Andrieu[1] and G. Navarro[1,*]

[1]CEA, LETI and Univ. Grenoble Alpes, F-38000 Grenoble, France
[2]Univ Lyon, INSA Lyon, ECL, CNRS, UCBL, CPE Lyon, INL, UMR5270, 69621 Villeurbanne, France
[3]Univ. de Lyon, Ampere-UMR 5005, INSA Lyon, 69621 Villeurbanne, France.

Abstract—**In this paper, we present an innovative Multilayer SeAsGeSi-based Ovonic Threshold Switching (OTS) Selector targeting high reliability for Crossbar arrays. We compare our Multilayer (ML) OTS with SeAsGeSi-based bulk alloy (SAGS). We demonstrate the high thermal stability of the ML stack against the Back-End-of-Line (BEOL) thermal budget as well as the reduction of the device-to-device variability and reliable switching operations up to 300°C. We study by Raman and FTIR spectroscopy the integrity of the ML OTS material after an annealing of 3 hours at 400°C. SeAsGeSi Multilayer OTS delays crystallization mechanism along cycling. We finally report the successful co-integration of our ML with Phase-Change Memory technology.**

I. Introduction

Ovonic Threshold Switching (OTS) technology, based on amorphous chalcogenide materials, has been investigated in the last years as a reliable Back-End Selector solution because of its unique switching properties. OTS enables highly dense Resistive Non-Volatile Memory Crossbar arrays with agressive shrinkage of the single cell area down to $4F^2$. Therefore, OTS Selector becomes a key volatile device that combines the capability to provide at the same time a high current density when switched in the ON state for resistive memory programming, and an ultra-low leakage current when switched back to the OFF state. Indeed, Phase-Change Memory (PCM) and OTS Selector have been successfully integrated in 3D-stackable 1S1R cell (i.e. one selector and one resistance) [1]–[3]. However, one of the main concerns is the devices thermal stability. Devices should undergo a Back-End-of-Line (BEOL) thermal budget and ensure reliable behaviour at high operating temperatures around 160°C-180°C forecasting the introduction in advanced embedded circuits like in edge AI applications. Recent works show the possibility to improve the thermal stability of OTS alloys [1, 4], nevertheless material study should always be coupled to the investigation of the final devices thermal stability after the initialization step (i.e. called "firing") that could lead to an important evolution of the active layer. AsSe-based OTS selectors have been investigated for their high resistivity (with respect to SbSe-based OTS) and for their switching voltage compatibility

with OTS+PCM co-integration [1, 2]. However, AsSe-based OTS materials featuring the best high temperature stability performance are based on four or even more elements and their high complexity could lead to an intrinsic variability on their electrical parameters [4, 5]. In order to address such issue, we recently demonstrated that OTS Multilayer (ML) structure is an interesting method to decrease variability without affecting the main device characteristics [6, 7].

In this work, we investigate the benefit of ML structure in SeAsGeSi-based quaternary alloys to increase the material thermal stability of initialized (i.e. after firing) devices. The improved yield is verified by an enhanced endurance, a reduced variability of electrical parameters and a higher operating temperature. We gather Raman and FTIR spectroscopic studies on as-deposited and annealed OTS full-sheets for a thorough investigation of their structure and its evolution after annealing up to 400°C. The fabricated OTS devices are then characterized at high operating temperature up to 300°C and after a BEOL-like annealing of 3 hours at 400°C. Finally, we demonstrate for the first time a successful co-integration of ML OTS with PCM.

II. OTS Physico-Chemical Analyses

We deposited our materials using magnetron reactive sputtering from a SeAsGeSi (SAGS) single target. To achieve ML structure, we alternated the deposition of layers of SAGS and GeN, the latter obtained from a Ge target under constant N flow. Material samples were capped without air-break with 3.5 nm thick carbon layer to prevent surface oxidation. Annealing procedure was performed in a pre-heated chamber under N flow at a selected temperature and time.

In **Fig. 1**, we report the Raman spectra obtained for samples as-deposited and annealed at 400°C for 3 hours. As-deposited SAGS and ML exhibit the same three main features: As-Se, Si-Se and Ge-Se. The Ge-Se contribution decreases with annealing in SAGS while ML spectra are not strongly affected by annealing. SAGS spectrum shows an evolution after annealing, which could correspond to the beginning of material segregation. As shown in our previous work on SbSe-based OTS [6], the ML structure delays structural changes related to crystallization process in OTS layers. Indeed, the introduction of GeN layers in SAGS increases the Ge-Se bonds with respect to As-Se and Si-Se because of interlayer interactions leading

---

This work was partially funded by European commission, French State and Auvergne-Rhône Alpes region through ECSEL-IA 101007321 project StorAIge and French Nano2022 program.
*Corresponding author. E-mail: gabriele.navarro@cea.fr (Gabriele Navarro)

978-1-6654-8498-5/22 $31.00 © 2022 IEEE

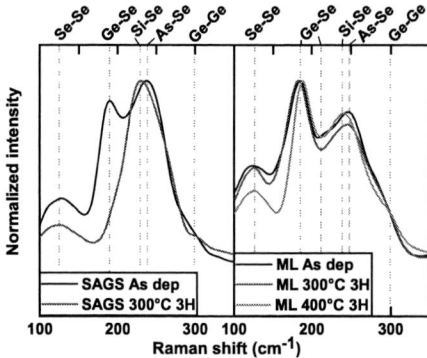

Fig. 1. Raman spectra of SAGS and ML as-deposited and after annealing for 3 hours at 300°C and 400°C. As-deposited, both materials show As-Se at 248 cm^{-1} [8, 9], Si-Se at 238 cm^{-1} [10] and Ge-Se features around 190 cm^{-1} (and 212 cm^{-1} for ML) [8, 11]–[13] with Ge-Ge shoulders at 300 cm^{-1}. Spectrum of SAGS annealed at 400°C is not reported because of the strong degradation of the sample after such annealing.

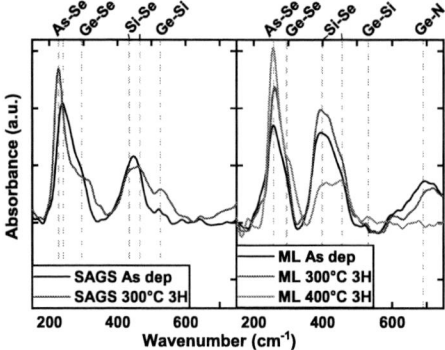

Fig. 2. FTIR spectra of SAGS and ML as-deposited and annealed at 300°C and 400°C during 3 hours. The feature at 230-300 cm^{-1} is a convolution of As-Se and Ge-Se vibrations [14, 15] and the one around 400 cm^{-1} is composed by at least two modes related to Si-Se vibrations [10]. In ML, a Ge-N feature is also observed at 690 cm^{-1} [16] at low temperature. Multiple features of difficult indexation are present around 830 cm^{-1} in ML after 400°C (not reported).

to Ge-Se bonds formation likely responsible for a higher layer stability.

In **Fig. 2**, we report the FTIR spectra for same samples. The spectra gather features linked to As-Se, Ge-Se, Si-Se, Ge-Si and Ge-N (only in ML) bonds vibrations. After annealing, we observe a thinning of the As-Se/Ge-Se convoluted feature, suggesting that the structural order is higher in annealed devices. Annealing modifies the shape of the double-mode Si-Se feature, favouring vibrations at higher frequencies. We observe Ge-N features in ML as-deposited and annealed at 300°C at 690 cm^{-1}. Despite the strong reduction of Ge-N features after annealing at 400°C we think that they should shift at higher frequencies (around 830 cm^{-1} [16]) due to their stabilization in Ge-poorer motifs. Unfortunately the analysis at such frequencies is perturbed by several other contributions in our samples preventing a correct indexation.

ML can undergo the annealing at 400°C without triggering crystallization nor segregation whereas bulk SAGS layer starts important structural changes. The increase of Ge-Se and addition of Ge-N features in ML delay crystallization phenomena

and preserve the amorphous nature integrity of the OTS layer.

III. OTS DEVICES ELECTRICAL CHARACTERIZATION

We integrated SAGS and ML layers in analytical OTS single devices based on a tungsten bottom electrode with a diameter of 300 nm. A thin carbon layer is inserted between the OTS and the titanium nitride top electrode to prevent Ti diffusion within the chalcogenide layer. When not differently specified, data are obtained using AC measurements (i.e. pulsed voltage/current) on populations of 30 devices.

A. High operating temperature effects

In this section, we compare the electrical parameters of SAGS and ML up to an operating temperature of 300°C to evaluate their functionality in high temperature environment.

In order to evaluate the change of the conduction properties in the studied materials after firing, we performed DC current-vs-voltage measurements (IV chracteristics) at several temperatures in the sub-threshold regime before (virgin) and after firing (initialized). From the conductivity dependency on the temperature, we interpolated the effective activation energy (E_m) for each voltage as reported in **Fig. 3** and compatibly with model reported in [17]. The extraction of the real activation energy (E_a) reveals that ML exhibits always a higher E_a with respect to SAGS (i.e. higher energy gap). The presence of additional Ge-Se features and the introduction of GeN [16], highlighted in material analyses, likely contribute to the increase of the energy gap in ML. Initialization is a process responsible of the creation of a channel in the OTS that likely has a different structural organization (and even composition) with respect to the virgin device. This is why the thermal stability of as-deposited layers could significantly differ from the one after firing.

We evaluated the functionality of the devices applying AC pulses inducing ON current in mA range. The initialization voltage (V_{fire}) and the threshold voltage (V_{th}) decrease when the operating temperature increases as shown in **Fig. 4**. SAGS devices do not show functionality above 250°C. At the same time, the leakage current (I_{off}) increases with operating temperature as already observed in [4, 6]. The saturation voltage

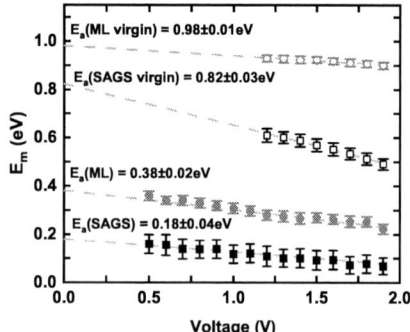

Fig. 3. Activation energy (E_a) extraction from E_m-vs-Voltage interpolations. E_m for each voltage is obtained by current vs temperature fitting (not reported) compatibly with [17] for virgin (before firing) and initialized devices (after firing). ML shows a higher activation energy (i.e. higher gap) with respect to SAGS before and after firing.

978-1-6654-8498-5/22 $31.00 © 2022 IEEE

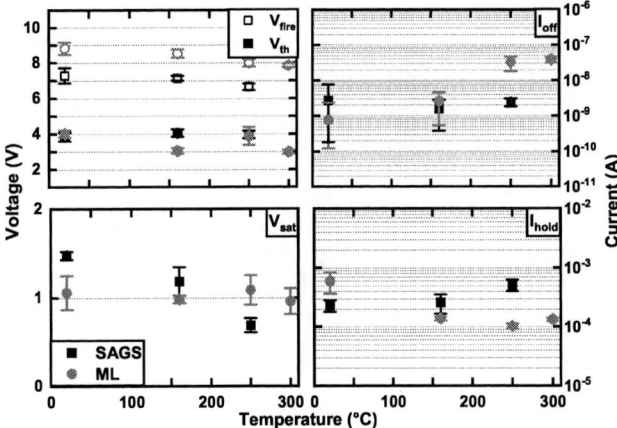

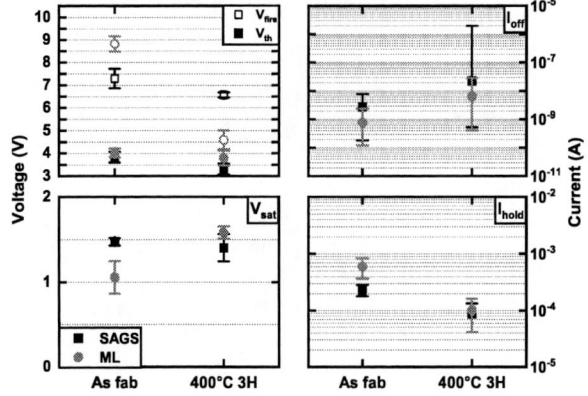

Fig. 4. Electrical parameters V_{fire}, V_{th}, V_{sat} and I_{hold} measured by AC protocols and I_{off} measured by DC voltage application (at $V_{th}/2$), from room temperature up to 250°C for SAGS and up to 300°C for ML. SAGS devices do not show functionality at 300°C.

Fig. 5. Electrical parameters measured in AC (at the exception of I_{off} measured in DC) on devices as-fabricated and annealed for 3 hours at 400°C. The close electrical behavior of ML and SAGS, is likely related to the same features (i.e. AsSe) responsible for the switching mechanism in both materials.

(V_{sat}), calculated as the voltage drop on the devices at zero current extrapolated from ON current-vs-voltage characteristics, is stable in temperature for ML devices while it decreases in SAGS. We suggest that, considering the impact of annealing on SAGS reported in Fig. 1 and Fig. 2, the active material should evolve at 250°C. This leads to a reduction of the saturation voltage and an increase of the holding current (i.e. I_{hold}, the current needed to maintain the OTS in its ON state). On the contrary, the decrease of I_{hold} at high temperature in ML is compatible with the temperature activated ON conduction in the device (i.e. less current required to achieve the same temperature in the material to sustain the conduction in the ON state).

SAGS and ML show a high compatibility in terms of electrical parameters. However, ML OTS allows functionality up to higher temperatures: the addition of GeN layers in ML delays crystallization phenomena without degrading devices electrical properties and increases the activation energy of the conduction (i.e. higher energy gap).

B. BEOL thermal budget effects

In this section, we study the influence of a thermal budget on the devices electrical parameters, V_{th} drift and cycling capabilities. Devices were annealed in a pre-heated chamber at 400°C during 3 hours and slowly cooled within the chamber. Such high thermal budget can be assimilated to a BEOL thermal budget that the devices should overcome during a complete fabrication process. In the following, we will address as *As-fab* the initialized (after firing) devices after fabrication and as *400°C 3H* the initialized devices after annealing. As reported in **Fig. 5**, the annealing has similar effects as the ones observed at high operating temperatures. The structural evolution (i.e. relaxation) and the formation of stable features in the two materials lead to similar trends. Thus, the presence of GeN layers in ML does not affect the switching capabilities of SAGS-based chalcogenide glass even after annealing.

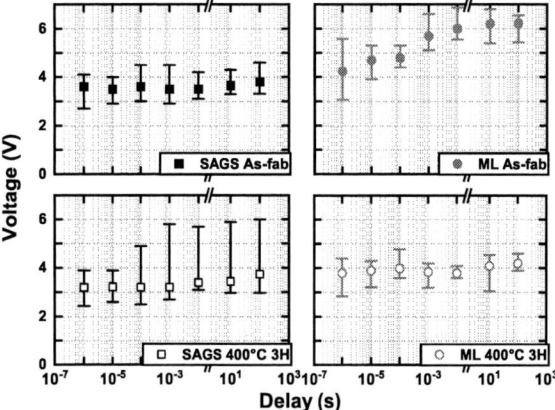

Fig. 6. V_{th} drift measured on as-fabricated and annealed devices. The defects related to Ge-Ge bonds owning to GeN layer in ML, present in *As-fab* devices, relax in time causing an important drift. After annealing, SAGS exhibits a median low drift affected by a 3 V spread while ML shows lower drift (with respect to *As-fab*) and low variability.

In **Fig. 6**, we report the V_{th} drift (i.e. the relaxation-induced changes after a given delay on the threshold voltage) in ML and SAGS devices as-fabricated and after annealing. Considering SAGS devices, the drift is low before annealing but after 3 hours at 400°C, device-to-device variability increases due to material degradation leading to a high drift in some devices. As-fabricated ML shows a high increase of V_{th} of about 2 V after 100 s from programming pulse, likely due to the presence of highly defective Ge-Ge bonds in GeN layers, which tend to relax after pulse application, as reported in [18]. Annealing, as observed in physico-chemical analyses, leads to the formation of stable GeN features and the increase of Ge-Se bonds, therefore to a reduction of Ge-Ge bonds in the system and a reduced drift in the device. Indeed, after annealing ML exhibits a reduced drift and low device-to-device variability.

Fig. 7 reports the maximum cycles number reached by SAGS and ML as-fabricated and after annealing together with the evolution of V_{th} along cycling on long and short time

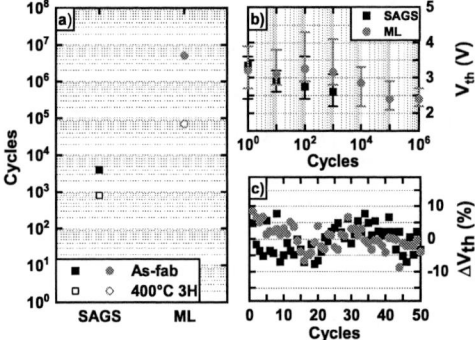

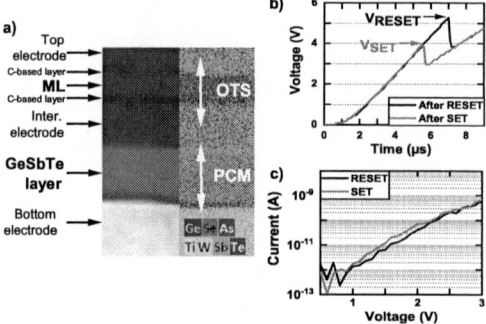

Fig. 7. a) Maximum cycles number reached preserving a selectivity between I_{on} and I_{off} of at least 10^3 in as-fabricated and annealed devices. The ML structure delays crystallization, enabling a higher endurance, however the annealing reduces cycling capabilities of both materials. **b)** V_{th} measured along cycling on as-fabricated devices. SAGS shows a V_{th} decrease since the first cycles while ML reaches 1000 cycles before showing a similar behavior. **c)** Cycle-to-cycle V_{th} variability during first 50 cycles: very low variation in the range of sensitivity used (about 0.2 V).

Fig. 8. a). TEM and EDX analyses of our ML OTS+PCM. **b)** V_{SET} and V_{RESET} switching voltages read by a voltage ramp applied on the OTS+PCM cell, after SET and RESET operations respectively. The reading window is higher than 1 V. **c)** Sub-threshold IV characteristics after SET and RESET pulses. A low leakage below 0.1 nA is ensured at half reading voltage (i.e. $(V_{SET}+V_{RESET})/2 \sim 2.2$ V).

scales for as-fabricated devices. The degradation of SAGS material begins within the first cycles. On the contrary the structure of ML delays crystallization preserving the switching capability for a higher number of cycles. As-fabricated ML devices reach 10^6 cycles with a V_{th} decrease of 0.6 V that begins after the first 1000 cycles. To be noticed that ON current is kept at high values of about 1 mA during endurance tests. The structural evolution observed after annealing, seems to induce a higher sensitivity of the layer to ON current, being the endurance reduced of about 10 times. Cycle-to-cycle variability evaluated on first 50 cycles shows a compatible result between the two materials.

ML OTS presents a lower device-to-device variability with respect to SAGS after BEOL-like thermal budget. Endurance capability is also enhanced in ML of at least 2 orders of magnitude.

C. Multilayer OTS and PCM co-integration

We successfully co-integrated ML OTS with PCM based on $Ge_2Sb_2Te_5$. In **Fig. 8**, we observe a reading window of more than 1 V between V_{SET} and V_{RESET} (respectively the switching voltages of the device when the PCM is programmed in SET and RESET state). Leakage current is kept below 0.1 nA at half of the reading voltage.

IV. CONCLUSIONS

We investigated Multilayer structure in SAGS-based OTS Back-End selectors to enhance thermal stability against BEOL-like thermal budget of 3 hours at 400°C and improve switching reliability. We have demonstrated that ML OTS enables functionality at high operating temperatures up to 300°C, higher endurance and lower device-to-device variability. Finally, we provide demonstration of successful co-integration of ML OTS with PCM.

REFERENCES

[1] H.Y. Cheng et al., "Si incorporation into AsSeGe chalcogenides for high thermal stability, high endurance and extremely low Vth drift 3D stackable cross-point memory", *in Proc. VLSI 2020*, pp. 1-2, 2020.

[2] D.C. Kau, "The pursuit of atomistic switching and cross point memory", *in Proc. VLSI-TSA 2021*, pp. 1-2, 2021.

[3] G. Navarro et al., "Innovative PCM+OTS device with high sub-threshold nonlinearity for non-switching reading operations and higher endurance performance", *in Proc. VLSI 2017*, pp. 1-2, 2017.

[4] D. Garbin et al., "Composition Optimization and Device Understanding of Si-Ge- As-Te Ovonic Threshold Switch Selector with Excellent Endurance", *in Proc. IEDM 2019*, pp. 35.1.1-35.1.4, 2019.

[5] H.Y. Cheng et al., "Optimizing AsSeGe chalcogenides by dopants for extremely low Ioff, high endurance and low Vth drift 3D crosspoint memory", *in Proc. IEDM 2021*, pp. 28.6.1-28.6.4, 2021.

[6] C. Laguna et al., "Multilayer OTS Selectors Engineering for High Temperature Stability, Scalability and High Endurance", *in Proc. IMW 2021*, pp. 1-4, 2021.

[7] S. Zhang et al., "A symmetric multilayer gese/gesesbte ovonic threshold switching selector with improved endurance and stability", *in Proc. ICTA 2021*, pp. 45-46, 2021.

[8] Y. Gan et al., "Analysis of Raman Spectra of GeAsSe Glass Using Different Peak-fitting Method", *in Proc. SPIE 2014*, vol. 9446, 2014.

[9] C.M. Schwarz et al., "Processing and fabrication of micro-structures by multiphoton lithography in germanium-doped arsenic selenide", *Opt. Mater. Express*, vol. 8, no. 7, pp.1902-1915, 2018.

[10] M. Tenhover et al., "Vibrational studies of crystalline and glassy SiSe2", *Solid State Communications*, vol. 51, no. 7, pp. 455-459, 1984.

[11] C. Zha et al., "Optical properties and structural correlations of GeAsSe chalcogenide glasses", *J Mater Sci: Mater Electron*, vol. 18, pp. 389-392, 2007.

[12] P. Němec et al., "Structure and properties of the pure and Pr3+-doped $Ge_{25}Ga_5Se_{70}$ and $Ge_{30}Ga_5Se_{65}$ glasses", *J. Non-Cryst. Solids*, vol. 270, no. 1, pp. 137-146, 2000.

[13] M. Olivier et al., "Structure, nonlinear properties, and photosensitivity of $(GeSe_2)_{100-x}(Sb_2Se_3)_x$ glasses", *Opt. Express*, vol. 4(3), pp. 252-540, 2014.

[14] P. Khan et al., "Investigation of Temperature Dependent Optical Modes in $Ge_xAs_{35-x}Se_{65}$ Thin Films: Structure Specific Raman, FIR and Optical Absorption Spectroscopy", *Thin Solid Films*, vol. 621, pp. 76-83, 2017.

[15] M. Munzar et al., "Far-infrared spectra and bonding arrangement in Ge–As–S–Se glasses", *J. Phys. Chem. Solid.*, vol. 61, no. 10, pp. 1647-1652, 2000.

[16] I. Chambouleyron et al., "Nitrogen in germanium", *J. Appl. Phys.*, vol. 84, no. 1, pp. 1-30, 1998.

[17] D. Ielmini et al., "Analytical model for subthreshold conduction and threshold switching in chalcogenide-based memory devices", *J. Appl. Phys.*, vol. 102, no. 054517, pp.1-13, 2007.

[18] S. Clima et al., "Ovonic Threshold-Switching GexSey Chalcogenide Materials: Stoichiometry, Trap Nature, and Material Relaxation from First Principles", *Phys. Status Solidi RRL*, vol. 14, pp. 1900672, 2020.

Characterization of reset state through energy activation study in Ge-GST based ePCM

Matteo Baldo
STMicroelectronics
Agrate Brianza, Italy
matteo.baldo@st.com

Lorenzo Turconi
DEIB, Politecnico di Milano
Milano, Italy
lorenzo.turconi@mail.polimi.it

Alessandro Motta
STMicroelectronics
Agrate Brianza, Italy
alessandro.motta@st.com

Elisa Petroni
STMicroelectronics
Agrate Brianza, Italy
elisa.petroni@st.com

Luca Laurin
STMicroelectronics
Agrate Brianza, Italy
luca.laurin@st.com

Daniele Ielmini
DEIB, Politecnico di Milano
Milano, Italy
daniele.ielmini@polimi.it

Andrea Redaelli
STMicroelectronics
Agrate Brianza, Italy
andrea.redaelli@st.com

Abstract—Embedded phase change memories (ePCM) based on Ge-rich $Ge_2Sb_2Te_5$ (Ge-GST) alloys are gaining increasing interest in recent years from the scientific community. The introduction of Ge-GST allows to meet the retention specifications of embedded nonvolatile memory (eNVM), where stored codes and parameters must remain unaffected by the high-temperature soldering reflow. This work presents a detailed study of the reset state of the ePCM. Resistance drift and the conduction mechanisms are assessed for different bake times and temperatures. For the first time, evolution of the activation energy for conduction is detailed for the Ge-GST compound. Finally the dependence of the activation energy and resistance value on the forming state is assessed.

Index Terms—ePCM, Amorphous state, Activation Energy, Structural Relaxation, Drift, Forming Process

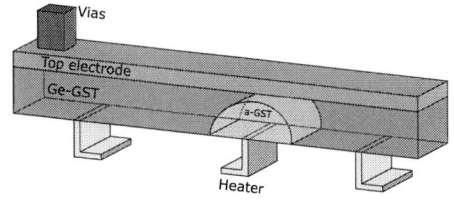

Fig. 1. Schematic representation of a Ge-GST line where three Wall ePCM cells are implemented. The CD of the Ge-GST line is 60 nm. Different colors represent distinct regions of the structure. The active material is highlighted in green and will be considered amorphous through the discussion.

I. INTRODUCTION

Stability of the amorphous, high resistive, state is one of the figure of merits of embedded phase change memory (ePCM). In ePCM, the resistance value tends to increase with time in a phenomenon called resistance drift [1]. Such event has to be considered for applications that need a large number of programming levels due to the instability in time of the absolute resistance value. Multilevel programming of PCM devices is required in many applications of growing importance such as hardware accelerators of neural networks (NN), artificial intelligence (AI) and machine learning (ML) [2]. Detailed characterization and understanding of the drift phenomena is essential for accurate and reliable in-memory computing accelerators [3].

This work addresses the drift of Ge-rich ePCM cells (Ge-GST in the following). The structure of the cell used in this work is the wall-type PCM [4] shown in Fig. 1. The time evolution of the reset state resistance is studied to validate and compare the obtained results with the previously studied behavior of $Ge_2Sb_2Te_5$ (from now on GST-225) [1], [5], [7]. For the first time, a detailed analysis on the activation energy for conduction on Ge-GST during the drift phenomena is performed. Correlations between different physical and electrical parameters are analyzed to gain further insight on

drift phenomenon and its impact on the cell properties. Finally, the most relevant amorphous characteristics are studied as a function of the forming operation. The work presents a comprehensive study of the conduction and drift properties of ePCM for both memory and computing applications.

II. CELL CHARACTERIZATION

In the following two groups of cells, belonging to different wafers, will be considered. Each batch is composed by 37 Ge-GST PCM cells, embedded in a Bipolar-CMOS-DMOS (BCD) process flow, located on the same wafer. Both groups were formed and reset with the same current pulses in order to provide an unique starting condition. The distributions of cells between the two groups after reset were compared to confirm the latter statement.

After these preliminary steps, each wafer underwent a series of sequential bakes at temperatures T_{bake}, either 150°C or 180°C, specific for each group. The bakes were stopped in order to measure the cells after 1, 10 and 100 hours of bake time. The measurement routine consists of acquiring each device current-voltage characteristic through a DC sweep. The threshold voltage of the cell was estimated to be around 1V, so the maximum value of the sweep was chosen to be 0.7 V in order not to induce crystallization or bias drift during the reading phase [5]. The minimum voltage applied was set to $V_{read} = 0.02$ V due to noise constraints. Each measurement

978-1-6654-8498-5/22 $31.00 © 2022 IEEE

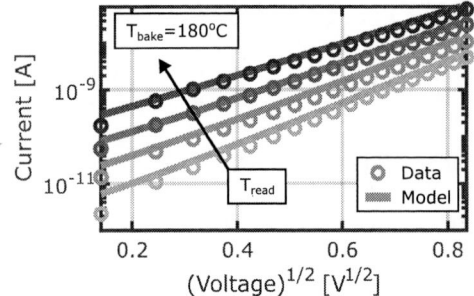

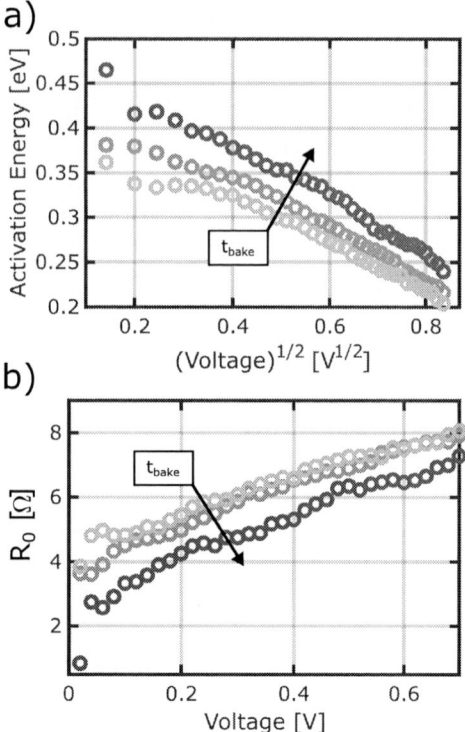

Fig. 2. IV measured after 1h of bake time at 180°C (gray shades) for the different temperature considered and the model used (red shades). The IV shape is well modeled by a Poole-Frankel behavior. This is highlighted by the linearity in the $\log(I) - \sqrt{V}$ axis.

was performed at four different temperatures T_{read} namely 25°C, 0°C, -20°C and -40°C. These temperature values are much lower then the bake ones in order to minimize the drift during acquisition.

Fig. 2 shows the measured current-voltage characteristic after 1h of bake at 180°C. The cells display a Poole-Frenkel current characteristic at all temperatures and bake times. This is highlighted by the linearity in the graph where the curves are represented in a $\log(I) - \sqrt{V}$ fashion. In order to describe with a unique conduction model for all operating conditions considered in this study, we used an analytical equation for Poole-Frenkel conduction given by [6]:

$$I = \frac{V \cdot G_0}{\sqrt{1 + (V/V_{sat})^2}} \exp\left(-\frac{E_g}{2k_b T_{read}}\right) \exp\left(\frac{\Delta U_{PF}}{k_b T_{read}}\right) \tag{1}$$

where G_0 is a constant, V_{sat} the saturation voltage, E_g the band-gap of the material, k_b the Boltzmann constant, T_{read} the measurement temperature and ΔU_{PF} represents the Poole-Frenkel barrier lowering that can be expressed as

$$\Delta U_{PF} = \sqrt{\frac{qV}{\pi \epsilon u_a}} \tag{2}$$

where $\epsilon = \epsilon_0 \epsilon_r$ is the dielectric constant of the material and u_a the amorphous region's thickness. This last parameter was fixed for all working conditions since all the cells were programmed with the same electrical pulses before the bakes. G_0 and V_{sat} were also kept constant. The latter of the two was set to 0.02 V. This low value is used to take into account high frequency of phonon scattering events inside the amorphous material. Extraction of the activation energy for conduction was made possible by repeated acquisition of the current-voltage characteristics at different temperatures. In order to extract the parameter, resistance value was considered to have the following Arrhenius expression

$$R = R_0 \exp\frac{E_a}{k_B T_{read}} \tag{3}$$

where R_0 is a pre-exponential factor, E_a is the activation energy for conduction, T_{read} is the temperature at which the

Fig. 3. Parameters extracted following the model of Eq.3, for the 180°C bake, in function of the read voltage. In a) activation energy for conduction (E_a). In b) the parameter R_0. Different shades of color represent the different time steps: lighter colors for short bake time (1h) darker for long (100h).

measure was performed and k_B is the Boltzmann constant. Having access to the full current-voltage curve, it is possible to extract for each reading voltage the parameters of Eq. 3 as shown by Fig. 3 for the 150°C bake. The horizontal axis of Fig. 2a is proportional to $\sqrt{V}$ in order to highlight the linearity of the curves at different bake times. This is in agreement with Eq. (1) since the parameter extracted from Eq. (3) merges all exponents of the function in one variable. E_a was extracted by interpolation from the curves at $V = 0$. The band-gap energy E_{gap} was considered to be two times E_a in agreement with [5]. The slope of the data can also be used to extract the sub-threshold slope (STS) of the current-voltage characteristic. One can notice that the activation energy, hence

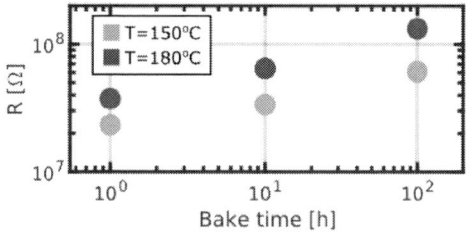

Fig. 4. Resistance evolution in time for the two bakes. In light shades Light shades where used for 150°C, dark shades for 180°C.

978-1-6654-8498-5/22 $31.00 © 2022 IEEE

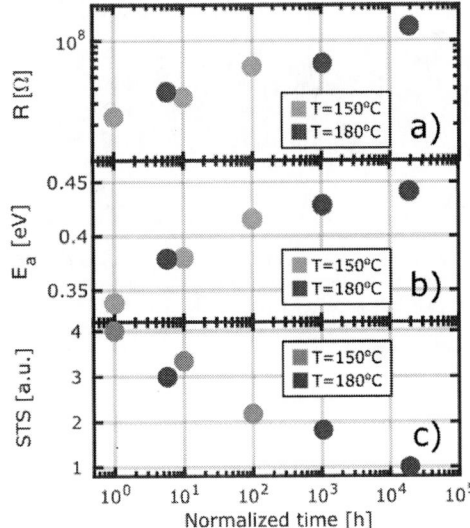

Fig. 5. Resistance (a), activation energy for conduction (b), and sub-threshold slope (c) evolution for both bakes on a normalized time axis with respect to 150°C. Light shades where used for 150°C, dark shades for 180°C. The first two parameters were considered at a reading voltage of 0.04 V.

E_{gap}, monotonically increases with time as a result of the defect annihilation due to structural relaxation (SR). Fig. 2b shows the linear dependence of R_0 on the voltage applied to the cell. This behavior is also in agreement with the pre-exponential factor of Eq. (1), further supporting the reliability of the approach. The resistance evolution in time for the two bake temperatures is shown in Fig. 4. In order to properly relate the two experiments, a time normalization was performed. To this purpose, the bake times at 180°C were transformed according to the [7] formula

$$t_{norm} = \tau_{00}^{1-\beta \frac{T_{150°C}}{T_{180°C}}} \cdot t_{150°C}^{\beta \frac{T_{150°C}}{T_{180°C}}} \qquad (4)$$

where $\beta = (T_{SR} - T_1)/(T_{SR} - T_2)$ with T_{SR} structural relaxation temperature. The latter was estimated from the same set of cells as $T_{SR} = 480$ K. This value is significantly lower than what reported for GST-225 in [7] where $T_{SR} = 760$ K. This difference underlines how the SR temperature is a parameter strongly dependent on material composition. Time normalization's results for different extracted parameters are reported in Fig. 5. All data follow the same universal trend as a function of normalized time, which confirms the validity of the normalization procedure also for Ge-GST materials. The STS decreases logarithmically with time while resistance and activation energy increase, in line with previous results [1]. Based on the dimension of the amorphous dome, it is possible to extract from the STS the value of the dielectric constant for each bake condition. In Fig. 6 the correlation between ϵ, the resistance value and the band-gap for different bakes and times is shown. Clear trends of correlation can be noticed for ϵ_r and R.

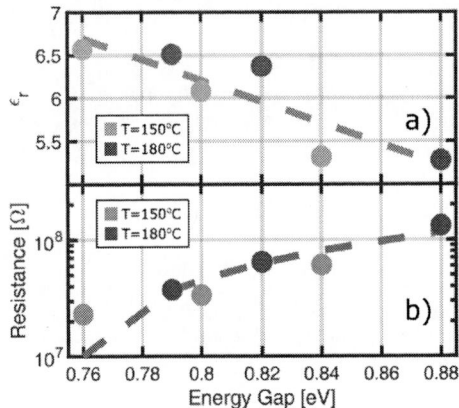

Fig. 6. Correlation plot of a) relative dialectic constant, b) resistance and the band-gap value of the two bakes for all the times considered. Light shades where used for 150°C, dark shades for 180°C. Dashed lines show the linear interpolation of the data. The horizontal axis was extracted from the activation energy following the relation $E_a = E_{gap}/2$. ϵ_r and R have opposite correlation behaviors with respect to E_{gap} at different bake times.

III. DRIFT COEFFICIENT

To extract the drift coefficient ν, the following relation [1] was used

$$\nu = \frac{\Delta E_a}{\Delta E_{SR}} \frac{T_{bake}(1 - T_{read}/T_C)}{T_{read}(1 - T_{bake}/T_{SR})} \qquad (5)$$

where T_C is the isokinetic temperature, T_{SR} is the SR temperature used in Eq. (4) and ΔE_a and ΔE_{SR} are the difference between the value after 1 and 100 hours of bake time of the activation energy of conduction and the SR energy. The value of the isokinetic temperature, extracted form the Arrhenius behavior described by Eq. (3), was determined to be $T_c \simeq 500$ K for Ge-GST. Fig. 7 shows the drift coefficient extracted for both bakes when the cells were read at $T_{read} = 25°C$. As

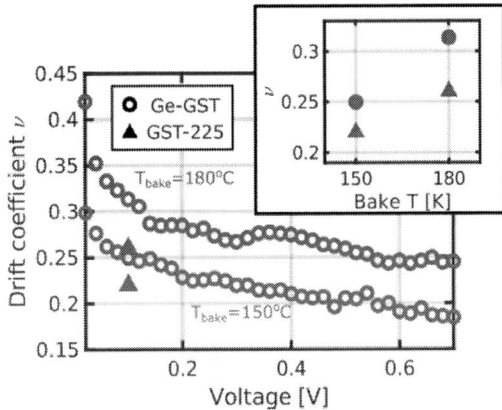

Fig. 7. Drift coefficient extracted for $T_{read} = 25°C$ dependence on the read voltage of the cell. Both bakes are analyzed showing a dependence on the temperature of the drift coefficient. Also data on GST-225 from [1] is shown both in the graph and in the inset in order to highlight the difference between the two materials.

978-1-6654-8498-5/22 $31.00 © 2022 IEEE

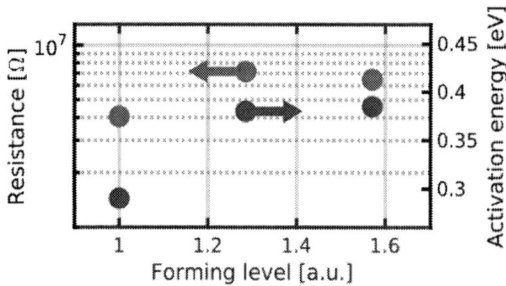

Fig. 8. Resistance value (in red axis on the left) and activation energy for conduction (in gray axis on the right) correlation with the initial forming level. At high forming level values a saturation behavior can be noticed for both parameters.

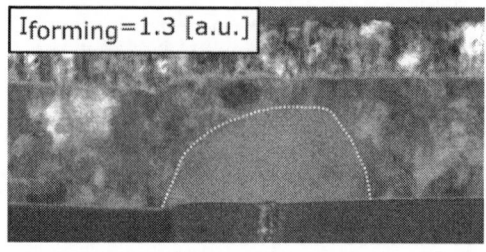

Fig. 9. Dark Field Scanning Transmission Electron Microscopy (STEM-DF) image of the cell with forming current of 1.3 [a.u.]. The amorphous region is highlighted by the white dotted line. Non-crystalline areas can be distinguished by the typical uniform and opaque contrast. The dome dimension are close to the thickness of the cell.

expected by Eq. (5) the drift coefficient increases with the bake temperature. The decrease along the read voltage can be explained by taking into account the dependence of the activation energy for conduction on the voltage signal shown in Fig. 3. In the inset to Fig. 7 is highlighted the comparison between the drift coefficient for GST-225 (data from [1]) and Ge-GST for a reading voltage of 0.1 V. It can be seen, as expected, that the drift phenomenon of the reset resistance is strongly accentuated in Ge-rich GST compositions. This enhancement can be attributed to the removal of Ge-Ge homopolar bond's chains [8].

IV. FORMING DEPENDENCE

A substantial difference between PCM cell based on GST-225 and Ge-GST is the need for the latter of an activation step at the end of the fabrication process. This first electrical signal applied to the cell, called forming [9], defines the active area of the material during the device's lifetime.

In order to investigate the influence of the activation step on the amorphous proprieties of the material three groups of 37 cells each were formed with different current peaks. Each group was then subjected to a reset pulse with a current peak equal to the one received with the forming step. This was done to ensure a fully amorphized active region. Temperature measures of the resistance value and extraction of the activation energy for conduction were carried out as described in Section II. Fig. 8 shows the correlation between reset resistance, activation energy for conduction and forming level. Both values increase as the activation current peak increases. This could be explained as a direct consequence of the growth of the active region with the forming pulse. Compositional differences between the three points could also be a contributing factor [9].

For the highest forming level the two parameters do not grow further but saturate to a maximum level. This can be related to the evolution of the active region's extension at high forming pulses. Fig. 9 shows a Dark Field Scanning Transmission Electron Microscopy (STEM-DF) image of one of the cell formed with current 1.3 [a.u.]. The active region, defined by the forming step, does not present any crystalline areas so it can be concluded it was fully amorphized, as desired by the experiment setup. The radius of the dome almost equals the entire thickness of the cell. Due to the thermal sink behavior of the top electrode (TE) the chalcogenide material in contact at the top of the cell is cooled down and will never reach melting [10]. Due to this phenomenon a saturation of the active region dimension at high forming values can be noticed. This justifies the plateau of the resistance at high forming, since this value is closely related to the dimensions of the dome.

V. CONCLUSION

In this work a fully consistent description of the amorphous state characteristics in time is presented. The conduction mechanism evolves coherently with trap annihilation due to SR. The energy activation for conduction was measured along the full current-voltage characteristic and at different bake times and temperatures. Correlations between various electrical parameters were provided in order to establish trends in the amorphous material evolution. The extraction of the drift coefficient was performed and compared with GST-225 values showing how the drift phenomenon is accentuated in Ge-GST devices. Finally a dissertation on how different forming pulses affect the resistance and activation of the cell is shown.

REFERENCES

[1] M. Boniardi and D. Ielmini, *Applied Physics Letters*, vol. 98, no. 24, p. 243506, Jun. 2011. doi: 10.1063/1.3599559

[2] I. Boybat, M. Le Gallo et al., *Nature Communications*, vol. 9, no. 1, p. 2514, Jun. 2018. doi: 10.1038/s41467-018-04933-y

[3] I. Giannopoulos, M. L. Gallo et al., in *2020 2nd IEEE International Conference on Artificial Intelligence Circuits and Systems (AICAS)*, 2020, pp. 286–290. doi: 10.1109/AICAS48895.2020.9074003

[4] F. Arnaud, P. Zuliani et al., in *2018 IEEE International Electron Devices Meeting (IEDM)*. San Francisco, CA: IEEE, Dec. 2018, pp. 18.4.1–18.4.4. doi: 10.1109/IEDM.2018.8614595

[5] D. Ielmini, D. Sharma et al., *IEEE Transactions on Electron Devices*, vol. 56, no. 5, pp. 1070–1077, May 2009. doi: 10.1109/TED.2009.2016397

[6] A. Calderoni, M. Ferro et al., *Electron Device Letters, IEEE*, vol. 31, pp. 1023 – 1025, 10 2010. doi: 10.1109/LED.2010.2052016

[7] D. Ielmini, M. Boniardi et al., *Microelectronic Engineering*, vol. 86, no. 7-9, pp. 1942–1945, Jul. 2009. doi: 10.1016/j.mee.2009.03.085

[8] S. Gabardi, S. Caravati et al., *Phys. Rev. B*, vol. 92, p. 054201, Aug 2015. doi: 10.1103/PhysRevB.92.054201

[9] E. Palumbo, P. Zuliani et al., *Solid-State Electronics*, vol. 133, 04 2017. doi: 10.1016/j.sse.2017.03.016

[10] M. Baldo, O. Melnic et al., in *2020 IEEE International Electron Devices Meeting (IEDM)*. San Francisco, CA, USA: IEEE, Dec. 2020, pp. 13.3.1–13.3.4. doi: 10.1109/IEDM13553.2020.9372089

Enhanced Thermal Confinement in Phase-Change Memory Targeting Current Reduction

C. De Camaret[1,2], G. Bourgeois[2], O. Cueto[2], V. Meli[2], S. Martin[2], D. Despois[2], V. Beugin[2],
N. Castellani[2], M.C. Cyrille[2], F. Andrieu[2], J. Arcamone[2], Y. Le-Friec[1] and G. Navarro[2,*]

[1]STMICROELECTRONICS, 38926 Crolles, France
[2]CEA, LETI and Univ. Grenoble Alpes, F-38000 Grenoble, France

Abstract—In this work, we present the extensive electrical characterization of 4kb Phase-Change Memory (PCM) arrays based on "Wall" structure and Ge-rich GeSbTe (GST) material, integrating a SiC dielectric with low thermal conductivity surrounding the heater element for enhanced cell thermal efficiency. We investigate the effects of the introduction of such dielectrics on the electrical performances of the device and we provide a promising path to achieve energy-efficient PCM cells supporting our results by electro-thermal TCAD simulations.

I. INTRODUCTION

Phase-Change Memory (PCM) is a non-volatile, scalable and mature memory technology, that has proven to be of great interest for embedded applications. In order to target automotive requirements, memory devices need to demonstrate good data retention in high temperature environment and low power consumption. The first requirement was achieved in PCM by material engineering, thanks to the introduction of Ge-rich GeSbTe (GGST) chalcogenide alloys, which provide great retention capability of more than 10 years at 185°C [1]. Thanks to this breakthrough, PCM became a suitable candidate, introduced in advanced CMOS technology nodes such as 28 nm and even 18 nm [2]. The programming current reduction to target portable and low power applications remains an important goal to reach and the main development axes are scaling [3, 4], material engineering [5], and thermal optimization [6]. Indeed, the perfect PCM cell would exploit adiabatically all the power generated during the heating of the chalcogenide layer, necessary to perform the phase change transition from the amorphous (highly resistive i.e. RESET) to the crystalline phase (lowly resistive i.e. SET). On the contrary, a significant part of this power is lost, and in particular through the dielectrics surrounding the heating element of the device [7]. In previous studies, it was demonstrated that replacing the dielectric encapsulating the PCM layer by an optimized one featuring low thermal conductivity can lead to a reduction of the programming current, as well as an improved reliability [8, 9]. However, in a heater-based "Wall" architecture (**Fig. 1**), recognized as a PCM cell solution with a great optimization potential [4], most of the heat is generated inside the heater

This work was partially funded by European commission, French State and Auvergne-Rhône Alpes region through ECSEL-IA 101007321 project StorAIge and French Nano2022 program.
*Corresponding author. E-mail: gabriele.navarro@cea.fr (Gabriele Navarro)

itself, hence the dielectrics surrounding this element should be carefully optimized. In this work, we study the effects of the introduction of a SiC dielectric presenting a low thermal conductivity around the heater of a "Wall" PCM structure to enhance its thermal isolation. Through electrical characterization, we analyze the programming characteristics, the endurance and the retention at statistical level in 4kb arrays and the results are compared to the ones obtained from standard SiN-based devices. Electro-thermal TCAD simulation is used to support our observations and our solutions for further device optimization. Despite some process challenges to be overcome, we present promising paths for next-generation thermally-optimized PCM cells.

II. SiC vs SiN-based 4kb PCM Electrical Characterization

We integrated GGST materials in "Wall" PCM devices fabricated in the Back-End-of-Line (BEOL) of CEA-LETI Memory Advanced Demonstrators (MAD) based on 130 nm CMOS technology, for single device analysis and 4kb statistical electrical parameters evaluation. The critical dimension (unless specified differently) corresponding to the width (w) of the heater element is 80 nm. The heater is surrounded by three dielectrics, namely A and B as reported in Fig. 1, and the encapsulation dielectric layer that covers up the sides of the entire device as described in [9]. We considered devices with encapsulation and dielectric B based on standard SiN, and varying the dielectric A fabricated either with SiN or SiC. We will address in the following the PCM devices with respect to the material integrated as dielectric A. Our SiC dielectric has a low thermal conductivity (k_{th}) of about 0.4 $Wm^{-1}K^{-1}$, while for SiN it is close to 1.4 $Wm^{-1}K^{-1}$ as measured in [8]. The heater element has a resistance of about 1 kΩ.

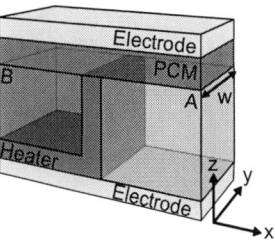

Fig. 1. Description of a "Wall" PCM device. The dielectrics integrated on the two sides of the heater are named respectively A (SiN or SiC) and B (SiN).

978-1-6654-8498-5/22 $31.00 © 2022 IEEE

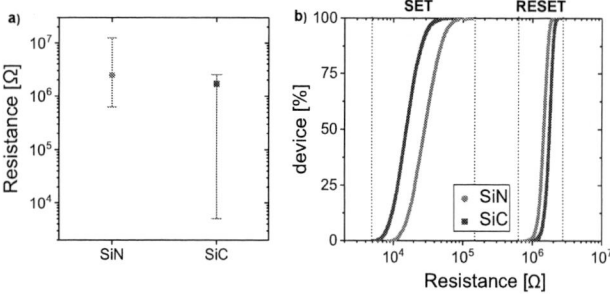

Fig. 2. a) Median virgin resistance with error bars defined as minimum and maximum values. b) Cumulative distributions in the 4kb array of the SET and RESET resistance states after initialization, for both SiC and SiN-PCM.

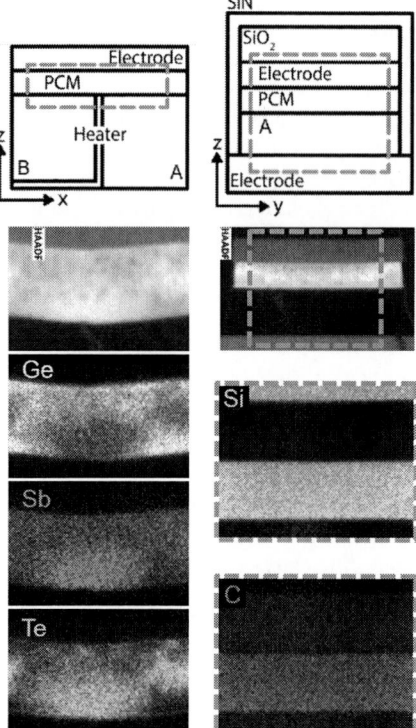

Fig. 3. Left: schematic view of a cross-section orthogonal to the heater width (z,x-plane) with High-angle annular dark-field scanning transmission electron microscopy (HAADF-STEM) and Energy Dispersive X-Ray (EDX) maps of the active volume of a SiC device (with w = 300 nm), showing a Ge enrichment at the edges of the active volume and increased Sb and Te concentrations compatibly with previous observations [1]. Right: description and HAADF-STEM image of a cross-section parallel to the heater width (z,y-plane), along with EDX images of the silicon and carbon distributions.

A. Programming characteristics

The virgin resistance of the devices measured after fabrication is reported in **Fig. 2a**. We can deduce from the spread of the resistances in SiC devices a likely crystallization of the GGST layer already triggered during the BEOL fabrication. After an initialization protocol to bring the cells into a stable SET state, we programmed the matrices in the RESET state using a sequence of pulses with increasing voltage amplitude

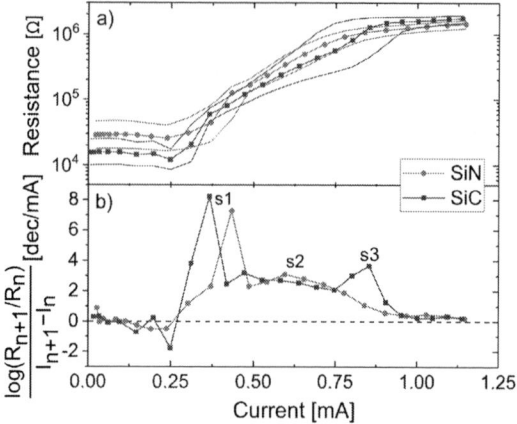

Fig. 4. RI characteristics for both SiN and SiC 4kb PCM arrays, obtained with staircase protocol (pulse width = 300 ns, fall-time = 10 ns). Median resistance with 15%-85% confidence interval is reported as a function of the programming current (a), associated with the evolution of the slope s of each curve (b).

(i.e. staircase protocol). We obtained the cumulative distributions of the SET and RESET states presented in Fig. 2b. For both SiN and SiC based devices a good functionality is achieved in the whole 4kb array, with a wider resistance window for SiC, mainly related to its capability to achieve a lower SET resistance. The HAADF-STEM analyses coupled to EDX maps on a SiC device after programming are reported in **Fig. 3**. The Si and C maps show the successful integration of SiC dielectric in the PCM device. Moreover, we observe an elemental segregation in the active volume of the cell [1] leading to a Ge accumulation outside of the active volume, expected in GGST-based PCM.

Resistance versus current (RI) characteristics for both SiC and SiN 4kb arrays are reported in **Fig. 4** with the associated derivative that was correlated to the thermal properties of the material surrounding the PCM cell [9] (i.e. $s = \Delta log R / \Delta I \propto 1/\sqrt{k_{th,eff}}$ with $k_{th,eff}$ the effective thermal conductivity of the materials surrounding the active volume). We observe three current ranges corresponding to different values of s. In the first one ($s1$), the heat is mainly generated inside the GGST layer. As soon as the current increases, the chalcogenide layer becomes extremely conductive and the heater element starts to have the highest resistive contribution, leading to a gradual displacement of the maximum temperature from the GGST layer to the heater element itself ($s2$). The effect of the presence of low thermal conductive SiC can be visible only at higher current values ($s3$), with an increase of s parameter in SiC devices that is not present in SiN.

In order to investigate the effect of the presence of SiC dielectric around the heater element on the SET performances of the device, we performed SET resistance cartographies as a function of the programming current and of the fall-time of the SET pulse (i.e. the duration of the falling edge of the pulse). The devices were pre-programmed in the RESET state before each SET pulse and the results are reported in **Fig. 5**. Confirming previous results, the SiC PCM devices present a

978-1-6654-8498-5/22 $31.00 © 2022 IEEE

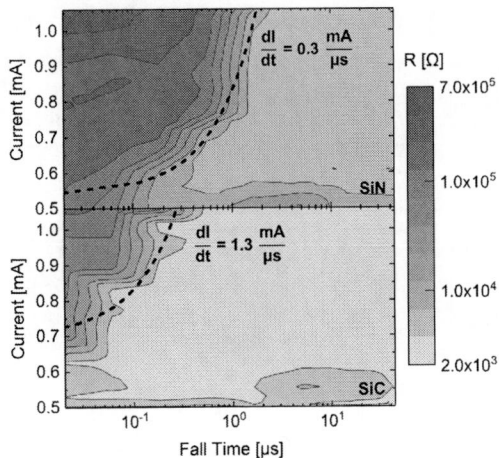

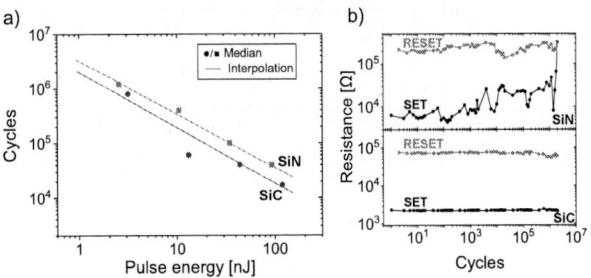

Fig. 6. a) Median (based on 25 devices) of the number of cycles until failure as a function of the total energy of the SET+RESET programming, modulated by varying the width of the SET and RESET combination of pulses. Lines are linear regressions of the data. b) Example of endurance tests performed with same programming conditions for SiN and SiC PCM.

Fig. 5. SET resistance cartographies as a function of the programming current and of the duration of the falling edge of the SET pulse (pulse width kept constant at 300 ns). Median values from 30 cells with 100 nm heater width are reported. Dashed lines represent the current over fall-time rate necessary to achieve the SET state in both SiC and SiN (SET state = 10% of the Resistance window).

faster crystallization as their resistance decreases already for short fall-times. This can be quantified by calculating from the graphs (dashed lines) the minimum current over time rate necessary to reliably SET the cells. We find a $4\times$ difference in the minimum falling edge slope necessary to crystallize SiC with respect to SiN PCM.

Combining the already presented results, we observe an enhanced crystallization in SiC PCM with respect to SiN even during BEOL fabrication, i.e. independently from programming operations (Fig. 2a). We think that the lower density of SiC could increase the surface roughness and enhance the oxidation propensity of SiC with respect to SiN. This could favor heterogeneous crystallization phenomena at the chalcogenide/SiC interface already at BEOL-like temperatures (i.e. 300°C-400°C) [10]. If these phenomena can also contribute to the higher SET capability of SiC devices, we should consider that the improved thermal isolation can induce a different behavior of the device during the initialization step (i.e. called "forming"). In particular, we expect a higher temperature reached in the case of SiC with respect to SiN, leading to a reduced Ge content in the active volume (i.e. higher Ge expulsion towards the edges). A lower Ge content is then compatible with an improved SET capability (Fig. 5). Moreover, the increased thermal resistance of the device thanks to the SiC introduction is responsible for a slower cooling (i.e. higher thermal time constant) that contributes to enhance crystallization even for short SET fall-times (Fig. 5). Such improved crystallization in SiC PCM can also explain the limited effect observed on the programming current reduction in our devices. Indeed, since the crystallization is favored in SiC PCM in the range of currents compatible with s2 (Fig. 4), the benefit from a higher thermal isolation that should lead to a higher amorphized volume with respect to SiN (i.e. higher resistance) is compensated by a reduction of the final effective

amorphized volume due to the enhanced crystallization during the falling edge of the applied RESET pulse. Only at high current (range s3) the thermal isolation improvement is visible, when current decrease rates at the falling edge are sufficiently low to avoid any unwanted crystallization.

B. Endurance and Data Retention

We compared endurance of both SiC and SiN devices using combination of SET and RESET pulses with increasing duration (i.e. increasing energy). In **Fig. 6a**, we report the maximum endurance achieved for a given combination of SET and RESET duration. Although the endurance of SiC devices seems slightly lower, we consider the difference insignificant as we usually compare orders of magnitude. On the contrary, we can observe the reliable resistance window achievable (without any smart programming protocol) up to more than 10^6 cycles, likely favored by the enhanced SET capability in these devices (Fig. 6b). Hence, the SiC does not seem to severely affect the endurance performance of GGST.

Data retention tests performed on 4kb arrays by isochronal annealing of one hour at increasing temperatures are reported in **Fig. 7**. As expected from programming characteristics analyses, SiC PCM exhibits a slightly lower data retention with respect to SiN. Indeed, SiC RESET (and SET) population shows a more advanced re-crystallization at 240°C, still preserving good retention performances.

III. PCM Device Thermal Optimization

We reported on the effects of the introduction of SiC in our "Wall" PCM to enhance the thermal isolation of the heater element. The electrical/thermal optimization of the PCM cell should preliminarily consider the matching of the heater resistance with the chalcogenide layer resistance in order to maximize the efficiency of the heating as previously reported in [11]. Such matching is responsible for a localization of the heating at the heater/chalcogenide material interface reducing the losses and the power needed to completely amorphize the active volume in the PCM. To further investigate how the thermal conductivity of the dielectric A in Fig. 1 can change the cell behavior, we performed 3D TCAD electro-thermal simulations using a PCM dedicated tool relying on Sentaurus

978-1-6654-8498-5/22 $31.00 © 2022 IEEE

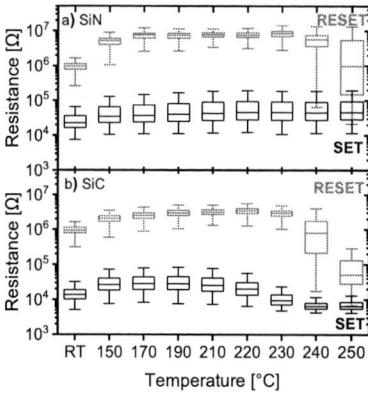

Fig. 7. Data retention tests performed by isochronal annealing of 1 hour at increasing temperatures for SiC and SiN devices.

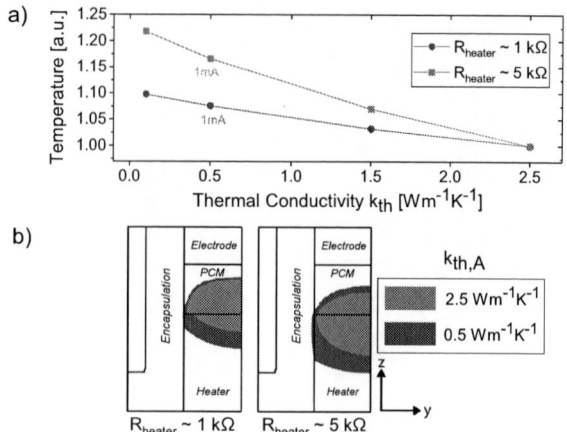

Fig. 8. a) Evolution of the maximum temperature reached inside the chalcogenide material in the PCM cell, depending on the thermal conductivity of the dielectric A ($k_{th,A}$) and for two different values of heater resistance. Temperature is normalized and is obtained from equivalent programming currents. b) Temperature profile (region that during RESET current application achieves a temperature higher than 930 K) achieved in the two devices for a low and a high thermal conductivity dielectric A. Based on the same maximum temperature obtained for $k_{th,A} = 2.5$ Wm^{-1}K^{-1}.

Device [12]. Our electro-thermal model considers a PCM material whose electrical conductivity depends on temperature (based on electrical data fitting). To study the role of the heater resistance, we considered two structures, with a heater resistance of 1 kΩ (like in our electrical tests) and 5 kΩ respectively. In **Fig. 8a**, we observe that lowering the thermal conductivity of dielectric A from 2.5 Wm^{-1}K^{-1} to 0.5 Wm^{-1}K^{-1} induces an increase of the maximum temperature inside the chalcogenide of about 8% for the low resistive heater and more than 15% for the highly resistive heater. Dielectric B (Fig. 1) replacement by SiC would enhance even more such gain (not reported). Moreover, we think that phenomena related to electric field activated conduction in the chalcogenide (not considered in the used tool), would contribute to an even higher gain from the heater thermal isolation on the thermal efficiency of the PCM cell [13]. Finally, we report in Fig. 8b the temperature profile comparison achieved for both 1 kΩ and 5 kΩ heater resistances, with a high and a low thermal conductivity simulated for dielectric A ($k_{th,A}$). Note that the low resistive heater exhibits bigger melted volume inside the PCM layer, which is expected since the heat is generated directly inside the chalcogenide. However, the required power to melt this zone is also higher. Therefore, here, we are only interested in the benefit from an improved thermal isolation of the heater element, which is higher in 5 kΩ device.

IV. CONCLUSIONS

We analyzed the electrical performances of a thermally optimized "Wall"-based PCM in 4kb arrays. We could evidence the effects of the integration of a SiC dielectric featuring low thermal conductivity on one side of the heater element. SiC PCM showed high SET speed and higher general propensity to crystallization, likely due to both interface and thermal related phenomena that contribute to reduce in such devices the effective gain on programming current reduction. We highlighted in RI characteristics analyses, how the higher thermal isolation in low resistive heater-based cells becomes interesting at high current, when the heating is mainly localized inside the heater.

Finally, thanks to 3D TCAD simulations, we showed how coupling SiC dielectric integration demonstrated in our PCM cells with an optimized heater resistance can lead to better thermal efficiency in future generations of PCM technology.

REFERENCES

[1] A. Redaelli et al., "Material and process engineering challenges in Ge-rich GST for embedded PCM", Materials Science in Semiconductor Processing, vol. 137, p. 106184, 2022.

[2] D. Min et al., "18nm FDSOI Technology Platform embedding PCM and Innovative Continuous-Active Construct Enhancing Performance for Leading-Edge MCU Applications", 2021 IEEE International Electron Devices Meeting, pp. 13.1.1–13.1.4, 2021.

[3] S. Lee et al., "Programming disturbance and cell scaling in phase change memory: For up to 16nm based 4f2 cell", 2010 Symposium on VLSI Technology, pp. 199–200, 2010.

[4] M. Boniardi et al., "Optimization metrics for Phase Change Memory (PCM) cell architectures", 2014 IEEE International Electron Devices Meeting, pp. 29.1.1-29.1.4, 2014.

[5] M. Boniardi et al., "Electrical and thermal behavior of tellurium poor gesbte compounds for phase change memory", 2012 4th IEEE International Memory Workshop, pp. 1-3, 2012.

[6] J. Wu et al., "A low power phase change memory using thermally confined tan/tin bottom electrode", 2011 International Electron Devices Meeting, pp. 3.2.1–3.2.4, 2011.

[7] S. W. Fong et al., "Dual-layer dielectric stack for thermally isolated low-energy phase-change memory", IEEE Transactions on Electron Devices, vol. 64, no. 11, pp. 4496–4502, 2017.

[8] A. L. Serra et al., "Outstanding improvement in 4kb phase-change memory of programming and retention performances by enhanced thermal confinement", 2019 IEEE 11th International Memory Workshop, pp. 1–4, 2019.

[9] A. L. Serra et al., "Phase-change memory electro-thermal analysis and engineering thanks to enhanced thermal confinement", Solid-State Electronics, vol 186, p. 108111, 2021.

[10] P. Noé et al., "Impact of interfaces on scenario of crystallization of phase change materials", Acta Materialia, vol. 110, pp. 142–148, 2016.

[11] U. Russo et al., "Modeling of programming and read performance in phase-change memories—part i: Cell optimization and scaling", IEEE Transactions on Electron Devices, vol. 55, no. 2, pp. 506–514, 2008.

[12] TCAD Tools, S-2021.06, Synopsys, Mountain View, CA, USA, 2021

[13] M. L. Gallo and A. Sebastian, "An overview of phase-change memory device physics", Journal of Physics D: Applied Physics, vol. 53, no. 21, p. 213002, 2020.

TiTe/Ge$_2$Sb$_2$Te$_5$ Bi-layer-based Phase-Change Memory Targeting Storage Class Memory

G. Lama, M. Bernard, J. Garrione, N. Bernier, N. Castellani,
G. Bourgeois, M.C. Cyrille, F. Andrieu and G. Navarro*
CEA, LETI and Univ. Grenoble Alpes, F-38000 Grenoble, France

Abstract—**In this work, we introduce an innovative Phase-Change Memory (PCM) based on a TiTe and Ge$_2$Sb$_2$Te$_5$ (GST) bi-layer stack that presents low resistance variability since the out-of-fabrication in 4 kb array. It allows creating reliably an intermixed system right from the first programming in the active volume of the device. TiTe/GST PCM exhibits higher speed, lower variability of intermediate resistance states and lower drift compared to standard GST. An endurance of more than 10^8 cycles can be achieved and we found a reduced cycle-to-cycle variability even after endurance stress. Such new TiTe/GST stack, based on our results, demonstrates to be a valuable candidate for PCM targeting Storage Class Memory applications.**

I. INTRODUCTION

In the last years, the memory hierarchy is facing a big challenge in covering the gap between DRAM and Flash memory. Indeed, DRAM features high speed and endurance, whereas Flash memory is characterized by high density and low cost, but limited speed and endurance [1]. This gap can be filled by a new category of memory called Storage Class Memory (SCM) that would bring non-volatility close to DRAM improving system performances in terms of speed and density and, moreover, reducing the power consumption. Among Non-Volatile Memories (NVM), Phase-Change Memory (PCM) is considered the best candidate for SCM for its high speed, high density and high endurance [2], also after having proved to be a mature NVM entering both standalone [3] and embedded market [4]. PCM working principle is based on a reversible phase transition from a crystalline low resistive phase (SET) to an amorphous high resistive one (RESET). Material engineering in PCM is the key method to improve speed and endurance performances to target SCM requirements. For example, Sb-rich GST has been recognized as a suitable phase-change material for SCM thanks to its ns range programming time [5] and a record endurance of 2×10^{12} [6]. However, these materials require a very good control of the stoichiometry, which otherwise could induce devices variability. Moreover, they present a really steep SET-to-RESET characteristic [5], which prevents the possibility of achieving intermediate states with low variability i.e. multi-level cell (MLC). Another example of phase-change material with reduced compositional and structural variability is the multilayer based on TiTe/Sb$_2$Te$_3$ stack that ensures also good speed and cyclability [7], however requiring a huge

stability of the multilayer structure that could result difficult along cycling.

For the first time, we introduce a relatively simple PCM structure based on TiTe and Ge$_2$Sb$_2$Te$_5$ (GST) bi-layer stack, which ensures low device-to-device variability (D2D) already at the out-of-fabrication thanks to a TiTe layer that features low resistivity and high stability in temperature. An initialization step drives a reliable intermixing of TiTe and GST. We explore the SET speed, the MLC capability, drift in temperature and in time as well as cycle-to-cycle (C2C) variability in 4 kb, making a comparison with standard GST based PCM. Results are supported by Transmission Electron Microscopy (TEM) and Energy-Dispersive X-ray spectroscopy (EDX) analyses. Finally, we analyze the behavior of TiTe/GST devices before and after endurance stress highlighting an extremely low variability of the resistance states even after cycling.

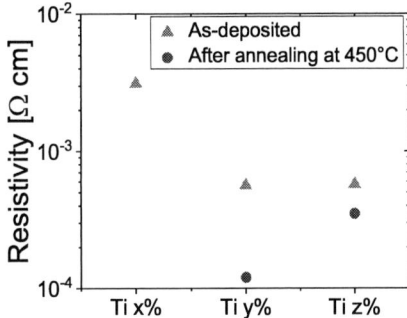

Fig. 1. Resistivity of TiTe samples with three different percentages of Ti (x, y and z with x<y<z) measured at room temperature before and after annealing at 450°C. The resistivity after annealing of the layer with Ti x % is not reported since the layer resulted degraded after the test.

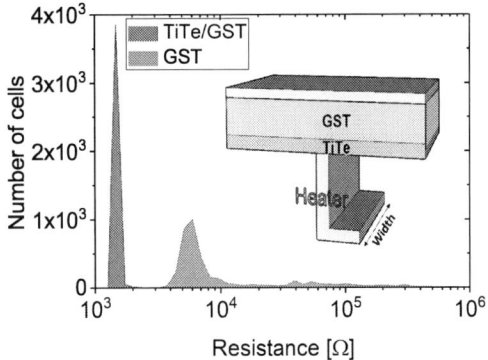

Fig. 2. As-fabricated resistance distributions of TiTe/GST and GST 4 kb arrays. Inset: simplified scheme of the studied "Wall" PCM device based on TiTe/GST bi-layer stack.

This work was partially funded by European commission, French State and Auvergne-Rhône Alpes region through ECSEL-IA 101007321 project StorAIge and French Nano2022 program.
*Corresponding author. E-mail: gabriele.navarro@cea.fr (Gabriele Navarro)

978-1-6654-8498-5/22 $31.00 © 2022 IEEE

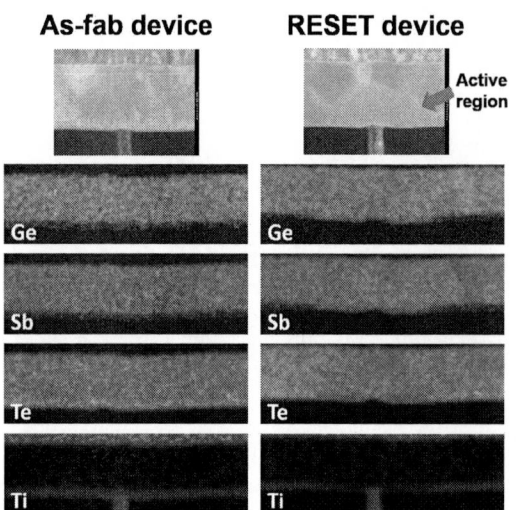

As-fab device **RESET device**

Active region

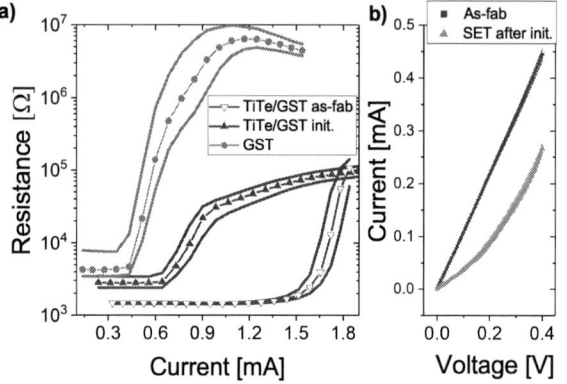

Ge Ge
Sb Sb
Te Te
Ti Ti

Fig. 3. TEM/EDX analyses performed on as-fabricated TiTe/GST device (left column) and on a device programmed in the RESET state after initialization (right column).

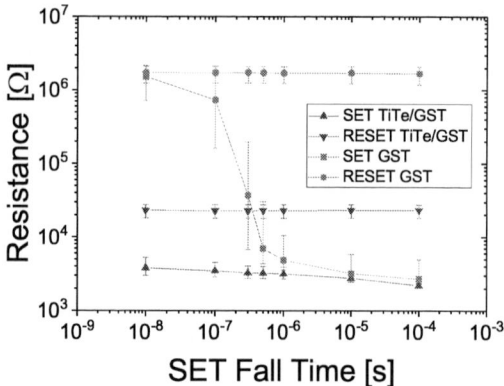

Fig. 5. SET speed test performed on TiTe/GST and GST 4 kb arrays. Data are obtained applying a SET pulse with optimized current, 300 ns width time and incremental fall time on arrays pre-programmed in RESET state before each SET pulse. RESET state resistance is reported as well.

a)

b)

Fig. 4. a) Resistance as a function of current measured in 4 kb array for GST and TiTe/GST devices starting in the SET state. We represent median values and 1σ intervals, and in particular before (as-fab) and after initialization (init.) for TiTe/GST. b) Current-Voltage (I-V) characteristics of five TiTe/GST PCM devices as-fabricated (as-fab) and after initialization in the SET state (SET after-init.).

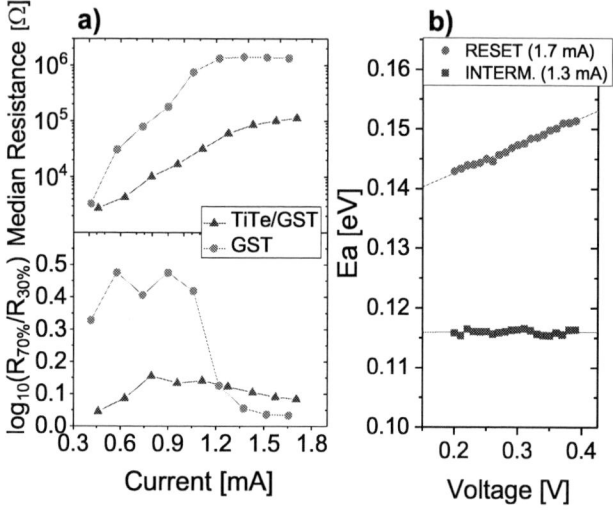

Fig. 6. a) Intermediate states obtained in GST and TiTe/GST applying 300 ns squared pulses with increasing current intensity pre-programming the array in the SET state before each pulse. Median (above) and variability (below) of the resistances in 4 kb arrays are represented. The variability was calculated as the ratio between the 70th and the 30th percentile of the 4 kb resistances. b) Activation energy of the conduction measured from I-V characteristics of TiTe/GST (not reported), for RESET state and an intermediate state, following the model from [8].

II. TiTe/GST Bi-Layer Development and Characterization

In order to ensure stability against Back-End-of-Line (BEOL) fabrication process thermal budget, we engineered the thin TiTe layer integrated in our PCM devices. For this purpose, we deposited three TiTe layers with different percentages of Ti ranging from ∼10 at.% to ∼50 at.% (addressed as Ti x, y, z% with x<y<z) and we measured at room temperature the resistivity of TiTe layers before and after annealing at 450°C by four-probe method (**Fig. 1**). The resistivity of TiTe layer decreases as the Ti content increases due to its metallic nature. The layer with Ti z% maintains a resistivity similar to the initial one after being annealed, showing a higher stability compared to the layers with less Ti content. For this reason, we selected Ti_zTe_{100-z} as thin layer to be co-integrated with a GST layer in "Wall" PCM devices, as described in the

inset of **Fig. 2** (this PCM will be addressed as TiTe/GST in the following). We performed electrical characterization on 4 kb arrays consisting of PCM devices with a heater width of 100 nm integrated into the BEOL of LETI Memory Advanced Demonstrator (MAD) based on 130 nm CMOS technology.

A. TiTe/GST Programming Characteristics

Resistance of as-fabricated devices is reported in Fig. 2 showing an extremely low dispersion in TiTe/GST with respect to GST. Such low resistance variability is ensured by the low resistive Ti_zTe_{100-z} layer, which offers a good temperature stability against fabrication process thermal budget. This is confirmed by TEM/EDX analyses of **Fig. 3** illustrating that the TiTe layer remains intact after the fabrication, while in

978-1-6654-8498-5/22 $31.00 © 2022 IEEE

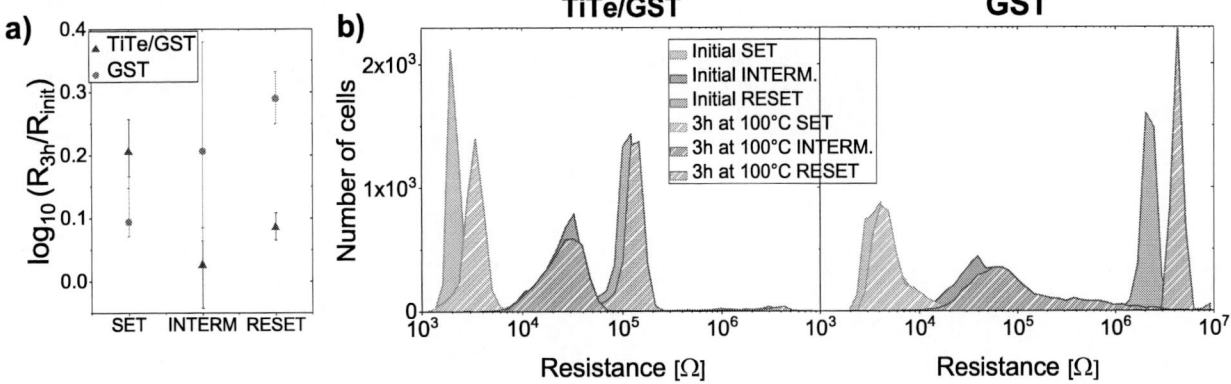

Fig. 7. Data retention evaluated after annealing at 100°C for the SET, intermediate and RESET states in TiTe/GST and GST 4 kb arrays. a) The resistance drift is quantified as the ratio between the resistance measured after a 3 hours annealing and the initial resistance. Median values and variability are represented. b) Resistance distributions of the three states before and after 3 hours annealing at 100°C.

a programmed TiTe/GST the TiTe layer is no longer visible in the active region due to the intermixing of TiTe with GST. The R-I curve in **Fig. 4a** evidences that as-fabricated TiTe/GST devices need an initialization step to give rise to the intermixing of TiTe with GST in the active region. Such initialization effect is also highlighted by current-voltage (I-V) measurements in Fig. 4b: as-fabricated TiTe/GST devices exhibit an ohmic behavior, while after the initialization and the first programming in the SET state, the I-V characteristic changes showing an exponential behavior. Resistance-Current (R-I) curve of TiTe/GST devices after initialization shows an extremely low variability and more gradual SET-to-RESET transition with respect to GST, despite a reduced resistance window. The SET speed is evaluated in (**Fig. 5**) indicating that GST needs pulses longer than 1 μs to obtain a reliable SET state, whereas in TiTe/GST the SET operation is achievable using pulses with a short fall time lower than 10 ns. MLC capability is analyzed in **Fig. 6a**, which reports the resistances obtained applying pulses of increasing intensity in arrays programmed in the SET state before each pulse. GST shows a higher variability, except for the RESET state. On the contrary, a low variability of all the programmed states is confirmed in TiTe/GST. The activation energy of conduction (E_A) of the RESET state and of an intermediate state in TiTe/GST is measured from subthreshold I-V realized at different temperatures (Fig. 6b) according to [8]. In common phase-change materials, such as GST, the amorphous phase shows a trap-limited transport described by Poole-Frenkel mechanism, where E_A decreases with increasing voltage [8]. In the case of TiTe/GST, E_A is not dependent on the voltage for the intermediate state and it increases with voltage for the RESET state. Moreover, TiTe/GST presents a low E_A (~ 0.1 eV i.e. small energy gap) compared to GST (0.37 eV) [8]. These results evidence a different conduction mechanism in RESET TiTe/GST that appears more close to a metallic behavior, with the presence of defects that generates scattering, increasing E_A value as the electric field increases.

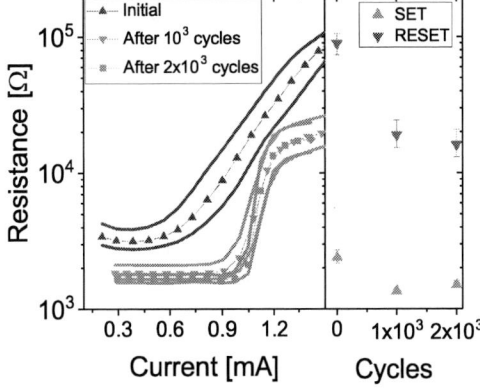

Fig. 8. Left: R-I characteristic at the beginning of the devices life and after endurance for TiTe/GST. Right: 4 kb TiTe/GST SET and RESET resistances evolution along cycles performed with pulses of 10 μs to accelerate and evidence the evolution.

B. Data Retention

Data retention in our PCM devices was investigated comparing the resistance states values before and after an annealing at 100°C for 3 hours. The resistance drift is quantified for each device of the array in **Fig. 7**. SET state drifts more in TiTe/GST than in GST devices, likely due to the presence of defects in the crystalline matrix of the material related to Ti inclusion. GST intermediate state shows a large spread in the behavior of the devices after annealing, that is suppressed in TiTe/GST. The RESET state is also more stable in TiTe/GST with respect to GST. We think that the (almost) metallic behavior of amorphous TiTe/GST observed in previous E_A analyses contributes to the improved stability in this material.

C. Endurance

In order to analyze the behavior of TiTe/GST during cycling, R-I curves and SET and RESET resistances before and after endurance stress are reported in **Fig. 8**. The cycles were executed with long pulses of 10 μs for accelerated aging. After 10^3 cycles, the R-I curve evolves and SET and RESET resistances decrease. Additional 10^3 cycles, applied with the same protocol, highlight that this evolution is not detrimental

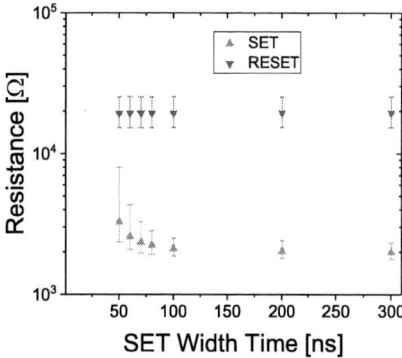

Fig. 9. SET speed test realized in TiTe/GST 4 kb array after endurance test performed in Fig. 8b, using SET pulses with increasing width time and short constant fall time of 10 ns.

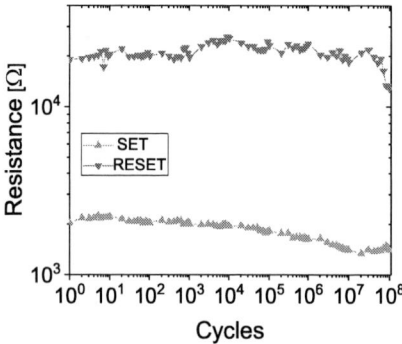

Fig. 10. Endurance evaluated in a TiTe/GST device with optimized SET and RESET pulses (width time = 50 ns, rise/fall time = 10 ns).

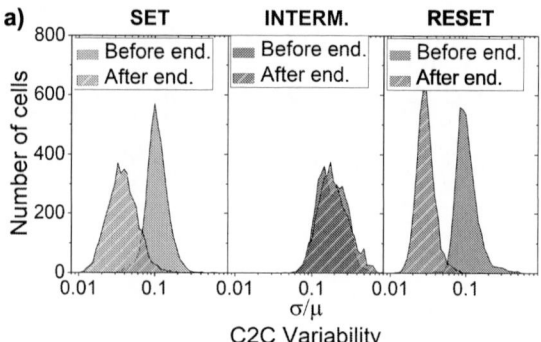

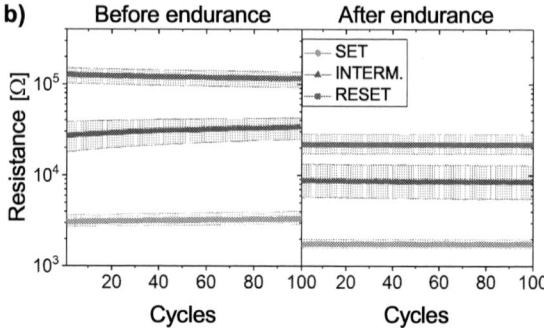

Fig. 11. Variability of SET, RESET and intermediate states evaluated along 100 programming cycles (SET/RESET pulses of 50 ns) in TiTe/GST 4 kb arrays, before and after an endurance test compatible with the one performed in Fig. 8. a) C2C variability distributions: for each device in the array the C2C variability was calculated as the ratio between the standard deviation of the resistance value and its median resistance along 100 cycles. b) D2D variability: median and standard deviation for the 4 kb resistance values along the 100 cycles.

and the devices achieve an extremely stable behavior. Indeed, we realized that the composition achieved in the active region of the PCM features unique stability properties. **Fig. 9** shows the programming SET time kept in tens of ns range and an endurance higher than 10^8 cycles without the use of any smart programming protocol (**Fig. 10**). The devices were still perfectly functional after all the stresses applied. The C2C and D2D variability in TiTe/GST 4 kb arrays along 100 cycles were evaluated for SET, intermediate and RESET state before and after the endurance stress (**Fig. 11**). The three states modify their resistance values after cycling, as shown in Fig. 11b, compatibly with previous results. Nevertheless, the three states are still perfectly achievable and distinguishable and the C2C variability of SET and RESET states is even reduced after cycling (Fig. 11a). Further analyses are ongoing to reveal the nature of this new composition featuring such striking stability and endurance performances.

III. CONCLUSIONS

We investigated through electrical characterization an innovative PCM device based on a TiTe/GST bi-layer stack. The TiTe layer was engineered to be stable after the fabrication process, as demonstrated by the low variable resistance distributions of as-fabricated 4 kb devices and by TEM/EDX analyses. The first programming into the high resistive state gives rise to a new alloy made by GST and TiTe, that showed improved

performances with respect to GST, such as higher speed and intermediate states featuring lower variability. Endurance up to more 10^8 cycles was demonstrated, nevertheless a modification of R-I characteristic and a reduction of SET and RESET resistances were found after aging. The obtained new alloy revealed striking stability along cycling, showing high SET speed and reduced C2C variability even in MLC mode. This new alloy, obtained from our TiTe/GST investigation, holds great promise for targeting DRAM-like performances for SCM applications.

REFERENCES

[1] R.Freitas et al., "Storage-class memory: The next storage system technology", *IBM J. of Research and Development*, 52, 4.5, pp 439-447,2008.

[2] SW. Fong et al., "Phase-change memory—Towards a storage-class memory", *IEEE TED*, 64,11, pp 4374-4385, 2017.

[3] A. Fazio, "Advanced technology and systems of cross point memory", *IEEE IEDM*, pp 24-1, 2020.

[4] P. Cappelletti et al., "Phase change memory for automotive grade embedded NVM applications", *J. Phys. D: Appl. Phys.*, 53,19,193002,2020.

[5] G.Navarro et al., "Highly Sb-Rich Ge-Sb-Te Engineering in 4kb Phase-Change Memory for High Speed and High Material Stability Under Cycling", *IEEE IMW*, pp 1-4, 2019.

[6] W. Kim et al., "ALD-based confined PCM with a metallic liner toward unlimited endurance", *IEEE IEDM*, pp 4.2.1-4.2.4, 2016.

[7] K. Ding et al., "Phase-change heterostructure enables ultralow noise and drift for memory operation", *Science*, 366, 6462, pp 210-215, 2019.

[8] D. Ielmini et al., "Evidence for trap-limited transport in the subthreshold conduction regime of chalcogenide glasses", *Applied Physics Letters*, 90, 19, 192102, 2007.

Design exploration of IGZO diode based VCMA array design for Storage Class Memory Applications

M. Gupta, M. Perumkunnil, A. Fantini, S. A. Chamazcoti, W. Kim, M.G. Bardon, G.S. Kar, A. Furnémont,

imec, Leuven, Belgium, email: Mohit.Gupta@imec.be

Abstract—**This paper presents the Design Technology Co-optimization (DTCO) study of the 1-diode 1-Voltage controlled magnetic anisotropy (1D-1VCMA) stack, which functions as Storage Class Memory (SCM) to bridge the gap between DRAM and flash memory. The dual requirement of low sneak current and high non-linearity for bidirectional selectors in 1S-1R crossbar memories is extremely challenging to achieve practically. Moreover, the IR drop due to parasitic resistance (R_{PAR}) results in significant degradation in the voltage across the memory element (ME) and causes write disturbance for the 1S-1R crossbar. 1D-1VCMA solves the above-mentioned issues by having low write current (I_W), (thus reducing IR drop) and increasing the number of DINs /DOUTs to improve energy/bit. Sneak current and non-linearity are also significantly improved due of the diode selector. Based on our VCMA and IGZO diode technology data and an extensive DTCO, we are able to achieve a write energy consumption of 880fJ/bit (at a delay of 40.5ns and V_{DD} of 2V), and read energy consumption of 414fJ/bit (at a read delay of 39ns), thus showing a significant improvement over similar SCMs.**

Keywords—DTCO, Storage class memory, VCMA, diode, IGZO, DRAM, 1D-1VCMA

I. INTRODUCTION

Performance and efficiency requirements from storage data have exploded over the last years due to 5G and AI. This has exacerbated the need for a change in the traditional memory hierarchy via SCMs that can fit in between DRAM and flash memory, and thereby enhance system performance (Fig. 1). 1OTS-1PCM crossbar variants [1-3] have targeted this gap and some have even moved to commercial production with both new and existing protocols [2-3]. However, the required voltage for read/write (R/W) operations remains high, leading to high energy consumption along with endurance, and reliability issues.

Thanks to its low voltage operation (~0.4-0.8V), fast operation speed (>100MHz), and high endurance (>10^9), STT-MRAM is one of the best candidates for nonvolatile memory [4]. Therefore, the crossbar memory using STT-MTJ (Fig. 2) can be an attractive memory for SCM [4]. For the 1(S)elector-1MTJ crossbar array to function effectively, the selector should be bi-directional, have low off current, high active current density and low on-resistance [4]. However, there is still no consensus on how to achieve these requirements at the same time [5-6]. To this end, we used a 1(D)iode-1VCMA [7-8] based crossbar, where the VCMA is employed as a ME and a diode is used as the selector. The selector requirements in this case are relaxed since the VCMA is a voltage-based unipolar device. The diode closes the sneak current path by reverse biasing, and it has a higher ON/OFF current ratio >10^5, which will solve the non-linearity issue. As the required I_W for VCMA switching is much lesser (compared to STT), the diode can easily provide this current. Along with the selector requirements, the IR drop induced by the R_{PAR} is also an issue with the 1S1R

crossbar, which results in degradation in the voltage across ME for the 1S1R crossbar. This IR drop can be reduced using two possible solutions: 1) By opting for lower switching current (I_{SW}), which impacts the read current and hence read window and 2) having lesser number of cells on the wordline (WL) and bitline (BL), which impacts the area efficiency as periphery area increases. The IR drop worsens when writing into more than one cell in an array as the total I_{SW} increases linearly with the number of cells. A 1D-1VCMA cell can solve this issue without impacting area efficiency as the VCMA works on a lower I_W ~10uA (reduces the IR drop) making it possible to increase DIN (number of cells to be written) /DOUT (number of cells to be read) width and improve energy.

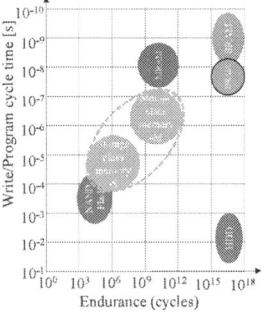

Fig 1. Typical memories positioning w.r.t program time and endurance.

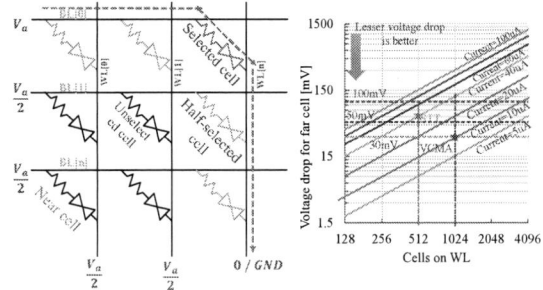

Fig 2. Typical crossbar array which has memory element and Bi-directional selector.

II. EXPERIMENTAL DATA

In VCMA, applying a voltage across the MTJ stack (Fig. 3) results in an electric field through the MgO oxide and reduces the energy barrier [9]. Here, the resulting resistance states [Parallel state (P) and Antiparallel state (AP)] depend on the orientation of the Free layer magnetization. Switching from P to AP or vice versa depends on the previous state and is achieved by simply applying the required voltage pulse in one direction. Hence, it is a 'Unipolar device' or 'toggle memory'. Figure 3 shows the measured data for a VCMA device. The MTJ stack of the VCMA device basically consists of MgO barrier, CoFeB

978-1-6654-8498-5/22 $31.00 © 2022 IEEE

free layer and Ta capping layer, along with synthetic antiferromagnetic structure underneath, and is also BEOL-compatible with 400C annealing. This device data is used to model the VCMA behavior.

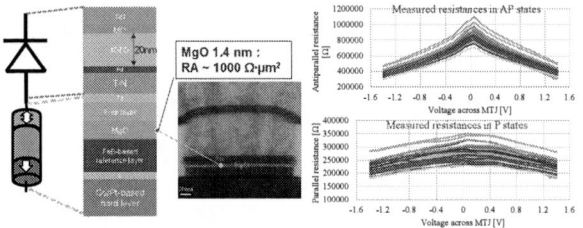

Fig 3. VCMA device specifications and cross-section TEM showing device pillar alongside.

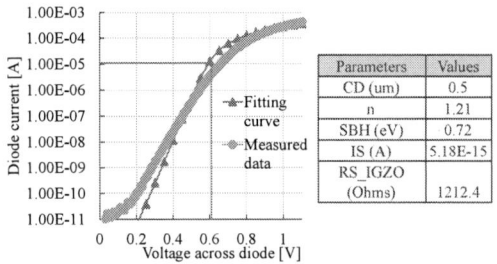

Fig 4. The measured I-V characteristics of IGZO diode which is used for the calibration of diode compact model.

An indium-gallium-zinc oxide (a-IGZO) based Schottky diode with a Pd\IGZO\Mo electrode is utilized as the selector (Fig. 3) for the 1D-1VCMA cell. The diode is composed of a 20nm thick IGZO layer with contact electrodes. It has an ideality factor n of 1.21 and a Schottky barrier height (φSBH) of 0.72 eV [6]. The turn-on/off time of the selector element is negligible due to the absence of minority carriers in the Schottky diode. Figure 4 shows the measured data of the IGZO diode, that is used to calibrate the device model employed for DTCO analysis.

It should be noted that our work (based on the perpendicular MTJ) is different from the in-plane MTJ device related prior work (in [7][8]). Moreover, the time scale in these papers is in the order of ms with STT effect, which means it is to use the static response by VCMA effect. The shortest time scales considered, even with H_{eff}, is the range of ~10ns compared to timescales in the range of a few ns to sub ns in our work (precessional dynamics of magnetization in play for perpendicular MTJ).

III. DTCO OF 1D-1VCMA

For an efficient memory design, accurate technology assumptions and extensive DTCO is important. This section discusses about 1D1VCMA assumptions and basic operations related to 1D1VCMA.

A. Array considerations

The VCMA device serving as the non-volatile ME with a diode as a selector (Fig. 5), makes the read and write operation much simpler compared to other 1S1R stacks [2-3]. Fig. 5 shows the bias points of the diode and VCMA for R/W operations. A very high subthreshold slope (72 mV/dec, n=1.21) ensures that the current ratio of the IGZO diode matches the VCMA resistance ratio and thus maintains the

VCMA TMR ~200%. For the R/W operations, only V_{DD} and V_{SS} supplies are required to bias both BL and WL. A separate $V_{DD}/2$ supply is not needed (compared to 1S1MTJ), thus saving area and energy (since no dc-dc converter is required). In inactive banks, BLs are biased to V_{SS}, and WLs are biased to V_{DD}, whereas, in selected banks, BL transitions from V_{SS} to V_{DD} and WL transitions from V_{DD} to V_{SS}.

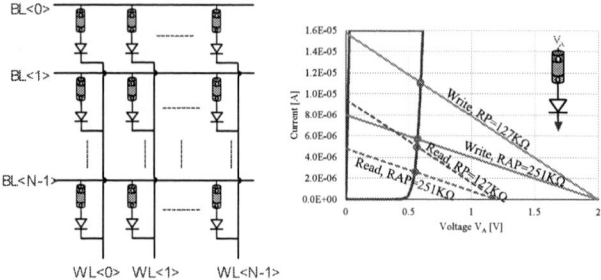

Fig 5. The 1D1VCMA crossbar array which uses IGZO diode as a selector and VCMA as a memory element. The bias points of IGZO diode and VCMA is shown in right figure.

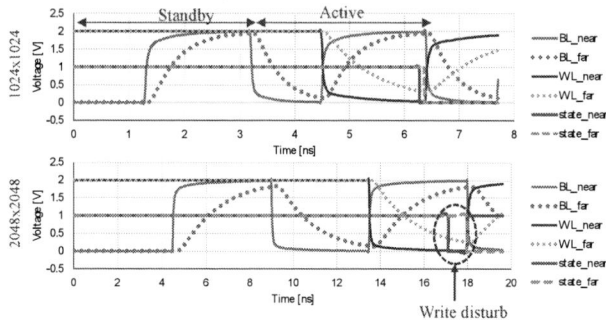

Fig 6. The 1D1VCMA write operation for the 1024x1024 and 2048x2048 array sizes. The delay between the near cell (red colour) and far cell (green colour) switching increases with bigger array for 1024x1024 array size. However, to write the far cell (green colour) in 2048x2048 array, the near cell (red colour) switches two times (i.e., creating the write error) considering the fixed write driver strength in an array.

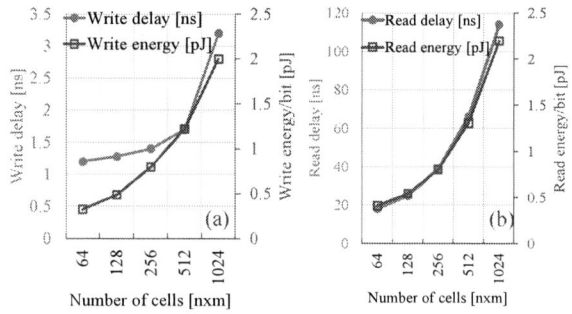

Fig 7. (a) The 1D1VCMA write delay and energy versus different array size. For 1Mb 1D1VCMA array, the write delay and energy are 3.2ns and 2pJ/bit, respectively. (b) The 1D1VCMA read delay and energy versus different array size. For 1Mb 1D1VCMA array, the read delay and energy is 114ns and 2.25pJ/bit, respectively.

B. Write operation

The 1D-1VCMA write operation for an array having 1024 cells on the BL and WL each is highlighted in Fig. 6. If the number of cells is increased from 1024 to 2048 on BL and WL to achieve higher memory density/efficiency, then the difference between the near cell and far cell DL_{WR} increases due a corresponding increase in parasitic

978-1-6654-8498-5/22 $31.00 © 2022 IEEE

capacitance (C_{PAR}). This can lead to 'Write Disturb' issues. For example, in a 2Kx2K array (Fig. 6), when we switch the far cell, the near cell flips back to the initial state (switches twice due to the DL_{WR} discrepancy between and leading to a write error). On the other hand, if we optimize DL_{WR} for the near cell, then far cell switching is bound to be affected (degraded slew rate) due to parasitic capacitance (C_{PAR}). Hence, the maximum number of cells that can be connected on the BL and WL is 1K. Fig. 7 (a) shows the E_{WR} and DL_{WR} for the different array sizes (64 to 1024). For the 1Mb 1D-1VCMA array (1024 cells on WL and 1024 cells on BL), DL_{WR} and E_{WR} are 3.2ns and 2pJ/bit, respectively.

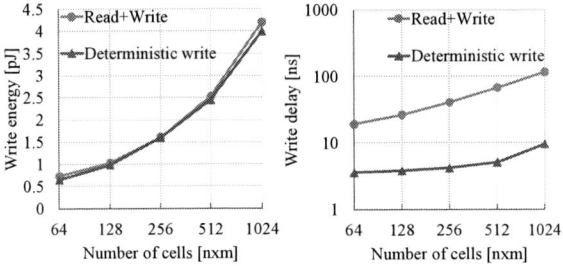

Fig 8. Due to the toggling nature of VCMA device, reading before write is necessary. As a result, overall write energy is summation of the read and write energy. However, deterministic switching is possible using the method presented in [8]. The total write delay is summation of the read and write delay. Deterministic write can drastically reduce the write time as no read before write is needed.

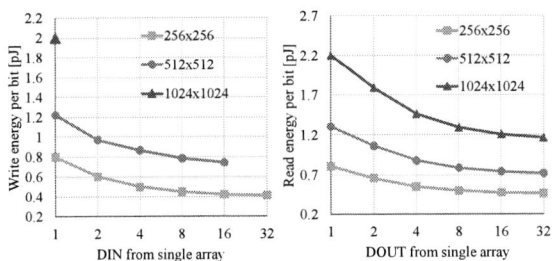

Fig 9. The write energy/bit improves with DINs. 1024x1024 would not work with more than 1 DIN (near cell and far cell cannot work simultaneously). The read energy/bit improves with DINs. ~48% energy saving can be achieved with 32 DOUTs.

C. Read operation

The reading of 1D-1VCMA is similar to other resistive memories [11]. The active current source provides current to the stack, and the voltage across the stack is compared with a reference voltage (V_{REF}). Since VCMA is a voltage-based device, the clamp keeps the voltage across VCMA low enough to avoid the read disturb (or overwrite the cell data). Figure 7(b) shows that the DL_{RD} is 15 to 20 times of DL_{WR} (due to the high VCMA resistance). For the 64x64 and 1024x0124 array, the DL_{RD} is 20ns and 114ns, respectively. In a 1Mb array, the E_{RD} is 2.25pJ/bit and similar to the E_{WR} as capacitive loading contributes most to the energy consumption. A higher RA for VCMA is good for the write operation (lower I_W) but the read becomes much worse. The DL_{RD} can be improved by opting for a reduced voltage margin between the read voltage (due to the ME) and the V_{REF}. This reduced voltage margin does however increase sense amplifier complexity as it has to sense smaller voltage

differences. Here, V_{REF} will be half of the voltage difference between the two states.

D. Read-before-write operation

Since the VCMA device is a toggle memory, to ensure reliable writing, reading the ME state before write becomes necessary. For the 1Mb 1D-1VCMA array, the DL_{WR} and E_{WR} are 117.2ns and 4.2pJ/bit, respectively (for a read + write operation), (Fig. 8). However, deterministic switching is also possible using the method presented in [10] and reduces the DL_{WR} drastically as no read before a write is needed (Fig. 8).

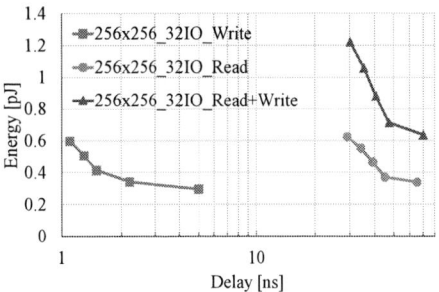

Fig 10. The energy/bit can be improved by reducing the V_{DD} supply. 1.8V is the optimum voltage w.r.t energy/bit and performance.

E. Multibit write/read in an array

Due to the diode and a voltage based unidirectional VCMA device, it is possible to R/W into multiple cells in an array, unlike other 1S1R crossbar array architectures. The E_{WR} reduces with an increase in the number of DINs (Fig. 9) because of the amortization of capacitive load energy over all the DINs. However, the energy improvement saturates as the energy contribution from the switching current increases linearly with DINs. The number of DINs also depends upon the array size (Fig. 9). For the 512x512 array, the maximum number of cells that can be written in a cycle is 16, whereas, for the 256x256 array size, it is 32. The reason is the 'Write Disturb' issue (near cells switch twice while writing into the far cell), which is exacerbated with more DINs as the write current increases, thus raising the voltage drop due to R_{PAR}. Following the same reasoning, the 1024x1024 array would not work with more than 1 DIN. In our multibit 256x256 array design, for a 1-bit write, the write driver is shared among 8 cells, and the 8:1 multiplexer helps in selecting one BL out of 8. As a result, the E_{WR} improves by 48% for 32 DINs. The E_{RD} also improves with an increase in DOUTs and follows the same reasoning as write energy (Fig. 9). It is evident from the Fig. 9 that the 1024x1024 array works perfectly with 32 DOUTs, but only with one DIN. A DOUT of 32bits is possible due to the two reasons: 1) A lower read current (<<write current) resulting in a lower voltage drop and 2) having enough time for the read voltage generation and settlement.

VDD scaling is one of the techniques that is employed to improve the Energy/bit at the cost of degraded delay. V_{DD} is scaled is from 2.4V to 1.6V in-order to improve the Energy/bit at the cost of degraded delay (Fig.10). Below 1.6V, the DL_{RD} degrades too much, and above 2.4V, the number of 'Write Disturbs' increase. For the 256x256 array with 32 DINs/DOUTs, 1.8V is the optimum voltage w.r.t energy/bit and performance. The energy/bit improves by 23% at the expense of an increase in the DL_{RD} (by 7ns). The write energy/bit @1.8V is 712fJ with the DL_{WR} of 47ns, and the RE/bit @1.8V is 370fJ with the DL_{RD} of 45ns.

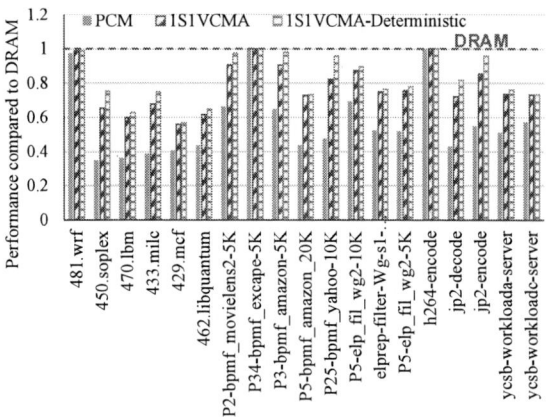

Fig 11. System level benchmarking of 1D1VCMA and PCM (compared to DRAM). Core configuration: X86, 1 core, 128-entry instruction window, 4-wide issue, 3.2 GHz. DRAM configuration: DDR4_2Gb_x8 , DDR4_2400U. Simulator used: SEAT (modified gem5 + Ramulator) [14].

IV. SYSTEM BENCHMARKING

In this work, we are targeting the gap between DRAM and Flash. To fill the gap between this gap, we need a memory that should be denser than DRAM and have performance better than Flash. As a result, we have compared 1D1VCMA with DRAM and PCM (as PCM is already filling this gap). Table II show the comparison of 1D-1VCMA with the other memories like DRAM, PCM and MRAM. From a process technology point of view, 1D-1VCMA can be located at the backend like a crossbar array with silicon circuitries beneath it. This can increase the cell area efficiency for the Memory Macro to as high as >90%, like in the case of 1OTS-1PCM based memory in the Intel 3D-Xpoint product [12]. The read and write energy/bit are impressive compared to other standalone memories. However, the read performance remains a concern with 1D-1VCMA (due to the high resistance VCMA). While the resistance of the VCMA stack can be reduced by improving the RA, this would impact the write operation (due to increasing current flowing through the device and the increased resulting voltage drop). Finally, compared to other memories (Fig. 1) positioned between DRAM and NAND flash, the 1D-1VCMA has good endurance due to the thicker oxide barrier. Figure 11 highlights the system-level benefits of using the 1D-1VCMA memory over a 1OTS-1PCM one (similar to Intel 3D-Xpoint [2]) as a SCM with DDR interface for different benchmarks. On an average, a performance improvement of ~43% is observed over the 1OTS-1PCM option. Moreover, these benefits are enhanced while opting for deterministic switching for the VCMA (average performance improvement of ~53%). In our simulations, the 1OTS-1PCM and 1D-1VCMA stacks have a CD of 20nm and 100nm, respectively, which hints at more

Table II: Comparison of 1D1VCMA with other memories

Characteristics	DRAM*	PCM*	MRAM	1D1VCMA_256x256_32 DIN/DOUT @ 2V	
				Write	Read + Write
Bitcell area	$6F^2$	$4F^2$	$18F^2$	$4F^2$	
Read latency	15ns	48ns	5ns	**39ns**	**39ns**
Write latency	15ns	150ns	20ns	**1.5ns**	**40.5ns/4.5ns#**
Addressability	Byte	Byte	Byte	**Byte**	**Byte**
Volatile	Yes	No	No	**No**	**No**
Read Energy/bit	1.17pJ	2pJ	0.1pJ	**0.465pJ**	**0.465pJ**
Write Energy/bit	0.39pJ	13.5pJ	1.2pJ	**0.414pJ**	**0.88pJ**
Endurance	$>10^{12}$	$>10^6$	$>10^9$	$\mathbf{\sim 10^{12}}$	$\mathbf{\sim 10^{12}}$

*Data is taken from [13]

room for optimization for 1D-1VCMA.

V. CONCLUSION

1D-1VCMA can be a better option for the crossbar SCM designs. Thanks to the voltage based and unidirectional ME, the selector requirements can be relaxed. A diode can be used as the selector, which improves sneak current, non-linearity and eliminates the need for $V_{DD}/2$ supply voltage at the macro level. The write operation is very fast, but the read determines overall latency as read before write is needed for ensure reliable operations. R/W operations are possible in <100ns cycle time at an energy of 880fJ/bit for 256x256_32IO. Finally, 1D-1VCMA shows ~43% average speed-up over 1OTS1PCM at the system level.

REFERENCES

[1] Geoffrey W. Burr et al., "Access devices for 3D crosspoint memory," Journal of Vacuum Science & Technology B 32, 040802 (2014).

[2] https://www.intel.com/content/www/us/en/products/details/memory-storage.html

[3] https://www.anandtech.com/show/15029/micron-finally-announces-a-3d-xpoint-product-micron-x100

[4] H. Yang et al., "Threshold switching selector and 1S1R integration development for 3D cross-point STT-MRAM," 2017 IEDM, 2017, pp. 38.1.1-38.1.4.

[5] G. Molas et al., "Crosspoint Memory Arrays: Principle, Strengths and Challenges," 2020 IEEE IMW, 2020, pp. 1-4.

[6] S. H. Sharifi et al., "Sub-μm a-IGZO, Fully integrated, Process improved, Vertical diode for Crosspoint arrays," 2020 IEEE IMW.

[7] R. Dorrance et al., "Diode-MTJ Crossbar Memory Cell Using Voltage-Induced Unipolar Switching for High-Density MRAM," in IEEE Electron Device Letters, vol. 34, no. 6, pp. 753-755, June 2013.

[8] J. G. Alzate et al., "Voltage-induced switching of nanoscale magnetic tunnel junctions," 2012 IEDM, pp. 29.5.1-29.5.4

[9] R. Carpenter et al., "Demonstration of a Free-layer Developed With Atomistic Simulations Enabling BEOL Compatible VCMA-MRAM with a Coefficient ≥100fJ/Vm," 2021 IEEE IEDM, 2021.

[10] Y. C. Wu et al., "Deterministic and Field-Free Voltage-Controlled MRAM for High Performance and Low Power Applications," 2020 IEEE Symposium on VLSI Technology, 2020, pp. 1-2.

[11] M. Gupta et al., "High-density SOT-MRAM technology and design specifications for the embedded domain at 5nm node," 2020 IEEE IEDM, 2020, pp. 24.5.1-24.5.4.

[12] https://www.techinsights.com/blog/memoryselector-elements-intel-optanetm-xpoint-memory

[13] Benjamin C. Lee, et al, "Architecting Phase Change Memory as a Scalable DRAM Alternative," ISCA '09: Proceedings of the 36th annual ISCA, June 2009 Pages 2–13.

[14] M. Perumkunnil et al., "System exploration and technology demonstration of 3D Wafer-to-Wafer integrated STT-MRAM based caches for advanced Mobile SoCs," 2020 IEEE International Electron Devices Meeting (IEDM), 2020, pp. 15.4.1-15.4.4.

Device optimization for 200V GaN-on-SOI Platform for Monolithicly Integrated Power Circuits

Olga Syshchyk
IMEC
3001 Leuven, Belgium
olga.syshchyk@imec.be

Deepthi Cingu
IMEC
3001 Leuven, Belgium
deepthi.cingu@imec.be

Karen Geens
IMEC
3001 Leuven, Belgium
karen.geens@imec.be

Stefaan Decoutere
IMEC
3001 Leuven, Belgium
stefaan.decoutere@imec.be

Thibault Cosnier
IMEC
3001 Leuven, Belgium
thibault.cosnier@imec.be

Dirk Wellekens
IMEC
3001 Leuven, Belgium
dirk.wellekens@imec.be

Pavan Vudumula
ESAT, KULeuven
3001 Leuven, Belgium
pavan.vudumula@imec.be

Tian-Li Wu
National Yang Ming Chiao Tung University, NCYU
Taiwan
tlwu@nycu.edu.tw

Zheng-Hong Huang
National Yang Ming Chiao Tung University, NCYU
Taiwan
zhenhong.icst08g@nctu.edu.tw

Anurag Vohra
IMEC
3001 Leuven, Belgium
anurag.vohra@imec.be

Urmimala Chatterjee
IMEC
3001 Leuven, Belgium
urmimala.chatterjee@imec.be

Benoit Bakeroot
Center for Microsystems Technology, IMEC and Ghent University
9052 Ghent, Belgium
benoit.bakeroot@imec.be

Abstract—The device performance of monolithically integrated power Schottky barrier diodes (SBDs) and depletion-mode (D-mode) MIS HEMTs is studied in relation to the thickness of the gate dielectric, the gate-edge termination (GET) layer and AlGaN barrier. Special attention is paid to the turn-on voltage (V_{TON}), ON-resistance (R_{ON}), device dispersion, leakage current and breakdown voltage (V_{BD}) of SBDs and D-mode MIS-HEMTs. Based on the current design, SBDs show the lowest dynamic R_{ON} for devices with 7.5-9.5 nm AlGaN barrier thickness and 25-35 nm GET thicknesses. The best performance of the D-mode MIS-HEMTs is observed for devices with 5.5 nm AlGaN barrier thickness and 45 nm gate dielectric thickness.

Keywords— *GaN, GaN-ICs, Schottky barrier diodes, enhancement mode HEMT, depletion mode HEMT, monolithic integration.*

I. INTRODUCTION

GaN power integrated circuits (GaN-ICs) unlock the full potential of the fast-switching speed and high operating frequency of GaN power devices [1]. Monolithic integration of power devices, drivers and control/protection circuits allows to obtain a better control over the gate signal, a lower gate-voltage stress, and improve efficiency by reducing power loop size, interconnect parasitics and fabrication costs.

In [2], we have successfully integrated depletion-mode HEMTs and Schottky barrier diodes into an enhancement-mode (E-mode) GaN HEMT baseline process on Silicon-on-Insulator (SOI) substrates yielding a 200V GaN-IC platform. The depletion mode HEMTs allow to replace the resistive load from the RTL (Resistor/Transistor Logic) circuits [3] with a current source (Direct Coupled FET Logic or DCFL). This allows more compactness and flexibility in circuit design and improves the speed while reducing the power dissipation of the circuits [3]. Offering SBDs in this GaN-IC

platform allows designing clamping circuits, level shifters, protections and freewheeling diodes.

This paper reports the impact of the AlGaN barrier thickness and GET dielectric thickness on the device performance of SBDs and D-mode MIS-HEMTs monolithically integrated with E-mode pGaN HEMTs.

II. GaN POWER INTEGRATION TECHNOLOGY

A. Epitaxy

Monolithically integrated devices are fabricated using 1070 μm thick 200 mm SOI substrates with a 1.5 μm SiO_2 buried layer, and a 1.5 μm Si (111) top layer. The epitaxial stack with a total thickness of 4 μm was grown using metalorganic chemical vapor deposition (MOCVD). This stack includes a 200 nm AlN nucleation layer, a 100nm $Al_{0.75}Ga_{0.25}N$ transition layer, 2.4 μm (Al)GaN supper-lattice buffer layer, a 1 μm carbon doped GaN back-barrier and a 200 nm unintentionally doped GaN channel layer, a 14 nm $Al_{0.23}Ga_{0.77}N$ barrier layer, and an 80 nm Mg-doped p-GaN layer with a doping concentration of ~3×10^{19} cm^{-3}. Channel thickness, AlGaN barrier thickness, composition and Mg doping are similar to what is used in our p-GaN E-mode GaN power devices [4].

B. Device fabrication

Fig.1 shows a schematic representation of the fabrication steps with SEM images of critical steps. The device fabrication starts with the patterning of the TiN/p-GaN stack to form the gate of the p-GaN HEMTs (Fig.1.a). The key steps for SBDs and D-mode HEMTs include the simultaneous patterning and partial recess (Fig.1.b) of the AlGaN barrier using atomic layer etching (ALE), which impacts the SBD turn-on voltage and D-mode HEMT

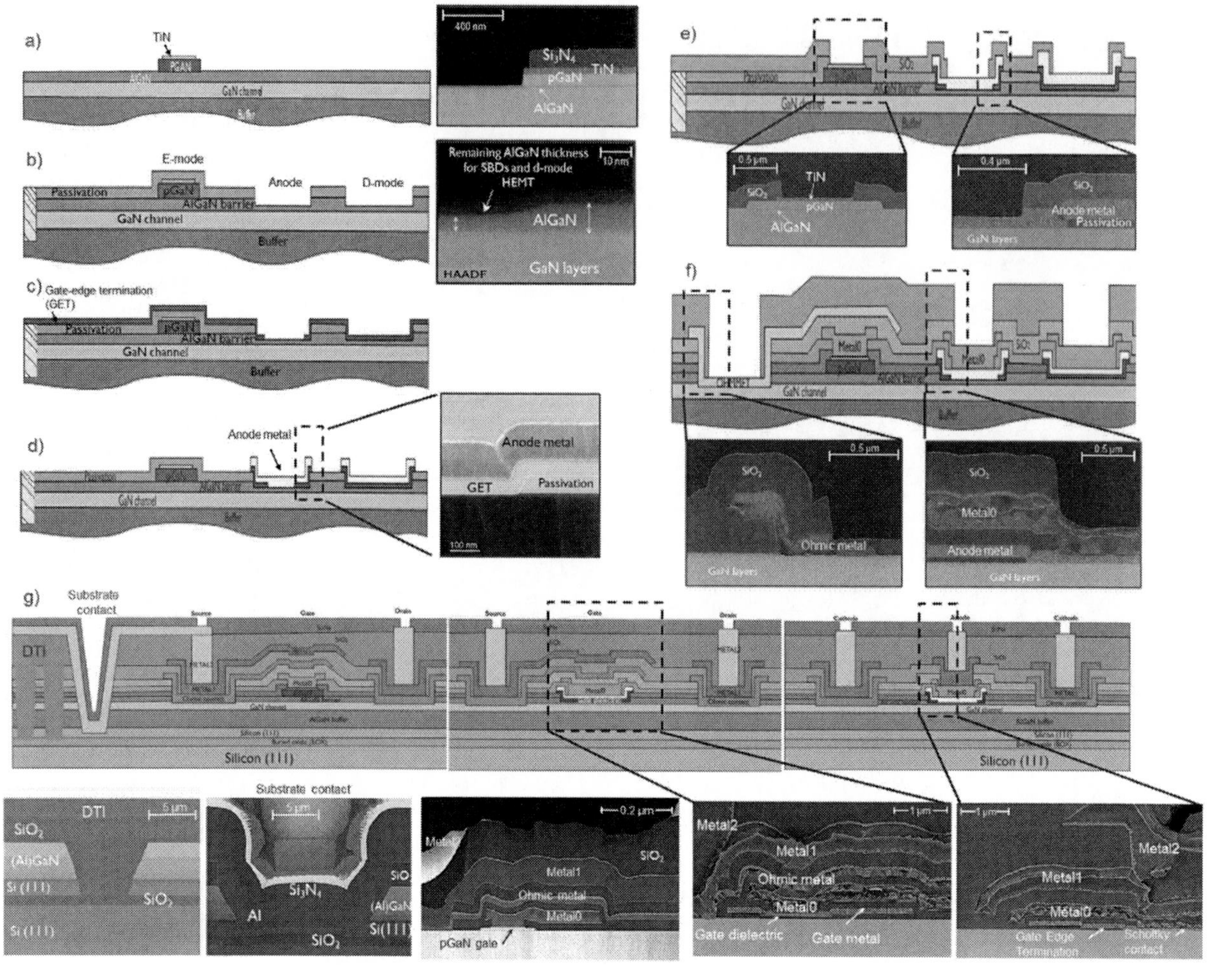

Fig.1. Schematics of the co-integration process steps (a)-(g) of E-mode pGaN HEMTs, D-mode HEMTs and Schottky barrier diodes for IMEC's 200 V GaN-on-SOI platform with cross-section SEM images of the processed devices.

threshold voltage. An Al_2O_3/SiO_2 stack is used as gate dielectric for the D-mode HEMT and Gate-Edge-Termination (GET) dielectric for the SBDs [5] (Fig. 1.c, d). Further processing includes the formation of source, drain and cathode ohmic contacts with different field plate configurations (Fig.1.e and 1.f). A 100 nm thick TiN layer is used as a Schottky contact for the SBDs and Ti/Al/TiN - based stack for ohmic contact formation using a low-temperature anneal. In addition, two interconnect levels are formed by Al-based metal depositions. PECVD SiO_2 layers are used as inter metal dielectric finished by a Si_3N_4 passivation, final polyimide passivation and thick Cu plated redistribution layer. To avoid back-gating effects for the high-side HEMT during switching, and crosstalk between the power devices and the sensitive analog/logic areas of the GaN-IC, the GaN-on-SOI platform includes oxide filled deep trench isolation (DTI) in combination with the buried oxide as shown Fig. 1.g.

III. RESULTS AND DISCUSSION

Fig.2 compares the transfer characteristics and sub-threshold slope of discrete GaN-on-Si and co-integrated GaN-on-SOI p-GaN HEMTs, with a threshold voltage of 2.5 V. The observed difference in ON-resistance is related to a difference in sheet resistance of the two-dimensional electron gas (2-

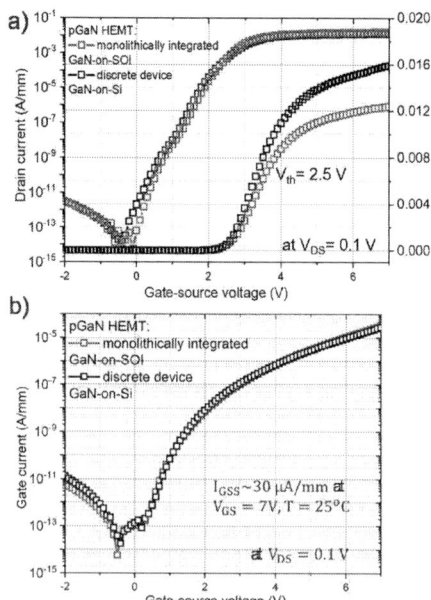

Fig.2. Transfer characteristics (a) and gate leakage currents (b) of a 200 V monolithically integrated p-GaN HEMT and a discrete GaN-on-Si HEMT at 25°C.

DEG) channel at the AlGaN/GaN interface. The forward gate current in Fig. 2.b remained unaffected up to $V_{GS} = 7$ V.

The thickness of the AlGaN barrier in the gate/anode regions after recess and the thickness of the dielectric for the gate/GET determine the D- mode HEMT and GET-SBD characteristics, shown in Fig. 3 to 5. For the GET-SBDs, the turn-on voltage (i.e., V_{AC} extracted at 1 mA/mm) weakly depends on the recessed AlGaN barrier thickness (0.65V to 0.75V at T=25°C) and no significant impact on ON-resistance and dynamic R_{ON} is observed as shown in Fig 3. The leakage current is below 1 nA/mm at T=25°C and below 1 µA/mm at T=150°C. The breakdown voltage is well above 200V. These results show that a large process window was obtained for the target AlGaN barrier thickness underneath the anode region. For the SBDs

shown in Fig.3, the metal field plate configuration has been optimized. However, SBDs with other FP configurations show higher than 20% dynamic R_{ON} and lower V_{BD} for the thinnest 5.5nm AlGaN layer. Therefore, optimum AlGaN thickness is 7.5 nm or 9.5 nm. This also leads to a better ALE uniformity of the remaining AlGaN layer over the full 200 mm wafer. A dielectric thickness of 2.5 nm Al_2O_3 and 35-55 nm SiO_2 forms a good compromise between the GET-SBD parameters and the depletion mode characteristics. SBDs with different GET layers do not show significant impact on the turn-on voltage and ON-resistance. However, dynamic R_{ON} increases above 20% for the devices with GET thickness above 45 nm. The leakage current of SBDs with different GET thicknesses is 10^{-9}-10^{-10} A/mm at T=25°C and ~10^{-7} A/mm at T=150°C, as shown in Fig. 4.d.

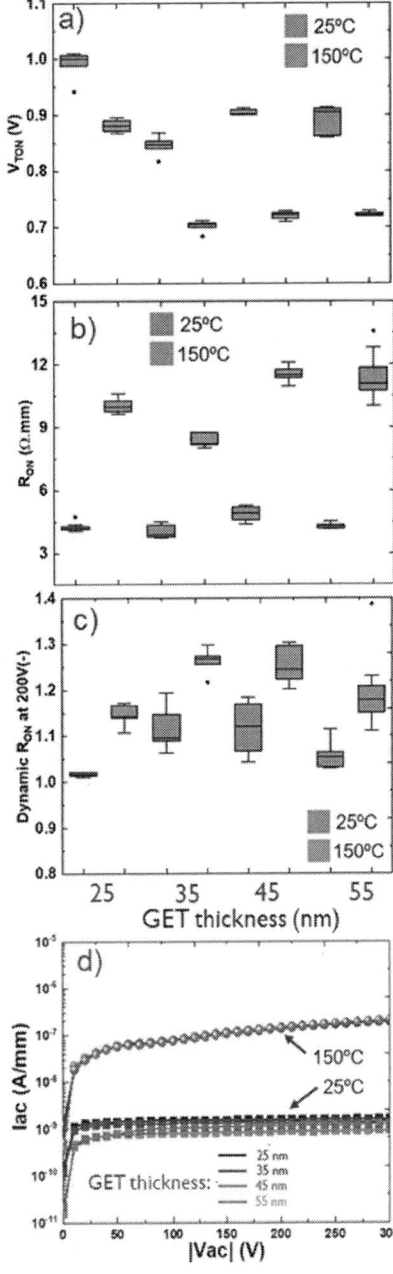

Fig.3. Turn-ON voltage V_{TON} (a), ON-resistance R_{ON} (b), dynamic R_{ON} (c) and medium curves of the reverse characteristic (d) for high-voltage GET-SBDs with anode-to-cathode distance $L_{AC} = 7$ µm and Schottky contact opening $L_{SC} = 4$ µm for different AlGaN thickness at 25°C and 150°C.

Fig.4. Turn-ON voltage V_{TON} (a), ON-resistance R_{ON} (b), dynamic R_{ON} (c) and medium curves of the reverse characteristic (d) for high-voltage GET-SBDs with $L_{AC} = 7$ µm and $L_{SC} = 4$ µm for different thickness of gate-edge termination layer at 25°C and 150°C.

978-1-6654-8498-5/22 $31.00 © 2022 IEEE

As displayed in Fig 5.a and 5.c the threshold voltage of D-mode MIS-HEMTs can be tuned by the thickness of AlGaN barrier and/or GET dielectric. The thinnest AlGaN layer and gate dielectric lead to lower negative threshold voltage around -4 V. AlGaN thickness has no significant impact on the leakage current and breakdown voltage and hence V_{BD} is well above 200 V. However, D-mode HEMTs with gate dielectric below 45 nm exhibit V_{BD} around 200V. Therefore, a thicker gate dielectric layer leads to a higher V_{BD}. The same tendency remains for the low-voltage devices with $L_G = 1$ μm, $L_{GD} = 1.5$ μm and $L_{SG} = 0.5$ μm. These results show that a change in AlGaN thickness leads only to a change in V_{th} while a change in the gate dielectric thickness affects both the V_{th} and V_{BD} simultaneously.

IV. CONCLUSION

High performance Schottky barrier diodes and D-mode HEMTs have been co-integrated in a 200V GaN-on-SOI GaN-IC platform without impacting the performance of the baseline p-GaN HEMTs. The results show a process window of 5.5-9.5 nm AlGaN thickness for the turn-ON voltage and ON-resistance of SBDs. However, SBDs with thicker AlGaN layer (7.5 nm and 9.5 nm) show better performance in dynamic R_{ON} for devices with different FP configurations. In order to obtain a low dynamic R_{ON}, the GET thickness should be between 25 nm - 35 nm for the optimum thickness for current diode design. Therefore, the best performance is observed for SBDs with 7.5-9.5 nm AlGaN thickness and 35 nm of GET thickness. A significant impact of AlGaN and gate dielectric thicknesses on V_{th} and V_{BD} is observed for D-mode HEMTs. The thinnest AlGaN and gate dielectric layers show $V_{th} = -4V$ and V_{th} linearly increases with the increase of AlGaN and gate dielectrics. In addition, reducing the gate dielectric layer for D-mode HEMTs results in breakdown limitations for 200V applications. Therefore, a gate dielectric ≥ 45nm is necessary to obtain V_{BD} above 250V. The best performance for DCFL applications with $V_{th} = -4.5$ V is observed for d-mode HEMTs with 5.5 nm AlGaN and 45 nm of gate dielectric. The extended functionality of the GaN-IC platform described in this paper opens perspectives for design innovation for high performance, compact power systems.

REFERENCES

[1] X.Li et al., "GaN-on-SOI: Monolithically Integrated All-GaN ICs for Power Conversion ", IEEE Intern. Electr. Dev.Meet. (IEDM), vol.39, no. 7, 2018, doi: 10.1109/LED.2018.2833883.

[2] T.Cosnier et al., "200V GaN-on-SOI Smart Power Platform for Monolithic GaN Power ICs", IEEE Intern. Electr. Dev.Meet. (IEDM), 2021, doi: 10.1109/IEDM19574.2021.9720591.

[3] O. Trescases et al., "GaN Power ICs: Reviewing Strengths, Gaps, and Future Directions", IEEE Intern. Electr. Dev. Meet. (IEDM), 2020, pp. 27.4.1-27.4.4, doi: 10.1109/IEDM13553.2020.9371918.

[4] N.Posthuma et al., "An Industry-Ready 200 mm p-GaN E-mode GaN-on-Si Power Technology", IEEE Intern. Symp. Pow. Semicon. Dev. & ICs (ISPSD), 2018, doi: 10.1109/ISPSD.2018.8393658.

[5] E. Acurio et al., "Realibility Improvement in AlGaN/GaN Schottky Barrier Diodes with a Gate Edge Termination", IEEE Transactions on Electron Devices, vol. 65, no. 5, 1765-1770, 2018, doi: 10.1109/TED.2018.2818409.

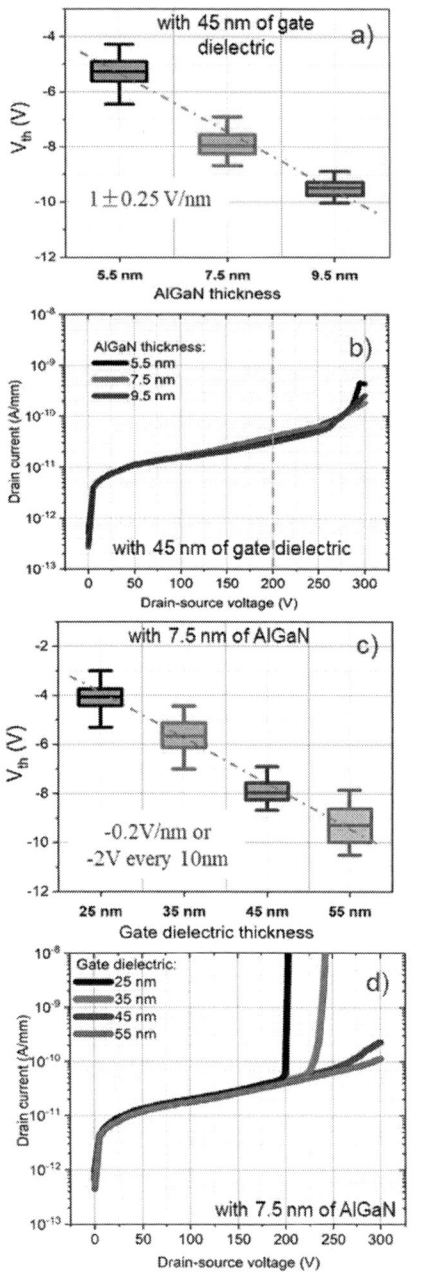

Fig.5. Threshold voltage V_{th} (a) and (c) of low-voltage logic/analog D-mode MIS-HEMTs with $L_G = 1$ μm, $L_{GD} = 1.5$ μm and $L_{SG} = 0.5$ μm and medium curves of the high-voltage drain leakage characteristics (b, d) of devices with $L_G = 1.3$ μm, $L_{GD} = 6$ μm and $L_{SG} = 0.75$ μm with different AlGaN thicknesses and gate dielectric thicknesses at 25°C.

Effect of Post Annealing on the Electrical Characteristics and Deep Level Defects of Ga₂O₃/SiC Heterojunction Diodes

Dong-Wook Byun
Department of Electronic Materials Engineering
Kwangwoon University
Seoul, Republic of Korea
byun1994@kw.ac.kr

Min-Yeong Kim
Department of Electronic Materials Engineering
Kwangwoon University
Seoul, Republic of Korea
alsdud9971@kw.ac.kr

Soo-Young Moon
Department of Electronic Materials Engineering
Kwangwoon University
Seoul, Republic of Korea
justin825@kw.ac.kr

Myeong-Cheol Shin
Department of Electronic Materials Engineering
Kwangwoon University
Seoul, Republic of Korea
smc0753@kw.ac.kr

Michael. A Schweitz
Department of Electronic Materials Engineering
Kwangwoon University
Seoul, Republic of Korea
michael.schweitz@schweitzlee.com

Sang-Mo Koo*
Department of Electronic Materials Engineering
Kwangwoon University
Seoul, Republic of Korea
smkoo@kw.ac.kr

Abstract— **Ga₂O₃/SiC heterojunction were fabricated by RF magnetron sputtering. The influence of nitrogen atmosphere annealing on the electrical and deep level traps in heterojunction diodes was investigated by current density-voltage (J-V), capacitance-voltage (C-V) characteristics, and as well as deep level transient spectroscopy (DLTS). The leakage current and barrier height of post annealed device were 10^{-11} A/cm² and 0.98 V in J-V. And carrier concentration of Ga₂O₃ was improved by 5 % compared with non-annealed diode. It means that post annealing process under nitrogen could be made n-type doping. Based on the DLTS measurement results, the trap related to Fe$_{Ga}$ or Co$_{Ga}$ was detected in all samples at 0.519 eV with concentrations of 4.5 x 10^{13} cm⁻³. Moreover, the N₂-annealed diode exhibited two deep level traps at 1.087 and 1.213 eV related to V$_O$ (normally generated in the dopant epi layer with n-type conductivity Ga₂O₃ such as Ge dopants). It could suggest that post annealing process under nitrogen may cause improving n-type doping.**

Keywords— *Silicon carbide, gallium oxide, post annealing, DLTS*

I. Introduction

Gallium oxide (Ga₂O₃) are used widely in high voltage, high temperature, and high-speed switching applications due to their excellent electrical properties and power efficiency.[1] It is a promising candidate for deep-ultraviolet (DUV), photodetectors, high power, and radio frequency (RF) electronics thanks to its UWBG (Eg = ~ 4.8 eV)[2], strong radiation hardness, and high theoretical electrical breakdown field of ~ 8 MV/cm.[3] The availability of high quality, low cost substrate grown from the edge-defined film-fed growth (EFG) and the floating zone (FZ) method are complemented by the ability to control net free-electron concentrations in a very wide range: from semi-insulating up to 10^{20} cm⁻³ using shallow donors and deep acceptors.[4-7] Additionally, it has been broadly explored with efforts as growth optimization, a large various doping engineering, heterostructure development, device characterization, and defect investigations.[8-11]

However, challenges still remain for the application of Ga₂O₃ in commercial devices. It has a low thermal conductivity(range of 11 − 27 W/m·K).[12] Therefore, it is rather important to realize Ga₂O₃ thin film devices on other semiconductor materials with better thermal characteristics and low lattice mismatch with Ga₂O₃. For example, the heterostructures of Ga₂O₃ on SiC and on Si have been suggested for improving the low thermal parameters, in particular, Ga₂O₃ on SiC with low lattice mismatch (~ 1.3 %).[13]

Despite various efforts of researches in Ga₂O₃, the lack of controlling defects and dopants hinders the development of high-performance devices. Deep level traps can negatively affect the performance of devices by trapping charge carriers, resulting in reduced minority carrier lifetime and increased leakage current.[14] Hence, it is essential to get an understanding of the nature of defects and others such as intrinsic vacancies (V$_{Ga}$, V$_O$),[15] impurities, interstitial or substituted atoms and their complexes. Identifying and controlling deep level defects especially is important in devices used for high-power applications, which frequently contain wide-bandgap semiconductor materials such as Ga₂O₃.

In this work, Ga₂O₃/4H-SiC heterojunction diodes were fabricated by depositing using RF sputtering on SiC substrates, and post-annealing using tube furnace under N₂ atmospheres. The electrical properties were explored from

978-1-6654-8498-5/22 $31.00 © 2022 IEEE

current density-voltage (J-V) and capacitance-voltage (C-V) characteristics such as on/off ratio, barrier height, and carrier concentration. The effect of the annealing atmosphere on deep level traps level, concentration and capture cross section in both diodes was investigated by measuring using deep level transient spectroscopy (DLTS).

II. EXPERIMNETAL

The inset of Figure 1 shows the schematic cross section of a Ga_2O_3/SiC heterojunction diode. The starting material was a highly doped 4H-SiC substrate (N_D = 1.0 × 10^{19} cm^{-3}) with an epitaxial SiC layer (N_D = 1.0 × 10^{16} cm^{-3}). substrate. The 150 nm of nickel (Ni) was deposited onto the back side of substrate using electron-beam (E-beam) evaporation, followed by rapid thermal annealing (RTA) at 1050 °C for 1 min to create an Ohmic contact cathode. Ga_2O_3 thin films were uniformly deposited onto the 4H-SiC wafers using radio frequency (RF) magnetron sputtering. Films of 450 nm were grown under 25 mTorr with a sputtering power of 120 W. The samples were then annealed at 950 °C for 1 h under nitrogen (N_2) gas atmospheres using a tube furnace. Finally, an anode electrode with Ni (100 nm) was formed on the Ga_2O_3 thin film layer using E-beam evaporation.

III. RESULATS AND DISCUSSION

A. Electrical Properties

Figure 1 shows the J-V characteristics of heterojunction diodes. The values of leakage current with both diodes at −5 V were found to be approximately 10^{-8}, and 10^{-11} A/cm^2, respectively. The diode annealed under N_2 condition has excellent rectifying behavior, which improved its on/off ratio at −5 and 5 V by approximately 10^7 times compared with non-annealed sample. It could be attributed to the improved crystallinity of the film increased by the post-annealing process.[16]

Table 1 shows the ideality factor and barrier height for the diodes from their J-V characteristics based on the thermionic emission theory. And barrier height is the given by:[17,18]

$$I_s = AA^{**}T^2 \exp(-q\emptyset_B/k_BT) \quad (1)$$

where A the is contact area (0.00785 cm^2), A^{**} is the Richardson constant, and $\emptyset_B$ ($\emptyset_{B(I-V)}$) is the Schottky barrier height in the I-V curve. For Ga_2O_3, the Richardson constant can be calculated to be 33.65 A/cm^2·K^2.[19]

The value of barrier heights were deduced to be 1.03, and 0.98 V, respectively. The N_2 annealed diode exhibited the lowest barrier height and had an ideality factor nearest to 1 of any of the devices. This may be implied that the Ga_2O_3 annealed with nitrogen was able to improve electrical properties by filling /residing the oxygen vacancy in Ga_2O_3 when they annealed in nitrogen.[20]

Figure 2 shows A^2/C^2 measured in the reverse voltage bias applied at a pulse frequency of 50 kHz. Based on Anderson's theory of heterojunction diodes,[21] the capacitance of the depletion layer can be expressed as follows:[22]

$$A^2/_{C^2} = [2(\varepsilon_n N_D + \varepsilon_p N_A)/q\varepsilon_n\varepsilon_p N_D N_A](V_{bi} - V) \quad (2)$$

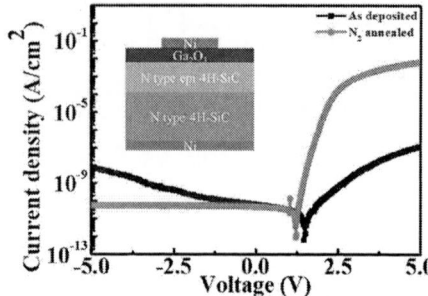

Fig. 1. J–V characteristics of Ga_2O_3/4H-SiC heterojunction diodes. The inset is a schematic section view of the Ga_2O_3/SiC heterojunction diode. (The as deposited (non-annealed), and N_2, post-annealed samples.)

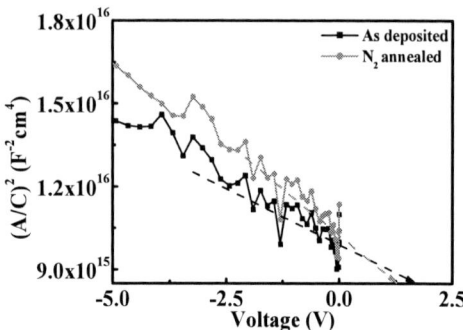

Figure 2. A^2/C^2-V characteristics of the diodes

TABLE I. Structural Parameters of the devices

	As deposited	N_2 annealed
On/Off ratio	17.6	11.2x10^7
η	2.15	1.66
$\emptyset_{B(I-V)}$ (eV)	1.03	0.98
$\emptyset_{B(C-V)}$ (eV)	2.33	1.92
N_A (cm^{-3})	1.48 x 10^{14}	1.55 × 10^{14}

where ε_n (= 9.7)[23] and ε_p (= 10)[21] are the relative dielectric constants of 4H-SiC and Ga_2O_3, respectively, N_D (= 1.0 x 10^{16}) and N_A are the carrier concentrations of 4H-SiC epi layer and Ga_2O_3, respectively, and V_{bi} is the built-in potential. The N_A of the both diodes can be obtained from the slope of the A^2/C^2 curves, and are approximately 1.48 × 10^{14}, and 1.55 × 10^{14} cm^{-3}. The values of the built-in potential can be deduced to be 1.78, 1.34 V, respectively, from the V-axis intercept. Using the extracted built-in potential, the experimental barrier height can be determined as follows:[24]

$$\emptyset_{B(C-V)} = V_{bi} + k_B\ln(N_C/N_D) \quad (3)$$

where N_D is the Ga_2O_3 carrier concentration and N_C = $2(2\pi m^* k_B T/h^2)^{3/2}$ is the conduction band density of states from m^* = 0.28m_0 is the electron effective mass in

Ga_2O_3.[25] The conduction band density of state can be calculated to be $N_C = 2.67 \times 10^{18}$ cm^{-3}.

The barrier height from capacitance – voltage characteristics were determined to be 2.33, and 2.18 V, respectively, which were similar to the barrier heights of the I-V curves. However, the barrier heights from the C-V characteristics ($\emptyset_{B(C-V)}$) were slightly larger than those from the I-V ($\emptyset_{B(I-V)}$) characteristics, as shown in Table 1. It can be seen in the existence of excess capacitance and Schottky barrier height inhomogeneities. Moreover this could be caused by metal trap states and semiconductor interface layers.

B. Deep Level Properties

Figure 3 (a) shows the DLTS spectra obtained from the non-annealed, and N_2 post-annealed samples. The measurements were conducted for temperatures ranging from 80-700 K. The as deposited sample's DLTS peak labeled E1 emerges before disappearing after annealing. However, the peaks in N_2 sample appear above 400 K. The broad DLTS peaks at 400 and 550 K seem to consist of peaks E2 and E3, with peak N4 being observed at approximately 600 K, as shown in the Figure 3 (b). The broad peak represents overlapping peaks, in which one of the overlapping peaks is small in amplitude but the overlap is strong. The trap parameters of a broad peak should be defined through an Arrhenius plot of the single scan analysis data.[26,27]

From the DLTS spectra, the trap parameters such as the trap level and the trap capture cross section can be easily determined. This follows from the following relationship:[26]

$$e_n = \sigma_n \gamma_n T^2 exp(-(E_C - E_V)/k_B T) \qquad (4)$$

where γ_n is a constant associated with the effective mass of the electrons, σ_n is the capture cross section, and k_B is the Boltzmann constant.

As shown in Figure 4, and $\Delta E = E_C - E_t$ can be extracted from the slope and σ_n can be deduced from the intersection of the plot. Using Eq. (4), the trap parameters can be extracted from the overlapping E2 and E3 peaks. The E1 peak is generated at 0.519 eV, the traps being positioned at a shallow depth while other traps can be observed more deeply in N_2 post annealing. Deep level traps of Ga_2O_3/SiC heterojunction devices under N_2 atmospheres appear at 0.981 (E2), 1.087 (E3), and 1.213 (E4), respectively.

From the Arrhenius plots, the energy levels ($E_C - E_t$) and the capture cross sections (σ_n) of the observed traps were obtained, which are summarized in Table 2. The trap concentration (N_t) can be determined from the maximum of the DLTS-peak amplitudes.[24] Based on the trap level compared with the similarly reported deep level defect levels, the detailed behaviors of defects by their possible physical origin and Ga_2O_3 based bulk or epi layer materials were collected in the last two columns of Table 2. The E1

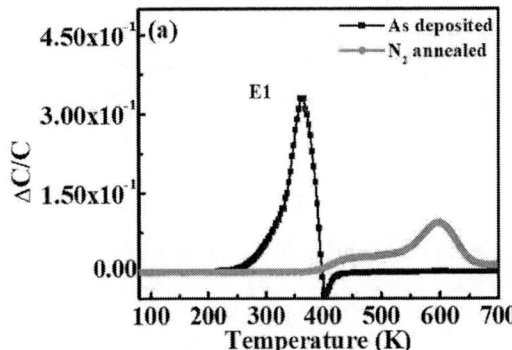

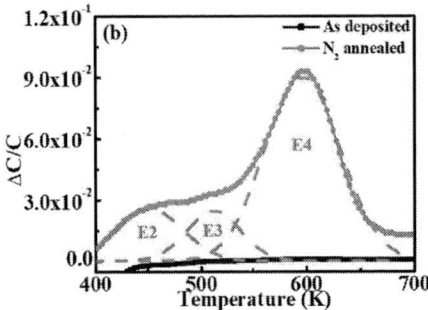

Figure 3. DLTS spectra of heterojunction diodes. (a) 80–700 K, and (b) detailed DLTS spectra of N_2 annealed diode.

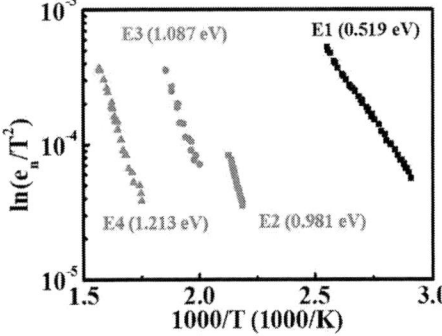

Figure 4. Arrhenius plot made from the DLTS spectra with the energy levels of the peaks.

trap in as deposited sample at 0.519 eV, with a trap concentration 4.5×10^{13} cm^{-3}, correlates with the Fe and Co traps in unintentionally doped (UID) bulk Ga_2O_3.[25] This may be estimated for part of the deep electron traps transition metal impurity since that is extracted from a preliminary investigation by electron paramagnetic resonance (EPR) spectroscopy.[25] The N_2 annealed device occurred three deep level traps below 0.9 eV. Labeled E2 trap (0.981 eV) was present deeper than non-annealed diode trap E1. This level is closed to approximately 1.0 eV. That reported it could generate when Ga_2O_3 grown methods by the edges-defined file-fed growth (EFG) and Czochralski (CZ) UID (n-type background) substrate materials.[18]

978-1-6654-8498-5/22 $31.00 © 2022 IEEE

TABLE II. Summary of Trap Information

Peak ID	$E_C - E_V$ (eV)	σ_n (cm^2)	N_t (cm^{-3})	Possible physical origin	Material
E1	0.519	2.1×10^{-14}	4.5×10^{13}	Fe$_{Ga}$, Co$_{Ga}$	UID bulk (CZ)[24]
E2	0.981	4.6×10^{-14}	2.2×10^{13}	Fe$_{Ga}$, Co$_{Ga}$	UID bulk (CZ, EFG)[24,25]
E3	1.097	1.3×10^{-13}	9.2×10^{12}	V$_O$	Sn-doped layer (MOCVD)[24,25]
E4	1.213	7.7×10^{-13}	4.4×10^{10}	V$_O$	Si-doped layer (MOCVD)[24,25]

In addition, the traps of E3 (1.087 eV) and E4 (1.213 eV) with 9.2×10^{12} and 4.4×10^{10} cm^{-3} trap concentration had similar level that oxygen vacancy (V$_O$) was observed in the dopant epi layer with n-type conductivity Ga$_2$O$_3$ materials such as Sn and Ge.[17] It suggests that the N$_2$ post annealing may cause improving n-type conductivity.

IV. CONCLUSION

The Ga$_2$O$_3$/4H-SiC heterojunction diodes were fabricated using RF sputtering. The effect of the post-annealing atmosphere under nitrogen condition on the electrical and deep level trap properties of the heterojunction diodes were investigated through I-V, C-V, and DLTS measurements. The reverse bias leakage currents density of annealed in N$_2$ had approximately 10^{-11} A/cm^2 is decreased by a factor of approximately 10^3 compared to the non-annealed sample. The estimated barrier heights from I-V and C-V in annealed diode were 0.98 and 1.92 V, respectively, lower in comparison with non-annealed diode. Additionally, the Ga$_2$O$_3$ carrier concentration was improved by 5 %. It means that post annealing process under nitrogen could be made n-type doping. By the DLTS investigation, the deep level traps were found to be related traps which are reported UID Ga$_2$O$_3$ substrate and epi layer. Moreover, traps of N$_2$ annealed device had 1.087 and 1.213 eV correlated with oxygen vacancy in n-type doped Ga$_2$O$_3$ materials.

ACKNOWLEDGMENT

This work was supported by the National Research Foundation of Korea grant (2021R1F1A1057620) funded by MSIT, the fostering global talents for innovative growth program through the KIAT grant (P0017308),and a Research Grant from Kwangwoon University in 2022.

REFERENCES

[1] Baliga. B. J, World Scientific Publishing Company., (2016).
[2] Tippins. H. H, Physical Review., 140(1A), A316 (1965).
[3] Tadjer. M. J, Mastro. M. A, Mahadik. N. A, Currie. M, Wheeler. V. D, Freitas. J. A, Greenlee. J. D, Hite. J. K, Hbart. K. D, Eddy. C. R, & Kub, F. J, Journal of Electronic Materials., 45(4), 2031–2037 (2016).
[4] Higashiwaki. M, Sasaki. K, Kuramata. A, Masui. T, & Yamakoshi. S, Applied Physics Letters., 100(1), 013504 (2012).
[5] Villora. E. G, Shimamura. K, Yoshikawa. Y, Aoki. K, & Ichinose. N, Journal of Crystal Growth., 270(3–4), 420–426 (2004).
[6] Kuramata. A, Koshi. K, Watanabe. S, Yamaoka. Y, Masui. T, & Yamakoshi. S, Japanese Journal of Applied Physics., 55(12), 1202A2 (2016).
[7] Tomm. Y, Reiche. P, Klimm. D, & Fukuda. T, Journal of crystal growth., 220(4), 510–514 (2000).

[8] Wagner. G, Baldini. M, Gogova. D, Schmidbauer. M, Schewski. R, Albrecht. M, Galazka. Z, Klimm. D, & Fornari. R, physica status solidi (a)., 211(1), 27–33 (2014).
[9] Rafique. S, Han. L, Tadjer. M. J, Freitas Jr. J. A, Mahadik. N. A, & Zhao. H, Applied Physics Letters., 108(18), 182105 (2016).
[10] Polyakov. A. Y, Nikolaev. V. I, Stepanov. S. I, Pechnikov. A. I, Yakimov. E. B, Smirnov. N. B, Shchemerov. I. V, Vasilev. A. A, Kochkova. A. I, Chernykh. A. V, & Pearton. S. J, ECS Journal of Solid State Science and Technology., 9(4), 045003 (2020).
[11] Ghadi. H, McGlone. J. F, Jackson. C. M, Farzana. E, Feng. Z, Bhuiyan. A. A. U, Zhao. H, Aehart. A. R, & Ringel. S. A, APL Materials., 8(2), 021111 (2020).
[12] Lin. C. H, Hatta. N, Konishi. K, Watanabe. S, Kuramata. A, Yagi. K, & Higashiwaki. M, Applied Physics Letters., 114(3), 032103 (2019).
[13] Yu. J, Nie. Z, Dong. L, Yuan. L, Li. D, Huang. Y, Zhang. L, Zhang. Y, & Jia. R, Journal of Alloys and Compounds., 798, 458–466 (2019).
[14] Nepal. N, Katzer. D. S, Downey. B. P, Wheeler. V. D, Nyakiti. L. O, Storm. D. F, Hardy. M. T, Freitas. J. A, Jin. E. N, Vaco. D, Yates. L, Gaham. S, Kumar. S, & Meyer. D. J, Journal of Vacuum Science & Technology A: Vacuum, Surfaces, and Films., 38(6), 063406 (2020).
[15] Wang. Z, Chen. X, Ren. F. F, Gu. S, & Ye. J, Journal of Physics D: Applied Physics., 54(4), 043002 (2020).
[16] Danno. K, & Kimoto. T, Journal of applied physics., 100(11), 113728 (2006).
[17] Kananen. B. E, Halliburton. L. E, Stevens. K. T, Foundos. G. K, & Giles. N. C, Applied Physics Letters., 110(20), 202104 (2017).
[18] Ahmadi. E, Koksaldi. O. S, Kaun. S. W, Oshima. Y, Short. D. B, Mishra. U. K, & Speck. J. S, Applied Physics Express., 10(4), 041102 (2017).
[19] Ingebrigtsen. M. E, Kuznetsov. A. Y, Svensson. B. G, Alfieri. G, Mihaila. A, Badstübner. U, Perron. A, & Varley. J. B, APL Materials., 7(2), 022510 (2019).
[20] Saikumar. A. K, Nehate. S. D, & Sundaram. K. B, ECS Journal of Solid State Science and Technology., 8(7), Q3064 (2019).
[21] Cheung, S. K., & Cheung, N. W. (1986). Extraction of Schottky diode parameters from forward current - voltage characteristics. Applied physics letters, 49(2), 85‐87.
[22] Farzana. E, Zhang. Z, Paul. P. K, Arehart. A. R, & Ringel. S. A, Applied Physics Letters., 110(20), 202102 (2017).
[23] Peelaers. H, & Van de Walle. C. G, physica status solidi (b)., 252(4), 828–832 (2015).
[24] Wang. Z, Chen. X, Ren. F. F, Gu. S, & Ye. J, Journal of Physics D: Applied Physics., 54(4), 043002 (2020).2
[25] Polyakov. A. Y, Smirnov. N. B, Shchemerov. I. V, Gogova. D, Tarelkin. S. A, & Pearton. S. J, Journal of Applied Physics., 123(11), 115702 (2018).
[26] Demircioglu. Ö, Karataş. Ş, Yıldırım. N, Bakkaloglu. Ö. F, & Türüt. A, Journal of Alloys and Compounds., 509(22), 6433–6439 (2011).
[27] Varley. J. B, Peelaers. H, Janotti. A, & Van de Walle. C. G, Journal of Physics: Condensed Matter., 23(33), 334212 (2011).

BEoL integrated hafnium zirconium oxide varactors for tunable mmWave applications

Sukhrob Abdulazhanov*, Dang Khoa Huynh*, Quang Huy Le*, David Lehninger*, Thomas Kämpfe*
and Gerald Gerlach[†]
*Center Nanoelectronic Technologies, Fraunhofer IPMS, 01109, Dresden, Germany
[†]Institute of Solid State Electronics, Dresden University of Technology, 01069, Dresden, Germany
*sukhrob.abdulazhanov@imps.fraunhofer.de, [†] gerald.gerlach@tu-dresden.de

Abstract—**In this work we demonstrate Back-End-of-Line (BEoL) integrated thin-film metal-ferroelectric-metal (MFM) varactors based on ferroelectric hafnium zirconium oxide. The varactors were developed for tuning a passive bandpass filter at millimeter wave frequencies. Upon the voltage sweep between -3 V and 2 V, the filter's center frequency shows a stable tuning between 39.1 GHz and 36.9 GHz, respectively, with a maximum insertion loss of 3.87 dB. The chip area of the device is 0.091 mm^2 excluding the pads. The experimental results were examined with the method-of-moment simulation, which has revealed that the varactor's permittivity shifts from 14.6 to 16.6, resulting in 12.5% tunability.**

Index Terms—**ferroelectric, MFM varactor, HZO, CPW, insertion loss, Back-End-of-Line**

I. INTRODUCTION

With the rapid development of wireless systems, communication standards move to higher frequency bands and wider bandwidths to provide greater channel capacity. In the meantime, there is a great deal of emphasis on the search for tunable materials and components that would provide better or comparable characteristics with conventional CMOS devices at mmWave frequencies. For instance, varactors based on microelectromechanical systems (MEMS) and ferroelectrics, like BaTiO$_3$ (BST), have lower dielectric losses than the MOS varactors, which is beneficial at frequencies above 20 GHz [1]. However, these materials have a significant drawback, namely their poor compatibility with conventional CMOS and BiCMOS integration processes. They usually require high tuning voltages and high annealing temperatures, exceeding the thermal budget of the Back-End-of-Line deposition processes, which hinders their integration into industrial RFICs. Recently there has been a large amount of interest in ferroelectric hafnium oxide - a fully CMOS and BEoL compatible material [2], that is mainly investigated as a candidate for nonvolatile memory applications [3] and neuromorphic computing [4]. It is also being investigated to be used as a varactor for microwave applications [5]–[9]. Within CMOS manufacturing methods, hafnium oxide can be structured by reactive-ion etching using optical lithography, which makes it possible to manufacture devices with a very small dimensions. Ferroelectric hafnium oxide layers are usually deposited by atomic layer deposition (ALD), using different precursors and varying deposition cycles to allow easy control of stoichiometry. The ferroelectric phase of HfO$_2$ can be achieved by doping with Si, Zr, La, .etc., and subsequent annealing. If doped with zirconium (Zr) with 1:1 doping ratio, the obtained Hf$_{0.5}$Zr$_{0.5}$O$_2$ (HZO) has the lowest crystallization temperatures of the ferroelectric phase, fitting well within the BEoL thermal budget [10]. Varactors, based on HZO have shown quite stable capacitance retention, tunability endurance, and decent Q-factors, when compared to the state-of-the-art ferroelectric varactors [11]. They also retain functionality at elevated temperatures up to 200°C [12], [13]. Passive microwave components, based on HZO have been proposed previously by our group [7], [8]. In this article we present the first experimental realization of a 40 GHz bandpass filter based on distributed HZO MFM varactors, fabricated in 180 nm CMOS technology.

II. DESIGN CONSIDERATIONS AND INTEGRATION FLOW

The bandpass filter design was developed in accordance with the 180 nm HV SOI CMOS technology. The BEoL stack consists of five metal layers [Fig.1(a)], where the HZO thin film of 10 nm thickness is capped by titanium nitride (TiN) electrodes and integrated between metal Met.#2 and Met.#3. The transmission electron microscopy (TEM) images of the stack is shown in Fig.1(b). Similar integration flow was implemented in [14].

Standart low-frequency C-V characteristics of the obtained thin-film varactor, measured at 10 kHz [Fig.2] shows that the HZO permittivity varies between 17.6 and 24.2 upon DC bias sweep between -3V and 2V, resulting in 27% capacitance tunability.

The bandpass filter itself is based on coplanar waveguide (CPW) delay-line with lumped shunting varactors (C_1) [Fig.3(a)]. It consists of 10 identical unit sections, each shunted to the ground from one side with MFM varactor (C_2) [Fig.3(b)]. By changing the number of sections, one can vary the resonance frequency. Two additional $4\,\mu m \times 4\,\mu m$ varactors (C_1) are connected in series for the low frequency rejection [Fig.3(c)]. Excluding the pads, the overall device area reaches 0.091 mm^2.

To exploit the maximum frequency possible in compliance with the technology design rules, the dimensions of the shunting varactors (C_2) set to minimum, which is

978-1-6654-8498-5/22 $31.00 © 2022 IEEE

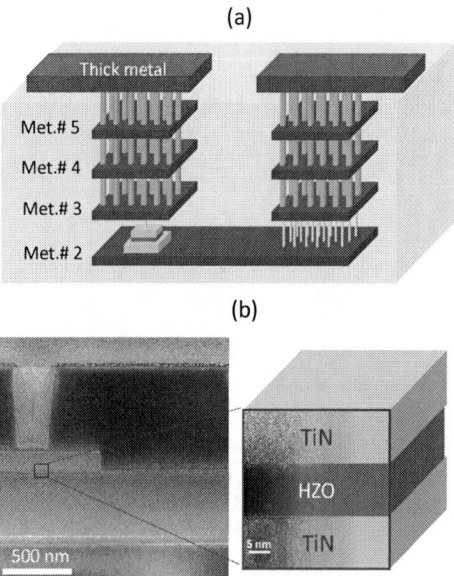

(a)

(b)

Fig. 1: Cross-section of the BEoL stack (a) and transmission electron microscopy images (b) of the MFM varactor integrated in 180 nm CMOS technology.

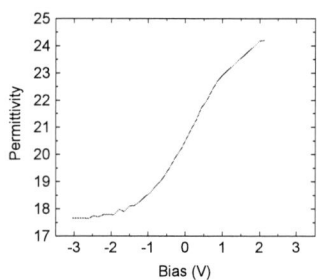

Fig. 2: C-V characteristics of HZO thin-film MFM varactor measured at 10 kHz.

$2\,\mu m \times 2\,\mu m$. This inevitably results in an increase of insertion loss due to the contact resistance (R_2), since only a single via with $0.26\,\mu m \times 0.26\,\mu m$ dimensions per varactor is allowed. To reduce the resistance of the bottom contact, as in [15], shunting varactors (C_2) are placed closer to the ground and are surrounded by the ground contact from three sides [Fig.3(b)]. The series inductance (L_1) and resistance (R_1) were minimized by using a wide ($35\,\mu m$) signal line. In contrast, the inductance of the shunt contacts (L_2) was increased by connecting thinner ($3\,\mu m$) metal lines to get a sharper peak and increase the Q-factor of the filter.

III. EXPERIMENTAL AND SIMULATION SETUP

The low-frequency C-V characterization was executed on Aixxact 3000 TF Analyzer. Microwave measurements were performed on a Keysight N5247B PNA-X vector

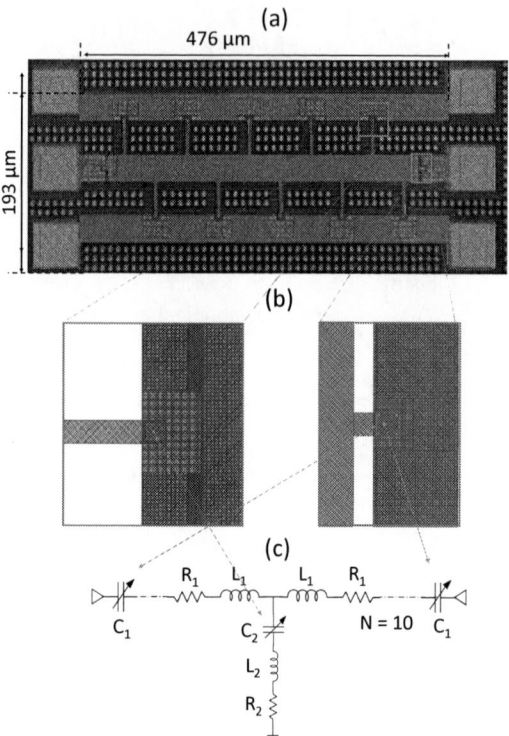

Fig. 3: Microscopy image of the bandpass filter with 10 unit sections (a), each shunted to the ground by $4\,\mu m^2$ MFM HZO varactors (b) and two $16\,\mu m^2$ varactors connected in series (c) for low frequency rejection.

network analyzer in the frequency range between 10 MHz and 67 GHz applying a -15 dBm signal. For contact with the pads, the $100\,\mu m$ pitch Ground-Signal-Ground Infinity probes were used. The capacitance was tuned using the DC bias, that was applied upon RF signal by an SMU B2902A through the bias tee. In addition, the performance of the device was investigated with method-of-moment simulations using the Cadence AWR AXIEM EM simulator to verify the permittivity change of HZO. In the simulations, the permittivity was set constant for the whole frequency range. The loss tangent ($\tan\delta$) of the HZO was set to 0.05, which was experimentally obtained in [16] at 20 GHz.

IV. RESULTS AND DISCUSSION

In Fig.4(a) the insertion loss S_{21} and return loss S_{11} are plotted versus frequency. In the inset, the shift of the center frequency is better visualized. Upon increasing bias, the center frequency f_0, defined as an average value between the two cut-off frequencies at 3 dB attenuation, is shifting from 39.1 GHz towards 36.9 GHz, exhibiting a frequency tunability of 9.3%. The maximum insertion loss increases from 3.51 dB to 3.87 dB and return loss decreases from 15.71 dB to 14.95 dB. The fraction bandwidth Δf, defined as a difference between cutoff frequencies at 3 dB, divided

978-1-6654-8498-5/22 $31.00 © 2022 IEEE

TABLE I: State-of-the-art tunable passive band-pass filters

Ref.	Technology	IL_{max} (dB)	f_0 (GHz)	Δf (%)	V_{max}(V)	FoM(dB^{-1}/kV)	Q-factor	Area (mm^2)
[17]	MEMS	2.8	57.7 – 63	3.6 – 4.3	140	3.44	23.3 – 27.9	40
[17]	MEMS	4.8	61.9 – 65	4.5 – 5	150	3.25	20 – 22.27	96
[18]	MEMS	3.2	63.2 – 67.4	3.4 – 8.3	27.3	9.52	12 – 29.4	2.53
[19]	MEMS	4.2	20 – 40	1.9–4.7	180	20.44	21.3–52.6	–
[20]	MEMS	3.09	23 – 35	1.3–3.6	140	15.75	27.7 –76.6	112
[21]	BST	6.9	29–34	9.48–12.3	30	3.72	8.13–10.54	6.3
This work	Hf$_{0.5}$Zr$_{0.5}$O$_2$	3.87	36.6 – 39.1	66.4–67.1	3	3.03	1.5	0.091

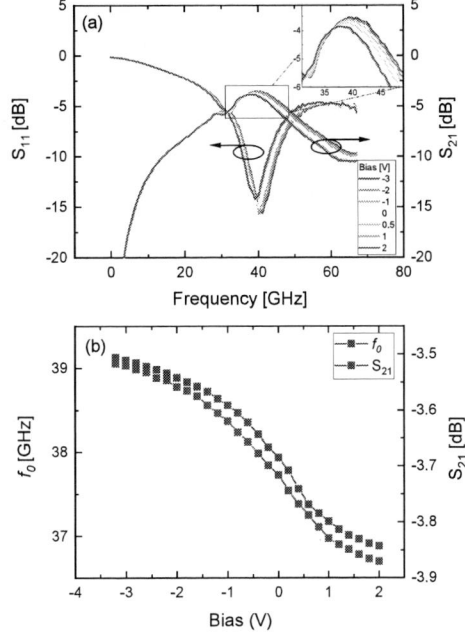

Fig. 4: S-parameters (a), center frequency and maximum insertion loss (b) variation of the bandpass filter upon applied bias voltage.

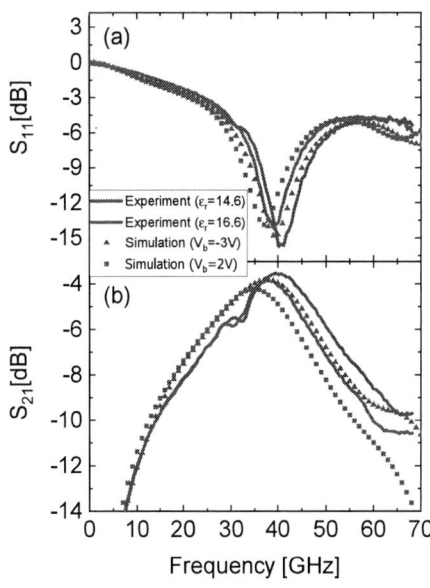

Fig. 5: The comparison of return (a) and insertion (b) losses of the experimental data at maximum and minimum bias (V_b) and the simulated data at maximum and minimum HZO permittivity (ϵ_r).

by f_0, shrinks from 67.1% to 66.53%. In Fig.4(b), the tuning of f_0 and S_{21} is plotted versus bias voltage, where it can be seen that the tuning has a good linearity, suitable for analog applications.

In Fig.5 the simulation results of the return loss [Fig.5(a)] and insertion loss [Fig.5(b)] are plotted versus frequency. When the permittivity is varied between 14.6 and 16.6, the corresponding peaks and the overall profiles of the S-parameters agree well with the experimental data, indicating the 12.5% capacitance tunability. This also indicates that the permittivity values are stable throughout the entire frequency range. However, the permittivity and tunability values are lower than those measured at 10 kHz [Fig.2]. This can be explained by the reduction of ferroelectric switching in GHz range, caused by the domain wall relaxation. The decrease of tunability in the MHz range was observed in our previous work [9], but was attributed to the effect of electric field distribution on TiN electrodes.

Based on the experimental results, the figure of merit FoM of the band-pass filter was calculated by using the relation [22]:

$$FoM = \frac{f_0^{max} - f_0^{min}}{V_{max} * \sqrt{\Delta f^{max} * \Delta f^{min}} * \sqrt{IL^{max} * IL^{min}}}$$

where V_{max} is the maximum tuning bias; IL^{max}, IL^{min}, f_0^{max}, f_0^{min}, Δf^{max} and Δf^{min} are maximum and minimum values of insertion loss, center frequency, and fractional bandwidth, respectively.

The FoM of our device is equal to 3.03 dB^{-1}/kV. The results are summarized in the Table I, where they are compared with the state-of-the-art passive tunable bandpass filters, known to the author. From the table, it is evident, that MEMS and BST bandpass filters have a higher frequency tunability and quality factor, which is defined as the ratio of the center frequency to fractional frequency. However, due to the high tuning voltages their figure of merit is comparable to our device. Moreover, our bandpass filter is much better in terms of miniaturization because it has a much smaller chip area. It should also be noted that the insertion loss and quality factor are mostly influenced by the via resistance and can be further reduced if the device is designed in a different technology where the design rules would allow implementation of interconnects with lower resistance.

V. CONCLUSION

In this work, we presented the first experimental characterization of 40 GHz tunable bandpass filter, based on distributed ferroelectric hafnium zirconium oxide varactors, integrated into the BEoL of 180 nm CMOS technology. Experimental measurements have shown that upon bias sweep between -3 V and 2 V, the filter's center frequency shifts from 39.1 GHz to 36.6 GHz and its insertion loss increases from 3.51 dB to 3.87 dB, showing a good linearity, appropriate for analog applications. The method-of-moment simulation has shown a good agreement with experimental results. It was revealed that upon bias sweep the varactor's permittivity varies between 14.6 and 16.6, showing 12.5% tunability, which is lower than 27% tunability at 10 kHz, which can be explained by the domain wall relaxation in GHz range. Compared to state-of-the-art passive tunable bandpass filters, our filter has a lower Q-factor and frequency tunability, yet a much smaller chip area and a comparable figure of merit due to the low tuning voltages.

VI. ACKNOWLEDGMENT

This work was financed by project OCEAN12 from the Electronic Component Systems for European Leadership Joint Undertaking and supported by X-FAB foundry group. The authors would also like to thank Hannes Maehne for the support during the design and fabrication.

REFERENCES

[1] S. Gevorgian, *Ferroelectrics in microwave devices, circuits and systems: Physics, modelling, fabrication and measurements*, ser. Engineering materials and processes. London: Springer, 2009.

[2] T. S. Böscke, J. Müller, D. Bräuhaus, U. Schröder, and U. Böttger, "Ferroelectricity in hafnium oxide thin films," *Applied Physics Letters*, vol. 99, no. 10, p. 102903, 2011.

[3] T. Ali, P. Polakowski, S. Riedel, T. Büttner, T. Kämpfe, M. Rudolph, B. Pätzold, K. Seidel, D. Löhr, and R. Hoffmann, "Silicon doped hafnium oxide (HSO) and hafnium zirconium oxide (HZO) based FeFET: A material relation to device physics," *Applied Physics Letters*, vol. 112, no. 22, p. 222903, 2018.

[4] M. Lederer, R. Olivo, D. Lehninger, S. Abdulazhanov, T. Kämpfe, S. Kirbach, C. Mart, K. Seidel, and L. M. Eng, "On the Origin of Wake–Up and Antiferroelectric–Like behavior in Ferroelectric Hafnium Oxide," *physica status solidi (RRL) – Rapid Research Letters*, 2021.

[5] M. Dragoman, M. Aldrigo, M. Modreanu, and D. Dragoman, "Extraordinary tunability of high-frequency devices using $Hf_{0.3}Zr_{0.7}O_2$ ferroelectric at very low applied voltages," *Appl. Phys. Lett.*, vol. 110, no. 10, p. 103104, 2017.

[6] M. Dragoman, M. Modreanu, I. Povey, S. Iordanescu, M. Aldrigo, A. Dinescu, D. Vasilache, and C. Romanitan, "2.55 GHz miniaturised phased antenna array based on 7 nm-thick $Hf_xZr_{1-x}O_2$ ferroelectrics," *Electron. Lett.*, vol. 54, no. 8, pp. 469–470, 2018.

[7] S. Abdulazhanov, Q. H. Le, D. K. Huynh, D. Wang, G. Gerlach, and T. Kämpfe, "A mmWave Phase Shifter Based on Ferroelectric Hafnium Zirconium Oxide Varactors," in *2019 IEEE MTT-S International Microwave Workshop Series on Advanced Materials and Processes for RF and THz Applications (IMWS-AMP)*. IEEE, 72019, pp. 175–177.

[8] ——, "A Tunable mmWave Band-Pass Filter Based on Ferroelectric Hafnium Zirconium Oxide Varactors," in *2019 IEEE MTT-S International Microwave Workshop Series on Advanced Materials and Processes for RF and THz Applications (IMWS-AMP)*. IEEE, 72019, pp. 46–48.

[9] S. Abdulazhanov, Q. H. Le, D. K. Huynh, D. Wang, M. Lederer, R. Olivo, K. Mertens, J. Emara, T. Kämpfe, and G. Gerlach, "RF-Characterization of HZO Thin Film Varactors," *Crystals*, vol. 11, no. 8, p. 980, 2021.

[10] D. Lehninger, R. Olivo, T. Ali, M. Lederer, T. Kämpfe, C. Mart, K. Biedermann, K. Kühnel, L. Roy, M. Kalkani, and K. Seidel, "Back–End–of–Line Compatible Low–Temperature Furnace Anneal for Ferroelectric Hafnium Zirconium Oxide Formation," *physica status solidi (a)*, vol. 217, no. 8, p. 1900840, 2020.

[11] S. Abdulazhanov, M. Lederer, D. Lehninger, C. Mart, T. Ali, D. Wang, R. Olivo, J. Emara, T. Kampfe, and G. Gerlach, "Tunability of Ferroelectric Hafnium Zirconium Oxide for Varactor Applications," *IEEE Transactions on Electron Devices*, vol. 68, no. 10, pp. 5269–5276, 2021.

[12] S. Abdulazhanov, M. Lederer, D. Lehninger, T. Ali, J. Emara, R. Olivo, and T. Kämpfe, "Influence of antiferroelectric-like behavior on tuning properties of ferroelectric HZO-based varactors," *MRS Advances*, vol. 6, no. 21, pp. 530–534, 2021.

[13] S. Abdulazhanov, M. Lederer, D. Lehninger, T. Ali, R. Olivo, and T. Kämpfe, "The effect of temperature on the ferroelectric properties of Hafnium Zirconium Oxide MFM thin-film varactors," in *2021 IEEE International Symposium on Applications of Ferroelectrics (ISAF)*, 2021, pp. 1–4.

[14] S. Abdualzhanov, D. K. Huynh, Q. H. Le, D. Lehninger, T. Kämpfe, and G. Gerlach, "Investigation of BEoL integrated ferroelectric thin-film HfO_2 for mmWave varactor applications," 2022, IEEE International Symposium on Radio-Frequency Integration Technology (RFIT 2022).

[15] M. Norling, D. Kuylenstierna, A. Vorobiev, and S. Gevorgian, "Layout optimization of small-size ferroelectric parallel-plate varactors," *IEEE Transactions on Microwave Theory and Techniques*, vol. 58, no. 6, pp. 1475–1484, 2010.

[16] T. T. Vo, T. Lacrevaz, B. Flechet, A. Farcy, Y. Morand, S. Blonkowski, J. Torres, and E. Defay, "In-situ microwave characterization of medium-k HfO_2 and high-k STO dielectrics for MIM capacitors integrated in back-end of line of IC," in *2007 Asia-Pacific Microwave Conference*. IEEE, pp. 1–4.

[17] M. Abdelfattah, D. Psychogiou, Z. Yang, and D. Peroulis, Eds., *V-band frequency reconfigurable cavity-based bandpass filters: 2016 IEEE/ACES International Conference on Wireless Information Technology and Systems (ICWITS) and Applied Computational Electromagnetics (ACES)*, 2016.

[18] D. Psychogiou, D. Peroulis, Y. Li, and C. Hafner, "V-band bandpass filter with continuously variable centre frequency," *IET Microwaves, Antennas & Propagation*, vol. 7, no. 8, pp. 701–707, 2013.

[19] Z. Yang and D. Peroulis, "A 23–35 GHz MEMS tunable all-silicon cavity filter with stability characterization up to 140 million cycles," in *2014 IEEE MTT-S International Microwave Symposium (IMS2014)*. IEEE, 62014, pp. 1–4.

[20] ——, "A 20–40 GHz tunable MEMS bandpass filter with enhanced stability by gold-vanadium micro-corrugated diaphragms," in *2016 IEEE MTT-S International Microwave Symposium (IMS)*. IEEE, 52016, pp. 1–3.

[21] K. Choi, S. Courreges, Z. Zhao, J. Papapolymerou, and A. Hunt, "X-band and Ka-band tunable devices using low-loss BST ferroelectric capacitors," in *2009 18th IEEE International Symposium on the Applications of Ferroelectrics*. IEEE, 082009, pp. 1–6.

[22] R. de Paolis, S. Payan, M. Maglione, G. Guegan, and F. Coccetti, "High-Tunability and High-Q-Factor Integrated Ferroelectric Circuits up to Millimeter Waves," *IEEE Transactions on Microwave Theory and Techniques*, vol. 63, no. 8, pp. 2570–2578, 2015.

978-1-6654-8498-5/22 $31.00 © 2022 IEEE

In-depth electrical characterization of deca-nanometer InGaAs MOSFET down to cryogenic temperatures for low-power quantum applications

Francesco Serra di Santa Maria, Christoforos Theodorou,
Francis Balestra, Gerard Ghibaudo
Univ. Grenoble Alpes, Univ. Savoie Mont Blanc, CNRS
Grenoble INP, IMEP-LAHC
Grenoble, 38000, France
francesco.serra-di-santa-maria@grenoble-inp.fr

Eunjung Cha, Cezar B. Zota
IBM Research GmbH Zürich Laboratory
Säumerstrasse 4
CH-8803 Rüschlikon, Switzerland

Abstract—This work presents a detailed electrical characterization of planar InGaAs on Silicon MOSFETs from room temperature down to cryogenic temperatures (10 K). The main temperature-dependent electrical parameters of MOSFET operation (threshold voltage V_t, low-field mobility μ_0, and subthreshold swing SS) were extracted in both linear and saturation regions through the consolidated Y-function method, for gate lengths down to 10 nm. The extracted parameters are first analyzed versus temperature and length and then compared against a more mature technology such as Silicon FDSOI MOSFETs. The work provides insight into the cryogenic operation of III-V MOSFETs and indicates a competing advantage versus Si CMOS for low-power cryogenic quantum computer applications.

Keywords— *III-V, MOSFET, cryogenic, Y-function, characterization, InGaAs*

I. INTRODUCTION

As the research on quantum computing (QC) advances, the need for active electronic devices that operate at deep cryogenic conditions will increase. This necessity is better understood when considering the qubit readout electronics [1]. In order to minimize signal transmission delays and noise amplification due to different temperatures amongst signal stages, readout electronics in QC have to be as close as possible to the qubits, therefore operating in the temperature range of a few units of Kelvins. Similar holds true for the qubit control side. The cryogenic environment places stringent requirements on the supporting device technology, as the circuits must be dense and operate with extremely low power and noise at gigahertz frequencies. While it is currently an open question how many qubits cryogenic Si CMOS circuits will be able to support, other device technologies, such as III-Vs, may in the future be better suited. Thanks to their enhanced mobility, III-V MOS devices can provide the same ON current at lower power supply voltages, and by turn reduced power consumption and heat dissipation, crucial at QC operation temperature [2]. Thus, the precise identification of their electrical parameters' behavior with temperature and channel length is required for reliable modeling and circuit design. On the other hand, as this technology is not as mature as its Silicon-based counterparts (Bulk, FDSOI, and FinFETs), the need of full electrical characterization down to deep cryo-temperatures, becomes challenging and critical, particularly in view of the emerging technology of QC.

To date, Si CMOS has been extensively studied including at cryogenic temperatures [3-6]. Similar studies on InGaAs MOSFETs are currently lacking. Therefore, in this work, we

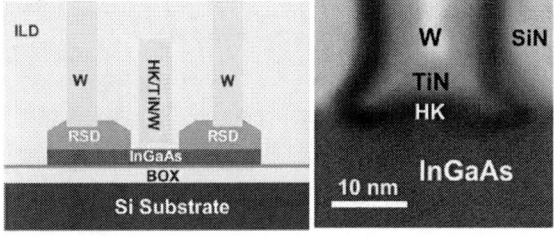

Fig.1 **(Left)** Schematics of studied III-V MOSFETs. (Right) TEM image of a device with nominal 10 nm gate length. Courtesy of IBM Zurich [2,7]

study scaled InGaAs MOSFET devices down to cryogenic temperatures, to determine their temperature- and size-dependent properties and compare them with Si CMOS. In particular, we compare the physical limits of the subthreshold swing at cryogenic temperatures. Finally, key cryogenic device properties are benchmarked with Si CMOS. The work provides valuable understanding of the cryogenic operation of III-V MOSFETs, and the results indicate that these devices are highly promising for cryogenic low-power quantum computer applications.

II. DEVICES UNDER STUDY AND EXTRACTION METHODOLOGY

The devices under test were fabricated by IBM Zurich [7], based on a III-V on insulator tehnology, incorporating a 20 nm InGaAs film, insulated by a buried oxide and integrated on Silicon substrates through direct wafer bonding (Fig.1). The fabrication process is CMOS-like with replacement metal gate, raised source/drain regrowth and a high-k metal gate. The devices share a common gate width value of W = 1 μm and channel lengths spanning from L = 300 down to 10 nm. Fig. 1 also shows a TEM image of a L = 10 nm device. Note that this is the nominal value, the real L varies up to 14 nm. Henceforth, we will refer to the nominal value. The measurements were carried out at wafer level down to 10K using a HP 4155A parameter analyzer and a SussTech 300mm Cryo probe station.

In order to extract the main parameters that define the MOSFET performance, we utilized the Y-function method, expressed through $Y(V_g) = I_d/\sqrt{g_m} \approx \sqrt{\beta}(V_g - V_t)$ [8], where $\beta = W\mu_0 C_{ox} V_d / L$. This function suppresses any access resistance effect and thanks to its linear behavior versus V_g at strong inversion, the linear fit can provide the threshold voltage, V_t, as well as the low-field mobility, μ_0, value through the x-axis intercept and slope, respectively.

978-1-6654-8498-5/22 $31.00 © 2022 IEEE

III. EXPERIMENTAL RESULTS AND DISCUSSION

A. Preliminary observations

The measured I_d-V_g input characteristics in linear regime are plotted in Fig. 2a, for the two temperature limits (10 K and 300 K). At first glance, we can already observe how the threshold voltage, V_t, is affected by both channel length and temperature: on one hand, for a shorter channel V_t shifts downwards ("V_t roll-off" short channel effect), while on the other hand, it increases for a decreasing temperature.

Moreover, we can notice some kind of a shoulder/hump in the high V_g region, in both the I_d-V_g and the Y-V_g (Fig. 2b) curves: this effect is most likely to be related to the onset of conduction in the L valley [9] of III-V. What is most important for our study, however, is that this hump does not impact the overall linear behavior of the Y function in strong inversion, allowing us therefore an easy and reliable extraction of the main intrinsic parameters. ·

B. Analysis of extracted parameters in ohmic region

When going down to low temperatures (LT), several effects are taking place. First, due to the Boltzmann statistics,

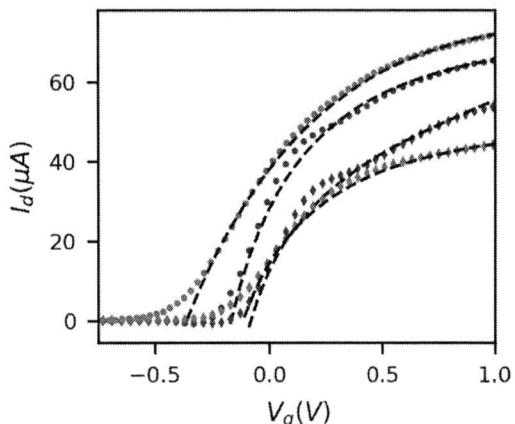

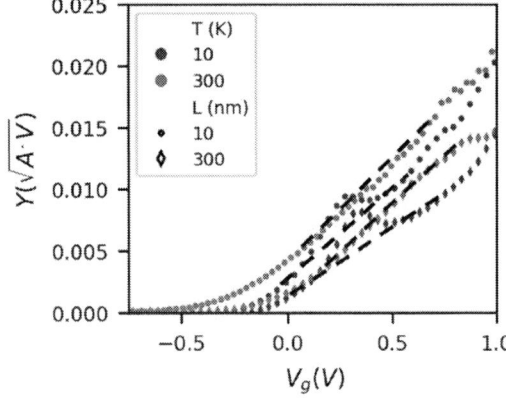

Fig. 2. Drain current (a) and Y-function (b) versus gate voltage at Vd = 30 mV for L = 10 and 300 nm at T = 10 and 300K. The dashed lines represent the fitting of the plots, based on Y function and linear regression respectively.

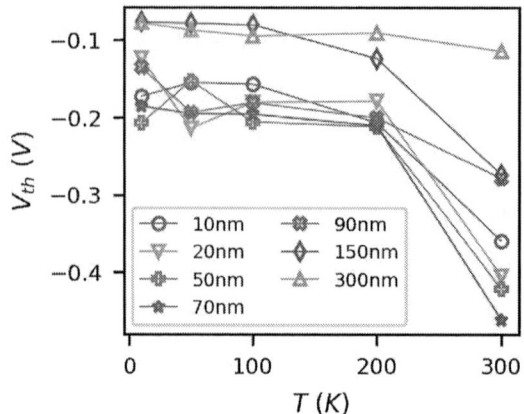

Fig. 3. Extracted threshold voltage versus temperature for Vd = 30 mV for various channel lengths

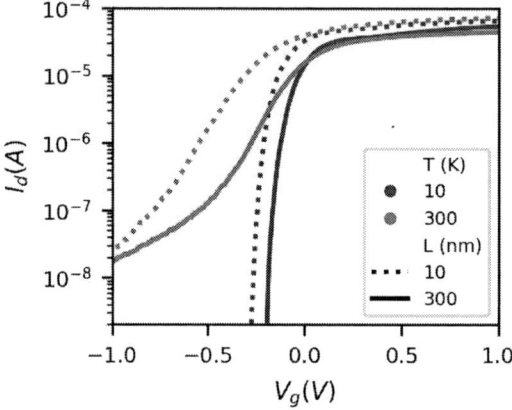

Fig. 4. Drain current versus gate voltage at Vd = 30 mV for L = 10 and 300 nm at T = 10 and 300K

as the temperature decreases, fewer and fewer electrons are promoted to the conduction band for the same gate voltage. This is reflected in an increase in threshold voltage as going towards lower temperatures, as shown in the extracted V_t values (plotted versus temperature for all measured gate lengths in Fig. 3), up to settling around 200 K due to the semiconductor statistics becoming degenerate past that temperature [10].

Moreover, the transition between OFF and ON states becomes sharper (Fig. 4), yielding a consequent decrease of the subtreshold swing, SW, for low temperatures, down to a settling value of about 10 mV/dec, which is attributed to the exponential band tails of states [3]. This behavior of SW is confirmed for the measured III-V devices, as shown in Fig. 5, where the extracted values of SW are plotted versus temperature for certain channel lengths.

Conversely, as the temperature lowers, there is less and less vibrational energy in the lattice, causing a decrease in phonon scattering, which allows an increase in the low-field mobility, μ_0, as shown in Fig. 6, where the extracted values of μ_0

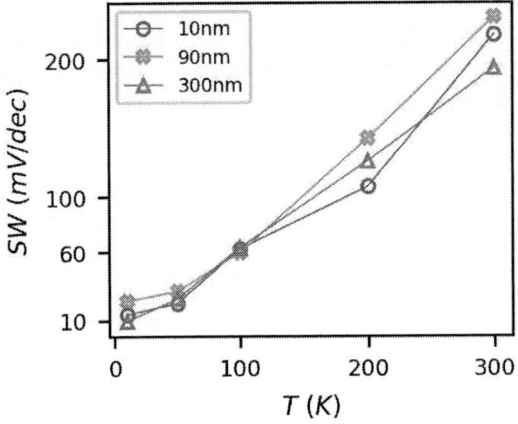

Fig. 5. Extracted subthreshold swing at V_d = 30 mV versus temperature for different lengths from 10 to 300 nm

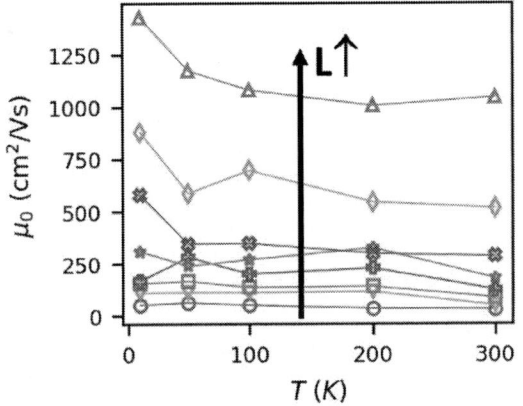

Fig. 6. Extracted low-field mobility at Vd = 30 mV versus temperature for different lengths from 10 to 300 nm

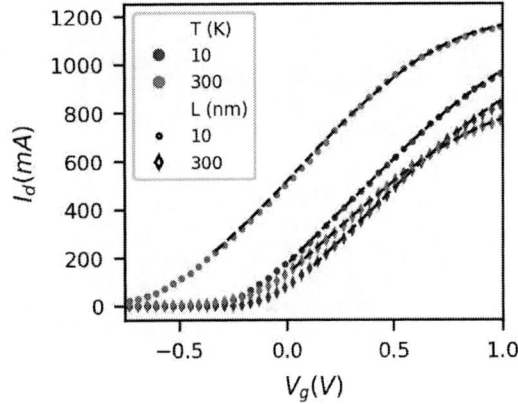

Fig. 7. Drain current versus gate voltage at Vd = 1 V for L = 10 and 300 nm at T = 10 and 300K. the dashed line shows the fit reconstructed with the parameters extracted thanks to the Y function

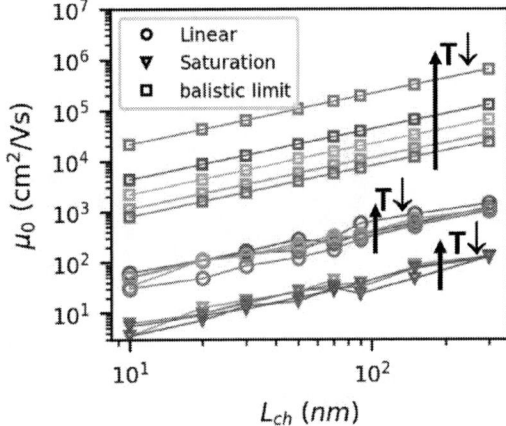

Fig. 8. Extracted low-field mobility versus channel length for linear and saturation regions along with theoretical ballistic limit

are plotted versus temperature for all measured gate channel lengths. This increase is evidently more pronounced in longer channel devices, whereas, in shorter channels, defect (neutral impurity) scattering is prevailing due to source/drain regions proximity, yielding a generally lower mobility and also a less significant increase at lower temperatures [6].

As it can be shown through Poisson-Schrodinger simulations [9], the effect of L valley carrier population becomes only visible below 100K, giving rise to specific structure in I_d-V_g and Y-V_g characteristics (Fig. 2).

C. Behavior in saturation region

As we proceed towards the analysis of the device parameters in saturation region of operation (V_d = 1 V), we observe that both 10 and 300 nm length devices, plotted in Fig. 7, show no significant variation in the ON region as going to low temperatures, except for an increased V_t for L_g = 10 nm at 10K.

From the consequent extraction, the variation of the low-field mobility with respect to temperature in the saturation region is very small.

On the contrary, we notice that in saturation region, the curvature created by the takeover of L valley is no longer visible at low temperature. This can be explained when considering that the drain current is obtained by integration along the channel from source to drain [11]: close to the drain region the influence of high V_d does not allow the quasi Fermi level to fill the L valley, thus attenuating its effect.

When extracting the μ_0 values through Y function, plotted in Fig. 8, compared to the linear region, the devices present a μ_0 reduction due to velocity saturation (v_{sat}) effect [4]. Moreover, it is worth noticing how the extracted mobility never reaches the ballistic limit [5], revealing that the exponential behaviour of μ_0 with respect to channel length is in fact scattering-related. The extracted values of v_{sat} in saturation region are plotted in Fig. 9, showing a good stability with respect to temperature and a slight increase for a decreasing channel length due to overshoot effect.

978-1-6654-8498-5/22 $31.00 © 2022 IEEE

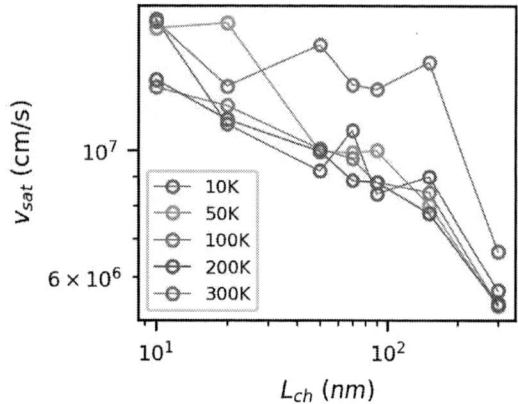

Fig. 9. Extracted saturation velocity versus channel length for various temperatures

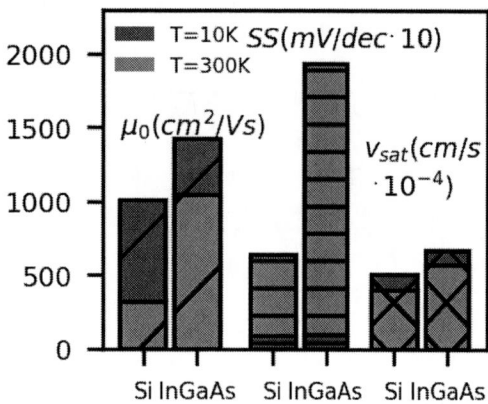

Fig. 10. Comparison of extracted parameters between III-V and Si [5] for a common channel length of 300 nm.

D. Benchmarking III-V against Si for Cryo

Fig. 10 illustrates the extracted parameters of μ_0, SW and v_{sat} along with respective extractions done for Si channel Fully Depleted (FD) SOI MOSFETs [6]. When comparing the III-V MOSFET devices to this more mature, industrial-like FDSOI technology (Fig. 10) [6], we note that although SW is much higher for III-V at 300K, both technologies reach the lowest limit value at 10K. Moreover, despite the higher interface trap density at the Al_2O_3/InGaAs interface, III-V shows higher μ_0 and v_{sat} at all temperatures, revealing a great III-V potential for cryo-related applications with further technology developments and the chance to overperform Si FDSOI in certain cases.

IV. CONCLUSION

We have performed a detailed electrical characterization of scaled planar InGaAs-on-Silicon MOSFETs from room temperature down to deep cryogenic temperatures (10K). The main MOSFET parameters (threshold voltage V_t, low-field mobility μ_0, and subthreshold swing, SS) were extracted in linear region of operation using the consolidated

Y-function method for gate lengths down to 10 nm, despite the possible presence of L-valley conduction, and were benchmarked to Si CMOS. The results build on the understanding of the operation of cryogenic III-V MOSFETs and indicate that this technology may be promising for future low-power cryogenic quantum computer applications. The saturation velocity was also extracted and analysed for all lengths and temperatures. The extracted parameters of the III-V devices follow the expected behavior with temperature as in Si, while demonstrating competing advantages as compared to Si MOSFETs, particularly when going down to cryogenic temperatures.

ACKNOWLEDGMENT

This work was supported by the European Union H2020 program SEQUENCE (Grant – 871764).

REFERENCES

[1] J. M. Hornibrook, J. I. Colless, I. D. Conway Lamb, S. J. Pauka, H. Lu, A. C. Gossard, J. D. Watson, G. C. Gardner, S. Fallahi, M. J. Manfra, and D. J. Reilly, "Cryogenic Control Architecture for Large-Scale Quantum Computing," Phys. Rev. Applied **3**, 024010 – Published 23 February 2015

[2] C. Convertino et al., "InGaAs-on-Insulator FinFETs with Reduced Off-Current and Record Performance," 2018 IEEE International Electron Devices Meeting (IEDM), 2018, pp. 39.2.1-39.2.4, doi: 10.1109/IEDM.2018.8614640.

[3] A. Beckers, F. Jazaeri and C. Enz, "Theoretical Limit of Low Temperature Subthreshold Swing in Field-Effect Transistors," in IEEE Electron Device Letters, vol. 41, no. 2, pp. 276-279, Feb. 2020

[4] C.Diouf, A.Cros, S.Monfray, J.Mitard, J.Rosa, G.Gloria, G.Ghibaudo, " 'Y function' method applied to saturation regime: Apparent saturation mobility and saturation velocity extraction," Solid-Satet Electronics, Volume 85, July 2013, Pages 12-14

[5] Y. Liu, M. Luisier, A. Majumdar, D. A. Antoniadis and M. S. Lundstrom, "On the Interpretation of Ballistic Injection Velocity in Deeply Scaled MOSFETs," in IEEE Transactions on Electron Devices, vol. 59, no. 4, pp. 994-1001, April 2012

[6] F.Serra di Santa Maria, L.Contamin, B.Cardoso Paz, M.Cassé, C.Theodorou, F.Balestra, G.Ghibaudo, "Lambert-W function-based parameter extraction for FDSOI MOSFETs down to deep cryogenic temperatures," Solid-State Electronics, Volume 186, December 2021, 108175

[7] C. Zota, C. Convertino, D. Caimi, M. Sousa, L. Czornomaz, "Effects of Post Metallization Annealing on InGaAs-on-Insulator MOSFETs on Si," 2019 Joint International EUROSOI Workshop and International Conference on Ultimate Integration on Silicon (EUROSOI-ULIS).

[8] G. Ghibaudo, "New method for the extraction of MOSFET parameters," Electronics Letter, Volume 24, Issue 9, 28 April 1988, p. 543 – 545.

[9] T. P. O'Regan, P. K. Hurley, B. Sorée, and M. V. Fischetti, "Modeling the capacitance-voltage response of In0.53Ga0.47As metal-oxide-semiconductor structures: Charge quantization and nonparabolic corrections," Appl. Phys. Lett. 96, 213514 (2010).

[10] R. Maurand, X. Jehl, D. Kotekar-Patil, A. Corna, H. Bohuslavskyi, R. Lavieville, L. Hutin, S. Barraud, M. Vinet, M. Sanquer and S. De Franceschi, A CMOS silicon spin qubit, Nature Commun., 7, 13575 (2016).

[11] Y. Taur, T. H. Ning, Fundaments of Modern VLSI Devices, 2nd ed., Cambridge University Press, 1998

III-V HBTs on 300 mm Si substrates using merged nano-ridges and its application in the study of impact of defects on DC and RF performance

A.Vais[1], S. Yadav[1], Y. Mols[1], B. Vermeersch[1], K. V. Kodandarama[1], M. Baryshnikova[1], G. Mannaert[1], R. Alcotte[1], G. Boccardi[1], P. Wambacq[1], B. Parvais[1,2], R. Langer[1], B. Kunert[1], and N. Collaert[1]

[1]imec, Leuven, Belgium, [2]VUB, Brussels, Belgium, Email: Abhitosh.Vais@imec.be

Abstract— **In this paper, we demonstrate III-V HBTs fabricated on GaAs/InGaP layers realized by merging the nano-ridges to create a bulk-like stack on a 300 mm Si substrate. The emitter-base and base-collector diodes show an ideality factor of ~1.2 and ~2.0, respectively. A maximum DC current gain of ~120 and breakdown voltage, BVCBO, of 10 V is achieved at Ft ~ 17GHz. A direct correlation between threading dislocation density (TDD) and various device metrics is shown using DC, RF and reliability measurements. Furthermore, 3D Monte Carlo simulations were done to model and understand the impact of different types of merged structures on the thermal performance of the device. With this work, we show the potential of merged nano-ridges, in enabling an efficient hybrid III-V/CMOS technology for mm-wave applications, as a material-independent tool to understand the impact of defects on the performance of III-V devices.**

Keywords—III-V, HBT, III-V-on-Si, 5G

I. Introduction

It is envisaged that future technologies for RF applications, such as 5G and beyond, would require RF performance significantly better than contemporary devices at significantly lower cost, flexibility in circuit design, reduced power losses, and smaller chip footprint [1,2]. All these requirements can be achieved by the integration of III-V compound semiconductor-based Heterojunction Bipolar Transistors (HBTs) on Si substrates to enable hybrid III-V/CMOS technology [1,2]. Several techniques have been explored to achieve this: strain relaxed buffer (SRB) layers [3,4], confined epitaxial lateral overgrowth (CELO) [5], Silicon on Lattice engineered substrates (SOLES) [6], and aspect ratio trapping (ART) inside narrow trenches [7,8]. In [9], we had shown GaAs/InGaP HBTs grown on a 300 mm Si substrate using III-V nano-ridge engineering (NRE). HBTs fabricated on this stack showed an electrical performance considerably better than GaAs(P) devices fabricated on a Si substrate with SRB layers, without any need to grow thick (>1-10 μm) buffer layers.

In this work, we demonstrate III-V HBTs fabricated on GaAs/InGaP layers formed by merging these nano-ridges (NRs) to create a bulk-like stack on a 300 mm Si substrate. In order to show the applicability of merged nano-ridges (mNR), these devices are then used to understand the impact of defects on their electrical performance in terms of DC current gain, β, cut-off frequency, Ft, breakdown voltage, BV$_{cbo}$, and reliability.

II. Experimental Details

The growth is done on 300 mm Si (001) wafers (n = 1×10^{19} cm^{-3}) with an STI-based trench-pattern as shown in figure 1. The shallow trench isolation (STI) SiO$_2$ layer is 400 nm thick with a pattern of narrow trenches with a Si {111}-faceted V-groove [11] at the bottom. The trench width is about 100 nm and the trench separation varies from 100 nm to 300 nm (Table 1). The patterned 300 mm Si substrate is transferred into a 300 mm III-V chamber where metal organic vapor phase epitaxy is performed applying standard liquid group III and V precursors with carbon tetrabromide and Silane as dopant sources [11]. Using selective area growth, III-V material is deposited inside the narrow trenches where all misfit defects initiated by the large III-V/Si lattice mismatch are confined close to the trench bottom due to high aspect ratio trapping [10]. This ensures growth of TD-free material out of the narrow trenches. NRE is then applied to widen the nano-ridges in order to maintain a flat (001) top facet. If the trench spacing is large, NRE results in a large box like shape.

In Fig. 1a, a patterned Si substrate with a second PECVD template oxide was applied to avoid sidewall deposition and

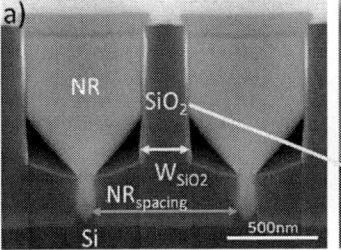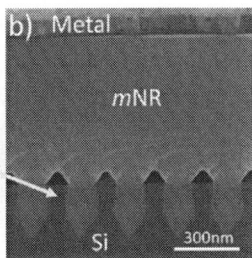

Fig. 1. XSEM image of a) isolated, and b) merged nano-ridges. While isolated NRs are separated by oxide template with large spacing in the STI trench, mNRs have smaller spacing and no oxide wall

to realize isolated NR HBT devices [11]. However, if the trench spacing, T$_s$, is sufficient small, continuous deposition leads to the merging of the NRs and to the generation of misfit defects along this fused interface. Once the nano-ridges are merged, the growth happens on a two-dimensional layer comparable to the growth on a blanket substrate. In this investigation no efforts were taken to improve the crystal quality of the merged region as the growth optimization was focusing on the isolated NR HBT structure. Most likely the TD density (TDD) can be reduced by controlling the merging NR interfaces and/or applying annealing treatments but in this investigation the presence of a variety of trench spacing, T$_s$,

978-1-6654-8498-5/22 $31.00 © 2022 IEEE

allowed us to study its impact on the quality of merged device layer (Table 1).

Table. 1: Description of variation in trench spacing and corresponding TDD

Cell no.	Trench spacing (μm)	Trench Width (μm)	TDD (x10^8 cm^-3)
C18	0.3	0.1	3.85
C19	0.2	0.1	5.1
C20	0.1	0.1	7.3

Table. 2: Details of the HBT stack used in this study

Material	Doping (dopant)	Thickness (nm, /cm^3)
GaAs	n(Si)	50,1e19
GaAs	n(Si)	20, 1e19
GaAs	n(Si)	30, 5e18
InGaP	n(Si)	50, 5e17
GaAs	p(C)	20, 7.5e19
GaAs	n(Si)	400, 5e16
GaAs	n(Si)	180, 1e19

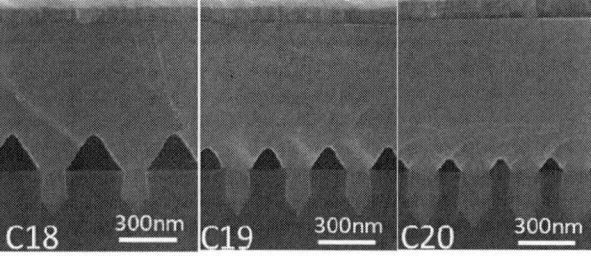

Fig. 2. XSEM images of mNRs with different trench spacing as specified in Table 1. Due to difference in the spacing, merging happens at different time of growth and hence, different height. The wave-like features are artifacts due to cleaving.

As shown in figure 2, the height at which merging occurs varies with T_s but the merging remains uniform across all NRs. It should be noted that the merging of NRs already happens during the n+ layer growth indicating that the subsequent growth of remaining n+ layer and of the corresponding HBT stack happens on top of the merged region. Details of the stack composition are given in Table 2. The electron channeling contrast imaging (ECCI) technique was used to examine the quality of the grown stack. An ECC image for C20 with trench spacing of 0.1 μm is shown in figure 3 as an example. The defect density, extracted from ECC images, is provided in Table 1.

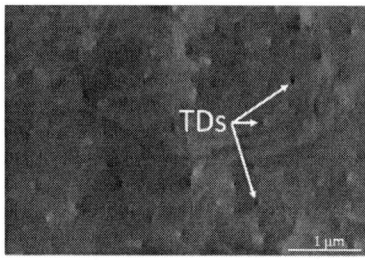

Fig. 3. ECCI image of C20 at the base layer depicting the presence of TDs

A TD density in the order of 1×10^8 cm^{-3} was observed for all 3 trench spacings. It should be noted that this is almost 2 orders of magnitude higher than the density observed for isolated NRs [12]. As the spacing increases, the TDD decreases. This is primarily because the number of merging interfaces decreases with an increase in T_s, in turn, leading to a higher TDD. In order to explore the crystal quality of the epitaxial layers and its impact on the device performance, HBT devices were fabricated using a simplified lab-based process flow on smaller pieces (5x5 cm^2), cut from the full 300 mm wafer after the III-V mNR growth [9]. GaAs and InGaP were etched using a combination of dry and wet etching. Ti/Pd/Au metal stack was used for metallization.

III. RESULTS AND DISCUSSION

A. DC and RF characterization of mNR-HBTs

Figures 4(a,b) show a representative Gummel plot and output characteristics for an mNR HBT on C18 with an area of 35x35 μm^2. The ideality factor of collector and base current is ~1.1 and ~2.0 respectively. The device showed a maximum DC current gain of ~120 at V_{be} = 1.65 V. Fig. 5 depicts a Gummel plot comparison between mNR HBT and a similar reference device fabricated on the same stack on a 2'' GaAs substrate.

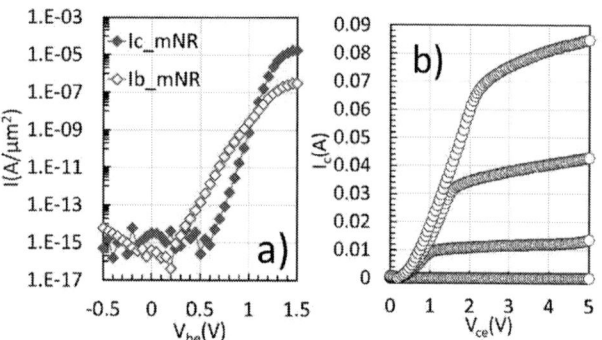

Fig. 4. a) Gummel plot of the measured mNR HBT device, b) Output characteristics of the mNR HBT device

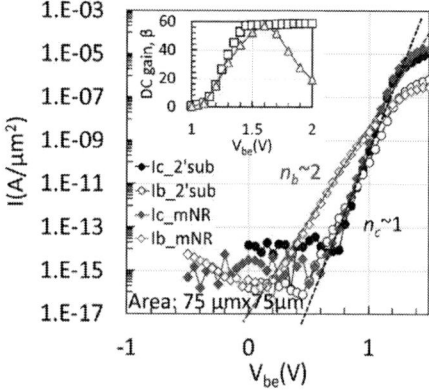

Fig. 5. A Gummel plot comparison between HBT devices fabricated on a stack grown on GaAs substrate and mNRs. Inset: A comparison of DC current gain obtained on an HBT device on mNR and GaAs substrate. The saturation is due to tool's compliance limit and hence just a measurement artifact.

The mNR HBTs show a collector current ideality factor and density similar to that of a reference HBT. The difference in the ideality factor of base current between the 2 devices can be explained by significantly higher TDD observed in case of mNR stack. A comparison of DC current gain, β, of 2 similar devices on mNR and 2'' substrate, shown in the inset of figure 5, demonstrates that there is no significant difference in the peak current gain of 2 devices. A comparison of DC current gain, β, breakdown voltages and output characteristics for different trench spacings is shown in Figures 6-7. All the above parameters undergo an improvement as the trench spacing increases which, as shown in fig. 7, can be traced to the reduction in TDD in the active area of the device. The hypothesis was also confirmed by theoretical calculations of current gain in the presence of TDs [13] as shown in fig 8. In fig. 9a, the RF performance, in terms of the cut-off frequency, F_t, of mNR is compared to that of reference devices on a 2'' GaAs substrate and isolated NRs. The difference in the F_t

978-1-6654-8498-5/22 $31.00 © 2022 IEEE

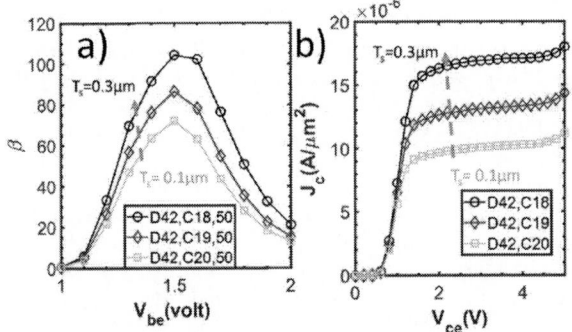

Fig. 6. Impact of trench spacing on a) the measured DC current gain, b) the output characteristics of the device

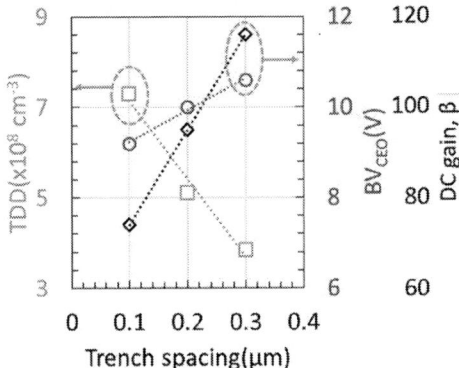

Fig. 7. The variation of DC gain, β, and breakdown voltage with the trench spacing and TDD.

performance between the 2 devices can be explained by the difference in their device structure as shown in fig. 9(b,c). The HBTs on 2" GaAs show a higher F_t at lower Jc as compared to mNR and i-NR HBTs mainly due to a self-aligned 'base all-around' design (Fig. 9b) which leads to a reduced emitter-crowing effect and results in a better F_t performance. Due to 'side-base' design of i-NRs and mNRs, the length of the NR becomes the effective width of the HBT, thereby leading to worse current crowding and device parasitics, and hence, degraded slope of F_t vs J_c curve as seen in fig. 9a. Figure 9a also shows that the HBT devices on Si (mNR and i-NR) show similar F_t performance owing to similar device layouts with base contact on the side (Fig. 14b).

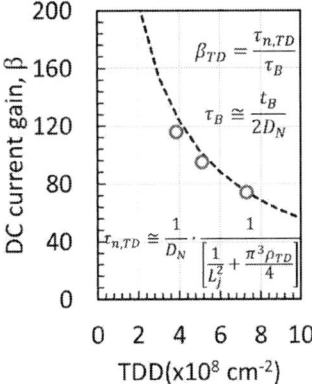

Fig. 8. A comparison of theoretical and measured β versus TDD. Inset shows the equations used for theoretical calculations.

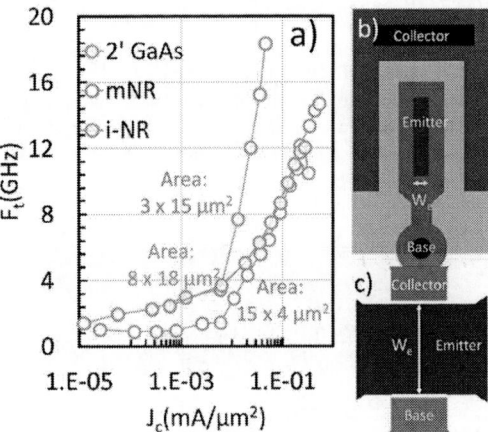

Fig. 9. A comparison of RF performance of mNR HBTs with HBTs on isolated NRs and 2' GaAs substrate, and core device layout of HBTs fabricated on b) 2' GaAs substrate, c) mNRs and isolated NRs. Note the difference in the coverage of base and collector around emitter in both cases

B. Reliability characterization

The stress measurements were also performed on mNR devices at high temperatures. The device is stressed at V_{ce} = 5.5V, I_b = 300μA for 10000 s till 125 °C. No appreciable degradation in transfer characteristics or output current was observed. The result of the stress measurement on mNR at 200 °C is shown in fig. 10. No significant degradation in output current was observed even in the presence of high TDD. In general, only a minimal degradation (at high T) in DC/RF performance is seen in mNR HBTs for long duration stress indicating a good reliability even with higher TDD.

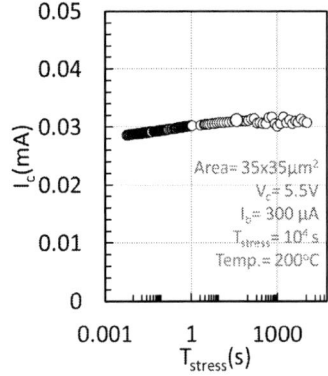

Fig. 10. Variation of mNR output current during the stress versus time at 200 °C.

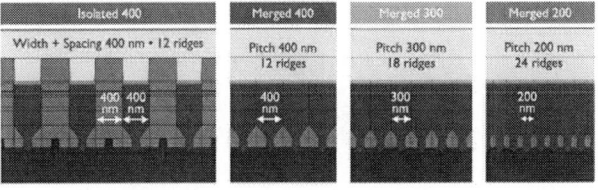

Fig. 11. A 2D cross-section along the emitter for 4 configurations: HBT structure with fixed trench width (100 nm), ridge height (600 nm) and total area (3.36 μm²) but realized in 4 different ridge configurations: i-NR at 400 nm spacing, and mNR at pitches of 400, 300 and 200 nm (as described in Table 1)

978-1-6654-8498-5/22 $31.00 © 2022 IEEE

C. *Thermal modelling of III-VHBTs on Si*

NR(Nano-ridge) merging and spacing/pitch also have implications on thermal performance and device self-heating. To account for non-diffusive heat transport effects, we carried out thermal modelling through 3D Monte Carlo simulations [14] with first-principles phonon dispersions and scattering rates [15]. A 2D cross-section of the structures used for the simulation is shown in fig. 11.

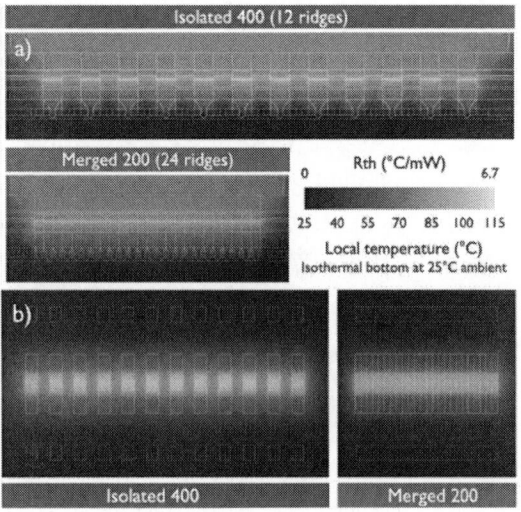

Fig. 12. A cross-section of simulated thermal resistance and local temperature for case 1 and 4 from Fig. 20: a) Front-view cross-section along the emitter, b) Top-view cross-section along base. Due to isothermal boundary assumptions, the junction temp. should be treated as lower-bound estimates. Actual temp. might be slightly higher

All simulations were carried out for the same power density (4 mW/μm²) with adiabatic conditions at the top side of the simulation domain, isothermal heat sink (25°C) at the bottom, and *transparent* side walls. The cross-sections of the simulated temperature field are shown in Fig. 12. At 400 nm ridge width, merging the isolated ridges carries an 11% thermal resistance (R_{th}) penalty (Fig. 13). Even though both cases have identical NR shapes, self-heating in the merged configuration is higher because the same total power is now dissipated across a smaller footprint. By reducing the pitch to 300 nm and further to 200 nm, however, merged ridges can lead to lower R_{th} compared to can their isolated counterparts (Fig. 13). This R_{th} reduction arises because under constant ridge height, smaller pitch means a larger GaAs to dielectric ratio (Fig. 11), thereby improving the effective thermal conductivity and easing the thermal bottleneck due to ridge necks.

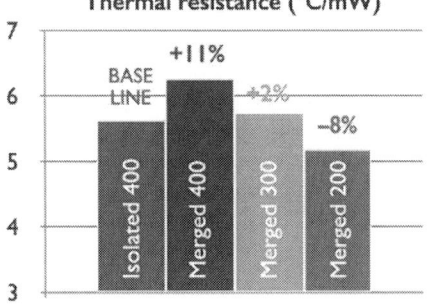

Fig. 13. Simulated impact of ridge merging and pitch on HBT self-heating

IV. CONCLUSIONS

With the results shown above, using HBTs fabricated on merged NRs, we have demonstrated the potential of merged nano-ridges, in enabling an efficient hybrid III-V/CMOS technology for mm-wave applications, as a material-independent tool to understand the impact of defects on the performance of III-V devices. By using various DC, RF and reliability characterization methods, we not only showed the electrical performance of HBTs on mNRs but in addition, we used this information to analyze the impact of TDs on the device performance. Furthermore, thermal modelling shows merged NRs at small pitch reduce device self-heating.

ACKNOWLEDGMENT

The authors gratefully acknowledge the support of imec members from 300mm fab, III-V lab and the amsimec lab.

REFERENCES

[1] E. Fitzgerald, et al. "Monolithic III-V/Si Integration." *ECS Transactions* 19, no. 5 (2009): 345.

[2] N. Collaert, et al. "Fabrication challenges and opportunities for high-mobility materials: From CMOS applications to emerging derivative technologies." In *Advanced Etch Technology for Nanopatterning VIII*, vol. 10963, pp. 13-21. SPIE, 2019..

[3] M. L. Huang, et al. "High performance In 0.53 Ga 0.47 as FinFETs fabricated on 300 mm Si substrate." In *2016 IEEE Symposium on VLSI Technology*, pp. 1-2. IEEE, 2016.

[4] C. Heidelberger,et al., "GaAsP/InGaP HBTs grown epitaxially on Si substrates: Effect of dislocation density on DC current gain." *Journal of Applied Physics* 123, no. 16 (2018): 161532.

[5] L., E. Czornomaz, et al., "Confined epitaxial lateral overgrowth (CELO): A novel concept for scalable integration of CMOS-compatible InGaAs-on-insulator MOSFETs on large-area Si substrates." In *2015 Symposium on VLSI Technology (VLSI Technology)*, pp. T172-T173. IEEE, 2015.

[6] J. R. LaRoche, et al. "Monolithically integrated III-V and Si CMOS devices on silicon on lattice engineered substrates (SOLES)." In *Proceedings of the International Conference on Compound Semiconductor Manufacturing Technology (CS MANTECH)*, pp. 2-5. 2009.

[7] N. Waldron, et al. "Gate-all-around InGaAs nanowire FETS with peak transconductance of 2200μS/μm at 50nm Lg using a replacement Fin RMG flow." In *2015 IEEE International Electron Devices Meeting (IEDM)*, pp. 31-1. IEEE, 2015.

[8] B. Kunert, et al., "How to control defect formation in monolithic III/V hetero-epitaxy on (100) Si? A critical review on current approaches." *Semiconductor Science and Technology* 33, no. 9 (2018): 093002.

[9] A. Vais, et al. "First demonstration of III-V HBTs on 300 mm Si substrates using nano-ridge engineering." In *2019 IEEE International Electron Devices Meeting (IEDM)*, pp. 9-1. IEEE, 2019.

[10] B. Kunert, et al., "Integration of III/V hetero-structures by selective area growth on Si for nano-and optoelectronics." *Ecs Transactions* 75, no. 8 (2016): 409.

[11] B. Kunert, et al. "III/V nano ridge structures for optical applications on patterned 300 mm silicon substrate." *Applied Physics Letters* 109, no. 9 (2016): 091101.

[12] M. Baryshnikova, et al. "Nano-ridge engineering of GaSb for the integration of InAs/GaSb heterostructures on 300 mm (001) Si." *Crystals* 10, no. 4 (2020): 330.

[13] Wan Khai Loke, et al., "MOCVD growth of InGaP/GaAs heterojunction bipolar transistors on 200 mm Si wafers for heterogeneous integration with Si CMOS." *Semiconductor Science and Technology* 33, no. 11 (2018): 115011.

[14] Jean-Philippe et al., "An alternative approach to efficient simulation of micro/nanoscale phonon transport." *Applied Physics Letters* 101, no. 15 (2012): 153114.

[15] J. Carrete, et al., "almaBTE: A solver of the space–time dependent Boltzmann transport equation for phonons in structured materials." *Computer Physics Communications* 220 (2017): 351-362.

Co-integration Process Compatible Input/Output (I/O) Device Options for GAA Nanosheet Technology

[1]Gautam Gaddemane, [2]Krishna K. Bhuwalka, [1]Philippe Matagne, [3]Gerhard Rzepa, [1]Maarten Van de Put, [1]Sybren Santermans, [3]Oskar Baumgartner, [2]Hao Wu, [1]Geert Hellings

[1]imec, Belgium, [2]Huawei Technologies R&D, Belgium, [3]Global TCAD Solutions, Austria

Abstract— **We benchmark possible device options for gate-all-around input/output devices which are process compatible with core logic gate-all-around (GAA) nanosheet devices. We consider a partial gate-all-around device (without metal filling between the sheets) as well as a device with a thin metal gate between the sheets, resulting in work-function mismatch between side- and inner-gates. The results are benchmarked against ideal GAA I/O devices to understand the performance impact for each case. The partial GAA device performs well under certain geometric conditions ($W_{NS} < 30$nm). However, the device with thin metal between the sheets shows excellent performance even for large W_{NS} with large work-function deviation assumptions.**

Keywords—Gate-all-around, nanosheet, input-output devices

I. INTRODUCTION

Gate-All-Around (GAA) Nanosheet (NS) transistors have enabled the scaling of logic devices beyond FinFET [1, 2] by providing excellent gate-control and superior drive-current per footprint, while enabling contact poly-pitch (CPP) scaling [3]. A key component defining the introduction of new logic technology is convergence, whereby satellite components, like SRAM cells and input-output (I/O) devices, are fabricated using the same process flow used for the core logic devices, barring a few changes and/or addition of masks. Thus, the compatibility and ease of fabricating an I/O device is essential. Since I/O devices generally operate at relatively high operating voltages (1.5-1.8 V as compared to sub-0.75 V for core), a thick dielectric (4-5 nm) is required to meet reliability constraints. For core logic on the other hand, a key requirement is that the vertical pitch of nanosheets should be scaled to ~15 nm to meet capacitive and performance targets [4]. At this pitch, vertical spacing between sheets is roughly 10 nm which leaves little or no room to have a full I/O device gate stack, including dielectric and work-function defining metals (Fig. 1). To this effect, multiple I/O device concepts, including a Si/SiGe superlattice FinFET [5], and GAA NSFET [6], have already been proposed. While in the superlattice FinFET, the process itself is simplified to a great extent, performance degradation, especially in PFET is observed due to residual tensile stress in PFETs as well as band-offsets between the Si and SiGe layers [7]. In Ref. [6], the GAA NSFET I/O devices were fabricated by selective oxidation of the silicon sheets (in addition to the deposited oxide) to accommodate the thick dielectric layers needed for I/O devices. This however, results in reduced sheet thickness which has adverse thin-body effects like mobility degradation.

The motivation for this work is twofold: (1) to understand performance limitations if an I/O device is formed following the core process at aggressive vertical pitch dimensions (15 nm); (2) define the nanosheet geometry range at which these devices are competitive to reference (hypothetical) I/O devices. In this study, the reference device is defined as one which provides a complete GAA behavior, despite the lack of space in between the sheets. This is achieved by inserting a thin metal layer between the sheets, connecting the side gates. The case study devices include (1) partial GAA device (with no inner-gates) and (2) a device with different work-function for the inner-gates as compared to the side-gates. This work-function change results from the thin metal layer due to limited vertical pitch. Finally, all device concepts are benchmarked against a conventional I/O FinFET following the same ground rules (layout, process).

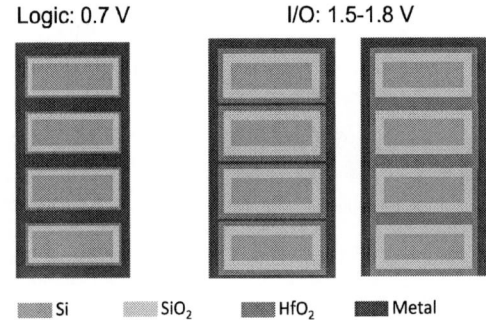

Fig. 1: Schematic cross-section of Logic and I/O NSFET devices. Due to the requirement of thick dielectric layers for I/O devices, we can observe there is only a thin metal between the sheets or no metal between the sheets.

II. TCAD STRUCTURE SET-UP & CALIBRATION

All our simulations were performed using a semi-classical device simulator [8]. A realistic geometry, including layered BEOL structures, as well as tapered trenches, vias, and fins, are generated using the layout-based structure generation tool [9]. A realistic junction profile of 2.25 nm/dec. is used. The TCAD device simulations are performed by solving the drift-diffusion/density-gradient (DD/DG) equations coupled to the Poisson equation and current relations in the interconnects. The DD/DG equations make use of empirical models for mobility and quantum-correction potentials. The mobility models used here are the Lombardi, Philips, and Ballistic models. The empirical mobility parameters are fitted to characterization data of a reference node [10] as well as preliminary characterizations of stacked nanowire transistors [11]. Further information on the calibration is provided in [12]. Device dimensions used throughout the study are listed in Table.1.

978-1-6654-8498-5/22 $31.00 © 2022 IEEE

Parameters (nominal)	NMOS
Gate length L_G (nm)	100
Sheet width W_{NS} (nm)	10 – 60 (16 for logic)
Sheet thickness T_{si} (nm)	5
Sheet spacing T_{SP} (nm)	10
Number of sheets N	4
Dielectric thickness (nm)	4 – 5 (3 nm SiO$_2$ and 1 – 2 nm HfO$_2$)
Supply voltage (V_{DD})	1.8
I_{OFF} (pA/μm)	10

Table. 1: The device dimensions used throughout the study. Only NMOS is considered.

The structure generated using the structure generation tool is shown in Fig. 2. To maintain process compatibility with its logic counterpart, dimensions, such as sheet thickness (T_{si}), sheet spacing (T_{SP}), number of sheets (N), and doping profiles for source/drain and extensions, are retained as given in Ref. [12] for core logic GAA NS device. However, longer gate length (L_G), thicker dielectrics ($T_{OX} + T_{HK}$), and larger supply voltage (V_{DD}) are used. The off-current (I_{OFF}) is fixed at 10pA/μm and device operating temperature is set at 300K.

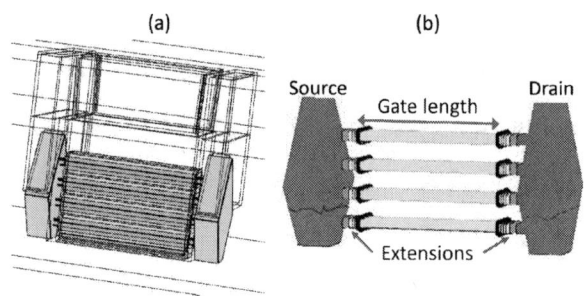

Fig. 2: (a) schematic of NSFET obtained from the structure generation tool and (b) zoom-in of the structure.

III. RESULTS

A. Comparison between the parital GAA device and reference GAA devices with varying high-k oxide thickness

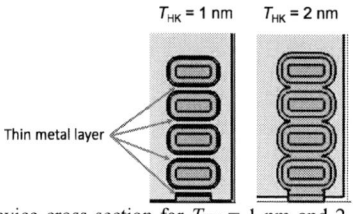

Fig. 3: I/O device cross section for T_{HK} = 1 nm and 2 nm with and without high-k pinch-off scenarios.

We first evaluate the transition from a reference GAA device (thin metal layer connecting side-gates) to a partial GAA device (with no inner-gates). This is done by varying the high-k oxide thickness (T_{HK}) from 1 nm (corresponds to 1 nm thickness of metal between the sheets) to 2 nm (no metal between the sheets). For this study, we take a constant sheet width, W_{NS} = 16 nm, which corresponds to the sheet width

used in the logic devices. The cross section of the reference device (T_{HK} = 1 nm) and partial GAA device (T_{HK} = 2 nm) is shown in Fig. 3.

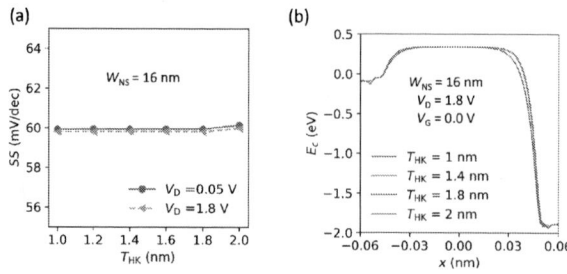

Fig. 4: (a) Sub-threshold slope (SS) versus high-k oxide thickness (T_{HK}), and (b) conduction band profile (E_c), plotted along the channel for different T_{HK}.

In Fig. 4(a), we show the sub-threshold slope (SS) obtained for different T_{HK}. We observe negligible deterioration of SS going from T_{HK} = 1 nm to T_{HK} = 2 nm. By plotting the conduction band profile (E_c) at off-state (Fig. 4(b)), we observe barrier lowering for the case without inner-gates (T_{HK} = 2 nm). However, because the gate is relatively long (L_G=100 nm), the barrier lowering has negligible effect on the sub-threshold behaviour.

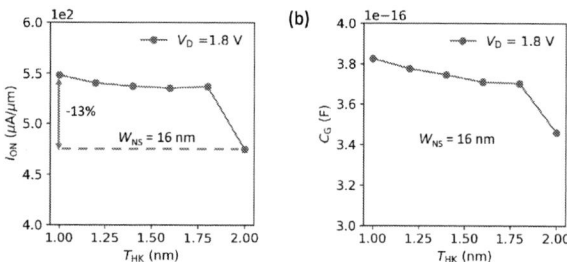

Fig. 5: (a) On-current (I_{ON}) versus high-k oxide thickness (T_{HK}), and (b) gate capacitance (C_G) for different T_{HK}.

In Fig. 5(a), we show the on-current (I_{ON}) obtained from our simulation at V_D = 1.8 V for different T_{HK}. The I_{ON} is normalized to the effective width (W_{EFF}), which is given by: $W_{EFF} = 2 \times (T_{si} + W_{NS}) \times N$. We observe that I_{ON} drops only by 13% when going from T_{HK} = 1 nm to T_{HK} = 2 nm. The drop in I_{ON} is almost entirely occurs when there is no metal left. By looking at the gate capacitance (C_G) in the on-state (Fig. 5(b)), we observe that the drop in C_G is also small going from T_{HK} = 1 nm to T_{HK} = 2 nm, which explains the small drop in I_{ON}.

B. Sheet witdh (W_{NS}) variation study

In the previous section, we performed our simulations by taking W_{NS} = 16 nm, and we found that there is negligible deterioration in SS and a small drop in I_{ON}, when transitioning from a reference GAA device to a partial GAA device. In this section, we show results of the study we performed by varying W_{NS} from 10 nm to 60 nm for two cases: T_{HK} = 1.5 nm (reference GAA device with 0.5 nm thickness of metal between the sheets) and T_{HK} = 2.0 nm (partial GAA device with no metal between sheets). The cross sections of the partial GAA devices with $W_{NS} = 20$ nm and $W_{NS} = 60$ nm are shown in Fig. 6.

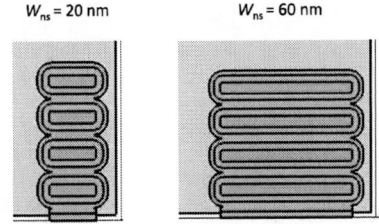

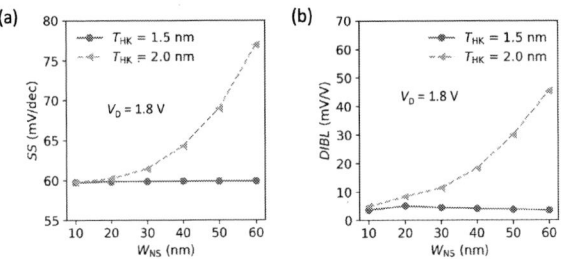

Fig. 6: The cross section of the partial GAA device for the case of $W_{NS} = 20$ nm and $W_{NS} = 60$ nm.

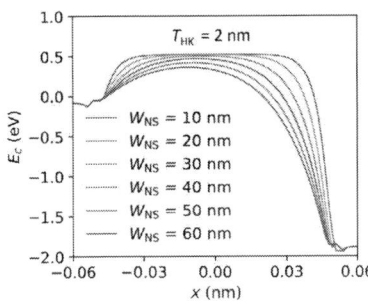

Fig. 7: (a) Sub-threshold slope (SS), and (b) drain induced barrier lowering ($DIBL$), versus sheet width (W_{NS}) for reference and partial GAA devices.

In Fig.7, we show the sub-threshold slope (SS) and drain induced barrier lowering ($DIBL$) obtained from our simulations, at $V_D = 1.8$ V, for different W_{NS}, both for reference and partial GAA devices. We observe that for the reference GAA device, there is excellent gate control for the entire range of W_{NS} used in our study (ideal SS and negligible $DIBL$). However, for the partial GAA device, the SS starts to deviate from ideality for $W_{NS} \geq 30$ nm. As W_{NS} increases, the side-gates start to lose control over the channels in the partial GAA device, leading to an increase in SS and $DIBL$ (Fig. 8 shows the S/D conduction band profiles (E_c) of the partial GAA devices at $W_{NS}/2$ for varying widths).

Fig. 8: S/D conduction band profiles (E_c) of the partial GAA devices at $W_{NS}/2$ for varying widths, emphasizing the barrier lowering as the channel width increases.

In Fig. 9(a), we show the threshold voltage (V_T) in the linear regime, plotted at different sheet width, W_{NS}. For the reference GAA device, V_T remains constant with increasing W_{NS}. However, for the partial GAA device, V_T increases sub-linearly with W_{NS} due to poor sub-threshold behaviour. The V_T shift is positive because we have set a fixed I_{OFF}.

In Fig. 9(b), we show the on-current, I_{ON} plotted for different W_{NS}, calculated at $V_D = 1.8$ V. For the reference GAA device, I_{ON} essentially remains constant for the entire range of W_{NS}. However, for the partial GAA device, I_{ON} starts deteriorating with increase in W_{NS}, with a 33% decrease for $W_{NS} = 30$ nm and a 61% decrease for $W_{NS} = 60$ nm.

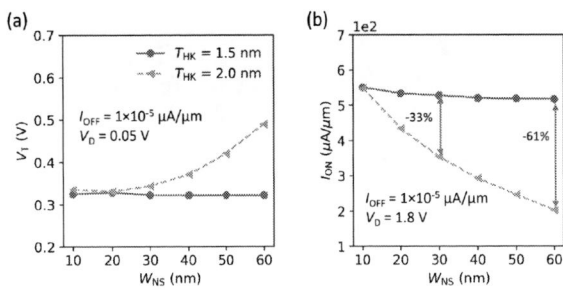

Fig. 9: (a) Threshold voltage (V_T) in linear regime, and (b) on-current (I_{ON}), versus sheet width (W_{NS}) for reference and partial GAA devices.

C. Benchmarking with repect to I/O FinFET

In this section, we benchmark I_{ON} of partial GAA I/O devices ($T_{HK} = 2$ nm) and reference GAA I/O devices ($T_{HK} = 1.5$ nm) to conventional I/O FinFETs. Parameters, such as L_G, T_{OX}, T_{HK}, and doping profiles for the FinFET are the same as those taken for NS devices. The fin width (W_{FIN}) and fin height (H_{FIN}) taken for this simulation are 5 nm and 55 nm, respectively. In this study, I_{ON} for all the devices is normalized with respect to the footprint (NSFET: $W_{NS}+21$ nm and FinFET: $W_{FIN}+21$ nm).

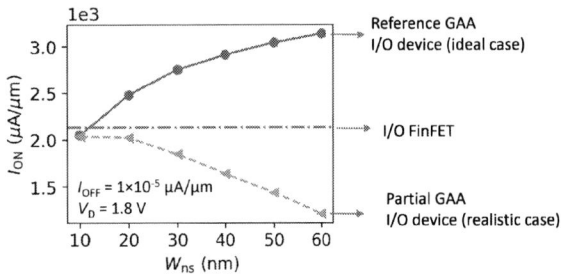

Fig. 10: Benchmarking I_{ON} of partial GAA I/O device and reference GAA I/O device to conventional I/O FinFET.

From Fig. 10, we can see that the partial GAA device slightly underperforms compared to the FinFET for $W_{NS} < 20$ nm, and the drop in I_{ON} increases for $W_{NS} > 20$ nm. However, the reference GAA I/O device outperforms the FinFET for $W_{NS} > 10$ nm.

D. Work function (between the sheets) variation study

From the results of the previous sections, we learn that the partial GAA I/O device shows good performance for certain geometric conditions ($W_{NS} \leq 30$ nm). However, this device does not exceed the performance of conventional I/O FinFET.

Nevertheless, the reference GAA device out-performs the I/O FinFET over a large range of sheet width values. In this section, we consider a GAA device, like the reference device (Fig. 11(a)), however, we consider that the WF of the inner-gates is different from the WF of the side-gates (Fig. 11(b)). This WF mismatch is a result of the thin metal layer between the sheets.

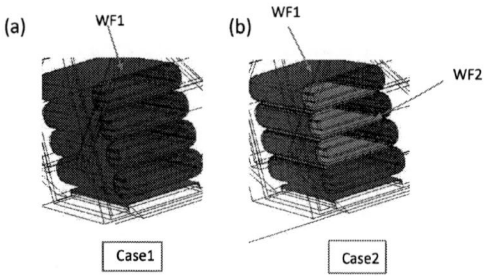

Fig. 11: (a) Schematic of the reference GAA device, and (b) GAA device with WF mismatch between the inner-gates and side-gates.

We show results of simulations performed by varying the WF of the inner-gates by $\pm$ 200 meV from the WF of the side-gates. We compare these results to the reference GAA device. For this study, we take T_{HK} = 1.5 nm in our simulations, which corresponds to 0.5 nm metal between the sheets, and we perform this study for W_{NS} = 16 nm and 40 nm. I_{ON} is normalized to W_{EFF} for all the devices.

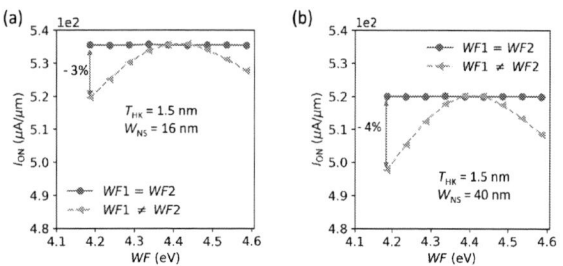

Fig. 12: On-current (I_{ON}) plotted for different WF values of the inner-gates at (a) W_{NS} = 16 nm and (b) W_{NS} = 40 nm.

In Fig. 12, we show I_{ON} plotted for different WF values of the inner-gates at two different sheet width, W_{NS} = 16 nm (Fig. 12(a)) and W_{NS} = 40 nm (Fig. 12(b)). In both the cases, I_{ON} drops with respect to the reference devices. Due to different WFs present in the device, the device behaves like multiple devices, with different V_T, connected in parallel. As a result, all the devices might not be fully on at V_{DD}, causing a drop in I_{ON}. The drop is larger when the WF of the inner-gates is lower than the WF of the side-gates (-0.2 eV). This is because, the major surface area of the gates is occupied by the side-gates, and the devices operating predominantly by the inner-gates are turned on fully compared to the devices operating with the side-gates in the on-state. However, from our simulations, we see that the drop in I_{ON} is rather limited ($-$ 3% at W_{NS} = 16 nm and $-$ 4% at W_{NS} = 40 nm).

IV. CONCLUSIONS

In this work, we studied possible I/O devices which are compatible with core logic GAA NS technology. We consider

two main device options: (1) A partial GAA I/O device (without metal filling between sheets), and (2) a device considering thin metal gate with a WF mismatch between side- and inner-gates. Both these devices are compared to a hypothetical full GAA device with thin conducting metal of same WF as side gates. The partial GAA device performs close to the reference device for shorter sheet widths ($W_{NS} < 30$ nm). However, its performance deteriorates for larger sheet widths. The partial GAA device is also benchmarked with respect to a conventional I/O FinFET, and a similar degradation of the partial GAA device is seen for larger sheet widths. However, the performance of device (2) is close to the reference GAA device. The drop in I_{ON} is negligible with respect to the reference device for a mismatch in WF ranging $\pm$ 200 meV. Even for larger W_{NS}, the drop in I_{ON} is limited to 4%. Therefore, if small W_{NS} suffice, the partial GAA device is viable choice. However, the addition of a thin metal gate between the nanosheets, even with a mismatched in WF, is a viable option for a wide range of W_{NS}, outperforming the conventional I/O FinFET.

ACKNOWLEDGMENTS

The authors would like to thank Fabian Bufler and Geert Enamen, for their valuable feedback on this work.

REFERENCES

[1] Bae, Geumjong, et al. "3nm GAA technology featuring multi-bridge-channel FET for low power and high-performance applications." *2018 IEEE International Electron Devices Meeting (IEDM)*. IEEE, 2018.

[2] Loubet, N., et al. "Stacked nanosheet gate-all-around transistor to enable scaling beyond FinFET." *2017 Symposium on VLSI Technology*. IEEE, 2017.

[3] Zhang, Jingyun, et al. "High-k metal gate fundamental learning and multi-V t options for stacked nanosheet gate-all-around transistor." *2017 IEEE International Electron Devices Meeting (IEDM)*. IEEE, 2017.

[4] Bardon, M. Garcia, et al. "Power-performance trade-offs for lateral nanosheets on ultra-scaled standard cells." *2018 IEEE Symposium on VLSI Technology*. IEEE, 2018.

[5] Hellings, Geert, et al. "Si/SiGe superlattice I/O FinFETs in a vertically-stacked gate-all-around horizontal nanowire technology." *2018 IEEE Symposium on VLSI Technology*. IEEE, 2018.

[6] Bhuiyan, M., et al. "Gate-Last I/O Transistors based on Stacked Gate-All-Around Nanosheet Architecture for Advanced Logic Technologies." *2021 IEEE International Electron Devices Meeting (IEDM)*. IEEE, 2021.

[7] Sanghoon Lee et al. SEMICONDUCTOR DEVICES. US11217695B2, United states Patent and Tademark Office, Jan 4 2022.

[8] http://www.globaltcad.com/minimos

[9] Stanojević, Z., et al. "Cell Designer-a Comprehensive TCAD-Based Framework for DTCO of Standard Logic Cells." *2018 48th European Solid-State Device Research Conference (ESSDERC)*. IEEE, 2018.

[10] Karatsori, T., et al. "Statistical characterization and modeling of drain current local and global variability in 14 nm bulk FinFETs." *2017 International Conference of Microelectronic Test Structures (ICMTS)*. IEEE, 2017.

[11] Mertens, Hans, et al. "Gate-all-around MOSFETs based on vertically stacked horizontal Si nanowires in a replacement metal gate process on bulk Si substrates." *2016 IEEE symposium on VLSI technology*. IEEE, 2016.

[12] Stanojević, Zlatan, et al. "Nano device simulator—a practical subband-BTE solver for path-finding and DTCO." *IEEE Transactions on Electron Devices* 68.11 (2021): 5400-5406

[13] Ryckaert, J., et al. "From design to system-technology optimization for cmos." *2021 International Symposium on VLSI Technology, Systems and Applications (VLSI-TSA)*. IEEE, 2021.

Cryogenic RF Characterization and Simple Modeling of a 22 nm FDSOI Technology

Hung-Chi Han*, Farzan Jazaeri*, Antonio D'Amico*, Zhixing Zhao[†], Steffen Lehmann[†],
Claudia Kretzschmar[†], Edoardo Charbon*, and Christian Enz*
*Ecole Polytechnique Fédérale de Lausanne (EPFL), Lausanne, Switzerland
[†]GlobalFoundries, Dresden, Germany
Email: hung.han@epfl.ch

Abstract—**This paper presents the RF characterization and modeling of a 22 nm FDSOI technology down to 3.3 K for quantum computing applications. The equivalent small-signal components are extracted analytically and automatically from the de-embedded two-port Y-parameters using an iteratively re-weighted least-squares method. The dynamic self-heating effect impacting $\Re(Y_{22})$ is characterized at different temperatures and bias points.**

Index Terms—**cryogenic electronics, cryo-CMOS, FDSOI, cryo-modeling, quantum computing, radio frequency, and self-heating**

I. INTRODUCTION

Cryo-CMOS, a standard CMOS technology operating at cryogenic temperatures, designs have recently been under the spotlight as a promising solution for large-scale solid-state-based quantum computers thanks to its mature scalability [1]. In this approach, the quantum processor consisting of many qubits at $T \leq 100$ mK is interfaced with CMOS front-end electronics, which is operating at 1-4 K for the larger cooling power. Particularly, the low-noise attenuated RF signal is sent to the qubits from a signal generator residing at room temperature. Besides, the fragile RF signal from the qubits, such as reflection coefficient [2] with using RF-reflectometry technique on CMOS qubits, is amplified after a multiplexer by the low-noise amplifier (LNA) [3]. It is a key building of a read-out line with a crucial design in terms of noise, gain, and power dissipation. Nowadays, many works have shown the Cryo-CMOS designs using a 22 nm FDSOI technology, which is one of the potential Cryo-CMOS candidates. Particularly, recent efforts show the promising approach of a fully monolithic integrated quantum processor, where qubits and front-end electronics are fabricated with such technology [4], [5]. However, the lack of a process design kit for cryogenic applications is an obstacle to having accurate Cryo-CMOS building blocks. Several works on the cryogenic characterization of FDSOI devices have already been published [6]–[9], little work has been done on the RF characterization and modeling at cryogenic temperatures. This paper presents new RF measurement results made at 3.3 K and successfully compares them to a simple small-signal model, the component of which are evaluated using a direct extraction technique.

This work was supported in part by the EU H2020 RIA project SEQUENCE under Grant No. 871764.

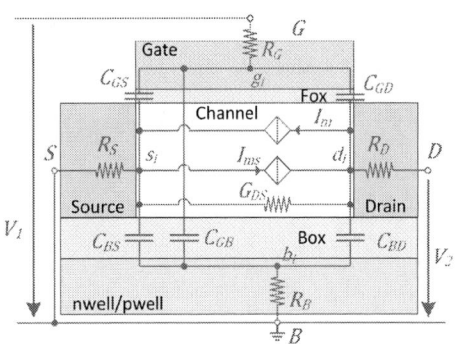

Fig. 1. Cross section of a FDSOI device with the quasi-static equivalent small-signal circuit as a transistor is in saturation regime.

II. EXPERIMENTS

The DC and RF characteristics of nMOS/pMOS transistors with 18 nm gate length (L_G) and multiple fingers (single finger width W_{fin} of 0.5 μm) are carried out from 300 K down to 3.3 K. Fig. 2(a, b) show the transfer and output characteristics at 300 K and 3.3 K; the saturation threshold voltage increases by 70 mV, the subthreshold swing reaches 22 mV/dec (nMOS) and 33 mV/dec (pMOS), and the output conductance G_{DS} in saturation remains about constant for the given $|V_{GS}| = 0.8$ V at 3.3 K. Besides, Fig. 2(c) describes the G_m^2/I_D (FoM used for the optimization of LNAs in terms of gain, noise figure, and power consumption [10]) w.r.t overdrive voltage V_{ov}. The FoM maximum moves deeper into the moderate inversion region at deep cryogenic temperature. The micro probe calibration is first performed, and open/short structure signals are obtained for each temperature. The calibrated de-embedded two-port S-parameters of the device-under-test (DUT) in strong inversion and saturation regime are acquired. Fig. 3 shows the de-embedded Y-parameters versus frequency (f), which is swept from 1 to 40 GHz.

III. MODELING AND EXTRACTION

The equivalent simplified small-signal circuit of a FDSOI MOSFET in saturation is shown in Fig. 1. The terms of I_m and I_{ms} correspond to the voltage-controlled current source, including the gate and source transadmittances given by $I_m =$

978-1-6654-8498-5/22 $31.00 © 2022 IEEE

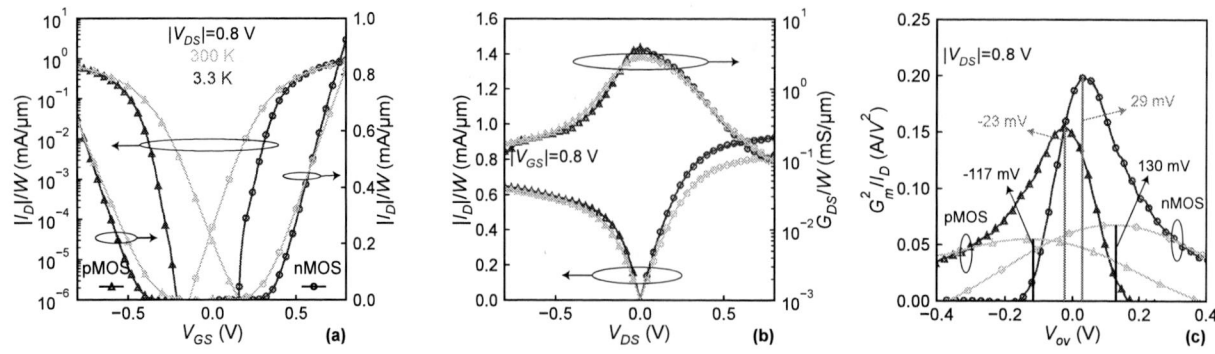

Fig. 2. DC characterization of RF FDSOI nMOS (circle) and pMOS (triangle) with $L_G = 18$ nm and $W_{fin} = 0.5\,\mu$m. (a) I_D-V_{GS} at 300 and 3.3 K in linear and log scales. (b) I_D-V_{DS} and G_{DS}-V_{DS} at 300 and 3.3 K. (c) FoM of LNA versus V_{ov} at 300 and 3.3 K, the peaks of FoM are annotated by the corresponding overdrive voltage.

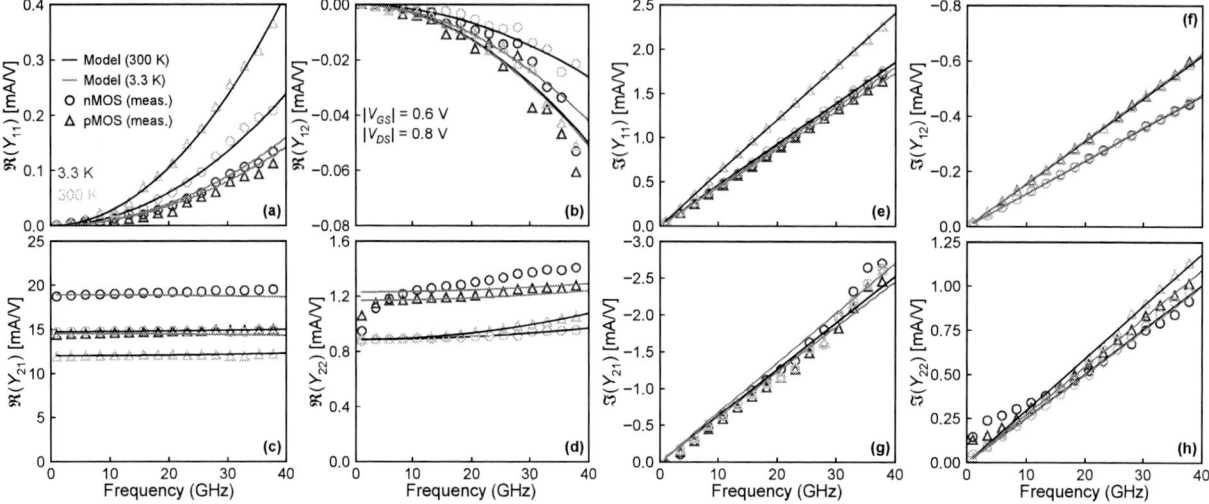

Fig. 3. Two-port de-embedded Y-parameters of nMOS (circle) and pMOS (triangle) with $L_G = 18$ nm and $W_{fin} = 0.5\,\mu$m at $|V_{GS}| = 0.6$ V and $|V_{DS}| = 0.8$ V, at 300 and 3.3 K. The markers and lines represent the experimental data and the analytical model, respectively. (a-d) and (e-h) are real and imaginary parts of Y-parameters, correspondingly.

$(G_m - j\omega C_m)(V_{gi} - V_{bi})$ and $I_{ms} = (G_{ms} - j\omega C_{ms})(V_{si} - V_{bi})$, respectively. A back-gate resistance R_B in the substrate network keeps the RF model of SOI structure simple, which allows parameters to be extracted analytically without using non-linear optimization. It is remarkable that this model is identical to the model that was proposed for bulk transistors [11], but the junction capacitances are replaced by the BOX capacitances and the substrate resistance are replaced by the back-gate resistance. The two-port Y-parameters of a transistor in saturation is therefore given by [12]

$$Y_{11} \approx \omega^2(C_{GB}^2 R_B + C_{GG}^2 R_G) + jwC_{GG} \tag{1a}$$

$$Y_{12} \approx \omega^2(C_{BD}C_{GB}R_B - C_{GD}C_{GG}R_G) - jwC_{GD} \tag{1b}$$

$$Y_{21} \approx G_m + \omega^2(C_{GB}R_B(C_{BD} - C_m + C_{ms}) \\ - C_{GG}R_G(C_{GD} + C_m)) - jw(C_{GD} + C_m) \tag{1c}$$

$$Y_{22} \approx C_{GD} + \omega^2(C_{BD}R_B(C_{BD} - C_m + C_{ms}) \\ + C_{GD}R_G(C_{GD} + C_m)) + jw(C_{BD} + C_{GD}) \tag{1d}$$

The simplified expression in (1) neglects the access resistances ($R_{S,D}$) due to the large device width, and the high-order terms, e.g., ω^2 and ω^4, are ignored from the Y-parameters. The capacitances, C_{GG}, C_{GD}, C_{GB}, and C_m, can be extracted from $\Im(Y)$ versus ω using a simple linear regression. However, extracting the RF components robustly and accurately from $\Re(Y)$ is challenging because they are noisier than $\Im(Y)$ and some of them consist of dynamic self-heating effect and DC component. Fortunately, $\Re(Y)$ is quadratic with respect to ω, they can be treated by a linear regression of $\Re(Y)$ versus ω^2 as described in Fig. 4. By adopting the iteratively re-weighted least-squares (IRLS) to extract ω^2 coefficients and DC elements, the outliers and data that are not linear to ω^2

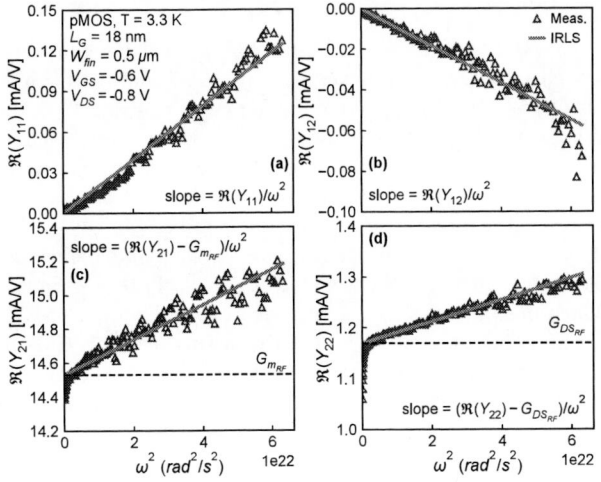

Fig. 4. Extraction with the IRLS method of f-dependent coefficients (slopes) and DC components from the real part Y-parameters of pMOS FDSOI with $L_G = 18\,\mathrm{nm}$ at 3.3 K

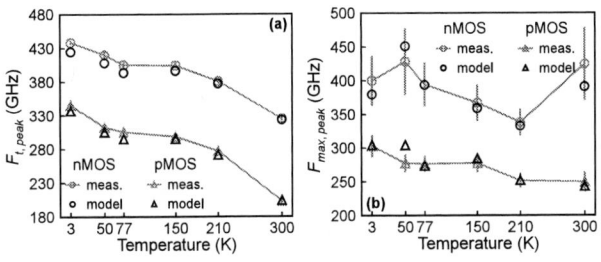

Fig. 5. The peak F_t and F_{max} from room temperature down to 3.3 K. The experimental data is marked in red with vertical error bar. The model based on extracted parameters is in black.

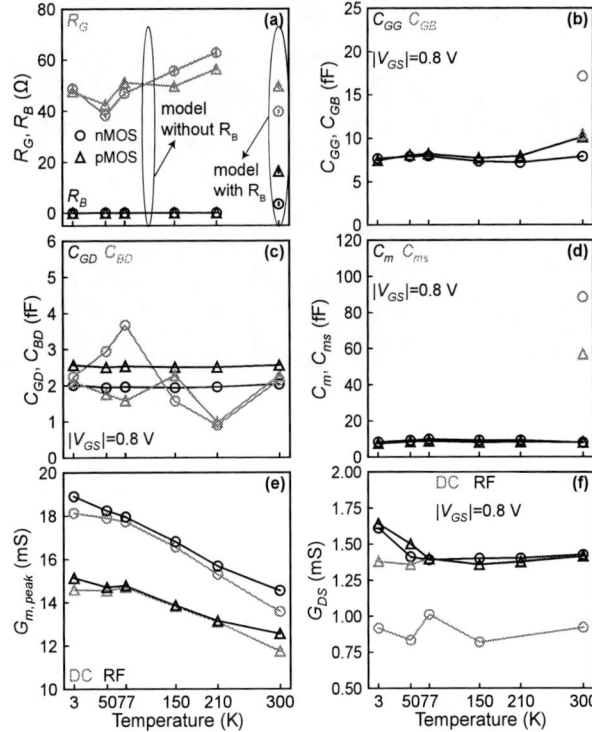

Fig. 6. Extracted small-signal components over a wide temperature range. The discontinuous lines of $R_{G,B}$ from 300 K to 210 K is due to the switch of the model. C_{GB} and C_{ms} are extracted at 300 K when R_B is different than 0. (e) shows the peak G_m for each temperature and (f) shows the G_{DS} extracted from DC (red) and RF (black) measurements.

are less weighted in the IRLS process, as shown by Fig. 4. Consequently, we could obtain the coefficient of the term ω^2 from the slope of the regression line. Besides, the intercept of the regression line tells the DC value, here we denote the $G_{m_{RF}}$ and $G_{DS_{RF}}$ that are extrapolated from the RF signal to differentiate $G_{m_{DC}}$ and $G_{DS_{DC}}$ obtained from the large signal. Finally, the remaining RF components, i.e., C_{GB}, R_B, R_G, and C_{ms}, embedding in $\Re(Y)$ are solved analytically from the ω^2 coefficients extracted from IRLS method. Fig. 3 shows that the model nicely correlates with the measurement for nMOS/pMOS at both 300 K and 3.3 K. It highlights the efficiency and accuracy of using the simple equivalent small-signal model and IRLS method to extract the RF parameters fully automatically.

IV. DISCUSSION

Fig. 5 shows the transit frequency F_t and maximum oscillation frequency F_{max} for each temperature, where a good agreement between the measurement and model is demonstrated. The peak F_t of nMOS/pMOS is increased by 1.34/1.70 times, compared to that at 300 K, and reaches 438/345 GHz at

3.3 K. On the other hand, the peak F_{max} fluctuates for nMOS, but it increases 1.2 times and reaches 302 GHz for pMOS.

In Fig. 6(a), R_B (nwell/pwell) is found small enough to be neglected for $T < 300\,\mathrm{K}$ due to phonon scattering reduction. Besides, R_B does not suffer from the dopant freeze-out effect at cryogenic temperatures, otherwise, R_B should be increased. This finding aligns to the temperature-independent back-gate coefficient presented in [7], [9]. Therefore, R_B is set to $0\,\Omega$ at low temperatures for the equivalent small-signal circuit and Y-parameters formulas in (1). As a result, a model with R_B is used at $T = 300\,\mathrm{K}$ and a model without R_B is used at $T < 300\,\mathrm{K}$. The change from room-temperature model to low-temperature model explains the increase in R_G from 300 to 210 K since R_G becomes the only term contributing to $\Re(Y_{11})$ when R_B is neglected. Fig. 6(b-d) shows capacitances that remain nearly constant over the temperature. It should be noted that C_{GB} and C_{ms} are coupled with R_B as shown by (1), they cannot be extracted when R_B is $0\,\Omega$ at $T < 300\,\mathrm{K}$. Fig. 6(e-f) plot the temperature dependency of G_m and G_{DS} that are extracted from DC and RF characteristics. The strong correlation between $G_{m_{DC}}$ and $G_{m_{RF}}$ highlights the effectiveness of the simple equivalent small-signal circuit and the IRLS method. Besides, it can be found that the dynamic self-heating effect

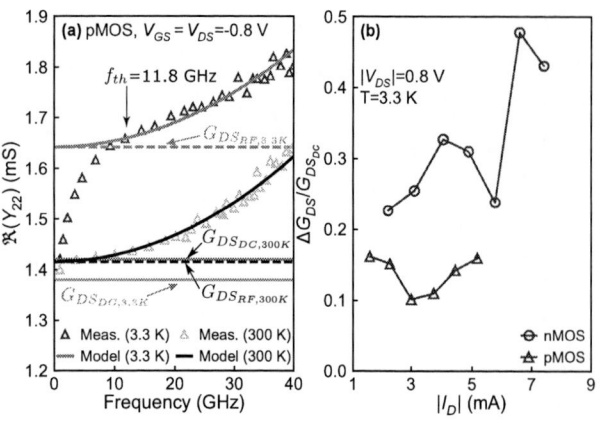

Fig. 7. Self-heating effect in RF FDSOI MOSFETs. (a) shows $\Re(Y_{22})$ at 3.3 K and 300 K with G_{DS} DC/RF extrapolation and model using IRLS method, a transition from f_{th} to DC at 3.3 K is due to dynamic self-heating effect. (b) presents the normalized ΔG_{DS} versus drain current at 3.3 K, where nMOS manifests stronger dynamic self-heating effect in terms of the power.

has less impact on G_m since the device is in saturation. In this case, the carrier velocity is saturated and show less temperature dependence [13].

However, a significant $\Delta G_{DS} = G_{DS_{RF}} - G_{DS_{DC}}$ is observed for nMOS from 300 K to 3.3 K, whereas pMOS shows the clear ΔG_{DS} below 77 K. The ΔG_{DS} is ascribed to the dynamic self-heating effect, in which the channel temperature cannot respond to the fast drain voltage oscillation, therefore, the dynamic self-heating vanishes at frequencies above the so-called thermal cutoff frequency (f_{th}) [14], [15]. Consequently, as shown by a RF pMOS device working at 3.3 K in Fig. 7(a), the self-heating results in the $\Re(Y_{22})$ having a decreasing transition from f_{th} to DC [16]. Moreover, the term ΔG_{DS} related to the dynamic self-heating effect is demonstrated for nMOS/pMOS by Fig. 7(b). Particularly, nMOS reaches 50% increase compared $G_{DS_{DC}}$ at 3.3 K.

In addition, Fig. 7(a) underlines the advantage of using the IRLS method in analyzing the self-heating effect. Conventionally, building a thermal network requires choosing the f_{th} that defines two frequency regions with and without the dynamic self-heating effect, respectively [17]. Although N. Rinaldi [15] proposed a technique that extracts f_{th} from $C_{21} = -\Im(Y_{21})/\omega$, where the capacitance changes the sign due to the delay in the heat transport [18], it turns out that $f_{th} = 1.9$ GHz with using the C_{21} technique for the RF pMOS at 3.3 K shown in Fig. 7(a). Evidently, $f_{th} = 1.9$ GHz locates at the region with the presence of AC self-heating, the C_{21} method underestimates the value of f_{th}. On the contrary, the IRLS method nicely captures the behavior of $\Re(Y_{22})$ at $f > f_{th}$ that is ascribed to the gate resistance, meaning that f_{th} can be obtained from the intersection point of the IRLS line and the measured $\Re(Y_{22})$. In the end, we obtain a reasonable $f_{th} = 11.8$ GHz. The IRLS method greatly improves the extraction process for RF components, but surprisingly locates f_{th} directly from $\Re(Y_{22})$ for the dynamic self-heating.

V. Conclusion

The RF nMOS and pMOS from a 22 nm FDSOI technology are characterized and modeled from 300 K down to 3.3 K for Cryo-CMOS designs. Using the simple equivalent small-signal model and the iteratively re-weighted least squares (IRLS) method allows the robust and automatic extraction of the small-signal circuit components directly from the de-embedded Y-parameters without any fitting. Moreover, the IRLS method is demonstrated as an accurate tool to define the thermal cutoff frequency from the dynamic self-heating effect, which finds a 50% increase in output conductance for nMOS in strong inversion at 3.3 K. Finally, biasing the Cryo-CMOS devices deeper into moderate inversion allows for keeping a maximum FoM and, at the same time, reducing the impact of dynamic self-heating.

References

[1] E. Charbon et al., "Cryo-CMOS for quantum computing," in IEDM, 2016.

[2] A. Crippa et al, "Gate-reflectometry dispersive readout and coherent control of a spin qubit in silicon," Nat Commun, vol. 10, 2019.

[3] B. Patra et al., "Cryo-cmos circuits and systems for quantum computing applications," JSSC, vol. 53, no. 1, pp. 309–321, 2018.

[4] M. J. Gong et al., "Design Considerations for Spin Readout Amplifiers in Monolithically Integrated Semiconductor Quantum Processors," in RFIC, 2019, pp. 111–114.

[5] R. B. Staszewski et al., "Cryo-CMOS for Quantum System On-Chip Integration: Quantum Computing as the Development Driver," SSCS Mag., vol. 13, no. 2, pp. 46–53, 2021.

[6] S. Bonen et al., "Cryogenic Characterization of 22-nm FDSOI CMOS Technology for Quantum Computing ICs," EDL, vol. 40, no. 1, pp. 127–130, 2019.

[7] B. Cardoso Paz et al., "Performance and Low-Frequency Noise of 22-nm FDSOI Down to 4.2 K for Cryogenic Applications," TED, vol. 67, no. 11, pp. 4563–4567, 2020.

[8] W. Chakraborty et al., "Characterization and Modeling of 22 nm FDSOI Cryogenic RF CMOS," JXCDC, vol. 7, no. 2, pp. 184–192, 2021.

[9] H.-C. Han et al., "Back-gate effects on DC performance and carrier transport in 22 nm FDSOI technology down to cryogenic temperatures," SSE, vol. 193, p. 108296, 2022.

[10] T. Taris et al., "A 60μW LNA for 2.4 GHz wireless sensors network applications," in RFIC, 2011, pp. 1–4.

[11] C. Enz, "An MOS transistor model for RF IC design valid in all regions of operation," TMTT, vol. 50, no. 1, pp. 342–359, 2002.

[12] M.-A. Chalkiadaki and C. Enz, "RF Small-Signal and Noise Modeling Including Parameter Extraction of Nanoscale MOSFET From Weak to Strong Inversion," TMTT, vol. 63, no. 7, pp. 2173–2184, 2015.

[13] R. Quay et al., "A temperature dependent model for the saturation velocity in semiconductor materials," Materials Science in Semiconductor Processing, vol. 3, no. 1, pp. 149–155, 2000.

[14] B. Tenbroek et al, "Self-heating effects in SOI MOSFETs and their measurement by small signal conductance techniques," TED, vol. 43, no. 12, pp. 2240–2248, 1996.

[15] N. Rinaldi, "Small-signal operation of semiconductor devices including self-heating, with application to thermal characterization and instability analysis," TED, vol. 48, no. 2, pp. 323–331, 2001.

[16] R. Singh et al., "Experimental Evaluation of Self-Heating and Analog/RF FOM in GAA-Nanowire FETs," TED, vol. 66, no. 8, pp. 3279–3285, 2019.

[17] S. Makovejev et al., "RF Extraction of Self-Heating Effects in FinFETs," TED, vol. 58, no. 10, pp. 3335–3341, 2011.

[18] A. Scholten et al., "Experimental assessment of self-heating in SOI FinFETs," in IEDM, 2009, pp. 1–4.

A Novel Approach to Modeling Insulator Wave-Function Penetration and Interface Roughness Scattering in MOSFETs

Zlatan Stanojević, Lee-Chi Hung, Chen-Ming Tsai, Markus Karner
Global TCAD Solutions GmbH, Vienna, 1010, Austria (e-mail: z.stanojevic@globaltcad.com).

Abstract—**We present novel models for insulator wave-function penetration and for interface roughness scattering. We review the intricate relationship between modeling of these two effects, with respect to usability, computational effort, and numerical stability. We demonstrate that our novel approach is capable of remedying all of the common issues of previous approaches at no additional computational cost.**

Index Terms—**TCAD, device simulation, Schrödinger-Poisson, roughness scattering, carrier mobility, CMOS**

I. INTRODUCTION

Surface roughness scattering (SRS) is one of the most important carrier scattering mechanisms in MOSFET devices based on silicon, germanium, and other semiconductors. SRS limits the drain current at high V_{gate}, low V_{drain} biases, and is thus the main contributor to R_{on}. In the semi-classical transport regime, the original treatment of SRS by Prange and Nee [1] has received numerous refinements over the decades, to include screening [2], extension to non-planar devices [3], and higher-order perturbations [4].

A somewhat less frequently considered effect in simulation of planar MOSFET devices was wave-function penetration into the insulator, that insulator being SiO_2 in most cases. Since SiO_2 typically represents a barrier of 2 to 3 eV, wave-function penetration is often neglected in Schrödinger-Poisson calculations, the wave-function is forced to zero by means of a Dirichlet boundary condition. Recently however, Fin-FET and nano-sheet (NSFET) based technologies commonly feature pi-gated of gate-all-around channels with thicknesses significantly below 10 nm, shifting the ground state energy to levels where wave-function penetration cannot be simply neglected.

As we shall see however, including wave-function penetrations is not as simple as merely solving the Schrödinger equation in both the semiconductor and insulator simultaneously. This is because the way the Si/SiO_2-interface is modeled in the Schrödinger equation has major consequences for the SRS rates and, consequently, the channel mobility.

II. MODELING WAVE FUNCTION PENETRATION AND ROUGHNESS SCATTERING

A. No Penetration

In the simplest approach, no wave-function penetration into the insulator is considered. The one-dimensional effective

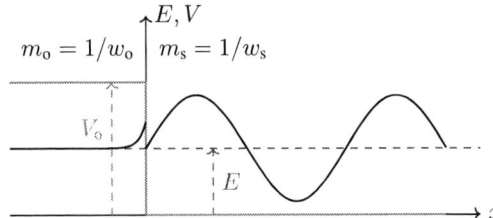

Fig. 1. Closed boundary solution of the Schrödinger equation for a carrier near an insulating barrier; the analytical extension of the wave-function into the insulator is also shown

mass Schrödinger equation is solved in the semiconductor region of a planar channel

$$\left(\frac{-\hbar^2}{2}\frac{d}{dx}w_s\frac{d}{dx} + V - E\right)\psi = 0, \quad (1)$$

where w_s denotes the inverse of the effective mass in the semiconductor, and the wave-function is terminated with a zero-value Dirichlet boundary condition at the semiconductor/insulator-interface as shown in Fig. 1. To model SRS, we need to calculate the SRS matrix element for small fluctuations of the interface position at $x = 0$ [3],

$$M_{i,j} = (V_o - V_s)\psi_i^*(0)\psi_j(0)$$
$$- \frac{\hbar^2}{2}(w_o - w_s)\frac{d\psi_i^*(0)}{dx}\frac{d\psi_j(0)}{dx}, \quad (2)$$

where w_o is the inverse of the oxide effective mass and $V_o - V_s$ the effective band edge barrier between semiconductor and oxide. We cannot obtain a meaningful value for the first term of $M_{i,j}$, since our wave-functions ψ are always zero at $x = 0$, hence we need to *extend* the wave-functions into the oxide by using an exponentially decaying function:

$$\psi_o = \psi_s(0)e^{\kappa x}, \quad \kappa = \frac{1}{\hbar}\sqrt{\frac{2(V_o - E)}{w_o}} \approx \frac{1}{\hbar}\sqrt{\frac{2V_o}{w_o}}, \quad (3)$$

since V_o is expected to be significantly higher than the state energy E. This wave-function is matched to the semiconductor wave-function in a way as if the Schrödinger equation was solved in both domains,

$$w_o\frac{d\psi_o}{dx} = w_s\frac{d\psi_s}{dx} \Rightarrow \psi_o(0) \approx \frac{w_s}{w_o}\hbar\sqrt{\frac{w_o}{2V_0}}\frac{d\psi_s(0)}{dx}. \quad (4)$$

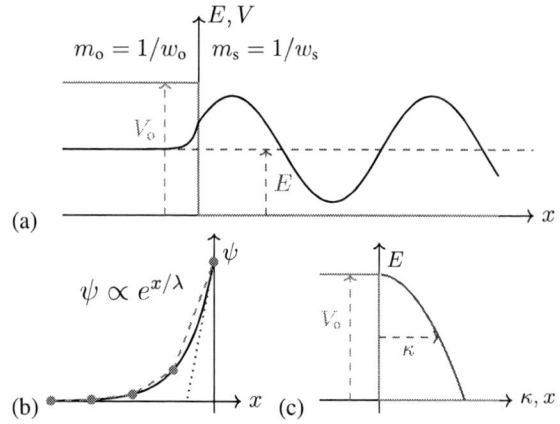

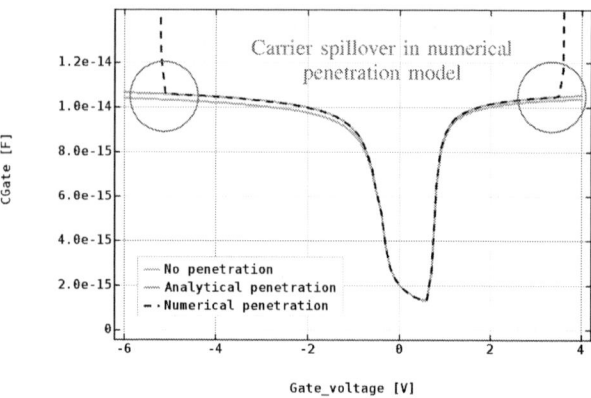

Fig. 2. Full solution of Schrödinger equation, including insulator (a); for a high barrier, inadequate insulator grid spacing leads to large numerical errors in the wave-function derivative (b); nonlinear dependence of κ on state energy (c)

Fig. 3. C/V-curves of a MOSCAP ($t_{\text{ox}} = 2\,\text{nm}, N_A = 2 \times 10^{17}\,\text{cm}^{-3}$) showing identical capacitance between analytical and numerical penetration as well as carrier spillover in the latter; without penetration, the apparent EOT increases

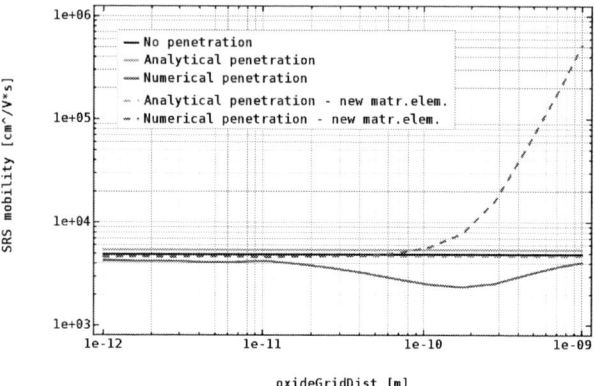

Fig. 4. Grid sensitivity of SRS-limited mobility for the different penetration models as well as the matrix elements in Eqs. (2) and (8)

With this we can simplify Eq. (2) to

$$M_{i,j} = \frac{\hbar^2}{2} w_{\text{o}} \frac{\mathrm{d}\psi_i^*(0)}{\mathrm{d}x} \frac{\mathrm{d}\psi_j(0)}{\mathrm{d}x}. \qquad (5)$$

It is interesting to note here that the insulator effective mass is a *hidden* parameter of the SRS model. This poses a challenge, since in insulators that are amorphous oxides the effective mass cannot be determined. It is common in literature to equate the insulator effective mass to the semiconductor effective mass, $w_{\text{o}} = w_{\text{s}}$.

B. Numerical Penetration Model

The seemingly straightforward way to include wave-function penetration into the oxide is to solve the Schrödinger equation in both materials as shown in Fig. 2 (a). Unless we are using open boundary conditions, the wave-function has to be terminated somewhere; for a MOSFET channel, this is typically done at the interface between the insulator and the metal gate, since after a quick decay in the insulator the wave-function will be very small there anyway.

This leads to the first problem. For strong enough inversion, the carrier states may spill over the insulator barrier and become trapped in the insulator near the metal gate, rendering the simulation useless outside a certain gate bias range (Fig. 3).

The second and much more serious problem relates to SRS. With the wave-function being non-zero at the semiconductor/insulator-interface, we can now readily evaluate all terms in Eq. (2). To do so, we require the value and first derivative of the incident and scattered wave-functions at the interface. While both are dependent on the discretization accuracy of the Schrödinger equation, the effects of numerical errors are greatly exaggerated for the first derivative as sketched in Fig. 2 (b), and the derivative can easily deviate from its true value by a factor of two. Numerical error is further amplified by the square dependence of $M_{i,j}$ on the wave-function derivatives and the square dependence of the SRS-limited mobility on $M_{i,j}$.

To get an understanding of the required grid resolution in the oxide, we can consult Eq. (3). For an energy barrier of 3 eV, the characteristic length $\lambda = 1/\kappa$ is approximately 1 Å; in practice, we need a value lower than that, whereas in the semiconductor a grid spacing of 0.5 nm is sufficient. This may not be a huge problem for one-dimensional Schrödinger-Poisson simulations, but a grid spacing below 1 Å renders two-dimensional simulation of non-planar devices prohibitively expensive.

It should be noted that the eigenenergies and densities are not very sensitive to numerical errors in the insulator, and the same is true for derived quantities such as sheet charge and capacitance. For mobility Fig. 4 shows a much more dire picture; we only see the value of the SRS-limited mobility stabilize at 0.1 Å grid spacing.

Users unaware of this effect might be lead to believe that insulator penetration necessitates a recalibration of the SRS model parameters, when attempting to fit simulation data with an inadequate insulator grid to experimental data.

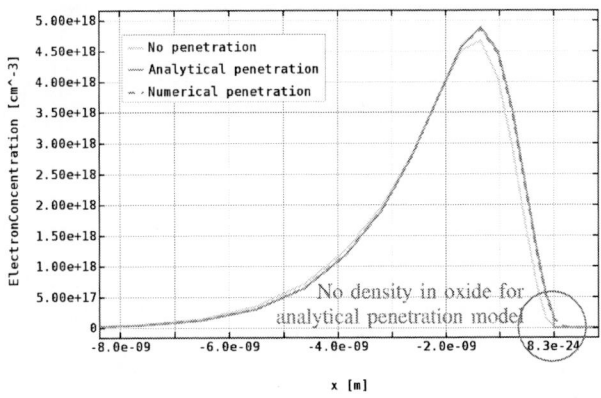

Fig. 5. Density profile of MOSFET in inversion for no, analytical, and numerical penetration

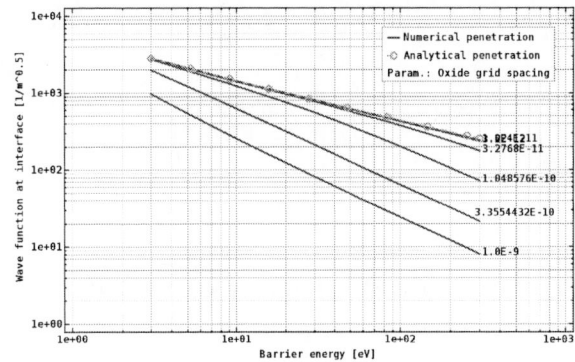

Fig. 6. Value of ground state wave-function at Si/barrier-interface for different barrier heights

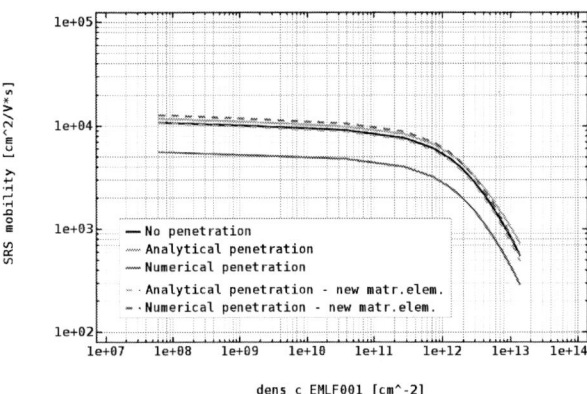

Fig. 7. SRS-limited mobility vs. inversion charge for the different penetration models as well as the matrix elements in Eqs. (2) and (8)

C. Analytical Penetration Model

To improve the numerical robustness of SRS in the presence of wave-function penetration, we have devised a novel scheme. Returning to Eqs. (3) and (4), we find that

$$w_\mathrm{s}\frac{\mathrm{d}\psi_\mathrm{s}}{\mathrm{d}x} = w_\mathrm{o}\frac{\mathrm{d}\psi_\mathrm{o}}{\mathrm{d}x} = w_\mathrm{o}\kappa\psi_\mathrm{s}(0)e^{\kappa x}, \qquad (6)$$

so at $x = 0$, we get

$$w_\mathrm{s}\frac{\mathrm{d}\psi_\mathrm{s}}{\mathrm{d}x} = w_\mathrm{o}\kappa\psi_\mathrm{s}(0) \qquad (7)$$

which is a Robin boundary condition for the Schrödinger equation. Due to the nonlinear dependence of κ on the state energy E in Eq. (3), which can be seen in Fig. 2 (c), this boundary condition turns the Schrödinger equation into a nonlinear eigenvalue problem, which we solve using approximative methods.

Since the insulator wave-function is eliminated from Eq. (7), the insulator grid is irrelevant for the result. The Schrödinger equation is not solved in the insulator at all, as can be seen in Fig. 5. Thus the analytical penetration model gives a constant SRS-limited mobility in Fig. 4.

Figure 6 shows that the analytical model is very accurate when it comes to representing the value of the wave-function at the semiconductor/insulator-interface over a wide range of barrier energies. The same cannot be said for the numerical penetration model with an insufficiently fine insulator grid.

III. REVISITING ROUGHNESS SCATTERING

Going back to the SRS matrix element Eq. (2), we can now replace the derivatives of the wave-functions in that expression using the relations in Eqs. (4) and (7). Doing so, we can simplify the matrix element to

$$M_{i,j} = \frac{w_\mathrm{o}}{w_\mathrm{s}}(V_\mathrm{o} - V_\mathrm{s})\psi_i^*(0)\psi_j(0). \qquad (8)$$

This allows us to completely eliminate the use of wave-function derivatives from the calculation of the SRS matrix element. This is significant because the numerical error of

ψ and E is $\mathcal{O}(\Delta x^2)$ while the accuracy of ψ-derivatives is $\mathcal{O}(\Delta x)$, with Δx being the grid spacing.

The usefulness of this formulation for the SRS matrix element goes beyond the analytical penetration model – it is also applicable to the numerical penetration model, as shown in Fig. 4, where it allows us to avoid the problematic ψ-derivative shown in Fig. 2 (b). While the error is even greater for coarser grids than for Eq. (2), the convergence is monotonic and predictable, and the limit coincides with the value obtained when using the analytical penetration model.

IV. DISCUSSION

A. SRS Parameter Recalibration

As can be seen from Figs. 4 and 7, the SRS-limited mobility obtained using the analytical penetration model and the SRS matrix element in Eq. (8) is largely consistent with what we obtained without wave-function penetration – our previous baseline. This is fortunate because we only need slight adjustments to our SRS parameters, which were previously calibrated to match the standard set of universal mobility curves from Takagi [5]. Table I shows the necessary adjustments while the mobility fits are shown in Fig. 8.

978-1-6654-8498-5/22 $31.00 © 2022 IEEE

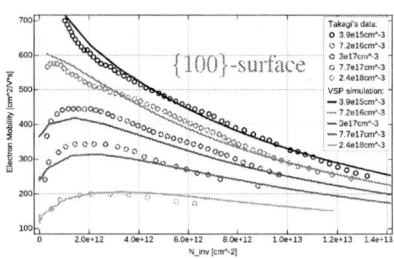 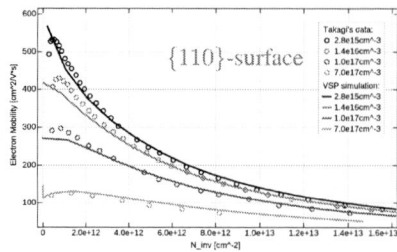 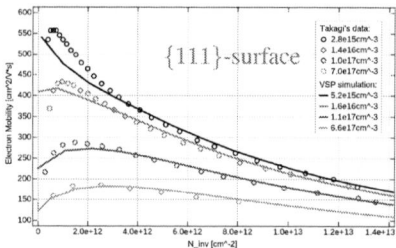

Fig. 8. Mobility curves simulated using analytical penetration and the SRS matrix element from Eq. (8) fitted to mobility data from Ref. [5]

TABLE I
CHANGES OF ROUGHNESS AMPLITUDE PARAMETER Δ_{RMS} TO MATCH
MOBILITY DATA FROM REF. [5] WHEN USING ANALYTICAL PENETRATION
AND THE SRS MATRIX ELEMENT EQ. (8)

Surface	No penetration	Analytical model
{100}	$\Delta_{\mathrm{rms}} = 3.0\,\text{Å}$	$\Delta_{\mathrm{rms}} = 2.8\,\text{Å}$
{110}	$\Delta_{\mathrm{rms}} = 4.5\,\text{Å}$	$\Delta_{\mathrm{rms}} = 5.0\,\text{Å}$
{111}	$\Delta_{\mathrm{rms}} = 2.9\,\text{Å}$	$\Delta_{\mathrm{rms}} = 3.6\,\text{Å}$
SiO_2 effective mass		
$m_{\mathrm{o}} = 0.5 m_e$		

B. Model Implementation

For the sake of simplicity and readability, we have only presented one-dimensional formulations of the effective mass Schrödinger equation and SRS matrix element in this work. However, the approach is generalizable to two and three dimensions and can be extended to $\mathbf{k} \cdot \mathbf{p}$-based effective Schrödinger equations. A generalized implementation of the models presented in this paper is available as part of the commercial simulation tools VSP and NDS from Global TCAD Solutions [6]–[9].

C. Model Limitations

The presented modeling framework is designed based on the assumption of high energy barrier formed by the insulating material of at least 1 eV. This applies to Si-based MOSFETs and MISFETs with high enough insulating barriers, such as GaN/AlN-based devices. It is not well-suited for devices, such as InGaAs/InP-based HEMTs, where the insulating barrier is on the order of a few 0.1 eV. In those cases however, λ is around 1 nm and the numerical penetration model can be safely used. Numerical penetration at low-energy barriers can also be mixed with the Robin boundary condition for analytical penetration at high energy barriers.

V. CONCLUSIONS

We have presented a novel model that allows us to consider wave-function penetration into the insulating barrier of MOSFET devices. We have shown that the approach of simply including the insulator in the Schrödinger equation is plagued by several problems, all of which our analytical penetration model addresses:

1) No carrier spillover at high inversion or accumulation biases

2) Numerically robust wave-function values and their first derivatives at semiconductor/insulator interface and thus stable surface roughness matrix element and mobility

3) No mesh density requirements in the insulating barrier and added computational cost w.r.t. simulations without penetration

We have further derived an alternative formulation of the surface roughness matrix element that does not require spatial derivatives of the wave-functions and is thus more numerically robust than its textbook counterpart.

We have demonstrated that the analytical and numerical penetration converge to the same energies, densities, ψ-values, and roughness-limited mobilities and we have shown that mobility parameters calibrated without wave-function penetration can be easily recalibrated for the analytical penetration model.

REFERENCES

[1] R. E. Prange and T.-W. Nee, "Quantum Spectroscopy of the Low-Field Oscillations in the Surface Impedance," *Phys. Rev.*, vol. 168, pp. 779–786, Apr 1968. http://link.aps.org/doi/10.1103/PhysRev.168.779. DOI: *10.1103/PhysRev.168.779*

[2] S. Jin, M. V. Fischetti, and T.-W. Tang, "Modeling of surface-roughness scattering in ultrathin-body soi mosfets," *IEEE Transactions on Electron Devices*, vol. 54, no. 9, pp. 2191–2203, 2007, DOI: *10.1109/TED.2007.902712.*

[3] Z. Stanojevic and H. Kosina, "Surface-Roughness-Scattering in Non-Planar Channels – the Role of Band Anisotropy," in *Intl. Conf. on Simulation of Semiconductor Processes and Devices*, 2013, pp. 352–355.

[4] O. Badami, E. Caruso, D. Lizzit, P. Osgnach, D. Esseni, P. Palestri, and L. Selmi, "An improved surface roughness scattering model for bulk, thin-body, and quantum-well mosfets," *IEEE Transactions on Electron Devices*, vol. 63, no. 6, pp. 2306–2312, 2016, DOI: *10.1109/TED.2016.2554613.*

[5] S. Takagi, A. Toriumi, M. Iwase, and H. Tango, "On the universality of inversion layer mobility in Si MOSFET's: Part II-effects of surface orientation," *IEEE T. Electron. Dev.*, vol. 41, no. 12, pp. 2363–2368, 1994, DOI: *10.1109/16.337450.*

[6] O. Baumgartner, Z. Stanojevic, K. Schnass, M. Karner, and H. Kosina, "VSP–a quantum-electronic simulation framework," *J. Comput. Electron.*, vol. 12, pp. 701–721, 2013. http://dx.doi.org/10.1007/s10825-013-0535-y. DOI: *10.1007/s10825-013-0535-y*

[7] Global TCAD Solutions. Vienna Schrödinger-Poisson. http://www.globaltcad.com/vsp

[8] Z. Stanojević, C.-M. Tsai, G. Strof, F. Mitterbauer, O. Baumgartner, C. Kernstock, and M. Karner, "Nano device simulator–a practical subband-bte solver for path-finding and dtco," *IEEE Transactions on Electron Devices*, vol. 68, no. 11, pp. 5400–5406, 2021, DOI: *10.1109/TED.2021.3079884.*

[9] Global TCAD Solutions. Nano Device Simulator. http://www.globaltcad.com/nds

On the convergence of the recurrence solution of McIntyre's local and non-local avalanche triggering probability equations for SPAD compact models

Dorian Saint-Pierre, Raphaël Clerc
Laboratoire Hubert Curien, UMR 5516 F-42023
Université de Lyon, UJM-Saint-Etienne, CNRS, Institut d'Optique
Graduate School, Saint-Etienne, France
raphael.clerc@institutoptique.fr

Rémi Helleboid, Denis Rideau
STMicroelectronics, 850 rue Jean Monnet, BP 16, F-38926
Crolles Cedex, France
denis.rideau@st.com

Abstract— **McIntyre's local and non-local equations are used to calculate the avalanche triggering probability, useful in the modeling of Dark Count Rates and Probability Detection Efficiency of Single Photon Avalanche Detectors (SPAD). Although non-linear, the *local* equations have a closed-form solution, while the more rigorous *non-loc*al history-dependent equations can be solved only by an iterative approach. However, the convergence of the latest approach is fairly slow, making the modeling of SPAD properties relatively time consuming. An alternative and numerically efficient method is proposed to solve the *local* equations, avoiding the use of the closed-form solution and its time-consuming root finding step. Several approaches are then discussed to improve the convergence of the *non-local* history-dependent equations.**

Keywords— *SPAD, avalanche triggering probability, history-dependent equations, McIntyre.*

I. INTRODUCTION

Single Photon Avalanche Diodes (SPAD) are present both in Silicon and III-V technologies and are commonly used to detect time dependent and small optical power signals. They are found in several technologies such as lidars, time of flight 3D imaging, fluorescence lifetime microscopy, optical communications, and particle detection [1] [2].

Biased above the breakdown voltage, electrons and holes pairs created by band-to-band absorption of few photons may trigger an avalanche, leading to a measurable electrical current. It is thus important to determine accurately the avalanche triggering probability versus device design. This question has been addressed at the early stage of the development of this technology in the seventies in the pioneering works of Oldham [3] and McIntyre [4]. In these works, a set of non-linear differential equations was derived [3] and solved [4] in order to calculate the probability $P_p(x)$, i.e. the probability that an electron or a hole generated in x will initiate an avalanche. This model has been successfully used by many different authors to calculate the SPAD figures of merit, such as the Dark Count Rates and Probability Detection Efficiency [5] [6] [7] [8].

As pointed out by Okuto and Crowell [3] [4], and then Hayat [11], the local nature of Oldham and McIntyre formalism ignores the fact that a carrier needs to gain a given threshold energy before being capable of triggering an impact ionization. The typical distance (sometimes referred as dead space [11]) for reaching this threshold energy being of the order of magnitude of ten nanometers (depending of the field and the material), it may not be negligible in highly doped

devices, where the space charge layer is typically hundreds of nanometers long. To address this issue, McIntyre has proposed in 1999 [12] an improved model to account for history-dependent impact ionization in the calculation of avalanche triggering probability, leading to a new set of non-linear integral equations that can be solved by an iterative procedure. This improved approach has also been extensively used in the field of SPAD modeling [13] [14] [15].

Despite their 1D nature, both local and non-local models are relatively time consuming, making the use of these models difficult for compact model applications. As recognized by McIntyre himself "It was found that the convergence is fairly slow close to the breakdown voltage but rapid well above the breakdown voltage" [12]. This work investigates and proposes solutions to accelerate the calculation of both models.

The paper is organized as follow. In section 2, the main models and solution algorithms are summarized. Results and discussion are reported in section 3.

II. NUMERICAL SOLUTION OF MCINTYRE LOCAL AND NON-LOCAL EQUATIONS

A. Hypothesis and definitions

For the sake of simplicity, the device was modeled as a simple NP junction with constant doping profile throughout this paper. The electrical field and potential were calculated using the total depletion approximation. We believe that a more rigorous solution of the Poisson equation would have only increased the computational burden, without affecting the overall conclusions.

The space charge layer extends from $x = 0$ to $x = W$, and consequently generated electrons are moving from W to 0 (negative x direction) and holes from 0 to W (positive x direction). Following Oldham [16], we define $P_e(x)$ (resp. $P_h(x)$) as the probability that an electron (resp. a hole), generated in x by any mechanism (light, trap, band to band or impact ionization), will initiate an avalanche (i.e. an infinity of successive impact ionization events). The probability that an electron (resp. a hole), generated in x, will not initiate an avalanche is referred as $P_{ne}(x)$ (resp. $P_{nh}(x)$), and is simply given as $P_{ne}(x) = 1 - P_e(x)$. The joint probability $P_p(x)$ that an electron or a hole generated in x will initiate an avalanche is thus given by:

$$P_p(x) = 1 - \big(1 - P_e(x)\big)\big(1 - P_h(x)\big) \qquad (1)$$

978-1-6654-8498-5/22 $31.00 © 2022 IEEE

B. McIntyre local model and its closed-form solution

According to [4] and [16], $P_e(x)$ and $P_h(x)$ are solution of the coupled non-linear differential equations given by:

$$\frac{dP_e}{dx} = \alpha_e(x)P_p(x)[1 - P_e(x)] \quad (2)$$

$$\frac{dP_h}{dx} = -\alpha_h(x)P_p(x)[1 - P_h(x)] \quad (3)$$

Where α_e (resp. α_h) are the electron (resp. hole) impact ionization coefficient. In this paper, the Reggiani's models for impact ionization for Silicon were used [17], [18].

If neither an electron generated at the entrance of the N region nor a hole at the entrance of the P region could initiate an avalanche, the boundary conditions are thus given by:

$$P_e(0) = 0 \quad (4)$$

$$P_h(W) = 0 \quad (5)$$

The conventional approach for solving these equations is detailed in [4] and leads to the following transcendental equation where $P_h(0)$ is the unknown variable:

$$1 = (1 - P_h(0)) \exp\left[\int_0^W \alpha_h(u) P_p(u, P_h(0)) \, du\right] \quad (6)$$

This equation can only be solved numerically, which may be computationally heavy. In the framework of the total depletion approximation, we use the Maserjian's procedure [19] to calculate analytically the first integral, but this simplifying method cannot be used in the general case, where $E(x)$ does not have an analytic expression. Moreover, for applied voltage close to the breakdown voltage, such numerical solution requires accuracy as $P_h(0)$ is close to zero, leading to floating point issues.

Once $P_h(0)$ is found, other parameters such as $P_e(W)$ and $P_e(x)$ can be easily deduced from the integration of equations (2) and (3) (see reference [4] for details).

C. An original iterative method to solve McIntyre local model:

The closed-form solution of McIntyre's local model remains numerically challenging, because of the root search step in Eq. (6). Inspired by the iterative approach of the McIntyre non-local model (see in the following for details), we suggest adopting a similar method to solve numerically the coupled equations (2) et (3). Using boundary conditions (4) and (5), these equations can be rearranged in the following integral form:

$$P_e(x) = F(P_e, P_h, x) = \int_0^x \alpha_e(u) P_p(u)[1 - P_e(u)] \, du \quad (7)$$

$$P_h(x) = G(P_e, P_h, x) = \int_x^W \alpha_h(u) P_p(u)[1 - P_h(u)] \, du \quad (8)$$

Such a set of equations can be solved by iteration. The first direct procedure (referred as "basic iteration") consists in starting from a guess value $P_{e,0}(x)$, $P_{h,0}(x)$ and deducing solutions by several iterative steps according to the recurrence relations:

$$P_{e,n+1}(x) = F(P_{e,n}, P_{h,n}, x) \quad (9)$$

$$P_{h,n+1}(x) = G(P_{e,n}, P_{h,n}, x) \quad (10)$$

However, this brute force method is generally diverging for the local model. To avoid this problem, two other approaches have been also investigated. The first one consists in re-injecting only a small fraction β of the solution (over-relaxation method) in the calculation of the n+1 step:

$$P_{e,n+1}(x) = \beta F(P_{e,n}, P_{h,n}, x) + (1 - \beta)P_{e,n}(u) \quad (11)$$

$$P_{h,n+1}(x) = \beta G(P_{e,n}, P_{h,n}, x) + (1 - \beta)P_{h,n}(u) \quad (12)$$

The second alternative method follows the principles of the well-known Newton-Raphson procedure. A first order power expansion is used:

$$P_{e,n+1} = F(P_{e,n}, P_{h,n}) + \frac{\partial F}{\partial P_e}(P_{e,n+1} - P_{e,n})$$
$$+ \frac{\partial F}{\partial P_h}(P_{h,n+1} - P_{h,n}) \quad (13)$$

$$P_{h,n+1} = G(P_{e,n}, P_{h,n}) + \frac{\partial G}{\partial P_e}(P_{e,n+1} - P_{e,n})$$
$$+ \frac{\partial G}{\partial P_h}(P_{h,n+1} - P_{h,n}) \quad (14)$$

Interestingly, each term of the Jacobian matrix can be analytically derived, leading to:

$$A = \frac{\partial F}{\partial P_e} = \int_0^x \left[\alpha_e \left(1 - 2P_{p,n}(u)\right)\right] du \quad (15)$$

$$B = \frac{\partial F}{\partial P_h} = \int_0^x \left[\alpha_e \left(1 - P_{e,n}(u)\right)^2\right] du \quad (16)$$

$$C = \frac{\partial G}{\partial P_e} = \int_x^W \left(\alpha_h(u)\left(1 - P_h(u)\right)^2\right) du \quad (17)$$

$$D = \frac{\partial G}{\partial P_h} = \int_x^W \left(\alpha_h(u)\left(1 - 2P_{p,n}(u)\right)\right) du \quad (18)$$

Equations (13) and (14) can be re-arranged by inverting the Jacobian matrix, leading to:

$$P_{e,n+1} = \frac{(D+1)F_n - BG_n + (A(D+1) - BC)P_{e,n} + BP_{h,n}}{A + D + AD - BC + 1} \quad (19)$$

$$P_{h,n+1} = \frac{(A+1)G_n - C F_n + CP_{e,n} + ((A+1)D - BC)P_{h,n}}{A + D + AD - BC + 1} \quad (20)$$

Results obtained using these three approaches to solve local McIntyre equations will be discussed in the "results and discussions" section.

D. Iterative solution of McIntyre non-local model:

McIntyre proposed a model to account for history dependent ionization probability in the calculations of avalanche triggering probability [12]. In this approach, the probability that an electron (resp. a hole), generated in x, will not initiate an avalanche is referred as $P_{ne}(x)$ (resp. $P_{nh}(x)$) is given by:

$$P_{ne}(x) = F(P_{ne}, P_{nh}, x)$$
$$= P_{se}(x|0) + \int_0^x p_e(x|u) P_{ne}(u)^2 P_{nh}(u) \, du \quad (21)$$

$$P_{nh}(x) = G(P_{ne}, P_{nh}, x)$$
$$= P_{sh}(x|W) + \int_x^W p_h(x|u) P_{nh}(u)^2 P_{ne}(u) \, du \quad (21)$$

978-1-6654-8498-5/22 $31.00 © 2022 IEEE

$p_e(x'|x)$ is the probability of first ionization by an electron and is given by:

$$p_e(x'|x) = \alpha(x'|x) \, P_{se}(x'|x) \qquad (22)$$

Where $P_{se}(x'|x)$ is the survival probability, i.e., the probability that the electron will travel from x' to x without undergoing an ionizing collision:

$$P_{se}(x'|x) = \exp\left(-\int_x^{x'} \alpha(x'|u)du\right) \qquad (23)$$

Several approaches have been proposed to deduce the non-local ionizing coefficient $\alpha(x'|x)$ from the local one, based on the dead space model [20] or the introduction of effective fields [12],[21]. Here we used the effective field definition proposed by C. Nichetti et al. [21], relying on the formalism of energy balance equations. The mean free path parameters λ_e for electrons and λ_h for holes have been taken constant and equal to 20 nm for simplicity. It has been numerically checked that local and non-local models give identical solutions when λ_e and λ_h tends to zero.

As in the previous section for the local model, three approaches (basic iteration, over-relaxation, and Newton Raphson) have been implemented to achieve the convergence of the iterative procedure. Regarding the Newton Raphson method, the same equations (19) (20) can be used, replacing the Jacobian matrix coefficients A, B, C, D by the appropriate expressions.

III. RESULTS AND DISCUSSION

In this section, the convergence of both the local and non-local iterative models are examined. The error between two consecutive steps is calculated with the r.m.s. error.

Fig. 1 shows the convergence of the three methods implemented for the iterative *local* equations for two different guess value. A moderately doped abrupt PN junction ($N_d = 7.\,10^{17}$ cm^{-3}, $N_a = 3\,10^{17}$ cm^{-3}) was considered. The first guess value is given by:

$$P_{e,0}(x) = \frac{x}{W} \quad \text{and} \quad P_{h,0}(x) = 1 - \frac{x}{W}$$

The second guess value is the exact solution obtained by direct solution of the local McIntyre equations (see section II.B).

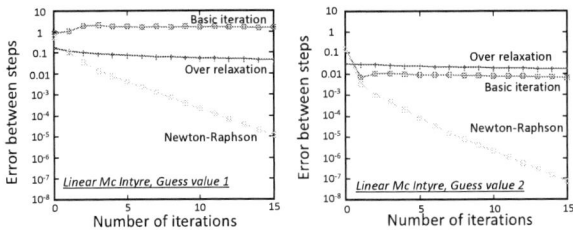

Fig. 1. Error between successive step versus number of iterations solving the local McIntyre equations, using a linear approximation as a guess value (Guess value 1, left figure) or the closed form solution (Guess value 2, right figure). Doping levels are $N_d = 7.\,10^{17}$ cm^{-3}, $N_a = 3\,10^{17}$ cm^{-3}, the excess voltage is 5 V, $\lambda_e = \lambda_h = 20$ nm.

The overall error at the initial step is lower in the second case (exact guess value) than in the first (linear approximation). The basic scheme is diverging in the first case, while the over-relaxation scheme ($\beta = 0.2$) is slowly

converging. The Newton Raphson scheme however shows a fast and smooth error decrease versus number of iterations, reaching an error of 10^{-5} after only 15 iterations in the first case, and 8 in the second. Similar results were obtained for all the doping and excess voltage considered, showing that the Newton-Raphson implementation is an extremely efficient way of solving the McIntyre local equations.

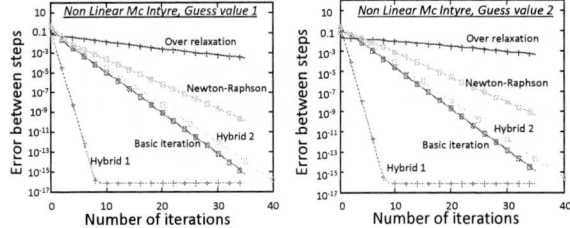

Fig. 2. Error between successive step versus number of iterations solving the non-local McIntyre equations, using a linear approximation as a guess value (Guess value 1, left figure) or the closed form solution (Guess value 2, right figure).

Conclusions significantly differ when *non-local* equations are considered (see Fig. 2). In this case, the basic iteration procedure is the most efficient scheme independently of the guess value. The same guess values than in the previous case were used, except that the local direct solution is no longer the exact solution of the non-local equations but only a rough approximation. Contrary to the local case, the change between each step of the non-local iterations is very small, preventing divergence. In this situation, the over-relaxation and Newton-Raphson only slow down the convergence. A "hybrid" method was also investigated, which consists in applying a Newton-Raphson step every 3 steps of basic iteration.

To compare, the correct way of counting a step is to count 4 (one Newton-Raphson step and 3 basic iteration steps) for one hybrid step: this counting method has been used for "Hybrid 2" in Fig. 2, while an incorrect counting (which consists in counting as one step for one Newton-Raphson step and 3 basic iterations steps) would artificially give the impression that the Hybrid method is more efficient ("Hybrid 1" in Fig. 2). Using the proper way of counting iterations ("Hybrid 2"), the hybrid method does not show any convergence improvement compared to the basic method.

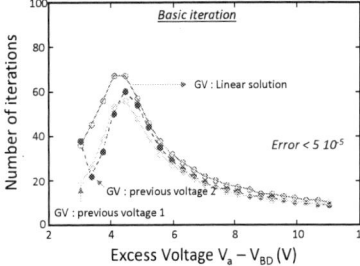

Fig. 3. Number of iterations required to reach an error lower than 5. 10-5 between two successive steps using the non-local model versus excess voltage : using the linear solution as a guess value (GV: Linear solution), or using the solution at the previous voltage as guess value, starting from the maximum voltage (GV: previous voltage 1) or starting from the minimum voltage, (GV: previous voltage 2).

As in most application, avalanche triggering probabilities are calculated for several applied voltages, we also investigate

978-1-6654-8498-5/22 $31.00 © 2022 IEEE

the possibility to use the result obtained at the previous applied voltage as guess value for the next one, instead of the solution of the linear equations, as usually done. It turns out that this approach is more efficient, as seen in Fig. 3. Moreover, in this case, it is recommended to start the simulation by the highest applied voltage, and not the lowest, as low voltage simulations, whatever the guest value, are much more time consuming.

IV. CONCLUSIONS

Several original findings about the numerical solution of the local and non-local McIntyre avalanche triggering probability in SPAD device have been discussed in this paper. First, it appears that the local equations can be very efficiently solved by a Newton-Raphson algorithm. The Jacobian matrix was calculated and inverted analytically, accelerating calculations. We believe that this approach is easier to implement and more efficient than the closed form solution derived by McIntyre.

Regarding the non-local equations, contrary to the local case, the basic implementation of the iterative solutions was found the most efficient approach compared to over-relaxation, Newton-Raphson or even hybrid method. Moreover, this implementation can be accelerated using the previous excess voltage as guess value (and not the local solution as it is usually done). In this case, it is recommended to start with the maximum voltage (where the local approximation is a good guess value) and not with the minimum voltage.

Finally, the local equations have been found to be a good approximation of the non-local one at high excess voltage. Further investigations are required to confirm if the local approximation is sufficiently accurate for all practical case of application in real device design.

ACKNOWLEDGMENT

The authors would like to thank A. Pilotto and P. Palestri from university of Udine (Italy) for helpful discussions. This work has been supported by the IPCEI Nano 2022 program and by the French ANR through the project GeSPAD under Grant ANR-20-CE24-0004.

REFERENCES

[1] M. Ghioni, A. Gulinatti, I. Rech, F. Zappa, and S. Cova, "Progress in silicon single-photon avalanche diodes," IEEE J. Sel. Top. Quantum Electron., vol. 13, no. 4, pp. 852–862, 2007, doi: 10.1109/JSTQE.2007.902088.

[2] G. S. Buller and R. J. Collins, "Single-photon generation and detection," Meas. Sci. Technol., vol. 21, no. 1, 2010, doi: 10.1088/0957-0233/21/1/012002.

[3] W. G. Oldham, R. R. Samuelson, and P. Antognetti, "Triggering Phenomena in Avalanche Diodes," IEEE Trans. Electron Devices, vol. 19, no. 9, pp. 1056–1060, 1972, doi: 10.1109/T-ED.1972.17544.

[4] R. J. McIntyre, "On the Avalanche Initiation Probability of Avalanche Diodes Above the Breakdown Voltage," IEEE Trans. Electron Devices, vol. 20, no. 7, pp. 637–641, 1973, doi: 10.1109/T-ED.1973.17715.

[5] T. C. De Albuquerque et al., "An analytical solution for McIntyre's model of avalanche triggering probability for SPAD compact modeling and performance exploration," Semicond. Sci. Technol., vol. 36, no. 8, 2021, doi: 10.1088/1361-6641/ac00d0.

[6] M. M. Vignetti, F. Calmon, P. Lesieur, and A. Savoy-Navarro, "Simulation study of a novel 3D SPAD pixel in an advanced FD-SOI technology," Solid. State. Electron., vol. 128, pp. 163–171, 2017, doi: 10.1016/j.sse.2016.10.014.

[7] A. Gulinatti, I. Rech, M. Assanelli, M. Ghioni, and S. Cova, "A physically based model for evaluating the photon detection efficiency and the temporal response of SPAD detectors," J. Mod. Opt., vol. 58, no. 3–4, pp. 210–224, 2011, doi: 10.1080/09500340.2010.536590.

[8] A. Panglosse, P. Martin-Gonthier, O. Marcelot, C. Virmontois, O. Saint-Pé, and P. Magnan, "Modeling, simulation methods and characterization of photon detection probability in CMOS-SPAD," Sensors, vol. 21, no. 17, pp. 1–17, 2021, doi: 10.3390/s21175860.

[9] Y. Okuto and C. R. Crowell, "Energy-conservation considerations in the characterization of impact ionization in semiconductors," Phys. Rev. B, vol. 6, no. 8, pp. 3076–3081, 1972, doi: 10.1103/PhysRevB.6.3076.

[10] Y. Okuto and C. R. Crowell, "Threshold energy effect on avalanche breakdown voltage in semiconductor junctions," Solid State Electron., vol. 18, no. 2, pp. 161–168, 1975, doi: 10.1016/0038-1101(75)90099-4.

[11] M. M. Hayat, W. L. Sargeant, and B. E. A. Saleh, "Effect of Dead Space on Gain and Noise in Si and GaAs Avalanche Photodiodes," IEEE J. Quantum Electron., vol. 28, no. 5, pp. 1360–1365, 1992, doi: 10.1109/3.135278.

[12] R. J. McIntyre, "A new look at impact Ionization-part I: A theory of gain, noise, breakdown probability, and frequency response," IEEE Trans. Electron Devices, vol. 46, no. 8, pp. 1623–1631, 1999, doi: 10.1109/16.777150.

[13] G. J. Rees and J. P. R. David, "Nonlocal impact ionization and avalanche multiplication," J. Phys. D. Appl. Phys., vol. 43, no. 24, 2010, doi: 10.1088/0022-3727/43/24/243001.

[14] J. S. Ng, C. H. Tan, and J. P. R. David, "A comparison of avalanche breakdown probabilities in semiconductor materials," J. Mod. Opt., vol. 51–9, no. 10, pp. 1315–1321, 2004, doi: 10.1080/09500340408235274.

[15] M. M. Hayat et al., "Breakdown probabilities for thin heterostructure avalanche photodiodes," IEEE J. Quantum Electron., vol. 39, no. 1, pp. 179–185, 2003, doi: 10.1109/JQE.2002.806217.

[16] W. G. Oldham, P. Antognetti, and R. R. Samuelson, "New method for breakdown voltage determination in p-n junctions," Appl. Phys. Lett., vol. 19, no. 11, pp. 466–467, 1971, doi: 10.1063/1.1653774.

[17] S. Reggiani, E. Gnani, A. Gnudi, M. Rudan, S. Member, and G. Baccarani, "Low-Field Electron Mobility Model for Ultrathin-Body SOI and Double-Gate MOSFETs With Extremely Small Silicon Thicknesses," vol. 54, no. 9, pp. 2204–2212, 2007.

[18] M. Rudan, R. Katilius, S. Reggiani, E. Gnani, and G. Baccarani, "Impact-ionization coefficient in silicon at high fields - A parametric approach," J. Comput. Electron., vol. 7, no. 3, pp. 151–154, 2008, doi: 10.1007/s10825-008-0184-8.

[19] J. Maserjian, "Determination of avalanche breakdown in pn junctions [4]," J. Appl. Phys., vol. 30, no. 10, pp. 1613–1614, 1959, doi: 10.1063/1.1735012.

[20] M. M. Hayat and B. E. A. Saleh, "Statistical Properties of the Impulse Response Function of Double-Carrier Multiplication Avalanche Photodiodes Including the Effect of Dead Space," J. Light. Technol., vol. 10, no. 10, pp. 1415–1425, 1992, doi: 10.1109/50.166785.

[21] C. Nichetti et al., "An improved nonlocal history-dependent model for gain and noise in avalanche photodiodes based on energy balance equation," IEEE Trans. Electron Devices, vol. 65, no. 5, pp. 1823–1829, 2018, doi: 10.1109/TED.2018.281

A self-sustaining Single Photon Avalanche Diode Model

S. Rink*, V. Quenette, J.R. Manouvrier
A. Juge, G. Gouget, D. Rideau, R.A.
Bianchi, D. Golanski, B. Mamdy
STMicroelectronics
Crolles, France
sven.rink@st.com

J.B. Kammerer, W. Uhring, C.
Lallement,
*ICube, CNRS & Université de
Strasbourg
Strasbourg, France

S. Pellegrini, M. Agnew, B. Rae
STMicroelectronics
Edinburgh, Scotland

Abstract—**This paper is describing structure of a Single Photon Avalanche Diode (SPAD) compact model. Using a loop architecture to describe the impact ionization phenomenon, this approach yields a closer to physics SPAD model using minimal fitting adjustments and an innovative approach, hence differentiating itself from the majority of available SPAD models.**

Keywords—SPAD, Verilog A, compact modelling, impact ionization, loop architecture, avalanche, breakdown, imager, CMOS

I. INTRODUCTION

The avalanche mode of photodiodes is well known and enable key applications to exist, such as laser ranging, LiDAR, 3D object detection etc. Consequently, there is an increasing demand for more precise device models to account for the increase in resolution and speed of improving applications. SPAD diode modelling has been growing in the last years. Most of the existing SPAD models rely on either a behavioural modelling approach or an electrical spice macro-model made of current sources, switches and passive devices [1][2][3]. Those models are calibrated with a set of parameters and mathematical corrections allowing to fit the measured characteristics. We focused our work on the development of a physics-centric modelling approach of the avalanche build-up mechanism, similar to Sadigov [4] in his iterative, and Inoue [5] in his time derivative approach. The key strengths of the compact model proposed in this paper are both a closed loop description of the avalanche phenomenon, calibrated on I(V) measurements and junction capacitance extraction on C(V) measurements. We propose the outline and basic principle of a VerilogA model description which, once it is coupled to this extraction methodology, is validated through a first comparison with experimental data.

II. IMPACT IONIZATION

A. Physical phenomenon

The model essentially describes the properties and mechanisms of the physical phenomenon of impact ionization [6], similar to the iterative model of Sadigov [4]. It describes the base mechanism of carrier multiplication inside the space charge region (SCR):

- An electron/hole pair is generated in the SCR of the SPAD through photoelectric effect.
- The carriers are then accelerated by the electric field present in the SCR, hence acquiring more and more energy.

- Once the electron/hole impacts the surrounding Si atoms with energy higher or equal to the ionization energy, a valence band electron is ejected in the conduction band leaving a hole in the valence band. These carriers are then accelerated and can in turn generate further electron/hole pairs on further impacts, resulting in an exponential increase of the number of carriers.

B. Ionization coefficients

The ionization coefficients $\alpha_{n,p}$, homogenous to the inverse of a distance (dead space = $\alpha_{n,p}^{-1}$), depend on the local field F_m and are defined by the following equations, [7]:

$$\alpha_{n,p} = A_{n,p} \cdot \exp(-B_{n,p}/F_m) \quad (1)$$

$$B_{n,p} = C_{n,p} + D_{n,p}T \quad (2)$$

Where $A_n = 4.43 \cdot 10^5$ cm^{-1}, $A_p = 1.13 \cdot 10^6$ cm^{-1}, $C_n = 9.66 \cdot 10^5$ V$\cdot$cm^{-1}, $C_p = 1.71 \cdot 10^6$ V$\cdot$cm^{-1}, $D_n = 4.99 \cdot 10^2$ V$\cdot$cm$^{-1} \cdot$K^{-1}, $D_p = 1.09 \cdot 10^3$ V$\cdot$cm$^{-1} \cdot$K^{-1} and T (in °K) the absolute temperature and F_m is the electric field inside the SCR [7].

III. LOOP MODEL

A. Model Structure

Since the underlying physical phenomenon is forming a loop, the compact model is also based on a loop structure as shown on Fig.1. It consists of two paths representing the electrons (green) and holes flows (red). This internal structure is only connected to the main model through 2 output nodes representing both endpoints of the SCR.

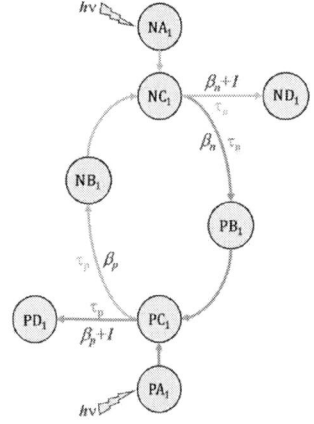

Fig. 1. Loop Model schematic principle

978-1-6654-8498-5/22 $31.00 © 2022 IEEE

NA and PA are the initial electron-hole pairs injection nodes used to model the photoelectric effect. It is worth noting that these entry points can also be used to inject charges induced by other phenomena such as the release of trapped carriers (Afterpulsing) or thermally induced electron-hole pair generation.

NC and PC are summing nodes. The purpose of those nodes is to add together, at each iteration, the currents coming from NA/NB or PA/PB.

NB and PB are multiplication nodes. Their purpose, as implied in their denomination, is to multiply the current of the previous iteration, using the extended ionization coefficients $\beta_{n,p}$ (see equation (3) in the next subsection).

ND and PD are the output nodes. The purpose of those nodes is to output the current of the running iteration and to inject it into the output terminals of the model.

B. Extended ionization coefficients

To better understand those extended coefficients, the definition of the impact ionization must be recalled:

After an electron or a hole has been accelerated along an average distance of $1/\alpha_{n,p}$ (Fig.2.), also called dead space, it collides and the excess energy accumulated generates an additional electron-hole pair. Consecutive collisions of both electrons and holes can trigger an avalanche breakdown of the junction.

Therefore, if one carrier generates an electron/hole pair after this distance $1/\alpha_{n,p}$, then, if a distance of W_{SCR} is considered, the multiplication of carriers will be $\alpha_{n,p}$ times stronger as in W_{SCR} as there is space for $W_{SCR}/(1/\alpha_{n,p})$ successive multiplications to happen. This coefficient that represents the number of successive multiplications that happen in W_{SCR} is called $\beta_{n,p}$ and its expression is given by :

$$\beta_{n,p} = \alpha_{n,p} \cdot W_{SCR} \quad (3)$$

A similar approach was used in a model developed by Oussaiti et al., using effective multiplication widths to describe the avalanche breakdown [8][9].

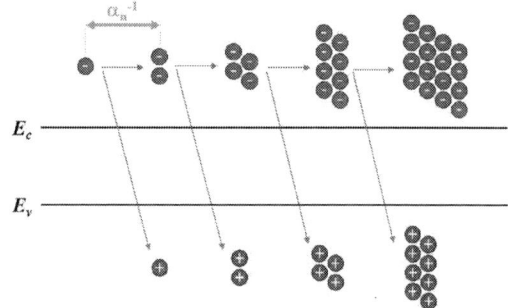

Fig. 2. Impact ionization mechanism, starting with electron generation

C. Transit time

The transit time $\tau_{n,p}$ across W_{SCR} corresponds to the time taken by carriers to pass through the SCR with a velocity equal to their saturation velocity v_{n,p_sat} :

$$\tau_{n,p} = W_{SCR}/v_{n,p\_sat} \quad (4)$$

D. Loop description and nodes equations

The description of the loop model is divided into distinct parts. The first part is the generation of carriers by photo-

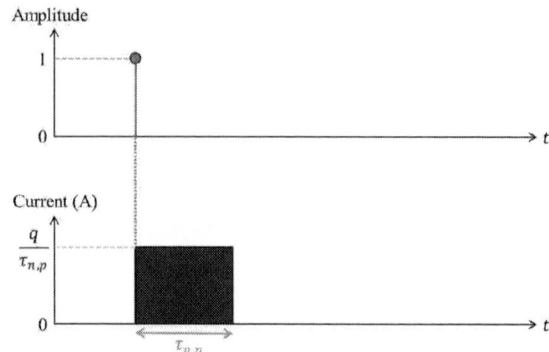

Fig. 3. NA/PA node signal generation explanation, trigger signal and time correlation

electric effect and the triggering of the avalanche. When a trapped electron is released or when a photon creates an electron/hole pair, a trigger signal is generated and detected by NA (electron injection) and potentially by PA (hole injection), therefore generating either an electron or an electron/hole pair. This event, shown on Fig. 3, produces a current step that has an amplitude of $q/\tau_{n,p}$, and a duration of $\tau_{n,p}$ so that the total injected charge corresponds to the elementary charge q.

The next part of this loop is the summing node (NC and PC). These nodes collect the currents generated in NA or PA. As the loop is repeated, in addition to possibly collecting photogenerated carriers, those nodes are also collecting the contribution of the NB and PB nodes:

$$NC = NA + NB \quad (5)$$

$$PC = PA + PB \quad (6)$$

The most important part of the loop resides in the multiplication of the current collected in the NC and PC nodes. This process is divided into two distinct contributions which are the multiplication nodes NB and PB, and the output nodes ND and PD.

- For NB and PB, the NC and PC nodes are multiplied by $\beta_{n,p}$ thus leading to equations (7) and (8):

$$NB = \beta_p \cdot PC \quad (7)$$

$$PB = \beta_n \cdot NC \quad (8)$$

- As NB and PB, the ND and PD nodes are also considered as multiplication nodes. Since the currents induced by the multiplication process are added to the currents already present, the equations associated with these nodes are slightly different:

$$ND = (\beta_n + 1) \cdot NC \quad (9)$$

$$PD = (\beta_p + 1) \cdot PC \quad (10)$$

Fig. 4 shows a set of different successive iterations of the impact ionization mechanism as the model executes them.

It is worth noting that if the multiplication factor is low or negligible, i.e. when the reverse voltage is lower than the breakdown voltage, this compact model will behave as a conventional photodiode. The more, when the photodiode is biased just below the breakdown voltage, the multiplication of the photocurrent is also correctly described by our compact model. (See Fig. 7. in IV.3.).

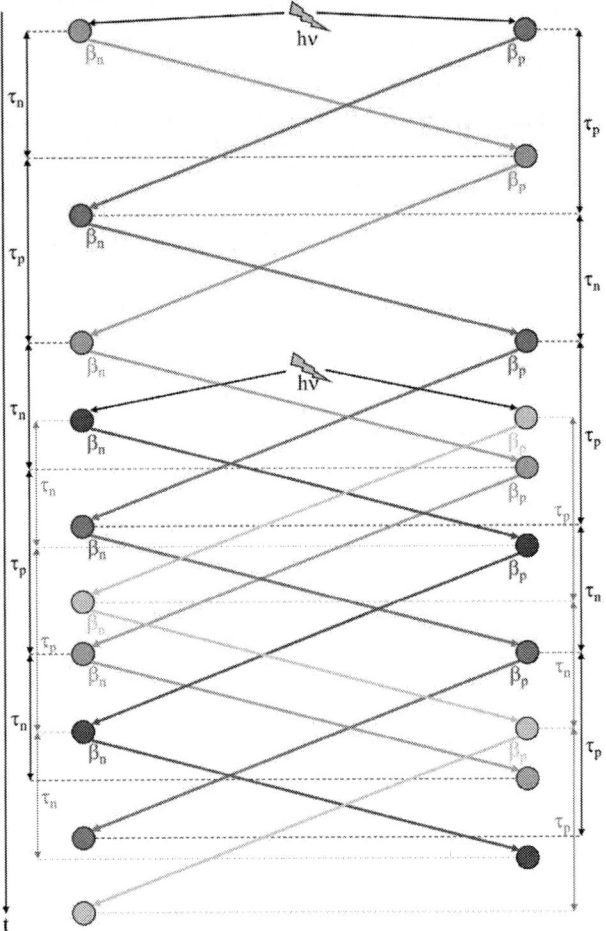

Fig. 4. Loop model avalanche breakdown principle over simulation time, decomposition into iterations. Multiple photon triggers are considered to represent accurately the device's behavior when consecutive photons are detected.

IV. PARAMETER EXTRACTION

The model uses a set of different parameters and variables from different sources, as shown on Fig. 5.

The necessary parameters of the loop model are obtained through both measurements and literature which in turn are

used to calculate the electric field F_m, necessary for the ionization coefficients $\alpha_{n,p}$. First, DC I(V) measurements are made on silicon dedicated test structure to extract the breakdown voltage V_{br}, which is then used in conjunction with AC C(V) measurements to calibrate F_m. Another set of physical parameters, mainly the impact ionization coefficients $\alpha_{n/p}$ and the carrier saturation velocities $v_{sat\_e/h}$, are given in the literature [7]. Finally, DC I(V) simulations finalize the set of parameters by extracting the adjustment parameter used in both the extended ionization coefficients $\beta_{n,p}$ and the transit times $\tau_{n,p}$ which is detailed in the following paragraph.

A. Adjustment parameter

The loop model requires an optimization parameter: W_{SCR} was introduced to account for effective avalanche multiplication width being different from the SCR width. This parameter was extracted from I(V) measurements to align the simulation with the measured V_{br}.

Consequently, the loop gain ($LoopGain = \beta_n \cdot \beta_p$) is equal to 1 when the applied voltage is at the breakdown voltage of the diode. When $LoopGain$ is above 1, avalanche is building up by self-sustaining carrier multiplication. If the loop gain drops below 1, i.e., when the voltage across the SCR drops below the breakdown voltage, then the avalanche current is decreasing, leading to the quenching of the avalanche process. Since the capacitance of the SCR is discharged by the avalanche current, the voltage across the SCR can automatically drop below the breakdown voltage depending on the electrical environment of the SPAD (quenching resistor, active quenching circuit, etc.).

V. SIMULATION RESULTS AND MEASUREMENTS

In this section, the simulation results of the loop model are compared to experimental results. Furthermore, transient simulation results are shown to support the fact that this model is fitted for implementation in a complete and realistic test circuit including quenching and readout circuitry. All simulation results obtained by the loop model are obtained without any further fitting correction, only using the physical equations of the impact ionization mechanism.

The testbench circuit, as seen on Fig. 6, was used for both simulations and measurements in the same conditions for accurate comparison. It is composed of a passive quenching circuit with a coupling capacitor and a digital readout circuit. This readout circuit produces digital pulses at its output when an avalanche occurs in the SPAD.

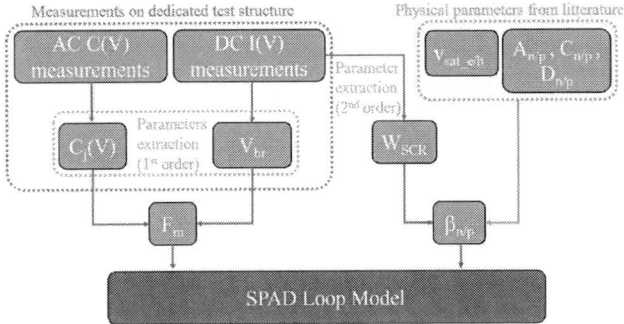

Fig. 5. Parameter extraction routine

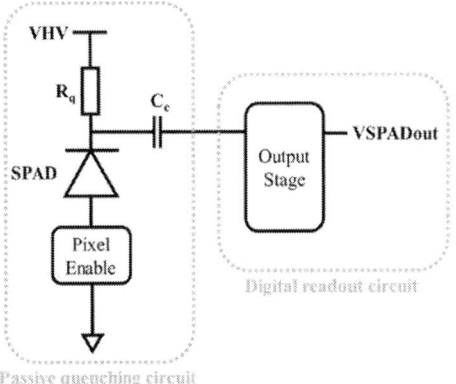

Fig. 6. Test bench circuit used for simulations and measurements. VHV is the biasing voltage, VSPADout the digital output

A. Transient response

To support the fact that the compact model is functional, two main transient output figures have been extracted: the output current flowing through the device, and the associated cathode voltage swing. Fig. 7. shows the behaviour of both the current swing at the output and the cathode voltage swing of the device. The transient response is triggered at 10ns (red arrows) and the model has been simulated at 60°C and an excess bias voltage of 5V. The SPAD avalanche current pulse has a rise and fall time of 50ps whereas the corresponding cathode voltage swing has similar rise time but considerably longer fall time, consistent with expectations given the dependency of the fall time on the quenching circuitry.

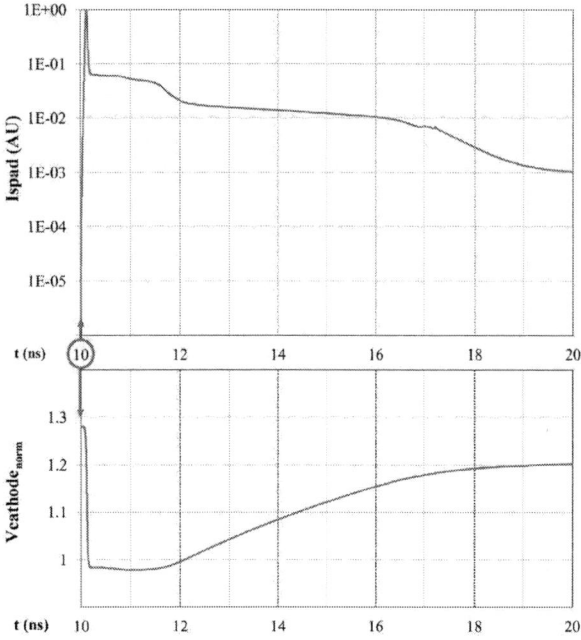

Fig. 7. Simulated transient response, SPAD avalanche current response (top), and the resulting normalized cathode voltage swing Vcathode$_{norm}$ = Vcathode/VHV0 (bottom)

B. Charge per pulse measurements

The figure of merit used to compare the model's simulation results to measurements is Charge Per Pulse (CPP). The definition of CPP is the amount of charge that is contained in one current pulse generated by the SPAD diode.

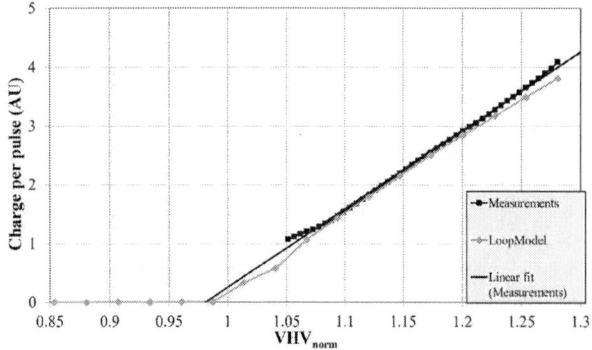

Fig. 8. CPP measurements compared to model simulation results at 60°C, V$_{norm}$ being the normalized biasing voltage VHV/VHV0, with VHV the biasing voltage and VHV0 the avalanche breakdown detection voltage

Fig. 8 is a comparison between the loop model and measurements of the device in the test bench environment. VHV0, used to normalize the biasing voltage, is defined as the voltage at which an avalanche breakdown of the SPAD is detected and a signal at the digital output VSPADout is visible. The model simulations provide a good fit to the experimental curve over most of the linear response, proving the ability of the model to predict power consumption in a complex circuit. When the excess voltage V$_{ex}$ is low, then, due to the measurement methods used, the measurement points in this region are not exploitable and are therefore not represented.

VI. CONCLUSION

The model developed and presented in this paper uses the base mechanism of impact ionization to accurately describe SPAD behaviour, The first set of simulations are yielding satisfying results, such as realistic transient rise and fall times as well as coherent avalanche current and cathode voltage swings. A confrontation of the model simulation results to measurements, made on silicon test circuits, shows good results for CPP, accurately predicting in circuit power consumption of the device. This confirms the model is fitted for implementation in a complete and realistic test circuit, including quenching and readout circuitry. The loop architecture used in this model description is a new and different approach on SPAD modelling. This loop-based approach, offers a broad set of possibilities in terms of capabilities such as the integration of photon absorption depth, impacting the avalanche breakdown characteristics in terms of avalanche duration, peak avalanche current, timing jitter etc.

REFERENCES

[1] H. Yang, X. Jin, X. Zhou, C. Chen, and J. Luo, "Verilog-A modeling of SPAD for circuit simulations," *Proc SPIE*, vol. 8908, pp. 12-, 2013, doi: 10.1117/12.2033037.

[2] J. López-Martínez, I. Vornicu, R. Carmona-Galan, and A. Rodriguez-Vazquez, "An Experimentally-Validated Verilog-A SPAD Model Extracted from TCAD Simulation," 2018, pp. 137–140. doi: 10.1109/ICECS.2018.8617962.

[3] F. Villa, Y. Zou, A. D. Mora, A. Tosi, and F. Zappa, "SPICE Electrical Models and Simulations of Silicon Photomultipliers," *IEEE Transactions on Nuclear Science*, vol. 62, no. 5, pp. 1950–1960, 2015, doi: 10.1109/TNS.2015.2477716.

[4] A. Sadigov, "An Iterative Model of Performance of Micropixel Avalanche Photodiodes," *IJARPS*, vol. 3, no. 2, Feb. 2016.

[5] A. Inoue, T. Okino, S. Koyama, and Y. Hirose, "Modeling and Analysis of Capacitive Relaxation Quenching in a Single Photon Avalanche Diode (SPAD) Applied to a CMOS Image Sensor," *Sensors (Basel, Switzerland)*, vol. 20, 2020.

[6] M. S. Hasan, S. Amer, S. Islam, N. Mcfarlane, and G. Rose, "Chapter Modeling Emerging Semiconductor Devices for Circuit Simulation," 2019.

[7] D. J. Massey, J. P. R. David, and G. J. Rees, "Temperature Dependence of Impact Ionization in Submicrometer Silicon Devices," *IEEE Transactions on Electron Devices*, vol. 53, no. 9, pp. 2328–2334, 2006, doi: 10.1109/TED.2006.881010.

[8] Y. oussaiti *et al.*, "Verilog-A model for avalanche dynamics and quenching in Single-Photon Avalanche Diodes," in *2020 International Conference on Simulation of Semiconductor Processes and Devices (SISPAD)*, 2020, pp. 145–148. doi: 10.23919/SISPAD49475.2020.9241648.

[9] G. J. Rees and J. P. R. David, "Nonlocal impact ionization and avalanche multiplication," *Journal of Physics D: Applied Physics*, vol. 43, no. 24, p. 243001, Jun. 2010, doi: 10.1088/0022-3727/43/24/243001.

Thermal Sensing Performances of Thin-Film Lateral PiN Diodes at 80 K and 300 K

Adrien Fournol, Jérémy Blond, Abdelkader Aliane, Hacile Kaya, Jérôme Meilhan and Laurent Dussopt
Univ. Grenoble Alpes, F-38000 Grenoble, France CEA, DRT, LETI, F-38054
Email: {adrien.fournol,jeremy.blond,abdelkader.aliane,hacile.kaya,jerome.meilhan,laurent.dussopt}@cea.fr

Abstract—The thermometric behavior of lateral 50-nm Silicon-On-Insulator PiN diodes are studied from ambient to liquid-nitrogen temperature. Prototypes were manufactured and their experimental performances are compared to theoretical models. Thanks to a temperature sensitivity reaching 25%/K and a low noise level at 80 K, a minimum thermal resolution of 0.1 mK is obtained. Such diodes represent an attractive solution for high performance thermal sensing in bolometric detectors.

Index Terms—Lateral diodes, PiN diode, SOI, thermal sensors

I. INTRODUCTION

Microbolometer detectors can be integrated into large pixel arrays commonly used in the infrared spectrum. Other fields such as Terahertz (THz) imaging being very promising for many applications such as biomedicine, non-destructive control, security or surveillance use this type of detector. Bolometric detection like other applications requires high-sensitivity thermal sensors. For example, the development of new thermometers is needed to meet the sensitivity requirements of THz passive imaging. These thermometers have to exhibit higher relative thermal sensitivity than classical semiconductor thermistors and a noise level as low as possible. Silicon diodes provide several advantages as thermal sensors over traditional semiconducting thermistors [1] [2]. In particular, they can easily be integrated in a CMOS foundry process and they turn out to be cheaper. The aim of this work is to assess the behavior of ultra-thin lateral PiN diodes from the Room-Temperature (RT) to cryogenic regime down to 80 K. A thickness as low as 50 nm is chosen with the goal of integration in a bolometer to limit the heat capacity of the suspended membrane. For this purpose, some diode prototypes have been fabricated on Silicon-On-Insulator (SOI) wafers. Their thermal sensitivity in low-injection forward regime is studied through their Temperature Coefficient of Current (TCC). Under a constant bias voltage, the diode output current is exponentially dependent upon the inverse of the temperature and provides the highest TCC value. Their Low Frequency Noise (LFN) also determines the minimal thermal resolution ΔT_{min} and is therefore also investigated. The experimental results presented hereafter demonstrate promising performances and a good agreement with theoretical models, which are used to provide insights for further optimizations.

II. DESIGN AND FABRICATION

A. Design

A lateral SOI PiN diode is composed of three regions in the active silicon layer. Two highly doped contact regions P+ and N+ are separated by the so-called intrinsic channel, which is in reality slightly doped. The P+ and N+ regions must have high dopant concentrations in order to limit the contact resistances R_c. Depending on the dopant type in the channel N or P, the PiN diode has a P+N or PN+ junction having a dynamic resistance r_d. The total series resistance R_S is the sum of R_c and the silicon resistance R_{Si} as shown in Fig. 1.a. The geometry of a rectangular PiN diode can be described by its channel length L_i, width W and thickness e_{Si}. The length of the diodes studied here range from 1 to 50 μm. The surface to volume ratio of the channel being high, the quality of the top interfaces with the buried oxide (BOX) and the passivation layer is essential.

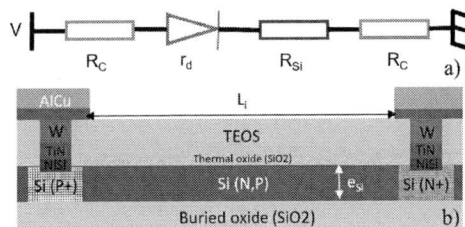

Fig. 1. Equivalent electrical circuit (a) and schematic cross-section (b) of a PiN diode.

B. Fabrication

Prototypes of PiN diodes are fabricated on 8-inch SOI wafers with an active silicon thickness of 70 nm originally doped by a 10^{15} at/cm³ boron concentration. First, the silicon layer is thinned down to 50 nm by thermal oxydation and etching. Next, each region is implanted selectively with Phosphorous (N) or Boron (P) through a 25-nm thermal SiO₂ protective layer. The implantation parameters are given in Tab. I. Then, a thermal annealing in an inert nitrogen atmosphere is performed for the dopant diffusion and activation in order to obtain uniform concentrations. After removing the protective oxide, a 10 nm passivation layer is formed by thermal oxidation followed by 300 nm of tetraethyl orthosilicate (TEOS) silicon dioxide (SiO₂). Then, the 300x300-nm² contact areas are

978-1-6654-8498-5/22 $31.00 © 2022 IEEE

opened by dry etching. NiSi contacts are formed by sputtering 10 nm of Ni followed by two silicidation annealings. Contact pads Ti/TiN/AlCu (10/40/440 nm) for electrical measurements are finally deposited. A final thermal annealing (30 min, 450°C, N_2/H_2) improves the interfaces quality as shown in Section. III. The final stack is shown in Fig. 1.b.

A silicon thickness of 46 nm was measured on prototype cross-sections. Two variants PPN and PNN, with the dopant concentrations given in the Tab. I, are fabricated. The access resistances of the contacts are of the order of 10^{-6} Ω.cm^2 thanks to high dopant concentrations in the N$^+$ and P$^+$ regions. These low values will allow to extract the thermometric performances of these diodes without a significant impact caused by the contact resistances.

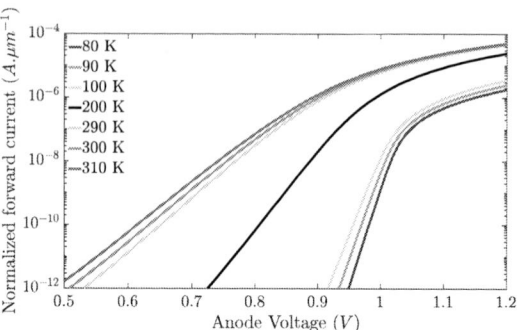

Fig. 2. Simulated I-V characteristics of a PNN diode with $L_i = 5$ μm.

TABLE I
DOPANT IMPLANTATION PARAMETERS AND CONCENTRATIONS

	Implantation Characteristic	Area		
		P$^+$	intrinsic	N$^+$
PPN	Energy (keV)	15	-	30
	Dose (cm^{-2})	$1,0.10^{15}$	-	$3,0.10^{15}$
	Concentration (cm^{-3})	$^a 3,6.10^{19}$	$^b 1,0.10^{15}$	$^a 2,2.10^{19}$
PNN	Energy (keV)	15	30	30
	Dose (cm^{-2})	$1,0.10^{15}$	$1,0.10^{12}$	$3,0.10^{15}$
	Concentration (cm^{-3})	$^a 3,6.10^{19}$	$^c 1,0.10^{17}$	$^a 2,2.10^{19}$

^{a}Characterized. ^{b}Initial SOI doping. ^{c}Simulated.

III. CURRENT-VOLTAGE RESPONSE

A. Modeling and Simulation

The current-voltage (I-V) response is simulated using SIL-VACO TCAD softwares [4]. The structures presented in Section. II are implemented into the electrical device simulator Victory for the calculation of I-V characteristics as a function of the temperature, as shown in Fig. 2. The choice of physical models valid over the full temperature range considered in our study is critical. The unified Klaassen model is chosen to describe the low-field mobility and the carriers lifetime as a function of temperature [5]. This model is coupled with bandgap narrowing, saturation velocity, impact ionization, band-to-band tunneling, Shockley-Real-Hall (SRH) and Auger recombination models. At cryogenic temperatures, the incomplete ionization of dopants must also be considered through the model of Altermatt [3]. The interfaces between the SiO$_2$ oxide layers and the channel are assumed to be trap less in the presented simulations.

The extraction of the ideality factors η corresponds to the slope of the straight part of $Log(I)-V$ curve in forward current :

$$\eta = (q/kT).(dV/d(Log(I))) \qquad (1)$$

Simulated I-V characteristics show η very close to 1 with a slight increase with L_i. Values of 1.05 at 300 K and 1.20 at 80 K are shown in Fig. 3. Also, the series resistance is dominated, as expected, by the channel silicon resistance R_{Si} in front of R_c (respectively 1 kΩ and 100 Ω for a PNN diode with L_i = 5 μm and $W = 2$ μm at 300 K). These results reveal a quasi-ideal forward behavior driven by a diffusion current.

B. Experimental Current-Voltage characteristics

The I-V response of the diodes were characterized at room temperature and in cryogenic regime. Measurements were done using an HP 4156 Semiconductor Parameter Analyzer at $300-330$ K and with a Keithley C6430 source measurement at $80-100$ K.

At RT, Fig. 3 shows really good uniformity tested on 50 prototypes per wafer and improved by a final N_2/H_2 annealing (LA). This annealing contributes to reduce the series resistance but above all to improve the quality of the interfaces. The ideality factors are very close to those simulated for lengths between 1 and 50 μm. At 80 K, the ideality factors are slightly higher but remain very close to 1 confirming a current driven by diffusion with weak Generation-Recombination (GR) current contribution.

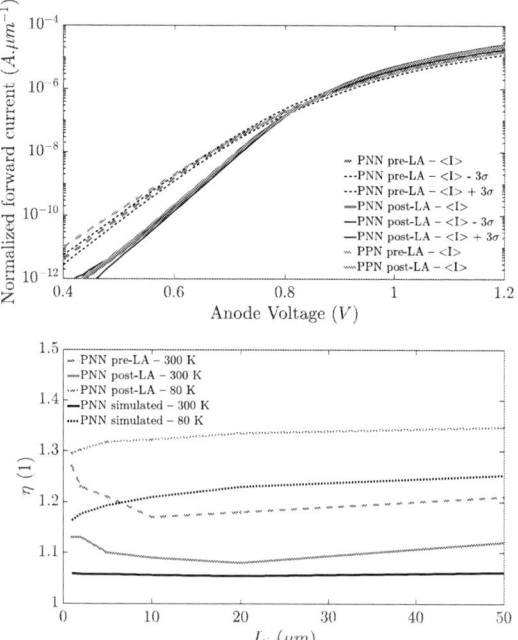

Fig. 3. Experimental and simulated I-V characteristics at 300 K (top) and ideality coefficients (bottom) for a diode with $L_i = 5$ μm.

C. Experimental TCC

At constant bias voltage, the TCC of the diode current is :

$$TCC = (1/I).dI/dT = d(Log(I))/dT \qquad (2)$$

The TCC is extracted from I-V characteristics taken every 5 K in a typical 25 K temperature range. The thermal response is higher at low forward currents and is increased by lowering the temperature down to 80 K. Fig. 4 indicates a factor higher than 3.5 between TCCs at RT and 80 K at 10^{-9} A.μm^{-1}, increasing from 7 to 25%/K. This trend is similar for the two diode variants and at each channel length. The TCC increases at short channel length and high normalized currents because of a smaller R_{Si}. The experimental results are in good agreement with the simulations particularly at RT. At 80 K, the slight differences in the low injection regime may come from a GR current not considered in simulation and which would explain also the larger experimental values of η. Another explanation may come from a dopant concentration in the channel slightly lower than in the simulation. Indeed, at low forward current, the TCC is improved by lowering the channel dopant concentration.

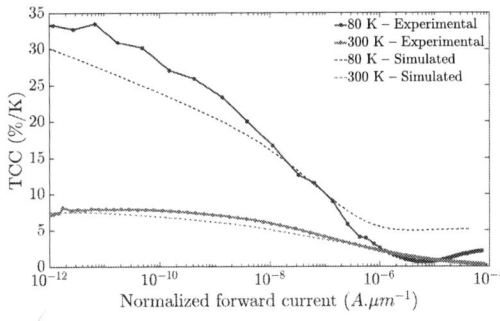

Fig. 4. Experimental and simulated TCC of a PNN diode with L_i = 5 μm.

IV. LOW FREQUENCY NOISE

A. Theoretical model

An analytical model of the current noise in PiN diodes is proposed here. It provides the Power Spectral Density of current (PSD) S_I in each forward conduction regime as a function of the frequency f. Using the superposition theorem to the equivalent circuit of a PiN diode (Fig. 1a), S_I is expressed as the sum of three noise contributions :

$$S_I = \left(\frac{r_d}{A}\right)^2 S_{I-PN} + \left(\frac{R_{Si}}{A}\right)^2 S_{I-R_{Si}} + \left(\frac{R_c}{A}\right)^2 S_{I-R_c} \qquad (3)$$

where the parameter $A = r_d + R_S$ depends on the current. Each term depends on the temperature but also on the current regime. Before the R_S regime, the noise contribution of the PN junction is predominant. Its PSD is the sum of the Schottky noise, characteristic of carriers crossing a potential energy barrier, and of a $1/f$ term (4) :

$$S_{I-PN} = 2qI\left(1 + \frac{\alpha_{H-PN}}{2\tau_{eff}f}\right) \qquad (4)$$

The effective lifetime of minority carriers in the channel τ_{eff} and the Hooge parameter α_{H-PN} are calculated according to the physical and geometrical parameters of the diode. Since previous results demonstrate a current dominated by diffusion with moderate GR, τ_{eff} is therefore equal to the diffusion time of the minority carriers in the channel, where $D_{n,p}$ is the diffusion coefficient of minority carriers :

$$\tau_{eff} = L_i^2/2D_{n,p} \qquad (5)$$

An expression of α_{H-PN} was proposed in [6] for short PiN diodes where the indices n and p depend on the types of minority carriers in the channel :

$$\alpha_{H-PN} = \frac{4\alpha_0}{3\pi} \frac{2q(V_{bi} - V) + 3kT}{m_{n,p}^* c^2} Log\left(\frac{n, p(0)}{n, p(L_i)}\right) \qquad (6)$$

$n, p(0)$ and $n, p(L_i)$ are the minority carrier concentrations at the two ends of the channel. V_{bi} is the buil-in voltage of the diode, α_0 the fine structure constant and m^* the effective mass of carriers.

The PSD for a resistor is the sum of the Johnson noise, coming from the carriers thermal fluctuations, and the $1/f$ noise :

$$S_{I-R_{Si}} = \left(\frac{4kT}{R_{Si}} + \frac{\alpha_{H-r}I^2}{N_{eff}f}\right), \quad S_{I-R_c} = \left(\frac{4kT}{R_c} + \frac{K_{1/f}I^2}{f}\right) \qquad (7)$$

For the channel contribution, the $1/f$ noise follows an Hooge interpretation where α_{H-r} is the Hooge parameter and N_{eff} the total number of carriers in the channel. For the contact resistance, the $1/f$ noise depends on the quality of the contact and is described by the coefficient $K_{1/f}$.

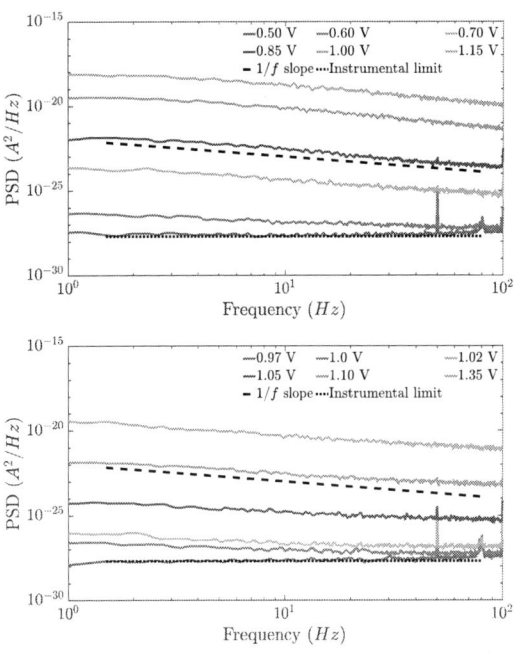

Fig. 5. Frequency behavior of the PSD of a PNN diode with L_i = 5 μm and W = 2 μm at 300 K (top) and 80 K (bottom).

B. Experimental Results

The noise spectral density measurements were done on samples wire bonded on ceramic holders. These ceramics were mounted into a cryostat under vacuum to perform the noise acquisitions using a Dynamic Signal Analyzer HP35670A with a noise floor at 2.10^{-28} $A^2.Hz^{-1}$.

The experimental results show a dominant $1/f$ noise, as expected for such thin silicon devices. Its appearance is more progressive at low currents at 80 K than at 300 K, as shown in Fig. 5. Fig. 6 compares the experimental and theoretical PSD for a short PNN diode as a function of the forward current. The experimental results clearly show the different noise regimes. At low currents, the PSD is proportional to the current and decreases at low temperature with a fairly good fit between the model and the measurements. Before the regime of series resistance, the noise is higher than predicted by the model particularly at 300 K; this comes from the calculation of τ_{eff} which is different in high injection. In the series resistance regime, S_I is proportional to the square of the current. So, the extraction of α_{H-r} and $K_{1/f}$ are done using measurements on several prototypes of a same variant. The values of α_{H-r} are around 10^{-3} at RT and almost one order of magnitude lower at 80 K. These values correspond to typical Hooge parameters for silicon resistances. $K_{1/f}$ is of the order of 10^{-5}, which is a sign of good contacts quality. These results show that the low-frequency noise of the diodes get closer to the Schottky limit at low temperature.

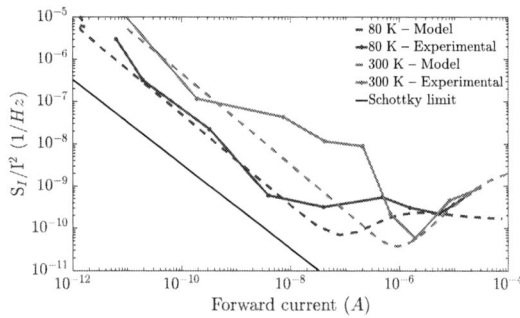

Fig. 6. Experimental and theoretical normalized PSD at 1 Hz of a PNN diode with $L_i = 5$ μm and W = 2 μm.

V. Thermal Resolution

The previous results allow to calculate ΔT_{min} defined as the ratio : $i_n/(|TCC|I)$. The total noise i_n is calculated for an integration time τ_{int} as $i_n = \sqrt{\int_0^{+\infty} S_I sinc^2(\pi f \tau_{int})df}$. Fig. 7 shows the thermal resolution of diodes with a length of 5 μm. ΔT_{min} is lower than 5 mK for a wide range of currents at 300 K, which compares favorably to the previous work presented in [7]. At 80 K, ΔT_{min} is much lower reaching values below 0.2 mK for the PNN variant. The difference between the PNN and PPN variants mainly comes from the noise level and is in agreement with the calculation of the α_{H-PN}/τ_{eff} ratio. Indeed, this ratio depends on the type of

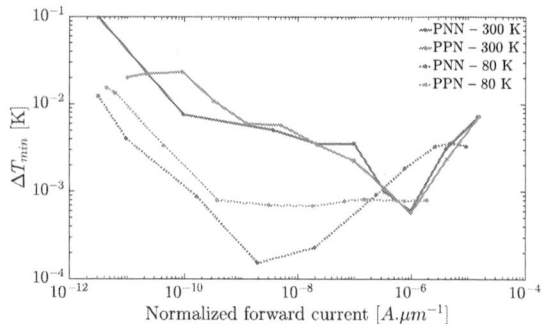

Fig. 7. Experimental ΔT_{min} of a diode with $L_i = 5$ μm and W = 2 μm for an integration time of 40 ms.

minority carriers and the channel doping level. The thermal resolution is a compromise between the TCC and the LFN. Therefore, the optimum range of current lies near the transition between the diffusion regime, where S_I is proportional to I, and the series resistance regime, where the TCC decreases.

VI. Conclusion

The thermometric performances of thin-film lateral PiN diodes have been studied. A simulation model enables to calculate the TCC of these diodes as a function of the temperature, the doping or the geometrical dimensions. Their LFN described analytically is in fair agreement with the experimental results. In future works, these models may allow further optimization of the thermal resolution. The presented diode prototypes have already very good resolutions close to 0.1 mK with moderate cryogenic cooling. Their integration in bolometers will enable the realization of passive THz imagers. Of course, these diodes can also be used as thermometers in SOI circuits operationg at RT and below.

Acknowledgment

The French Innovation Defense Lab partially funded this work.

References

[1] J. Blond, "Détecteurs térahertz hautes performances pour l'imagerie passive", PhD thesis, Université Grenoble-Alpes, France, 2021.

[2] M. de Souza, B. Rue, D. Flandre, et M. A. Pavanello, "Thin-film lateral SOI PIN diodes for thermal sensing reaching the cryogenic regime", *J. Integr. Circuits Syst.*, vol. 5, no 2, 2010.

[3] P. P. Altermatt, A. Schenk, et G. Heiser, "A simulation model for the density of states and for incomplete ionization in crystalline silicon. I. Establishing the model in Si :P", *Journal of Applied Physics*, 100(11) :(113714) 1–10, December 2006.

[4] Silvaco, Inc. Athena User's Manual - Process Simulation Software. Manuel d'utilisation, Santa Clara (CA-USA), 2017.

[5] D. B. M. Klaassen, "A unified mobility model for device simulation—I. Model equations and concentration dependence", *Solid-State Electron.*, vol. 35, no 7, p. 953–959, 1992.

[6] A. Van Der Ziel, "Unified presentation of 1/f noise in electron devices : fundamental 1/f noise sources", *Proceedings of the IEEE*, 76(3) :233–258, March 1988.

[7] D. Corcos, D. Elad, T. Morf, et U. Drechsler, "Sensitivity of SOI Lateral Diodes for Bolometric Sensing", In 2018 *43rd International Conference on Infrared, Millimeter, and Terahertz Waves (IRMMW-THz)*, pages 1–2, September 2018.

MEMS optical microphone based on light phase modulation

Niccolò de Milleri
Infineon Technologies Austria AG
Villach, Austria
niccolo.demilleri@infineon.com

Güclü Onaran
Infineon Technologies AG
Munich, Germany
abidingueclue.onaran@infineon.com

Dr. Andreas Wiesbauer
Infineon Technologies Austria AG
Villach, Austria
andreas.wiesbauer@infineon.com

Prof. Andrea Baschirotto
Dept. of Physics "G. Occhialini"
University of Milan-Bicocca
Milan, Italy
andrea.baschirotto@unimib.it

Abstract— **Acoustic sensing through optical transduction represents a promising alternative to the conventional capacitive sensing used in MEMS microphones, especially when aiming at ultra-low noise (i.e. high DR) applications. This paper reports the design and the modeling of the sensing elements of a MEMS optical microphone. The basic transduction mechanism is presented and the main design parameters and challenges are explained and analyzed with advanced modeling techniques.**

Keywords— *MEMS microphone, optical microphone, low noise, high signal-to-noise ratio (SNR), optical modeling*

I. INTRODUCTION

MEMS based silicon microphones are currently the most popular solution for consumer electronics due to their extraordinary miniaturization, relatively low production costs, and reliability. However, the constantly growing market of consumer electronics is continuously demanding for higher performance devices, one of the main trends is the increase of microphones' SNR (Signal-to-Noise Ratio), and this trend seems intrinsically impossible to be satisfied by the present MEMS technology. A potential option for satisfying such demand is the development of alternative technologies that allow to overcome the performance plateau that capacitive MEMS solutions have been facing in the last years. An increasingly popular audio sensing alternative is represented by optical-based solutions. Optical microphones solutions and design techniques were proposed in [1], [2], [3]. The initial approaches for the design of a MEMS-based optical microphone, including the optimum placement of the photodetectors with respect to the reflector and the method for calculating the optimum operating point of the device were illustrated in [4], [5]. The above aspects are here addressed and further developed and optimized with the final target of defining a mass-production aimed solution. In Section II the basic sensing concept with its main advantages is presented. In addition, the electrostatic modeling of the device is shown setting the basis for the subsequent modeling work. Section III introduces the initial modeling approach of the optical stage, based in an analytical description of the physical working principles of the device. In section IV a further development of the optics model is presented and the results are compared to the measurements.

II. TRANSDUCTION PRINCIPLE AND ELECTROSTATIC MODELING

The high-level diagram of the proposed sensing mechanism and the MEMS prototype are shown in fig. 1. The acoustic-to-mechanic transduction is obtained by a polysilicon moving diaphragm that includes a gold sputtered layer area in its central region that reflects back the diffracted LASER light towards the photodetectors. The variable position of the diaphragm reflector modifies the MEMS gap and consequently modulates the diffraction orders of the light generated at the diffraction grid (I_0, $I_{\pm 1}$), obtaining this way the mechanic-to-optical transduction. The final optical-to-electric signal conversion occurs at the photodetectors that collect separately the complementary diffraction beams in a differential readout that reduces the light source intensity noise [5], [6]. Both the VCSEL (Vertical Cavity Surface Emitting Laser) and the photodiodes dies are placed in the etched cavity of the MEMS facing the diffraction grid and the reflector.

Acoustic optical sensing can be classified according to the light modulation principle exploited (amplitude, phase or polarization). A common choice for MEMS-based optical microphones is the phase modulation of light as it requires no optical beam splitters, focusing lenses or waveguide structures for the construction of the optical system, resulting in a system that can be easier packaged and integrated in a large-scale production process [6]. Furthermore, the grating-based phase modulation method is advantageous as the periodic grating patterns can be achieved in modern silicon lithographic processes with high accuracy and reproducibility.

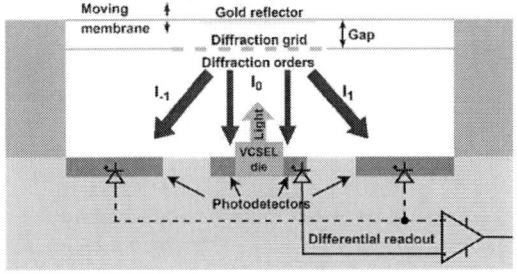

Fig. 1. MEMS prototype and conceptual diagram of the sensing mechanism

978-1-6654-8498-5/22 $31.00 © 2022 IEEE

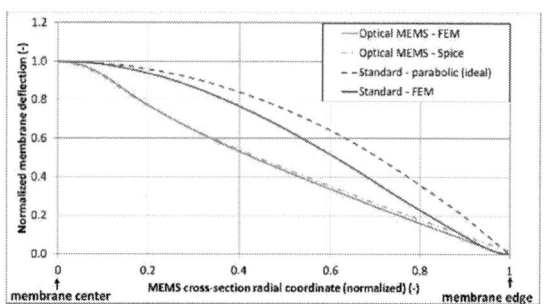

Fig. 2. Cross-sections of membrane displacement with applied bias, traditional MEMS vs. proposed structure

In contrast to traditional capacitive readout concepts, the performance of an optical-based acoustic sensing system is not limited by the capacitance of the MEMS. This allows to decouple the mechanic design from the electronic performance, which constitutes one of the main strengths of the acoustic optical sensing. In the proposed device the aggressive stator design is realized with few thin spokes that hold the central diffraction grid region (fig. 1). The mechanics of such design are almost noiseless and the resulting low device capacitance does not represent a limitation, as it does not need to be electrically sensed.

A correct modeling of the electrostatic behavior of the device is a key step, as the biasing of the MEMS is used to fine-tune the position of the reflector in order to obtain an optimum operating point for the optical system [1]. In addition, it is well known that for MEMS microphones the electrostatic biasing plays a key role on the acoustic-mechanic performance of the sensor, especially by the so called electrostatic softening effect that influences the acoustic compliance of the membrane [7]. In addition to the acoustic-mechanic behavior modeling, an accurate description of the MEMS membrane displacement shape is key for understanding the interaction of the membrane reflector with light. When applying a biasing voltage to the proposed optical MEMS capacitor structure, the electrostatic forces between the diaphragm and the stator act on the membrane in a non-uniform way resulting in a non-standard deflection shape of the diaphragm, that was modelled in Comsol and Spice, as shown in fig. 2, compared to a standard MEMS capacitor deflection with thin circular clamped membrane (approx. parabolic deflection).

III. ANALYTICAL OPTICS MODELING

The modeling strategies for the optical subsystem follow different paths that involve various FE tools and analytical propagation simulations, e.g. based on the scalar diffraction theories. In addition, also a mixed co-simulation that integrates analytical calculations with FE analysis is implemented. The following sections show the motivations and a comparison of the different modelling approaches.

The theoretic analysis of the working principle of the optical subsystem is based on waves full scalar propagation theory [8]. The models developed in Matlab rely on the angular plane-wave spectrum representation of optical diffracted fields, which has been proven to be equivalent to the first Rayleigh-Sommerfeld solution [2]. Such representation allows to describe the optical field by an infinite superposition of plane waves in the angular spectrum domain

and calculated at each point of a defined plane. Consequently, also the interaction with obstacles along the propagation field can be successfully captured. The main assumptions for such description of an optical field are: (I) a linear, isotropic and homogeneous diffracting medium, (II) apertures (or interfaces) with dimensions large with respect to the wavelength. On the other side the condition that both the diffraction and the calculated surfaces must be plane and parallel [8], can be overcome with proper discretization of the optical field especially when the problem is solved with computer simulations. The Matlab code is based on the Helmholtz equation in differential form:

$$\frac{\partial^2 U}{\partial z^2}(u,v;z) + k^2(1 - \lambda^2 u^2 - \lambda^2 v^2)\,U(u,v;z) = 0 \qquad (1)$$

Where $U(u,v;z)$ is the 2-D spatial Fourier transform of the field at a coordinate z, while k and λ are the wave number and the wavelength of the light respectively. A known solution of equation (1) is:

$$U(u,v;z) = U(u,v;0)e^{ikz\sqrt{1-\lambda^2 u^2 - \lambda^2 v^2}} \qquad (2)$$

Therefore, by knowing the field at a point z_0, it is possible to evaluate the propagated field at a generic point z ($z > z_0$) with very few approximations. The code takes into consideration the angular spectrum of the field at the planes relevant for the propagation (e.g. the diffraction grid), each term of the angular spectrum is multiplied by a z-linear phase factor, and transformed back using the inverse angular spectrum relation, or spatial Fourier transform. The developed code starts the propagation of a coherent LASER light source, modelled as a Gaussian beam [9] with divergence of 15° to reflect the typical commercial VCSEL devices, that propagates towards a diffraction grid plane (fig. 3 and fig. 4) located at 250μm distance.

Fig. 3. Silicon diffraction grating prototype

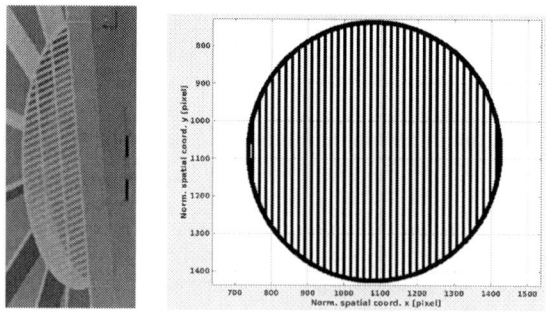

Fig. 4. FEM and Matlab diffraction grating models

At the grating plane the incident optical power is divided: a portion is reflected back, being also the grid coated with a sputtered gold layer, and the remaining optical power enters the interferometric cavity where it experiences a series of successive reflections that are described with an iteration loop in the simulation code, that ends when the desired percentage of optical field has finally left the cavity and propagates back towards the photodetectors. In fig. 5 (top) are visible the amplitude of the field and the 2D spatial FFT in a typical simulation case. The first diffraction order (central lobe) and several lateral orders are clearly visible, the optimum readout circuit makes use of the first and second orders [1].

After the propagation towards the photodetectors the optical field encounters and is shaded by the VCSEL die, stacked on top of the photodiodes chip, and then is finally integrated after reaching the photodetectors plane to estimate the collected optical power (fig. 5 - bottom).

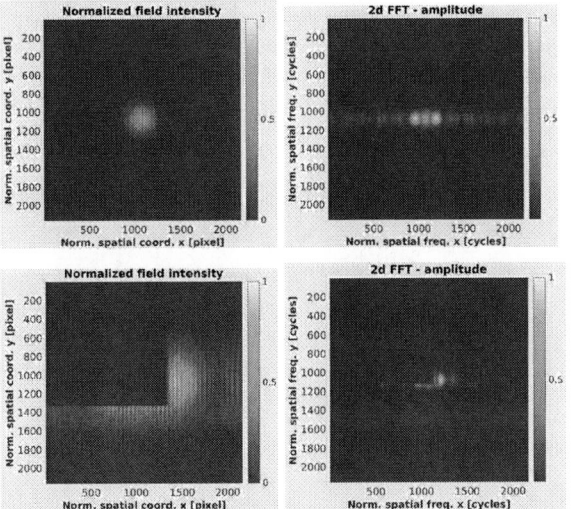

Fig. 5. Field amplitudes and 2D-FFTs at the output of the interferometric cavity (top) and at the photodetectors plane (bottom)

A typical simulated photocurrents plot vs the reflector motion (e.g. with incoming acoustic pressure) can be observed in fig. 6, being I_0 the 0-th order and $I_{\pm 1}$ the odd orders. It can be shown how the optimum operating point that maximizes sensitivity and linearity [5] [6] is obtained by tuning the resting position of the diaphragm at the zero-crossing of the differential photocurrent, where it has its maximum slope, by adjusting the bias voltage of the MEMS (see paragraph II).

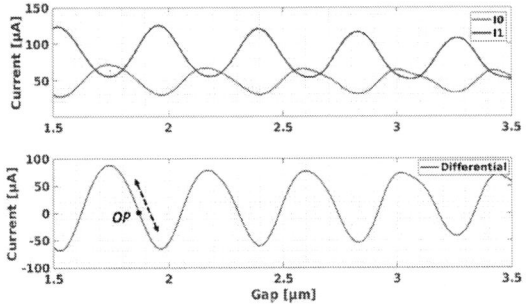

Fig. 6. Simulated collected photocurrents vs gap

For instance, the sensitivity of the differential readout at the optimum operating point (OP) g_{OP}=1.85µm can be calculated following the following equation [10]:

$$S_{diff} = R \frac{\partial(I_0 - \alpha I_1)}{\partial g}\Big|_{g=g_{OP}} = RI_{in}\frac{4\pi}{\lambda_0} \qquad (3)$$

that is equivalent to compute the slope of the differential current from fig. 6, yielding ~1mA/µm for a collected power by the photodetectors I_{in}~120µW (12% of the 1mW total power emitted by the VCSEL at a wavelength λ_0=850nm), assuming a photodiodes responsivity R=0.55A/W, and considering the measured responsivity masks shown in fig. 7.

Considering a typical MEMS membrane center compliance of 10nm/Pa, the overall projected sensitivity of the optical microphone would be 10µA/Pa. A system performance indicator is the Minimum Detectable Displacement (MDD), i.e. the ratio between the sensitivity in equation (3 and the noise portion of the system that depends on the detected currents, i.e. the photodiodes shot noise [4], [11]:

$$MDD = \frac{\sqrt{2q(I_0 + \alpha I_1)}}{RI_{in}\frac{4\pi}{\lambda_0}} \qquad (4)$$

being q the elementary charge. Considering the sensitivity values previously discussed and the current values in fig. 6, the MDD for the specific case would reach ~ 7fm/rtHz.

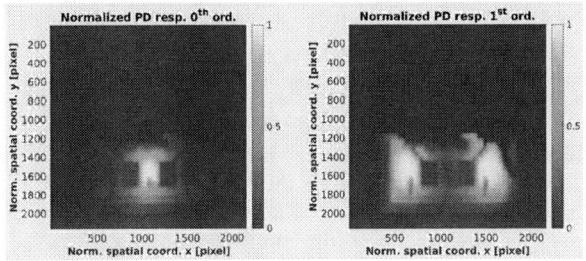

Fig. 7. PD normalized responsivity masks

IV. MEASUREMENTS AND MODEL EXPANSION

To introduce the second step of complexity added to the optical model, shown in section 2, it is useful to have a look to the first measurement results obtained in a simplified setup. Two different MEMS variants are evaluated, with and without gold coating applied as reflector on the membrane. Fig. 8 shows how the variant with gold coating shows unexpected kinks in the collected photo-currents plots (first and second orders), while the variant with no coating shows a cleaner periodic behavior as expected from the simulations.

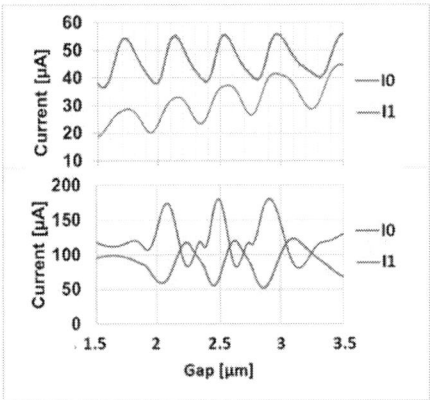

Fig. 8. Measured photocurrent, variant without (top) and with (bottom) gold coating

The optical model described in section III is enriched with a more detailed description of the optical cavity region and the addition of multiple reflections in the VCSEL to MEMS region, in order to describe the root cause of the undesired behavior. The resulting model comprises a propagation stage, VCSEL to diffraction grid, analytically solved in Matlab as described in section III, whereas the field in the optical cavity is co-solved with Comsol with the Gaussian beam approach, fig. 9. Such solver relies on Fourier optics theory where the light source is modelled as a Gaussian light beam. The current modulation plots obtained for the advanced model can be seen in fig. 10. The results (mid and bottom graphs) show kinks and strong non-idealities when the reflectivity of the VCSEL die is enabled, confirming that they are mainly generated by multiple reflections in the VCSEL-to-grating region. The comparison to the measurements (fig. 8) shows the improved matching achieved with the advanced model. The case with reduced VCSEL die reflectivity (top plot) shows a fundamental achievement of the present work, i.e. how the improved version of the system would look like when anti-reflective coating is applied to the VCSEL die, which could represent a trade-off with its output optical power. An alternative improvement could be obtained by designing an off-axis optical path, by simple tilting or with a reflections system, that avoids the direct interaction of the optical signal with the highly reflective VCSEL die.

Finally, in fig. 11 is shown another simulation case where the distance from VCSEL to the MEMS optical stage was varied within $\lambda/2$ tolerance in order to analyze a possible assembly tolerance. It is clearly visible the large dependence of the currents on the VCSEL-to-MEMS distance, highlighting the strong interdependence of multiple-reflections in the structure with the packaging tolerances and therefore the difficulty of an overall matching on a packaged system.

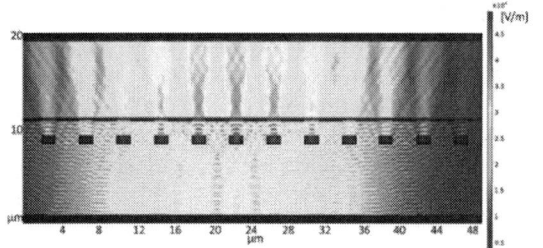

Fig. 9. Electric field in the optical cavity region - Gaussian beam method

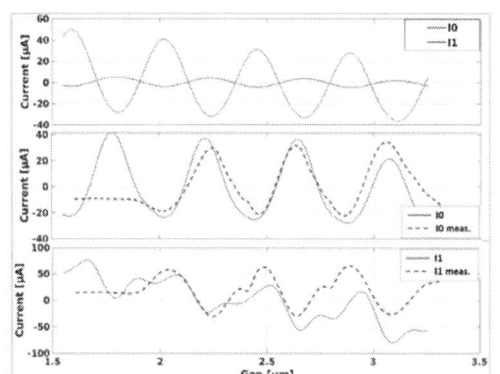

Fig. 10. Currents modulation w/out (top) and with (mid and bottom) VCSEL reflectivity vs measurements

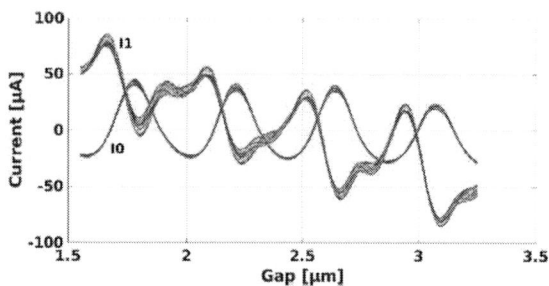

Fig. 11. Modulation currents with VCSEL-grating distance sweep – assembly tolerance example

V. CONCLUSIONS

The main challenges of the optical modeling and design of the MEMS are shown and solved. The main observed 2nd-order effect is studied, isolated and reproduced both in measurements and simulations. A fundamental design improvement is deduced, i.e. the reduction of the light reflections in the structure (especially in the Bosch cavity of the MEMS), e.g. with anti-reflective coating of the chips. A reliable and powerful modeling platform is developed that allows to predict the performance of the system also in the case of high-volume production with the unavoidable fabrication tolerances, aiming at a future statistical analysis.

REFERENCES

[1] N. A. Hall, "Micromachined Broadband Acoustic Transducers with Integrated Optical Displacement Detection," 2004.

[2] W. Lee, Diffraction-based integrated optical readout for micromachined optomechanical sensors., Georgia Institute of Technology, 2006.

[3] N. Bilaniuk, "Optical microphone transduction techniques," Appl. Acoust., vol. 50, no. 1, p. 35–63, Jan. 1997.

[4] C. T. Garcia, "Packaging and Characterization of MEMS Optical Microphones.," 2007.

[5] W. Lee, N. A. Hall, Z. Zhou and F. L. Deger, "Fabrication and characterization of a micromachined acoustic sensor with integrated optical readout," IEEE Journal of Selected Topics in Quantum Electronics, vol. 10, no. 3, pp. 643-651, May-Jun 2004.

[6] N. Hall, W. Lee and F. Degertekin, "Capacitive micromachined ultrasonic transducers with diffraction-based integrated optical displacement detection," IEEE Trans. Ultrason., Ferroelect., Freq. Contr., vol. 50, p. 1570–1580, Nov. 2003.

[7] D. Cattin, Design, Modelling and Control of IRST Capacitive MEMS Microphone, University of Trento, 2009.

[8] B. C. Kress and P. Meyrueis, "Appendix B: The Scalar Theory of Diffraction," in Applied digital optics, John Wiley & Sons, Ltd, 2009, pp. 587-595.

[9] Y. H. Lee, B. Tell, K. Brown-Goebeler, J. L. Jewell, C. A. Burrus and J. M. V. Hove, "Characteristics of top-surface-emitting GaAs quantum-well lasers," IEEE Photonics Technology Letters, vol. 2, no. 9, pp. 686-688, Sept. 1990.

[10] R. N. Miles and F. L. Degertekin, "Optical sensing in a directional MEMS microphone". United States Patent US 8503701 B1, 6 August 2013.

[11] K. Kadirvel, R. Taylor, S. Horowitz, L. Hunt and M. Sheplak, "Design and Characterization of MEMS Optical Microphone for Aeroacoustic Measurement," in 42nd Aerospace Sciences Meeting & Exhibit, Reno, NV, 2004.

978-1-6654-8498-5/22 $31.00 © 2022 IEEE

Filament Localization and Characterization in HfO₂ ReRAM Cells using Laser Stimulation

Franco Stellari[1], Ernest Y. Wu[2], Martin M. Frank[1], Leonidas E. Ocola[1], Takashi Ando[1], and Peilin Song[1]

[1] IBM T.J. Watson Research Center
1101 Kitchwan Rd.
Yorktown Heights, NY 10598 - USA
stellari@us.ibm.com

[2] IBM Research
1000 River St.
Essex Junction, VT 05452 - USA

Abstract— Optical Beam Induced Resistance Change (OBIRCH) laser stimulation is used to detect the spatial location of filaments inside large area Resistive Random Access Memories (ReRAMs). This technique allows one to detect filaments at very low bias voltages, down to ~10 mV, significantly improving previous results obtained using near-infrared Photon Emission Microscopy (PEM). This capability is leveraged to detect, for the first time, the location of filaments before they are fully formed. Multi-filaments and logic state detection are also demonstrated.

Keywords—Photon Emission Microscopy (PEM), Laser Stimulation, Optically Induced Resistance Change (OBIRCH), Resistive Random Access Memory (ReRAM), Hafnium Oxide.

I. INTRODUCTION

ReRAM is a type of Non-Volatile (NV) Random-Access Memory (RAM) that can be programmed by changing the resistance across a dielectric solid-state material. The virgin dielectric is insulating, but it can be made to conduct by *forming* a *filament* (*i.e.*, a conduction path) through the dielectric, after a sufficiently high voltage (or current) has been applied to it [1]. Afterwards, the filament may be *reset* (*i.e.*, broken, resulting in high resistance of the ReRAM cell) or *set* (*i.e.*, re-formed, resulting in lower resistance of the ReRAM cell) by applying negative/positive voltages (lower than the *forming* voltage). The formation of multiple current paths rather than a single filament in a single cell has also been theorized by [2] and experimentally reported for large 40 x 40 μm² cells in [3].

Often, the localization of the filament position inside a cell is a time-consuming process. For example, Conductive Atomic Force Microscopy (C-AFM) is a technique that allows to detect the presence of current paths in the dielectric material due to filaments [4]. This technique provides a very high spatial resolution but it is also slow, especially for large cells, thus making it costly for statistical analysis. Furthermore, it is not possible to monitor the filament over time, such as it is required in experiments involving multiple switching cycles, Constant Voltage Stress (CVS), Ramp Voltage Stress (RVS), *etc.*

In [3,5], we have shown that PEM can quickly localize the position of a filament inside large development crystalline HfO₂ ReRAM cells with sub-micron resolution. This allows for statistical analysis of a large number of devices. Furthermore, in [5], it was shown that PEM can monitor the filament position and characteristics over time, during both fast I-V sweeps and slow CVS/RVS reliability experiments.

In this work, we use the Optical Beam Induced Resistance Change (OBIRCH) laser stimulation technique [6] for the localization and characterization of filaments in large area HfO₂ ReRAM cells, spanning from 1 x 1 μm² to 100 x 100 μm². In this paper, we show that OBIRCH is effective in localizing formed filaments with voltage biases equivalent to normal sense voltages used for reading cells, significantly lower than what is required for previously reported measurements [3, 5]. OBIRCH spots are confirmed to point to the same filament location detected using PEM and can be effectively used to create filament maps and perform statistical analysis of the spatial distribution of the conduction paths. On the other hand, the slower acquisition frame rate of OBIRCH compared to PEM, imposes limitations to detecting fast changing phenomena during either I-V sweeps or long term CVS/RVS reliability measurements.

II. EXPERIMENT DESCRIPTION

Fig. 1 shows the ReRAM sample structure. The HfO₂ layer is deposited by Atomic Layer Deposition (ALD) using cycles of tetrakis(dimethylamido)hafnium (TDMAH) and H₂O, supplied alternatively. A temperature of 250 °C is used to obtain crystalline film, while 75 °C is used for amorphous film. Top and bottom 35 nm thick TiN electrodes provide electrical connections. Referring to the I-V characteristics in Fig. 2, a positive ramped voltage is first used to form a filament in the ReRAM cell (black line). This is achieved when the low cell current suddenly jumps higher, reaching a pre-determined compliance value of ~100 μA. At this point, a reverse/forward

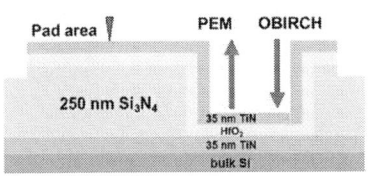

Fig. 1. Simplified diagram of the structure of a HfO₂ ReRAM cell.

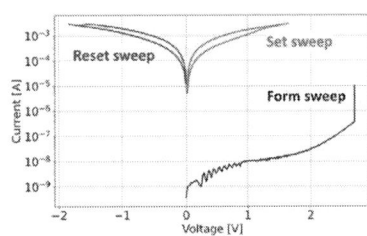

Fig. 2. I-V characteristics of an amorphous 10 x 10 μm² ReRAM cell acquired during forming, reset, and set voltage sweeps.

bias can be used to *reset/set* the cell state, respectively. This is indicated by the decrease/increase in current observed in the cell during the reverse/forward bias sweep, respectively. After the logic state of the cell has been written, it can be non-destructively read back by sensing its resistivity/current at a low bias voltage (*e.g.*, 100 mV).

Referring to Fig. 3, photon emission is acquired using either a NIR InGaAs camera or a shorter wavelength Silicon-Intensified CCD (SI-CCD), while OBIRCH is performed at ~1 μm laser wavelength. A 100X magnification air-gap lens with a Numerical Aperture (NA) of ~0.7 allows one to keep the entire area of the ReRAM cells (up to 100 x 100 μm²) in the field of view of the cameras. The top TiN contact layer covering the cell (see Fig. 1) is thin enough that both spontaneous photon emission and stimulating laser light can pass through.

III. OBIRCH VS. PEM MEASUREMENTS

PEM requires the application of a voltage/current bias to the cell under test and the observation of the spontaneous photon emission generated by the carrier flowing through the filament, in a completely non-invasive way. In [3], it was shown that the photon emission can be fitted using an electric-field model [7], where the intensity is linearly proportional to the filament current and exponentially dependent on the applied voltage (electric field). A single measurement can map the entire cell (up to 100 x 100 μm²) and the acquisition time can be very short, with the capability of acquiring movie sequences with up to ~90 frames per second for smaller cells. However, in most cases, a bias of ~1 V and a current of ~1 mA is required in order to detect the photon emission from the filament.

During OBIRCH measurements, a voltage/current bias is applied to the cell while a laser beam is raster scanned across the sample. For each laser location, a current/voltage measurement is quickly performed by an amplifier and its value is used to create the OBIRCH image, where color/intensity is related to the value of the measured resistivity at the pixel location. While the laser does not have any effect at any location where the dielectric is still intact, it is expected to alter the measured current/voltage of the device when it is impinging on the filament. Such change can be caused either by a localized slight increase in the temperature or by the localized creation of electron/hole pairs that are separated by the electric field. Given these premises, one can expect that the OBIRCH acquisition time is usually significantly longer than the PEM, especially

when averaging to reduce the noise is required. Additionally, if the scan speed is too fast, a shift in the filament position may be introduced. This effect may be due, for example, to the slow response of larger ReRAM cells that behave like capacitors. On the other hand, thanks to its much weaker dependence on the applied voltage, the OBIRCH technique can be used to detect filaments at very low bias voltages, down to 10 mV, which is much lower than even the sense voltage of 100 mV typically used to read a cell content.

Fig. 4 shows a comparison of the filament localization performed using (a) PEM and (b) OBIRCH. The false color spot represents the intensity of the PEM or OBIRCH signal, respectively, overlaid to the gray scale reflected light pattern image of the ReRAM cell. Both techniques can reliably determine the position of the filament at the same physical location inside the cell, thus making them equivalent for studying their spatial distribution statistics.

In summary, while PEM is uniquely suitable for detecting fast transitions, such as set/reset event during I-V sweeps, it does not allow for the localization of the filaments at low voltage (*e.g.*, read voltage) or before the filament has been formed. In this paper, we want to leverage the low voltage sensitivity capability of the OBIRCH technique to alleviate the potential shift in the electric characteristics of a ReRAM cell due to the need of applying elevated voltage to the cell to detect the PEM signal. Additionally, we want to investigate the use of OBIRCH to study filament forming, and potentially detect pre-formed filaments.

IV. SPATIAL DISTRIBTUIN OF THE FILAMENTS

In order to study the spatial statistics of the filaments in a given ReRAM technology and fabrication process, a plurality of nominally identical ReRAM cells is first electrically formed and then analyzed using the OBIRCH technique to precisely determine the location of the filament in relation to, for example, the cell center. Fig. 5(a) shows the filament location of 6 of the 45 individuals 10 x 10 μm² ReRAM cells measured during this experiment. By overlaying spot from each image, a map of the filament spatial distribution can be constructed, as shown in Fig. 5(b) for the entire dataset of 45 ReRAM cells. As discussed in [3] in regard to PEM spots, an Euclidian inter-filament distance can be computed and used to plot their Probability Density Function (PDF), as shown in Fig. 6. The

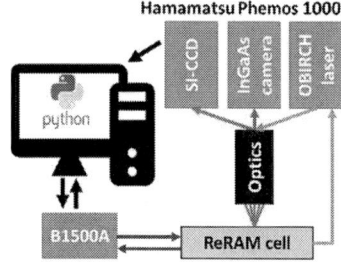

Fig. 3. Experimental setup can acquire both PEM and OBIRCH measurements.

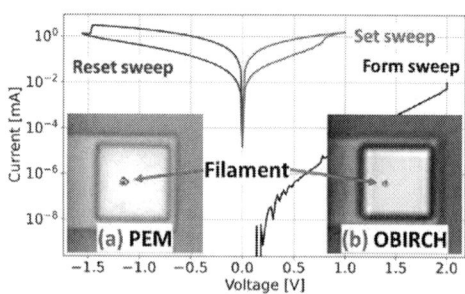

Fig. 4. Both the (a) photon emission spot acquired with PEM and the (b) OBIRCH spot acquired with the laser stimulation point to the same filament.

OBIRCH data (symbols) show a good agreement with the Poisson uniform spatial distribution model presented in [3].

Performing this analysis using the OBIRCH technique is typically slightly slower than that using the PEM methodology because of the need for a longer acquisition time. OBIRCH measurements are usually noisier than PEM images and the precise localization of the filaments benefits significantly from image averaging as well as better algorithms for interpolating the shape of the spot. At the end, though, the localization process yields very similar results with a reduced stress to the cells that do not need to be exposed to the elevated voltages required to measure the PEM signal.

V. CELL STATE DETECTION

The OBIRCH signal measured at a given (x, y) position of the image is proportional to the ReRAM cell current measured with the laser positioned at that location. The local temperature increase introduced by the laser spot at the location of the filament may produce an increase/decrease of the filament current, thus yielding a spot on the reconstructed image. Fig. 7 compares the signal acquired from a cell in three states: (a) immediately after forming, *i.e.* in a *set* logic state; (b) after resetting the logic state; (c) after setting the cell again. As expected, the filament current is the smallest when the cell is in the *reset* logic state with the highest resistance, as shown in Fig. 7(d). On the other hand, the OBIRCH signal normalized to the cell current is greater in the *reset* state, as shown in Fig. 7(e).

VI. MULTIFILAMENT OBSERVATION

Fig. 8 shows a comparison between (a) PEM and (b) OBIRCH measurements from the same 100 x 100 μm² HfO₂ ReRAM cell that was first reported in [3]. A detailed model of the spatial distribution of the multiple filaments can be found in [8]. Multiple spots are observed in the cell by both techniques. While the majority of locations is the same for both techniques

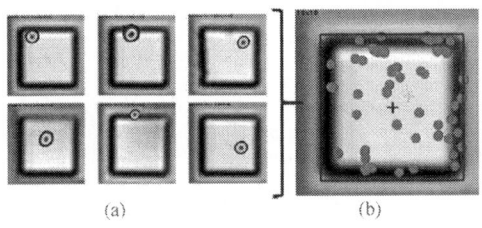

(a)   (b)

Fig. 5. (a) Individual filament locations acquired from different cells are combined to form a (b) map of the filaments.

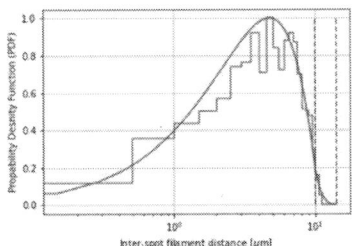

Fig. 6. The distribution of the inter-filament distances in Fig. 5(b) is in good agreement with the Poisson uniform spatial distribution model (red line).

(A), some locations are only present in the PEM image (B), and some are only present in the OBIRCH image (C), thus suggesting a potentially different behavior of the spots. To further support this hypothesis, one can also notice that the OBIRCH spot (A*) is false-colored in green instead of red, indicating that the cell current increases instead of decreasing when the laser impinges at its location. In the future, we hope to sufficiently develop this combined multi-dimensional approach to better understand the characteristics of different types of filaments and defects.

VII. FILAMENT FORMING INVESTIGATION

In [5], PEM was used to monitor the ReRAM filament behavior during I-V sweeps and long term RVS and CVS measurements. In most cases, photon emission measurements can be acquired quickly when the applied voltage and the filament current are sufficiently large. For this reason, they are very effective in monitoring the filament only after it has been completely formed. In this paper, we have shown that OBIRCH signal can be acquired at very low voltages and currents, thus allowing to minimize the impact of stress during measurements on the filament characteristics. In this section, we monitor the OBIRCH signal during long term measurements of unformed cells to understand if it is possible to detect the location of filaments before they are fully formed.

The formation of a filament depends on multiple different factors, including the applied voltage, the time spent at a given voltage, and the compliance current. In particular by reducing the applied voltage significantly below the typical forming

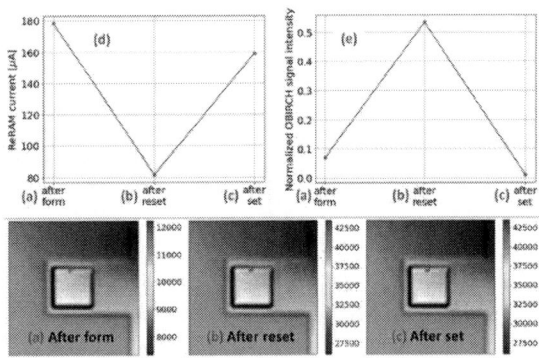

Fig. 7. A ReRAM cell current and OBIRCH signal values are compared after (a) form, (b) first reset, and (c) first set.

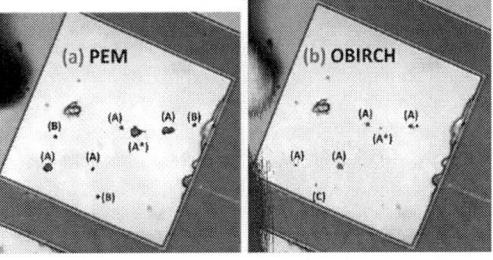

Fig. 8. A 100 x 100 μm² cell showing multiple concurrent filaments is imaged using both (a) PEM and (b) OBIRCH.

voltage, one can significantly slow down the speed of the formation process, thus allowing for a better chance to observe pre-formed filaments with OBIRCH.

Fig. 9 shows a 10 x 10 μm² crystalline ReRAM cell undergoing a CVS measurement performed at 1.6 V, much lower than the typical forming voltage for this ReRAM cells. The voltage is periodically lowered to 100 mV to perform a single OBIRCH measurement, as shown in Fig. 9(a). The cell current is continuously monitored during the stress to observe the formation of the filament corresponding to a rapid increase of the cell current, as shown on the right-hand side of Fig. 9(b). Referring to the OBIRCH signal amplitude shown in Fig. 9(c), one can observe that the intensity remains low during the initial part of the CVS measurement, up to the time corresponding to OBIRCH measurement A. The corresponding image at the bottom left of Fig. 9 confirms that no filament is detected. After measurement A has completed, the elevated 1.6 V is re-applied, and the cell current is observed to gradually increase, but without the usual large jump corresponding to a good forming of a filament. The following OBIRCH measurement B is then acquired, showing a weak signal corresponding to the filament being formed. Afterwards, when the elevated stress voltage is re-applied, the current continues to increase, finally jumping to the current compliance limit set to 10 μA. The following OBIRCH measurement C shows another weak spot corresponding to the same filament spatial location. After another period of time at 1.6 V, the OBIRCH signal D from the filament finally becomes strong and easy to detect. In summary, this experimental example shows that by carefully controlling and slowing down the filament formation process, one can potentially use the OBIRCH laser stimulation technique to detect the location of a filament before its full formation.

CONCLUSIONS

In this paper, we have discussed the use of the Optical Beam Induced Resistance Change (OBIRCH) laser stimulation technique to detect the spatial location of filaments inside large area crystalline and amorphous ReRAMs.

This technique is highly effective in observing filaments even at very low bias voltages, down to ~10 mV, which are even below the typical read voltage. This result represents a significant improvement over previous results obtained with near-infrared Photon Emission Microscopy (PEM) [3,5]. Furthermore, we show that OBIRCH can be used to determine the logic state of a cell, something that was not possible with the PEM technique.

In this paper, we show for the first time that the low voltage/current sensitivity of the OBIRCH technique can be leveraged in combination with controlled forming experiments to identify the spatial location of a filament that is about to form. This information can be used to stop the experiment before the filament is fully formed and potentially guide subsequent Physical Failure Analysis (PFA) explorations of the filament using, for example, a Scanning Electron Microscope (SEM) or a Transmission Electron Microscope (TEM).

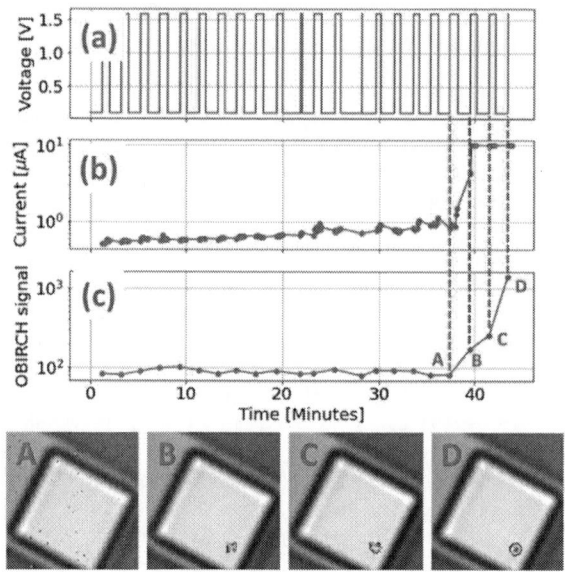

Fig. 9. OBIRCH signal is acquired during a long 1.6 V CVS to localize the filament before filament forming.

A key disadvantage of this technique is its relatively slower acquisition time compared to previous work using PEM. Therefore, this limitation does not allow high-frame rate measurement that may be required during I-V sweeps.

REFERENCES

[1] H.S.P. Wong, H.Y. Lee, S. Yu, Y.S. Chen, Y. Wu, P.S. Chen, B. Lee, F.T. Chen, and M.J. Tsai, "Metal–Oxide ReRAM", *Proceedings of the IEEE*, vol. 100, no. 6, 2012, pp. 1951-1970.

[2] W. Wu, H. Wu, B. Gao, N. Deng, S. Yu, and H. Qian, "Improving analog switching in HfOx-based resistive memory with a thermal enhanced layer", *IEEE Electron Dev Lett. (EDL)*, vol. 38, no. 8, 2017, pp. 1019-1022, doi: 10.1109/LED.2017.2719161.

[3] F. Stellari, E.Y. Wu, T. Ando, E. Cartier, M.M. Frank, C. Cabral, P. Song, and D. Pfeiffer, "Resistive Random Access Memory Filament Visualization and Characterization using Photon Emission Microscopy", *IEEE Electron Dev. Lett. (EDL)*, vol. 42, no, 6, 2021, pp. 828-831, doi: 10.1109/LED.2021.3071168.

[4] M. Lanza, "A review on resistive switching in high-k dielectrics: a nanoscale point of view using conductive atomic force microscope", *Materials*, vol. 7, no. 3, 2014, pp. 2155–2182, doi: 10.3390/ma7032155.

[5] F. Stellari, E.Y. Wu, T. Ando, M.M. Frank, and P. Song, "Photon Emission Microscopy of HfO2 ReRAM Cells", *Int. Symp. for Testing and Failure Analysis (ISTFA)*, 2021.

[6] E. Cole, "Laser-based defect localization on integrated circuits" Int. Electron Dev. Meeting (IEDM), 2006, pp. 1-4, doi: 10.1109/IEDM.2006.346968.

[7] F. Stellari, F. Zappa, S. Cova, and L. Vendrame, "Tools for non-invasive optical characterization of CMOS circuits", *Int. Electron Dev. Meeting Tech. Dig. (IEDM)*, 1999, pp. 487-490, doi: 10.1109/IEDM.1999.824199.

[8] E. Wu, F. Stellari, T. Ando, P. Song, and M. Frank, "Development of spatial nearest-neighbor analysis and Clustering/Gibbs statistical methodology for filament percolation in dielectric breakdown and forming process in ReRAM devices", *Int. Electron Dev. Meeting (IEDM)*, 2021, pp. 6.5.1-6.5.4, doi:10.1109/IEDM19574.2021.9720584.

Role of Conductive-Metal-Oxide to HfO_x Interfacial Layer on the Switching Properties of Bilayer TaO_x/HfO_x ReRAM

T. Stecconi[1], Y. Popoff[2], R. Guido[1], D. Falcone[1], M. Halter[2], M. Sousa[1], F. Horst[1], A. La Porta[1], B.J. Offrein[1] and V. Bragaglia[†,1]

[1]IBM Research GmbH-Zurich Research Laboratory, CH-8803 Rüschlikon, Switzerland
[2]ETH Zurich - Integrated Systems Laboratory, CH-8092 Zurich, Switzerland
Roberto.Guido@namlab.com ; mahalter@iis.ee.ethz.ch
{tec ; ypo ; dof ; sou ; fho ; alp ; ofb ; †vbr}@zurich.ibm.com

Abstract— **Filamentary bilayer ReRAMs based on conductive metal-oxide (CMO) / HfO_x have gained potential for applications in the field of analog in-memory computing. Compared to conventional metal / HfO_x based system, the resistive switching graduality improves and the switching stochasticity decreases. In this work we replace a standard Ti scavenging layer of an HfO_x-based ReRAM with a conductive metal-oxide such as TaO_x. We assess the material stack structural and electrical properties and identify an onset layer of oxidized metal-oxide at the interface between the dielectric and the CMO layer. We discuss its presence in relation to the increased forming voltage of the bilayer devices, the ON/OFF ratio, and the resistive switching window. Our scaled (<1 μm^2) bilayer ReRAM has excellent analog properties (at least 4 bits) and good retention measured for a minimum of 30 min. All materials and fabrication steps are compatible with complementary-metal-oxide-semiconductor (CMOS) and the back-end-of-the-line (BEOL) processes.**

Keywords—Bilayer ReRAM, analog memory, interface engineering, artificial synapse

I. INTRODUCTION

The advancements of non-volatile memories in device scaling with enhanced performance, as well as the establishment of the Tiki-Taka training algorithm that relaxes device requirement on the weight update, have increased the ReRAM potential to be scaled to large networks for training applications [1-3]. In recent years, it was demonstrated that the bilayer ReRAM concept based on conductive metal-oxide (CMO)/HfO_x shows reduced switching stochasticity [4], enhanced symmetry of the *set* and *reset* transition [5], and improved endurance [3,6] compared to conventional metal/HfO_x based systems [7]. The selection of the electrodes and CMO materials is of fundamental importance as they determine the energy of formation of the oxygen vacancies across the layers and their exchange upon switching [4]. For CMO, there are several established materials that can be found in literature such as TiO_x [8], AlO_x [5] and TaO_x [9]. From a material growth perspective, the choice and deposition of the CMO leads to different interfacial reactions as well as potential onset layer' growth that impacts various aspects of the resulting ReRAM devices, such as the forming voltage ($V_{forming}$), the ON/OFF ratio (i.e., the high resistive state (HRS) over low resistive state (LRS) value) and the writing reliability [4]. In this work we demonstrate a scaled (<1μm^2) ReRAM structure based on bilayer TaO_x/HfO_x stacked in between TiN electrodes with excellent analog properties and good multi-state state retention. We discuss the presence of an amorphous oxidized

CMO (TaO_x) and its effect on the device performances. A trade-off scenario is found between obtaining a good analog and low noise weight update at the expenses of an increased $V_{forming}$ as compared to a standard Ti/HfO_x-based ReRAM.

II. FABRICATION & LAYER STACK ASSESMENT

Fig.1 a) shows the schematic cross section of a typical CMO/HfO_x based ReRAM structure in a vertical configuration. To program the device, the Si n++ substrate is grounded, and the electrical signals are applied from the W/TiN top electrodes (TE). The bright field (BF) scanning transmission electron microscopy (STEM) image of a device scaled down to 200 X 200 nm^2 size is shown in Fig.1 b) with the corresponding layers highlighted on the right side of the image for convenience of the reader. The device fabrication is based on a process flow that is BEOL and CMOS compatible as summarized in Fig.1 c). The TiN bottom electrode (BE) and the HfO_x dielectric are deposited by atomic layer deposition (ALD) without breaking the vacuum to prevent oxidation of the TiN. The CMO layer is deposited by reactive sputtering and its sheet resistance is varied by more than one order of magnitude (from few kΩ up to 100 kΩ) upon changing the chamber pressure between 6 to 11 µbar [10,11]. The BF STEM micrograph in Fig.1 d) shows that most layers grow uniformly and distinct, with sharp interfaces. At the CMO/HfO_x interface a thin (~3 nm) of amorphous and more oxidized "a-CMO" is found (see bright contrast respect to upper CMO), followed by a thicker conductive crystalline "c-CMO" of ~16 nm.

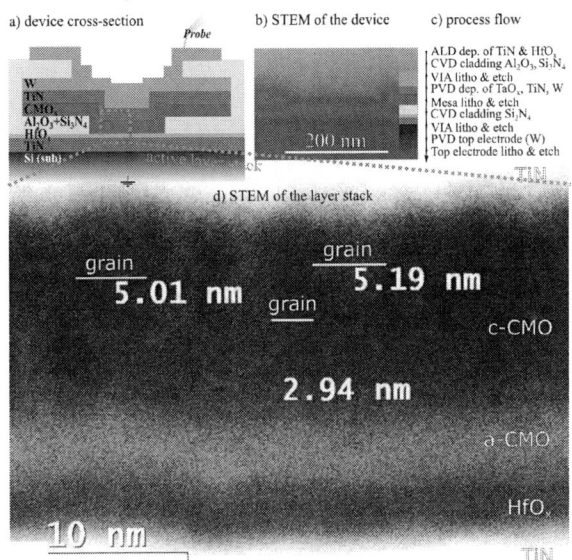

a) device cross-section   b) STEM of the device   c) process flow

d) STEM of the layer stack

Fig. 1. a) Sketch of the device cross-section and measurement configuration; b) STEM of a 200 x 200 nm² size device with highlighted layers for convenience of the reader; c) summary of the process flow. d) Cross-sectional bright field STEM micrograph of the ReRAM layer stack. The crystalline and conductive "c-CMO" layer grows on top of a thin onset of amorphous and more oxidized "a-TaO$_x$". Various grain' sizes in the c-CMO are highlighted for convenience of the reader.

A. X-ray reflectivity study

Table I summarizes the X-ray reflectivity (XRR) measurements and fitting based on two multilayer CMO-based stack. The extracted film thicknesses and densities refer to layer stacks deposited at chamber pressures of 6 and 10 µbar, resulting in CMO resistivities of 0.01 Ωcm and 0.25 Ωcm, respectively. The values of the density and thickness for the a-CMO are highlighted in yellow. Despite growing the CMO in different pressure conditions, the amorphous onset layer is not eliminated, and its thickness remains ~2.2 nm. Instead, the decreasing a-CMO density hints toward a reduced degree of oxidation between the two samples [12]. The presence of amorphous onset layers was already reported when growing a textured crystalline material on amorphous substrate [13]. The absence of a seed for the crystallization of the CMO, as well as presence of different strain conditions at the interface with HfO$_x$ can result in inhomogeneous CMO nucleation and crystallization across the CMO layer. This effect could be amplified by possible instabilities of the growth conditions at the onset of the deposition.

TABLE I. XRR SUMMARY

layer	P$_{chamb}$ 6 µbar		P$_{chamb}$ 10 µbar	
	Thick. [nm]	Density [g/cm³]	Thick. [nm]	Density [g/cm³]
c-CMO	25.2	11.6	26.6	10.7
a-CMO	2.2	9.3	2.3	8.6
HfO$_x$	5.7	9.0	5.7	9.0
TiN$_x$O$_y$	1.7	1.8	1.7	1.8
TiN	18.8	5.3	18.8	5.3
Sub	-	2.4	-	2.4

Fig. 2. Summary of thichkness ("thick.") and density values for two samples deposited at 6 and 10 µbar. The values of the density and thickness for the a-CMO are highlighted in yellow. The thickness does not vary, the density instead, decreases upon increasing the chamber deposition pressure.

III. ELECTRICAL PERFORMANCES OF CMO/HFO$_x$ RERAM

Various bilayer ReRAM devices were fabricated with the CMO parameters shown in Table I. Their performance was compared in terms of V$_{forming}$, resistive window range and ON/OFF ratio. A Ti/HfO$_x$/TiN baseline ReRAM was fabricated to compare the switching properties exhibited by the oxide bilayer stack, especially their excellent multilevel states.

A. DC characterization of CMO/HfO$_x$ ReRAMs versus Ti/HfO$_x$ ReRAM

We discuss the quasi-static DC characteristics of two bilayer ReRAMs with CMOs having resistivity ρ = 0.01 and 0.25 Ωcm and compare them to the Ti/HfO$_x$ based ReRAM. The impact of the a-CMO interlayer on the V$_{forming}$ of the bilayer stacks is clarified, as well as the overall CMO layer resistivity on the ON/OFF ratio of the devices.

1) Forming voltage comparison: Fig.3 a) shows the characteristic I-V forming curve of a Ti/HfO$_x$ based ReRAM. The soft breakdown of the HfO$_x$ dielectric occurs at 2.5 V. When the Ti metal is replaced with the CMOs, the V$_{forming}$ increases up to 5.5 V for the ReRAM based on CMO with ρ = 0.25 Ωcm, as shown in Fig.3 b). When a less resistive CMO (ρ = 0.01 Ωcm) is used in the ReRAM stack, the forming V is reduced to 4.2 V (see Fig.3 c)). The V$_{forming}$ distribution over more than 10 identical devices is reporeted in Fig.3 d) for the CMOs based ReRAM with different conductivities. Despite the difference in resistivities for the two CMO in the ReRAM stacks, they still act as electrodes, therefore the main contribution to the V$_{forming}$ increase is attributed to the oxidized a-CMO interlayers which act as a series resistor. The lower V$_{forming}$ for the device with ρ = 0.01 Ωcm is due to a slightly less oxidised state of this interlayer, as pointed out by the density decrease of the layers upon increasing the chamber pressure shown in Table I. The contribution of the interfacial CMO layer to the resistance of the pristine device is further discussed in Section IV using an impedance spectroscopy methodology.

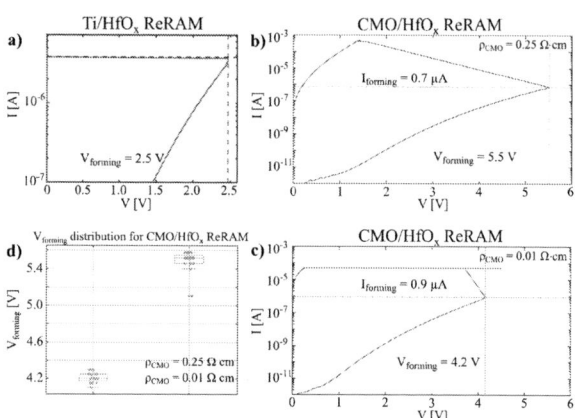

Fig. 3. DC I-V curve showing the forming operation for a 6 x 6 µm² a) Ti/HfO$_x$ ReRAM device; b) CMO/HfO$_x$ ReRAM device with ρ = 0.25 Ωcm and c) CMO/HfO$_x$ ReRAM device with ρ = 0.01 Ωcm. d) V$_{forming}$ distribution for the CMOs based ReRAM with different conductivities.

2) Impact of CMO resistivity on the ON/OFF ratio: in the bilayer devices, the tuning of the CMO resitivity, in addition to reduce the voltage at which the soft breakdown occurs, also offers a degree of tunability to increase the ON/OFF ratio (shown in Fig.4 a)) and the overall ON/OFF resistive window as displayed in Fig.4 b) and c). The increase of the resistive window is attributed to the overall increase of the full CMO resistivity and correlated stoichiometry, which affect physical parameters of the CMO such as thermal conductivity and availability of oxygen ions that can be exchanged with the filament in the HfO$_x$ layer [14,15]. It is agreed that a thin passive resistive layer in series to the ReRAM layer' stack increases the overall R of the devices but at the expenses of the ON/OFF range [16]. Our results show the opposite trend, since both the R window and the ON/OFF ratio increase for the sample with ρ$_{CMO}$ = 0.25 Ωcm (see Fig.2). Hence, also the a-CMO interlayer might undergo a soft breakdown during the

978-1-6654-8498-5/22 $31.00 © 2022 IEEE

forming operation. The a-CMO does not act anymore as a passive series resistor immediately after forming occurs.

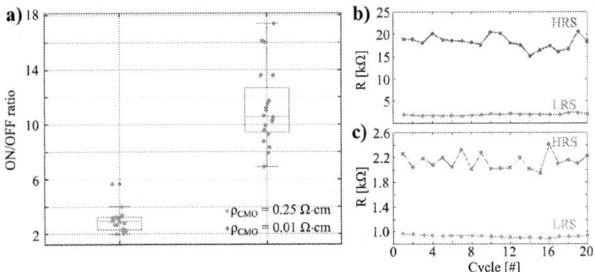

Fig. 4. a) ON/OFF ratio distribution for CMO-based ReRAM with $\rho_{CMO} = 0.25$ Ωcm (green data points) and $\rho_{CMO} = 0.01$ (blue data points); HRS and LRS cycle-to-cycle trends upon cycling 20 times for b) ReRAM device with $\rho_{CMO} = 0.25$ Ωcm and c) ReRAM device with $\rho_{CMO} = 0.01$ Ωcm.

B. AC characterization.

Major advantages of the bilayer structures are the increased number of states and the decreased stochasticity on the weight update [7]. Fig.5 a) shows a typical potentiation (lilac circles) and depression (purple circles) sequence for a 6 × 6 μm² bilayer ReRAM device with $\rho = 0.25$ Ωcm obtained by applying 200 pulses of amplitudes $V_{pot} = -0.95$ V and $V_{dep} = +1.25$ V, respectively. The pulse duration is 250 ns. The sequence was iterated 10 times to check the states' stability over cycles. Fig.5 b) shows the ten overlapping curves for both potentiation and depression, with median value highlighted in black. The average number of states ($\#states_{avg,pot}$) during potentiation is estimated as follows:

$$(G_{max} - G_{min})/ \sigma_G \qquad (1)$$

Where G_{min} is the average G value after the 1st potentiation pulse, G_{max} is the average G value after the 200th potentiation pulse, σ_G the median spread of the 200 G values over the 10 potentiation cycles. The same formula holds to extract the #states from the depression curve. For the bilayer ReRAM, the $\#states_{avg,pot} = 14.5$ while the $\#states_{avg,dep} = 21.3$. Fig.5 c) and d) shows a typical sequence of five potentiation (orange circles) and depression (sienna circles) traces for a baseline device programmed using $V_{dep} = -2.2$ V and $V_{pot} = +2.2$ V and pulse duration of 250 ns. To optimize the analog change of weights, the device is operated in subthreshold regime at small dynamic range and at low R. As opposed to the bilayer traces, the baseline shows a more abrupt conductance change, drift of the overall traces and a more stochastic conductance change. Applying formula (1) on the baseline device leads to $\#states_{avg,pot} = 3.4$ and the $\#states_{avg,dep} = 6.5$.

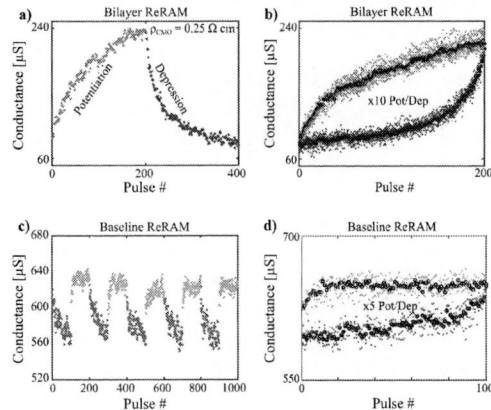

Fig. 5. a) Potentiation and depression trace for a CMO/HfOx ReRAM device of 6 × 6 μm²; b) overlap of ×10 potentiation (lilac cicles) and depression (purple circles) traces with highlighted median (black circles). c) Sequence of ×5 potentiation and depression traces for Ti/HfOx ReRAM. Noise and abruptness of the potentiation trace are evident in all three cases; d) overlap of ×10 potentiation (orange cicles) and depression (sienna circles) traces with highlighted median (black circles).

C. Scalability and Retention.

As scaling to sub-μm sizes of the ReRAM technology is an important requirement for future integration of large networks [1], in Fig.6 a) we show an example of DC I-V sweep for a 70 × 70 nm² bilayer device cycled 10 times. The curves overlap well as also confirmed by the HRS and LRS trend upon DC cycling shown in Fig.6 b). A retention tests of eight resistive states is also performed and shown in Fig.6 c). The device is set in various intermediate states varying the applied DC bias. Each state is read at 0.2 V with a sampling every 60 s for a total time of 30 min before setting the next state to read. All states remain distinct in the measured time and show no drift.

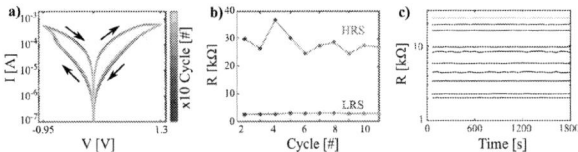

Fig. 6. a) Typical DC I-V sweep for a 70 × 70 nm² bilayer device cycled 10 times. The arrows indicate the polarity of the switching b) HRS and LRS cycle-to-cycle variations. C) Retention of multilevel states measured up to 30 min.

IV. EQUIVALENT CIRCUIT OF THE PRISTINE DEVICE STATE

The contribution of the interfacial a-CMO layer to the resistance of the device in pristine state is here further discussed in the context of an impedance spectroscopy experiment. A model physically representing a leaky dielectric between metallic electrodes was chosen to represent the device. The model parameters used to fit impedance data, measured between 100 Hz and 1 MHz, were the access resistance in series (R_{ser}), and the leakage of the dielectric in parallel (R_{par}) to the non-ideal capacitive contribution modelled as Constant Phase Element (CPE) Q_0 and n [17]. In addition to the device in pristine state, two more *reference* samples were fabricated and measured using the following layer stacks: a TiN/CMO/BE stack and a

TiN/HfO$_x$/BE one. The goal was to extract the model parameters of these *reference* samples, and then to compare them to the ones extracted from the full ReRAM stack in pristine state. Fig.7 b) shows that, for all device areas investigated, the measured R$_{par}$ of the full device (blue points) is systematically higher than the R$_{par}$ contribution of the stacks with CMO (purple) and HfO$_x$ layers alone (red points). The dashed grey line would be the expected value of R$_{par}$ for the bilayer ReRAM if it exactly corresponded to the sum of the R$_{par}$ of the TaO$_x$ and HfO$_x$ *reference* layer stacks (leakage contributions put in series). We attribute the higher resistance to the creation of the oxidized a-CMO interlayer during the growth of the crystalline CMO on the amorphous HfO$_x$ layer. The schematic in Fig.7 a) further depicts how the different parameters combine to represent the pristine stack model. This finding confirms that in the pristine state of the bilayer ReRAM, the a-CMO layer acts as an additional R$_{CMO-HfOx\ interface}$ increasing the R$_{par}$ resistance of the layer stack, leading to higher V$_{forming}$ compared to the baseline Ti/HfO$_x$ device.

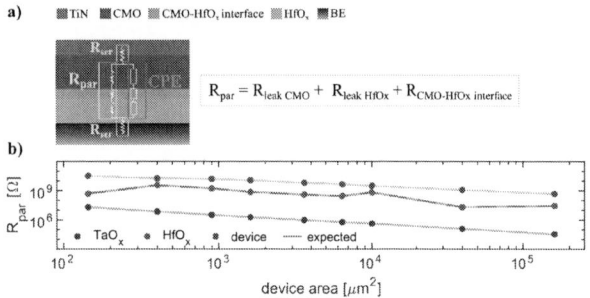

Fig. 7. a) Graphic representation of the expected model contributions for the bilayer ReRAM in pristine state. The R$_{par}$ of the device is the leakage of CMO and HfO$_x$ layers with an additional interface; b) R$_{par}$ values are extracted by fitting the impedance data with the model shown in a) assuming only the CMO layer between electrodes (purple points), only the HfO$_x$ layer between electrodes (red points), and the full device stack described in a) (blue points). The values are extracted for various device sizes ranging from 12 x 12 μm^2 to 400 x 400 μm^2 to better validate the model. The expected R$_{par}$ values for the device stack, assuming that they corresponds to the sum of the R$_{par}$ of the TaO$_x$ and HfO$_x$ *reference* layer stacks, are shown with the grey dotted line.

V. CONCLUSIONS

In summary we have demonstrated an analog multistate ($\geq$ 4 bits) bilayer TaO$_x$/HfO$_x$-based ReRAM technology that is scalable down to device sizes of 70 x 70 nm^2. During retention measurements, the states show no significant drift for up to 30 min. A structural assessment highlights the presence of an oxidized a-CMO interlayer between the CMO and the HfO$_x$ layer. The EEC built within an impedance spectroscopy experiment allows us to attribute the increased V$_{forming}$ compared to a baseline ReRAM to the R$_{ser}$ contribution of that a-CMO interlayer. Despite this, we have shown that varying the CMO' resistivity offers a degree of tunability in reducing the V$_{forming}$ of the bilayer devices and in increasing the ON/OFF ratio and resistive window by a factor of 10. Our bilayer ReRAM is fully CMOS- and BEOL-friendly.

ACKNOWLEDGMENT

The authors acknowledge the Binnig and Rohrer Nanotechnology Center (BRNC) at IBM Research Europe - Zurich. This work is funded by the European Union within the H2020 "MANIC" (grant ID: 861153), "MeM-Scales" (grant ID: 871371), "NEoteRIC" (grant ID: 871330), "NEBULA" (grant ID: 871658) and "PlasmoniAC" (grant ID: 871391) projects.

REFERENCES

[1] T. Gokmen, W. Haensch, "Algorithm for Training Neural Networks on Resistive Device Arrays," *Front. Neurosci*, vol. 14, no. 103, p. 1, Feb 2020, doi:10.3389/fnins.2020.00103.

[2] M. Onen, *et al.*, "Neural Network Training with Asymmetric Crosspoint Elements," *ArXiv*, abs/2201.13377, 2022.

[3] Y. -B. Kim *et al.*, "Bi-layered RRAM with unlimited endurance and extremely uniform switching," *2011 Symposium on VLSI Technology - Digest of Technical Papers*, 2011, pp. 52-53, Jan. 2011.

[4] W. Banerjee *et al.*, "Variability Improvement of TiO$_x$ /Al$_2$O$_3$ Bilayer Nonvolatile Resistive Switching Devices by Interfacial Band Engineering with an Ultrathin Al$_2$O$_3$ Dielectric Material." *ACS omega*, vol. 2, pp. 6888-6895. Oct. 2017, doi:10.1021/acsomega.7b01211.

[5] W. Jiyong *et al.*, "Improved Synaptic Behavior Under Identical Pulses Using AlO$_x$/HfO$_2$ Bilayer RRAM Array for Neuromorphic Systems." *IEEE Electron Device Letters*, vol. 37, pp. 994-997, June 2016, doi:10.1109/LED.2016.2582859.

[6] C. Y. Huang, T. L. Tsai, C. A. Lin, T. Y. Tseng, "Switching mechanism of double forming process phenomenon in ZrO$_x$/HfO$_y$ bilayer resistive switching memory structure with large endurance," *Appl. Phys. Lett.*, vol. 104, p. 062901, Feb. 2014, doi:10.1063/1.4864396.

[7] A. Hardtdegen, C. La Torre, F. Cüppers, S. Menzel, R. Waser and S. Hoffmann-Eifert, "Improved Switching Stability and the Effect of an Internal Series Resistor in HfO$_2$/TiO$_x$ Bilayer ReRAM Cells," in *IEEE Transactions on Electron Devices*, vol. 65, no. 8, pp. 3229-3236, Aug. 2018, doi: 10.1109/TED.2018.2849872.

[8] Cong Ye, *et al.*, "Enhanced resistive switching performance for bilayer HfO$_2$/TiO$_2$ resistive random access memory," 2016 *Semicond. Sci. Technol.*, vol. 31, no. 10, p. 105005, Sept. 2016, doi:10.1088/0268-1242/31/10/105005.

[9] Y. Zhao *et al.*, "Modeling and Optimization of Bilayered TaO$_x$ RRAM Based on Defect Evolution and Phase Transition Effects," in *IEEE Transactions on Electron Devices*, vol. 63, no. 4, pp. 1524-1532, April 2016, doi: 10.1109/TED.2016.2532470.

[10] T. Stecconi, et al., "Equivalent electrical circuit modelling of a TaO$_x$/HfO$_x$ based RRAM with optimized resistance window and multilevel states", unpublished.

[11] T. Stecconi, et al., "Filamentary TaO$_x$/HfO$_x$ ReRAM Devices for Neural Networks Training with Analog In-memory Computing;" unpublished.

[12] Density of Minerals III: Oxides and stoichiometry from L. Bruce Railsback' Some Fundamentals of Mineralogy and Geochemistry. Online at http://railsback.org/Fundamentals/FundamentalsCarbs.html (Accessed 12.04.2022).

[13] Redaelli, M. Boniardi, E. Varesi, R. Calarco, J. E. Boschker, "Textured memory cell structures". Patent No.: US 11,264,568 B2. Date of Patent: Mar. 1 , 2022.

[14] S. U. Sharath *et al.*, "Impact of oxygen stoichiometry on electroforming and multiple switching modes in TiN/TaO$_x$/Pt based ReRAM," *Appl. Phys. Lett.*, vol. 109, p. 173503, Oct. 2016, doi:org/10.1063/1.4965872.

[15] W. Wu, H. Wu, B. Gao, N. Deng, S. Yu and H. Qian, "Improving Analog Switching in HfO$_x$-Based Resistive Memory With a Thermal Enhanced Layer," in *IEEE Electron Device Letters*, vol. 38, no. 8, pp. 1019-1022, Aug. 2017, doi: 10.1109/LED.2017.2719161.

[16] F. Cüppers et al., "Exploiting the switching dynamics of HfO$_2$-based ReRAM devices for reliable analog memristive behavior," APL Mater. 7, 091105, Sept. 2019, doi:10.1063/1.5108654.

[17] J. Jacquelin, "A number of models for CPA of conductors and for relaxation in non-Debye dielectrics," *Journal of Non-Crystalline Solids*, 131, pp. 1080-1083, June 1991.

Impact of Gold Interconnect Microstructure on Electromigration Failure Time Statistics

Hajdin Ceric, Roberto Lacerda de Orio, and Siegfried Selberherr

Institute for Microelectronics, TU Wien

Vienna, Austria

Email: {ceric|orio|selberherr}@iue.tuwien.ac.at

Abstract—**Simulating the influence of microstructure on the electromigration reliability of a metallic interconnect is a challenging task due to the complexity of grain boundary physics and to the large number of grains. Crucial segments of gold metallization used for GaAs devices are susceptible to significant electromigration degradation and have a microstructure with a thousands of grains. In this work, a complete physics-based analysis of electromigration in gold is presented. A novel approach for the numerically efficient simulation of an interconnect containing a large number of grains is introduced. By building grain compounds containing hundreds of grains and equipping them with appropriate models, the dependence of statistical failure features on the variation of geometric properties is investigated. The experimentally observed dependence of the mean failure time and the associated standard deviation of the failure times on the interconnect geometry is well reproduced by our simulations.**

I. Introduction

Gold is the metal of choice for interconnects implemented on GaAs [1], because it forms a very low resistance ohmic contact, has a high melting temperature, and a low resistivity [2]. In the last decades numerous electromigration reliability studies have been performed on both aluminum and copper metalizations. In contrast, the number of studies on gold metallizations is very limited. The one-dimensional Korhonen model [3] is widely used today for electromigration analysis due to its main advantage which is to provide a closed equation for stress evolution. However, an important disadvantage is that it is not clear, how the evolution of the components of the stress tensor is affected by the geometry of the metallization or by more general external constraints imposed by barrier or passivation layers. In the pioneering work by Sarychev and Zithnikov [4], the general model for electromigration degradation of three-dimensional metallic interconnects is described. The Sarychev-Zithnikov model's generality makes it the optimal choice for studying electromigration degradation of gold metallization and in this work it is applied to study the experimental results published in [5]. As an extension of the original model we have introduced and applied a concept for modeling the impact of microstructure. The applied method is based on several previous works [6], [7], [8]. A number of grains are grouped into larger grain compounds and at the boundaries of these domains a detailed grain boundary model is applied [8]. The boundaries of the large grain compounds are chosen to comprise characteristic segments of the interconnect geometry, e.g., vias. Alternatively, these segments can also be divided into several grain compounds. The size statistics of grain compounds are set according to the experimental grain size distribution statistics. The lognormal distibution of grain sizes determines the duration of interconnect lifetime and it is also responsible for its lognormal distribution [9]. All simulations are performed with COMSOL multiphysics [10] and the background gold microstructure is generated using NEPER [11], a software package for polycrystal generation and meshing.

II. Modeling Electromigration

The degradation of metallic interconnect, until its complete failure, runs through the two distinctive phases [12], namely void nucleation and void evolution. Correspondingly, the time-to-failure for an interconnect structure, t_{TTF}, consists of a void nucleation time, t_N, and a void evolution time, t_E, corresponding to the two failure development phases.

$$t_{\text{TTF}} = t_N + t_E \tag{1}$$

The modeling of the first phase demands the solution of the vacancy balance equation together with the Laplace equation, the heat-transport equation, and the mechanical equations. The modeling of the second phase requires the solution of all equations from the first phase together with models describing the evolving void surface [12]. The central governing equations of the electromigration model are the vacancy flux equation (2) and the vacancy balance equation (3) [4]. The total vacancy flux consists of fluxes driven by the concentration gradient ($\vec{J}_v^c$), by the electromigration ($\vec{J}_v^e$), by the pressure gradient ($\vec{J}_v^s$), and by the temperature gradient ($\vec{J}_v^T$).

$$\vec{J}_v = \vec{J}_v^c + \vec{J}_v^e + \vec{J}_v^s + \vec{J}_v^T \tag{2}$$

$$\frac{\partial C_v}{\partial t} = -\nabla \cdot \vec{J}_v + G_{\text{eff}}(C_v) \tag{3}$$

$$\vec{J}_v^c = -D_{\text{eff}}\nabla C_v, \quad \vec{J}_v^e = D_{\text{eff}}\frac{C_v}{kT}|Z_{\text{eff}}^*|e|\rho\vec{j} \tag{4}$$

$$\vec{J}_v^s = \frac{D_{\text{eff}}C_v f\Omega}{kT}\nabla\sigma_{\text{Hyd}}, \quad \vec{J}_v^T = -\frac{D_{\text{eff}}C_v Q}{kT^2}\nabla T$$

D_{eff} is the local diffusivity, C_v is the vacancy concentration, σ_{Hyd} is the hydrostatic stress, ρ is the interconnect resistivity, $\vec{j}$ is the current density, Ω is the atomic volume, f is the atom-vacancy relaxation factor, Q is the heat of transport, G_{eff}

is the Rosenberg-Ohring recombination term, and Z_{eff}^* is the effective valence. The second phase of failure development starts, when the critical stress is reached at some point inside the interconnect or at some interface to surrounding layers. The second phase ends with the interconnect failure. Since there is no available data in the literature for the critical stress for gold interconnect embedded in nitride dielectrics, the critical stress in the present study was set according to the methodology previously used for copper interconnects [13].

III. ANALYSIS OF EXPERIMENTAL FINDINGS

The recent study by Hau-Riege and Yau [5] provided interesting insights in the development of electromigration induced failure in interconnect lines made of gold.

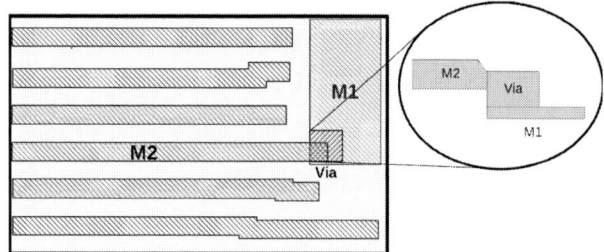

Fig. 1. Layout of the line-end of the investigated test structures.

The current density of $10\,\text{mA}/\mu\text{m}^2$ and the temperature of $361\,°\text{C}$ are used as test conditions in a MIRA EM module [14]. The gold interconnect, consisting of two metallization levels, M1 and M2, and the via is fully embedded in a dielectric (silicon nitride, SiN). Several types of test structures have been used (cf. Fig. 1). The electromigration lifetime distributions for all experiments were monomodal and lognormally distributed. It was expectedly observed that the test structures with smaller vias are more sensitive to the process and microstructural variations, leading to a larger variation in failure times. Some structures also exhibited multiple failure modes. The experimental results demonstrate that the area of square vias affects electromigration lifetime, until a via area of about 55 μm^2 is reached, therefter the lifetime becomes less sensible to the via size.

IV. SIMULATION RESULTS AND DISCUSSION

For our simulations test structures with square vias were chosen. In [5] two failure modes were observed: a) voiding at the M1/Via interface and b) voiding through the thickness of M2 at the Via/M2 interface. The layout for simulation, including all dimensions and the choice of materials, is constructed according to the structure used in the experiments (cf. Fig. 1). A key question is to which extent electromigration models previously used to analyze electromigration in copper and aluminum can explain such failures and whether simulations based on these models can reproduce the failure behavior observed in gold interconnect. Before conducting the simulation studies, it is necessary to parameterize the applied model. For the effective valence, Z_{eff}^*, a value calculated by an

ab initio method is used [15]. The values for the bulk and the grain boundary diffusivities are also available in the literature [16]. The mechanical parameters for gold and silicon nitride are set according to [17]. During the simulations the rise of tensile stress at the M1/Via interface was monitored. The time

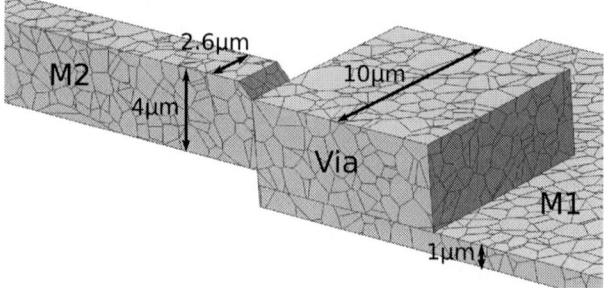

Fig. 2. An example of microstructure used for simulation. M2, via, and M1 are filled with a grain boundary network.

t_N needed for the tensile stress at the M1/Via interface to grow to the level of the critical stress threshold is assumed to be much larger (cf. Fig. 2 from [5]) than the subsequent time for void growth, t_E, and consequently the time-to-failure is determined as $t_{\text{TTF}} \approx t_N$. The polycrystalline background microstructure is generated with the generator NEPER [11]. All three segments of the interconnect, M1, Via, and M2 are filled by a numerically generated network of grain boundaries. The microstructure possessed a lognormal distribution of grain sizes with the average grain size of 0.95 μm and the standard deviation of 0.3 μm. These values are set according to the microstructural characterization measurements carried out in [14]. An example of the generated microstructure is presented in Fig. 2. A number of grains are grouped into grain compounds and at the boundaries of these compounds a detailed grain boundary model is applied [8]. For each of 6 different via sizes 10 microstructures were generated and for each via size and each microstructure a simulation was performed, until the critical stress threshold at the M1/Via interface is reached and the time-to-failure is determined.

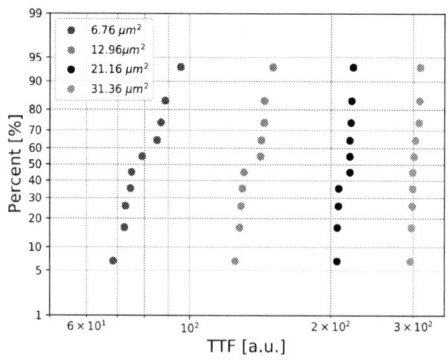

Fig. 3. Lifetime distribution for different via sizes.

In Fig. 3 the lifetime distributions for four different via sizes are presented. As one can see, all distributions are lognormal and monomodal, as observed in the experiments. With increasing via surface the mean-time-to-failure (MTTF)

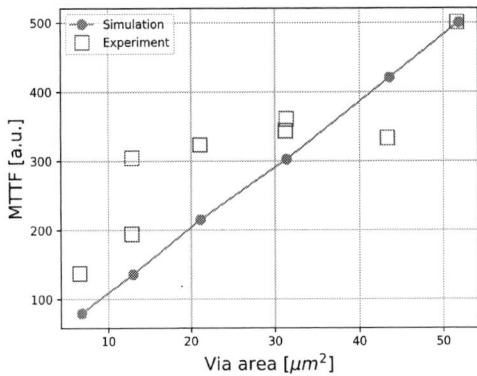

Fig. 4. MTTF for different via sizes. The maximal value of MTTF is adjusted to the experimental value from [5].

is also increasing, cf. Fig. 4. An explanation for an increased MTTF for larger via sizes is that due to a larger reservoir of atoms and wider via surface the stress build up is slower. The

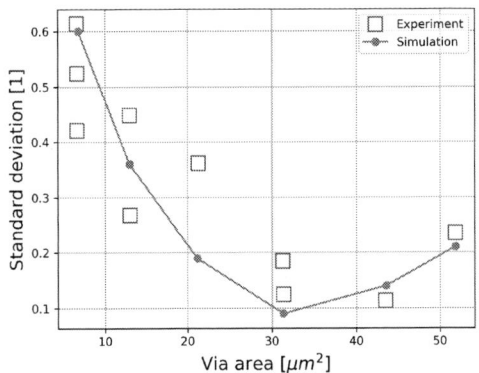

Fig. 5. Standard deviation for different via sizes. The maximal value of standard deviation is adjusted to the experimental value from [5].

corresponding standard deviation, displayed in Fig. 5, however, reduces up to a via surface of 31.36 μm^2, and then increases towards the via surface of 51.84 μm^2. While a reduction in the standard deviation is expected as a consequence of the smaller influence of the microstructure for larger vias, its subsequent increase calls for a more subtle explanation. The probable origin of this phenomena is the accumulation of standard deviation, analogously to the following case: If we assume that each of N grain sizes l_i behaves as an independent stochastic variable with the same statistical properties, the cumulative standard deviation (SD) can be expressed as:

$$\mathrm{SD}\left[\sum_{i=1}^{N} l_i\right] = \sqrt{N}\,\mathrm{SD}[l], \ \mathrm{SD}[l] = \mathrm{SD}[l_i], \ i = 1, \ldots, N \quad (5)$$

The properties of the vacancy flux, which determines the interconnect failure behavior, are influenced by both geometry and microstructure. The dimensions of the via apparently compete with a cumulative effect of grain size variations as the dominant factor in determining the standard deviation of failure times.

V. CONCLUSIONS

We applied state-of-the-art electromigration modeling to study the evolution of electromigration-induced degradation of gold interconnect. The model accounts for previously experimentally determined statistical properties of gold. As observed in the experimental reliability study, the MTTF increases with an increasing surface of the via, while simultaneously, the standard deviation exhibits both a decreasing as well as an increasing phase. This behavior is consistently reproduced by our simulations results.

REFERENCES

[1] N. A. Kulchitsky, A. V. Naumov, and V. V. Startsev, "Photonic and Terahertz Applications as the Next Gallium Arsenide Market Driver," *Mod. Electron. Mater.*, vol. 6, p. 77, 2020.

[2] S. Kasap and P. Capper, *Springer Handbook of Electronic and Photonic Materials*. Springer, 2017.

[3] M. A. Korhonen, P. Borgesen, K. N. Tu, and C. Y. Li, "Stress Evolution Due to Electromigration in Confined Metal Lines," *J. Appl. Phys.*, vol. 73, no. 8, pp. 3790–3799, 1993.

[4] M. E. Sarychev and Y. V. Zhitnikov, "General Model for Mechanical Stress Evolution During Electromigration," *J. Appl. Phys.*, vol. 86, no. 6, pp. 3068–3075, 1999.

[5] C. Hau-Riege and Y. Yau, "Electromigration of Gold Metallization," in *Proc. Intl. Symposium on the Physical and Failure Analysis of Integrated Circuits*, 2021, pp. 1–5.

[6] L. Filipovic, "A Method for Simulating the Influence of Grain Boundaries and Material Interfaces on Electromigration," *Microelectron. Reliab.*, vol. 97, pp. 38–52, 2019.

[7] K. E. Aifantis and S. A. Hackney, "Morphological Stability Analysis of Polycrystalline Interconnects under the Influence of Electromigration," *Rev. Adv. Mater. Sci.*, vol. 19, pp. 98–102, 2009.

[8] H. Ceric, H. Zahedmanesh, and K. Croes, "Analysis of Electromigration Failure of Nano-Interconnects through a Combination of Modeling and Experimental Methods," *Microelectron. Reliab.*, vol. 100-101, p. 113362, 2019.

[9] R. L. de Orio, H. Ceric, J. Cervenka, and S. Selberherr, "The Effect of Copper Grain Size Statistics on the Electromigration Lifetime Distribution," in *Proc. Simulation of Semiconductor Processes and Devices*, 2009, pp. 1–4.

[10] COMSOL Multiphysics, Version 5.6, 2021.

[11] R. Quey, P. Dawson, and F. Barbe, "Large-Scale 3D Random Polycrystals for the Finite Element Method: Generation, Meshing and Remeshing," *Comput. Methods. Appl. Mech. Eng.*, vol. 200, no. 17-20, pp. 1729–1745, 2011.

[12] H. Ceric and S. Selberherr, "Electromigration in Submicron Interconnect Features of Integrated Circuits," *Mater. Sci. Eng. R Rep.*, vol. 71, pp. 53–86, 2011.

[13] C. S. Hau-Riege, S. P. Hau-Riege, and A. P. Marathe, "The Effect of Interlevel Dielectric on the Critical Tensile Stress to Void Nucleation for the Reliability of Cu Interconnects," *J. Appl. Phys.*, vol. 96, no. 10, pp. 5792–5796, 2004.

[14] S. Kilgore, Ph.D. dissertation, Arizona State University, 2013.

[15] A. Lodder and J. P. Dekker, "The Electromigration Force in Metallic Bulk," *AIP Conf. Proc.*, vol. 418, pp. 315–329, 1998.

[16] H. Mehrer, *Diffusion in Solids, Fundamentals, Methods, Materials, Diffusion-Controlled Processes*. Springer Series in Solid-State Sciences, 2007.

[17] D. R. Lide, *CRC Handbook of Chemistry and Physics: A Ready-Reference Book of Chemical and Physical Data*. CRC Press, 2003.

Influence of Metal on Schottky Barrier Inhomogeneity in Ga_2O_3 Schottky Barrier Diodes

Min-Yeong Kim
Department of Electronic Materials Engineering
Kwangwoon University
Seoul, Republic of Korea
alsdud9971@kw.ac.kr

Geon-Hee Lee
Department of Electronic Materials Engineering
Kwangwoon University
Seoul, Republic of Korea
ghlee117@kw.ac.kr

Hee-Jae Lee
Department of Electronic Materials Engineering
Kwangwoon University
Seoul, Republic of Korea
tkrkek123@kw.ac.kr

Dong-Wook Byun
Department of Electronic Materials Engineering
Kwangwoon University
Seoul, Republic of Korea
byun1994@kw.ac.kr

Michael A. Schweitz
Department of Electronic Materials Engineering
Kwangwoon University
Seoul, Republic of Korea
michael.schweitz@schweitzlee.com

Sang-Mo Koo*
Department of Electronic Materials Engineering
Kwangwoon University
Seoul, Republic of Korea
smkoo@kw.ac.kr

Abstract—**A systematic study of Schottky barriers fabricated on Ga_2O_3 film is reported. The Schottky barrier heights (SBHs) and current transport mechanisms were estimated through the analysis of current-voltage (I-V), capacitance-voltage (C-V), current-voltage-temperature (I-V-T), and x-ray photoelectron spectroscopy (XPS) measurements. The estimated values of the SBH of Ni varied from 1.01 eV to 1.13 eV depending on the method of analysis. These values are close to the theoretical value derived from the Schottky-Mott relationship. In contrast, for Au, there is a lack of general agreement between SBH values determined by different methods. This appears to hint at the presence of spatial inhomogeneities affecting the estimated SBH values to varying degrees depending on the method of analysis. The dependence of the SBH and carrier transport mechanism on metals suggests that metals/Ga_2O_3 interfaces are not fully pinned, and this fact can affect device reliability.**

Keywords—*Schottky barrier height, Gallium Oxide, inhomogeneity, Schottky barrier diode, metal*

I. INTRODUCTION

Schottky diodes based on wide bandgap semiconductors like gallium oxide (Ga_2O_3) are being pursued for use in high power electronic systems [1]. Ga_2O_3 exists in five known polymorphs (α, β, γ, δ, and ε) as reported by Roy *et al* [2]. Among them, the β-Ga_2O_3 is the most thermodynamically and chemically stable [3]. β-Ga_2O_3 has an ultra wide bandgap of 4.7-4.9 eV [4], a high theoretical breakdown field of 8 MV cm^{-1}, which in turn gives it a much larger Baliga's figure of merit value than silicon carbide (SiC) and gallium nitride (GaN) [5,6]. Ga_2O_3 also has a relatively low thermal conductivity (10-29 W/m·K) [7,8], however, when compared to other semiconductors, including SiC (280 W/m·K) [9] and GaN (130 W/m-K) [10]. This material property can result in severe self-heating, with an associated deterioration of electrical performance and device reliability [11]. In order to mitigate thermal issues and increase device performance, Ga_2O_3 devices have been manufactured on a variety of heterogeneous substrates with high thermal conductivity [12,13].

Studies of Schottky contacts on Ga_2O_3 play a pivotal role which is a major requirement for superior device performance and reliability [14,15]. Schottky barrier heights (SBH) were derived from the measurements and are discussed in terms of predicted behavior according to the Schottky-Mott theory, which defines that the SBH is the difference in work function of the metal and the electron affinity of the semiconductor [16]:

$$\emptyset_M - \mathcal{X}_S = \emptyset_B \tag{1}$$

where $\emptyset_B$ is the SBH, $\emptyset_M$ is the work function of the metal and $\mathcal{X}_S$ is the electron affinity of the semiconductor.

However, the Schottky–Mott model gives grossly incorrect predictions for experimental values of SBHs [17]. To date, there are many systematic studies of SBH using different metals on Ga_2O_3 substrate [18-21], but studies of metals on deposited Ga_2O_3 thin films are still lacking. Furthermore, it is particularly important to develop an understanding of the inhomogeneity of metals on deposited Ga_2O_3 because of the need for using heterogeneous substrates to manage the self-heating effect. Therefore, we have investigated Schottky diode characteristics of devices fabricated with Ni and Au metal contacts on Ga_2O_3 films deposited on SiC substrates. Here, we present a comparison of the obtained material and device parameters, including Schottky barrier properties.

II. EXPERIMENTAL DETAILS

A. Device Fabrication

N-type 4H-SiC wafers (Fluoroware Inc., USA) (base substrate: $N_D = 1\times10^{19}$ cm^{-3}, n-type epitaxial layer: $N_D = 5\times10^{16}$ cm^{-3}) were soaked in acetone and methanol at 120 °C for 10 min, respectively, and subsequently rinsed in deionized water. Vertical Schottky diodes with either Ni or Au top metal Schottky contacts and large area Ni back metal ohmic cathode contacts were fabricated. The schematic of the fabricated Ga_2O_3 Schottky diodes is shown in the inset of Fig. 1. The Ni cathodes were deposited by electron beam evaporation of a Ni target in a vacuum ambient and then annealed at 1050 °C for 1 min in a rapid thermal annealing furnace in an N_2 atmosphere. Subsequently, Ga_2O_3 thin films were deposited on the n-type SiC substrates by a radio frequency magnetron sputtering (RF-sputtering) system

978-1-6654-8498-5/22 $31.00 © 2022 IEEE

using 99.99% pure Ga$_2$O$_3$ targets. The sputtering chamber was pre-evacuated to approximately 3.99×10^{-5} Pa. Argon gas (99.999%) was then introduced into the chamber to a pressure of 3.33 Pa which was maintained during the sputtering process. RF-sputtering was performed for 5 hours with the radio frequency generator operating at a power of 120 W. The thickness of the obtained Ga$_2$O$_3$ thin films was confirmed by ellipsometry to be 300 nm. We inserted the carrier with samples into the furnace and placed it in a constant temperature region when the furnace reached the equilibrium temperature of 900°C for 1 h. The Schottky contacts were subsequently deposited by thermal evaporation

B. Measurements and Characterizations

Bond structures were analyzed by X-ray photoelectron spectroscopy (XPS) using an ESCALAB 250 instrument with Al Kα (hν. = 1486.6 eV) as the X-ray source. Before measurements, all sample surfaces were cleaned by 200 eV Ar+ ion etching for 90 sec. All XPS spectra are calibrated by the C 1s peak (284.6 eV) to compensate for any residual charge. XPS analysis was carried out at a pressure of 4.8×10^{-7} Pa and the analyzed area was about 0.12 mm^2. A Keithley 4200 parameter analyzer was employed for the current-voltage (I-V) and capacitance-voltage (C-V) characterization. C-V characteristics were obtained for a frequency of 1 MHz. Temperature dependent I-V (I-V-T) measurements were performed for temperatures ranging from 298 K to 523 K, using a probe station with a temperature-controlled sample holder chuck.

III. RESULTS AND DISCUSSION

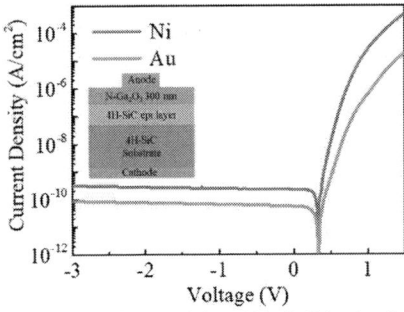

Fig. 1 Current-Voltage characteristics of Ga$_2$O$_3$ Schottky diodes measured at 298 K for two different metals, Ni and Au.

The 298 K I-V characteristics of Schottky diodes for Ni and Au metals are shown in Fig. 1. From I-V characteristics, some important device parameters like the distribution of the interface state between the metal and the semiconductor could be determined [22]. The excellent rectifying behavior with low leakage current and exponential current in the forward voltage range indicates that thermionic emission (TE) might be the dominant transport mechanism [23]. Therefore, the measured I-V characteristics were analyzed using the ideal TE theory demonstrated by Rhoderick and Williams [24]

$$I = I_S \exp[^{qV}/_{nkT} - 1], \qquad (2)$$

where

$$I_S = AA^*T^2\exp[^{q\phi_B}/_{kT}]. \qquad (3)$$

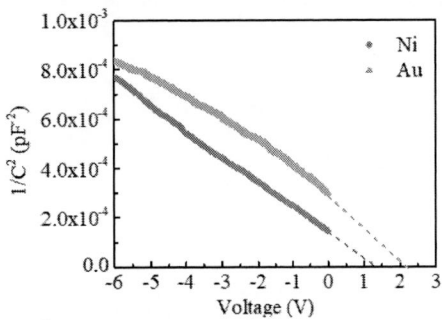

Fig. 2 1/C^2-V characteristics of Ga$_2$O$_3$ Schottky diodes measured for two different metals, Ni and Au.

Here, I$_S$ is the reverse saturation current, q is the electronic charge, A is the diode area, A* is the Richardson constant, V is applied voltage, n is the ideality factor, T is temperature, k is the Boltzmann constant and Φ$_B$ is the SBH. Ideality factors for Ni and Au Schottky diodes were extracted as 1.67 and 2.30, respectively. Using A* = 41 A/cm^{2}K^2 for Ga$_2$O$_3$ [22,23], the room temperature barrier heights for Ni and Au are obtained as 1.04 eV and 1.11 eV, respectively. The Schottky diodes were characterized by the C-V method in Fig. 2, to gain even further confidence in the extracted SBH values. The capacitance measurements were performed at a frequency of 1 MHz, C-V characteristics of Schottky diodes are expressed as

$$^1/_{C^2} = 2[q\phi_B - (E_C - E_F) - kT - qV]/q^2\varepsilon_s\varepsilon_0 N_D \qquad (4)$$

where E$_C$ is the conduction band minimum, E$_F$ is the Fermi level, ε$_0$ is the vacuum dielectric constant and relativity permittivity ε$_s$ = 10 for Ga$_2$O$_3$ [25]. For the ideal case, the intercept on the x-axis and slope of the 1/C^2-V plot provides the built-in voltage and doping concentration, respectively. The SBH can be determined using the doping concentration and built-in voltage V$_{bi}$ = Φ$_B$ − (E$_C$ − E$_F$)/q. Using an effective mass value for Ga$_2$O$_3$ of 0.34m [26], E$_C$ − E$_F$ was estimated to be 0.09 eV [26]. The SBHs of Ni and Au diodes estimated with C-V are 1.11 eV and 2.84 eV. Values of the SBH determined by C-V and I-V methods differ by 0.07 eV for Ni diodes, and 1.73 eV for Au diodes. Note that, SBHs determined from the C-V data are typically higher than SBHs determined from I-V. When using the I-V relationship, the SBH is obtained by the contact path in the vicinity of low barrier height, whereas the C-V derived SBH value may be different because it is essentially an averaged value [27]. Therefore, the fact that the difference between the values of SBH derived from I-V and C-V is significant can be related to the effect of slight such as spatially inhomogeneous SBH [27]. We accounted for the spatial distribution of Schottky barriers in our analysis of J-V-T measurements of the vertical (Ni and Au) device structures from 298K to 523 K, in 25 K steps. As such, J-V-T measurements provide a potential third method to determine SBHs for these diodes. According to the ideal TE model, the MS interface should be atomically flat and spatially homogeneous [28]. However, due to varying barrier height patches at the nanoscale [27,28], the current conduction at the interface is not the same in the overall temperature range. At low temperatures, charge carriers can get energies exceeding the barrier height in patches with low barrier height, while in the high temperature band, carriers can get energies exceeding the barrier height in patches with high barrier height. [29-31]. These temperature-dependent properties can be described as Gaussian distribution of

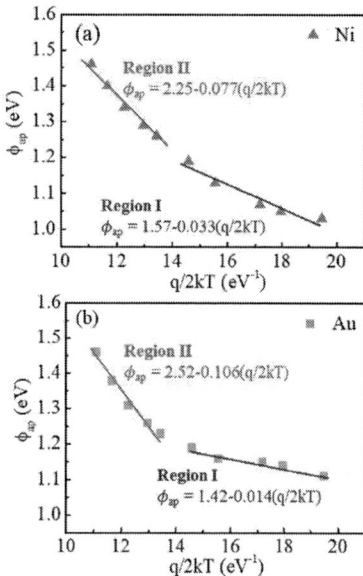

Fig. 3 Apparent barrier height versus q/2kT and linear fits for the (a) Ni/β-Ga$_2$O$_3$ SBDs and (b) Au /β-Ga$_2$O$_3$ SBDs according to the Gaussian distribution of the barrier heights.

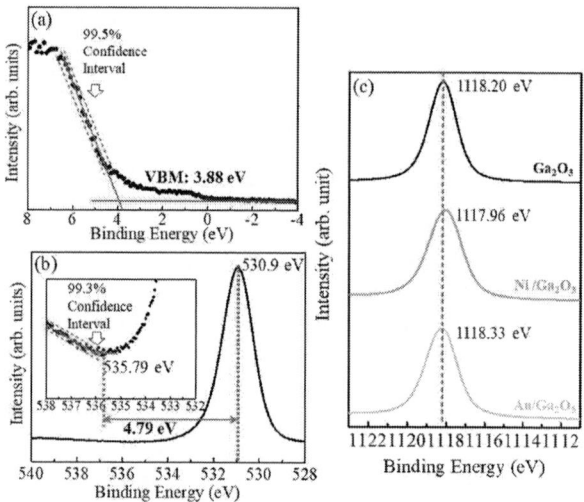

Fig. 4 Apparent barrier height versus q/2kT and linear fits for the (a) Ni/β-Ga$_2$O$_3$ SBDs and (b) Au /β-Ga$_2$O$_3$ SBDs according to the Gaussian distribution of the barrier heights. (c) Ga 2p$_{3/2}$ core levels spectra observed from Ga$_2$O$_3$ thin film, Ni/Ga$_2$O$_3$ interface, and Au/Ga$_2$O$_3$ interface.

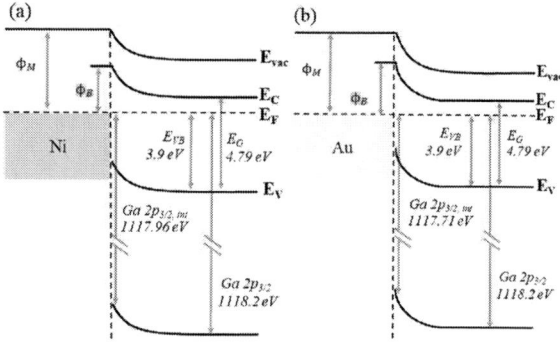

Fig. 5 Schematic band alignment diagrams of the (a) Ni/Ga$_2$O$_3$ and (b) Au/Ga$_2$O$_3$.

appropriate barrier height (using the analytic potentiation model proposed by Werner and Güttler) [28].The inhomogeneous barrier height may be described as

$$\emptyset_{ap} = \emptyset_{b0} - \frac{q\sigma_s^2}{2k_BT} \qquad (5)$$

Here, $\emptyset_{ap}$ is the apparent barrier height, $\emptyset_{b0}$ is measured experimentally as the mean barrier height and σ_s is the standard deviation which gives the level of barrier inhomogeneity at the MS interface. In Fig. 3, $\emptyset_{b0}$ and σ_s^2 can be extracted as the intercept on the y-axis and the slope, respectively. Fig. 3(a) and (b) show the Richardson plot of the Ni and Au samples, both of which have two linear regimes in temperature range I (298 - 398 K) and range II (423 - 523 K), each regime displaying its own Gaussian barrier height distributions, suggesting that more than one transport mechanism exists [22]. From these two Gaussian regions, the calculated values of $\emptyset_{b0}$ and σ_s^2 are 1.57 eV and 0.033 eV for region I, and 2.25 eV and 0.077 eV for region II respectively, as shown in Fig. 3 (a). The value of $\emptyset_{b0}$ increased by about 0.68 eV as the temperature increased from Region I to Region II, it could be inferred that the apparent barrier height was low in the high temperature region. Moreover, the value of σ_s^2 is larger in region I than in region II, it can be seen that the Schottky barrier height is relatively inhomogeneous at a high temperature. From Fig.3 (b), the estimated values $\emptyset_{b0}$ and σ_s^2 are 1.42 eV and 0.014 eV for region I and 2.52 eV and 0.106 eV for region II, respectively. From region I to II, the value of $\emptyset_{b0}$ increased by about 1.1 eV. The value of change in $\emptyset_{b0}$ was greater in the Au samples than in the Ni samples. This confirmed that the Schottky barrier was more inhomogeneous in the Au samples than in the Ni samples.

XPS analysis was performed to confirm the band alignment of the metals and Ga$_2$O$_3$. The Schottky barrier height can be derived by the following formula:

$$\emptyset_B = E_G - E_{VB} + (E_{Ga2p} - E_{Ga2p,int}) \qquad (6)$$

where E_{VB} is the core levels and valence band maximum (VBM) binding energies of Ga$_2$O$_3$, E_G is the bandgap of Ga$_2$O$_3$, E_{Ga2p} and $E_{Ga2p,int}$ are the core levels of Ga$_2$O$_3$ and the metals/Ga$_2$O$_3$ interfaces, respectively. Determined by linear extrapolation of the leading edge of the valence band (VB) spectra, as shown in Fig 4(a), the values of the VB maxima (VBM) of the Ga$_2$O$_3$ films is 3.88 eV. We can estimate the bandgap energy of Ga$_2$O$_3$ films by utilizing the inelastic energy loss spectrum of O1s peak as shown in Fig 4(b). The onset of the inelastic loss spectrum at lower kinetic energy (higher binding energy) relative to the core level peak corresponds to the energy bandgap [11,32]. An energy bandgap of 4.79 eV was determined from the separation between the threshold of the energy loss peak (defined by linearly extrapolating the peak onset shoulder to the background signal level) and the O 1s core level [32]. To quantify the uncertainty of the estimated Ga$_2$O$_3$ bandgap, the inset image in Fig 4(b) shows the 99.3% confidence intervals of the linear regression of the inelastic loss peak. Fig 4(c) shows the Ga 2p$_{3/2}$ as measured from the surface of the Ga$_2$O$_3$ thick film, Ni/Ga$_2$O$_3$ interface, and Au/Ga$_2$O$_3$ interface. From Fig 4(c), we detected that the binding energy of Ga2p appears at 1118.20 eV, and shifted by a binding energy 0.24 eV to negative with Ni deposited and 0.13 eV to positive with Au deposited. The shift of binding energy could be related to the number of oxygen vacancies increasing by the association of oxygen and Au atoms, which results in oxygen dangling bonds

978-1-6654-8498-5/22 $31.00 © 2022 IEEE

[33]. A schematic representation of the band alignment diagram produced using XPS for the heterojunction system studied is shown in Fig 5. From XPS data, the SBH of the Ni and Au diodes was determined to be 1.13, and 1.38 eV, respectively. These values differ somewhat from the values derived from Schottky-Mott for Ni SBDs (+0.13 eV) and Au SBDs (+0.28 eV). The barrier heights show little dependence on the respective metal work function. This is indicative of Fermi-level pinning [34].

IV. Conclusion

In conclusion, Schottky barrier heights and current transport mechanisms for two different metal/Ga_2O_3 Schottky diode junctions were characterized and compared. The electrical properties of Ni/Ga_2O_3 SBDs showed a general thermal emission transport dependence. Schottky barrier height values derived through I-V, C-V, I-V-T, and XPS fell within a fairly narrow range. On the other hand, the Au/Ga_2O_3 SBD junction displayed non-ideal thermal transport likely due to an inhomogeneous junction interface resulting in a wider distribution of local barrier height values. The spread of Schottky barrier height values across the metal work function suggests that Fermi-level pinning does not fully dominate the Schottky barrier height formation for SBDs. Therefore, the way to form a homogeneous Schottky barrier is related to the electrical characteristics, which requires further research.

Acknowledgment

This work was supported by Technology Innovation Program grant (20016102) by MOTIE, the fostering global talents for innovative growth program through the KIAT grant (P0017308), and a Research Grant from Kwangwoon University in 2022.

References

[1] B.K. Mahajan, Y.P. Chen, J. Noh, P.D. Ye, M.A. Alam, Electrothermal performance limit of β-Ga2O3 field-effect transistors, Appl. Phys. Lett. 115 (2019), 173508.

[2] Roy R, Hill V G and Osborn E F 1952 J. Am. Chem. Soc. 74 719.

[3] B.K. Mahajan, Y.P. Chen, J. Noh, P.D. Ye, M.A. Alam, Electrothermal performance limit of β-Ga2O3 field-effect transistors, Appl. Phys. Lett. 115 (2019), 173508.

[4] H. Zhou, J. Zhang, C. Zhang, Q. Feng, S. Zhao, P. Ma, and Y. Hao, J. Semicond. 40(1), 011803 (2019).

[5] B. J. Baliga, Semicond. Sci. Technol. 28, 074011 (2013).

[6] H. Zhang et al., "Progress of ultra-wide bandgap Ga2O3 semiconductor materials in power mosfets," IEEE Trans. Power Electron., vol. 35, no. 5, pp. 5157–5179, May 2020.

[7] Guo. Z., Verma. A., Wu. X, Sun. F, Hickman. A, Masui. T, Luo. T, (2015). Anisotropic thermal conductivity in single crystal β-gallium oxide. Applied Physics Letters, 106(11), 111909.

[8] M. Slomski, N. Blumenschein, P. P. Paskov, J. F. Muth, and T. Paskova, "Anisotropic thermal conductiYLW\RIü-Ga2O3 at elevated temperatures: Effect of Sn and Fe dopants,"-RXUQDORI$SSOLHG 3K\VLFVvol. 121, no. 23, p. 235104.

[9] Wei. R, Song. S, Yang. K, Cui. Y, Peng. Y, Chen. X, Xu, X. (2013). Thermal conductivity of 4H-SiC single crystals. Journal of Applied Physics, 113(5), 053503.

[10] Sichel, E. K. (1977). Thermal conductivity of GaN, 25-360 K. J. Phys. Chem. Solids, 38(3), 330-330.

[11] Kim. M. Y., Lee. H. J., Byun. D. W., Jung. S. W., Shin. M. C., Schweitz. M. A., & Koo. S. M. (2022). Modeling and Investigation of the Effect of Annealing on the Electrothermal Properties of Ga2O3/SiC Heterojunction Diodes. Thin Solid Films, 139200.

[12] T.C. Chen, T.C. Chang, T.Y. Hsieh, M.Y. Tsai, Y.T. Chen, Y.C. Chung, C.Y. Chen, Self heating enhanced charge trapping effect for InGaZnO thin film transistor, Appl. Phys. Lett. 101 (2012), 042101.

[13] Y. Chang, Y. Zhang, Y. Zhang, K.Y. Tong, A thermal model for static current characteristics of AlGaN / GaN high electron mobility transistors including self heating effect, J. Appl. Phys. 99 (2006), 044501.

[14] Jadhav. A., Lyle. L. A., Xu. Z., Das. K. K., Porter. L. M., & Sarkar. B. (2021). Temperature dependence of barrier height inhomogeneity in β-Ga2O3 Schottky barrier diodes. Journal of Vacuum Science & Technology B, Nanotechnology and Microelectronics: Materials, Processing, Measurement, and Phenomena, 39(4), 040601.

[15] Y. Yao, R. F. Davis, and L. M. Porter, J. Electron. Mater. 46(4), 2053 (2017).

[16] Liu. Y., Guo. J., Zhu. E., Liao. L., Lee. S. J., Ding. M., & Duan, X. (2018). Approaching the Schottky–Mott limit in van der Waals metal–semiconductor junctions. Nature, 557(7707), 696-700.

[17] Tung. R. T. (2014). The physics and chemistry of the Schottky barrier height. Applied Physics Reviews, 1(1), 011304.

[18] Farzana. E., Zhang. Z., Paul. P. K., Arehart. A. R., & Ringel. S. A. (2017). Influence of metal choice on (010) β-Ga2O3 Schottky diode properties. Applied Physics Letters, 110(20), 202102.

[19] Lyle. L. A., Jiang. K., Favela. E. V., Das. K., Popp. A., Galazka. Z., & Porter, L. M. (2021). Effect of metal contacts on (100) β-Ga2O3 Schottky barriers. Journal of Vacuum Science & Technology A: Vacuum, Surfaces, and Films, 39(3), 033202.

[20] M. Mohamed, K. Irmscher, C. Janowitz, Z. Galazka, R. Manzke, and R. Fornari, Appl. Phys. Lett. 101, 132106 (2012).

[21] K. Jiang et al., ECS Trans. 92, 71 (2019).

[22] Ertap. H., Kacus. H., Aydogan. S., & Karabulut. M. (2020). Analysis of temperature dependent electrical characteristics of Au/GaSe Schottky barrier diode improved by Ce-doping. Sensors and Actuators A: Physical, 315, 112264.

[23] Güler. G., Güllü. Ö., Karataş. S., & Bakkaloglu. Ö. F. (2009). Electrical Characteristics of Co/n-Si Schottky Barrier Diodes Using I–V and C–V Measurements. Chinese Physics Letters, 26(6), 067301.

[24] E. H. Rhoderick and R. H. Williams, Metal-Semiconductor Contacts, 2nd ed. (Clarendon Press, Oxford, 1988), p. 94.

[25] M. Passlack, E. F. Schubert, W. S. Hobson, M. Hong, N. Moriya, S. N. G. Chu, K. Konstadinidis, J. P. Mannaerts, M. L. Schnoes, and G. J. Zydzik, J. Appl. Phys. 77, 686 (1995).

[26] H. He, R. Orlando, M. A. Blanco, R. Pandey, E. Amzallag, I. Baraille, and M. Rerat, Phys. Rev. B 74, 195123 (2006).

[27] Gammon. P. M., Pérez-Tomás. A., Shah. V. A., Vavasour. O., Donchev, E., Pang, J. S., ... & Mawby, P. A. (2013). Modelling the inhomogeneous SiC Schottky interface. Journal of Applied Physics, 114(22), 223704.

[28] Sheoran. H., Tak. B. R., Manikanthababu. N., & Singh. R. (2020). Temperature-dependent electrical characteristics of Ni/Au vertical Schottky barrier diodes on β-Ga2O3 epilayers. ECS Journal of Solid State Science and Technology, 9(5), 055004.

[29] Tung. R. T. (1992). Electron transport at metal-semiconductor interfaces: General theory. Physical Review B, 45(23), 13509.

[30] Kumar, A., Heilmann, M., Latzel, M., Kapoor, R., Sharma, I., Göbelt, M., ... & Singh, R. (2016). Barrier inhomogeneities limited current and 1/f noise transport in GaN based nanoscale Schottky barrier diodes. Scientific reports, 6(1), 1-11.

[31] Hamri, D., Teffahi, A., Djeghlouf, A., Saidane, A., & Mesli, A. (2018). Temperature dependent transport characterization of iron on n-type (111) Si0. 65Ge0. 35 Schottky diodes. Journal of Alloys and Compounds, 763, 173-179.

[32] Nichols, M. T., Li, W., Pei, D., Antonelli, G. A., Lin, Q., Banna, S., ... & Shohet, J. L. (2014). Measurement of bandgap energies in low-k organosilicates. Journal of Applied Physics, 115(9), 094105.

[33] Yadav. M. K., Mondal. A., Das. S., Sharma. S. K., & Bag. A. (2020). Impact of annealing temperature on band-alignment of PLD grown Ga2O3/Si (100) heterointerface. Journal of Alloys and Compounds, 819, 153052.

[34] Kim, C., Moon, I., Lee, D., Choi, M. S., Ahmed, F., Nam, S., ... & Yoo, W. J. (2017). Fermi level pinning at electrical metal contacts of monolayer molybdenum dichalcogenides. ACS nano, 11(2), 1588-1596.

Compact Modeling of Phase Change Memory with Parameter Extractions

Feilong Ding[1], Xi Li[2], Yihan Chen[3], Zhitang Song[2], Runsheng Wang[4], Clarissa Cyrilla Prawoto[5], Mansun Chan[5], Lining Zhang*[1] and Ru Huang[4]

[1]School of Electronic and Computer Engineering, Peking University, Shenzhen 518055, China,
[2]Key Laboratory of Functional Materials for Informatics, Shanghai Institute of Microsystem and Information Technology, Chinese Academy of Sciences, Shanghai 200050, China,
[3]School of Humanities and Social Science, The Chinese University of Hong Kong, Shenzhen, Shenzhen, 518172, China,
[4]School of Integrated Circuits, Peking University, Beijing 100871, China,
[5]Department of Electronic and Computer Engineering(ECE),
The Hong Kong University of Science and Technology, Hong Kong, SAR, China.
*E-mail: eelnzhang@pku.edu.cn

Abstract—A compact model of mushroom-type phase change memory (PCM) with parameter extractions is reported in this work. General device physics of heating dynamics, crystallization kinetics and filaments formation are covered in four essential modules. The memory resistances are calculated with the conformal mapping technique, together with corrections for non-ideal geometries. A flow of model parameter extractions is then developed. With a procedure of seven steps, all model parameters are obtained with verifications from experimental PCM prepared with 40nm CMOS process. The model applicability in PCM designs are then demonstrated with Verilog-A implementation and simulation convergences of typical circuits.

Keywords—compact model, parameter extraction flow, phase change memory, mushroom type

I. INTRODUCTION

Phase change memory (PCM)[1-3] is one of the most promising candidates for next generation non-volatile memory technology, due to its outstanding scalability into nanoscale, CMOS process compatibility, and application maturity[4,5]. Phase change materials have two stable phase states with distinct resistivity at room temperature, crystalline state (low resistivity) and amorphous state (high resistivity). By utilizing joule heat generated by electrical pulse, a PCM cell can switch between high and low resistance states[6]. The transition mechanism involves the coupling among electrical, thermal and crystallization.

Models and simulations have been developed so far to capture the characteristics of PCM, including compact models[7] and finite element method numerical simulations [8]. For circuit simulations, compact models are preferred. Many compact models in form of macro model [9,10] and physics-based analytical models [11-16] with different approach have been proposed. The comprehensive PCM characteristics (e.g. resistance and switching) are gradually covered by the models, and model continuity with good convergence are becoming available. On the other hand, parameter extraction is a necessary step for the compact model to assist the PCM technology development, however, is discussed less often.

In this work, an improved compact model for fabricated mushroom-type PCM cell is described considering the non-ideality factors. A parameter extraction flow for this model is then developed for experimental data verifications. With the improved model and parameter extractions, the model capability to be used for PCM circuit simulations is finally demonstrated.

II. COMPACT MODEL DEVELOPMENT

The PCM considered is a mushroom-type PCM, with a TEM result of structure as shown in Fig. 1(a). TEC and BEC represent top and bottom electrodes. Phase change material is carbon-doped Ge$_2$Sb$_2$Te$_5$ (CGST). Geometry parameters and material parameters are displayed in Table I. A detailed description of the PCM process is seen in [17]. Applying an electrical pulse, the temperature profile from TCAD simulation is shown in Fig. 1 (b), which shows the mushroom-type active region.

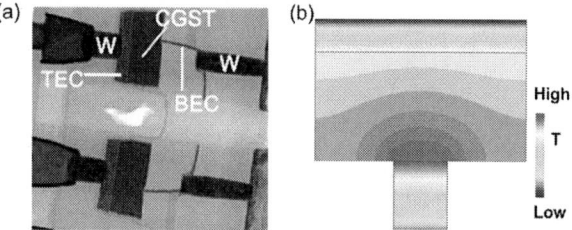

Fig. 1. (a) Cross section TEM view of the fabricated CGST-based PCM cell[17]. (b) Temperature profile from TCAD simulations of the PCM structure showing the mushroom-type active region.

A model based on the filament theory, including four parts, namely electrical module, thermal module, crystallization module and dynamic filament conductance module is developed. When voltage source or current source is applied to PCM cell, electrical module would generate Joule power to act as heat source to be used in the thermal module with a self-heating sub-circuit. Raise of temperature caused by joule heat leads cell to reach crystallization temperature (T_g) and melting temperature (T_m) which is used in the crystallization module. The temperature module is shown in (1):

$$C_{th} \cdot \frac{d(\Delta T)}{dt} = P_{th} - \frac{\Delta T}{R_{th}} \tag{1}$$

in which C_{th} is thermal capacitance, ΔT is the temperature rise, P_{th} is the heat power and R_{th} is thermal resistance. Equation (1) is implemented by sub-circuit as shown in Fig. 2 (a).

Crystallization happens where the cell temperature is between T_g and T_m, and this process is portrayed by the JMA equation as shown in (2):

$$C_f = 1 - \exp(-k_{gr} \cdot t) \tag{2}$$

in which C_f is crystal fraction which varies from 0 and 1 (0 represents fully amorphous while 1 stands for fully

978-1-6654-8498-5/22 $31.00 © 2022 IEEE

crystalline). k_{gr} is crystallization rate. Equation (2) can be realized in sub-circuit after derivations of (3):

$$\frac{dC_f}{dt} = k_{gr} \cdot (1 - C_f) = \frac{1}{C_{JMA}}(I_{JMA} - \frac{C_f}{R_{JMA}}) \quad (3)$$

where $C_{JMA} = 1$, $I_{JMA} = k_{gr}$ and $R_{JMA} = 1/k_{gr}$. Equation (3) is solved in sub-circuit as shown Fig. 2 (b).

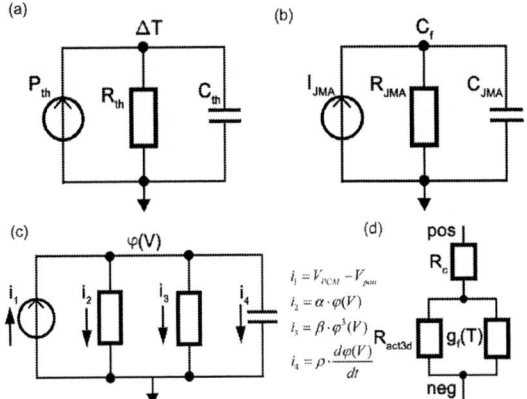

Fig. 2. Sub-circuits used in the PCM model.(a) Thermal module. (b) Crystallization module (c) Dynamic filament module. (d) Electrical module.

A dynamic filament conductance with hysteresis is described in (4), which is derived from the Landau-Khalatnikov equation originally for ferroelectric hysteresis.

$$V_{PCM} - V_{pan} = \alpha \cdot \varphi(V) + \beta \cdot \varphi^3(V) + \rho \frac{d\varphi(V)}{dt} \quad (4)$$

in which V_{PCM} is the voltage drop on PCM cell, V_{pan} represents voltage shift of hysteresis center, $\varphi(V)$ is the parameter that related to field-induced nucleation. What's more, α, β are the parameters that determine the shape of hysteresis curve. ρ is a parameter that used to enhance

TABLE I PCM INSTANCE AND MODEL PARAMETERS

Module	Symbol	Unit	Description
geometry parameters	W_g	nm	radius of top electrode
	W_{be}	nm	radius of bottom electrode
	H_g	nm	height of GST region
material parameters	T_g	K	crystallization temperature
	T_m	K	melting temperature
electrical module	R_{set}	Ω	resistance of fully SET state
	R_{reset}	Ω	resistance of fully RESET state
	α_{f_rst1}	Ω	fully RESET resistance multiplying factor
	α_{f_rst2}	1/nm	fully RESET resistance nonlinearity factor
	α_{f_rst3}	1	fully RESET resistance of constant
	S_{f_rst}	nm	smooth factor of transition of from SET to fully RESET
	R_{p_rst}	Ω	factor of controlling the maximum resistance partial resistance
	s_{p_rst}	nm	smooth factor of transition from SET to partial RESET
	r_{shift}	nm	radius shift factor of transition of from partial to fully RESET
thermal module	R_{th}	K/W	equivalent thermal resistance
	C_{th}	J/K	equivalent heat capacity
	T_0	K	ambient temperature
crystalline module	k_{gr}	1	crystallization rate
dynamic filament module	g_f	S	filament conductance
	E_t	V/nm	critical electrical field of threshold switching
	E_a	eV	active energy of temperature dependence of filament

robustness of simulation. Similarly, equation (4) is realized in sub-circuit in Fig. 2 (c).

For the electrical module, resistance calculation is illustrated with Fig. 2(d). R_{act3d} represents the resistance of active region. R_c stands for the resistance of outside active region. g_f is filament conductance, and temperature dependence is considered and formulated based on the active energy E_a. When there is no filament shown, total resistance is $R_c + R_{act3d}$. Otherwise, the total resistance is $R_c + R_{act3d}/(1+g_f \cdot R_{act3d})$. Model parameters are listed in Table I.

III. PARAMETERS EXTRACTION FLOW

To extract parameters for the above model, both DC and transient simulation are needed due to the non-volatility of memory devices. A proposed extraction flow is shown in Fig. 3 with seven steps in total.

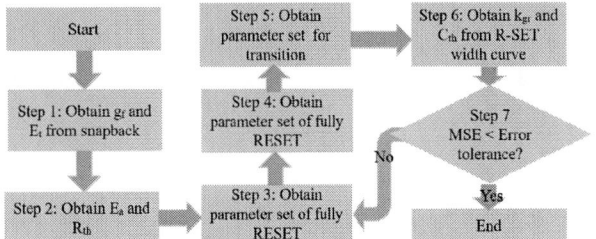

Fig. 3. Parameter extraction flow of the compact PCM model.

Step 1: Obtain g_f and E_t from snapback

PCM cells in the fully amorphous state are obtained when fabrication and initialization are completed. Measurement of snapback characteristics due to the threshold switching when the electrical filed goes beyond the critical value E_t is performed. With the active region size r_a, the threshold voltage is calculated. A filament is believed to form during the snapback process. After snapback, the filament conductance g_f leads to decrease in voltage drop on the cell.

From DC sweeping characteristics, the resistance of the status after snapback R_{after_snap} and the threshold voltage V_{OTS} can be obtained as shown in Fig. 4.

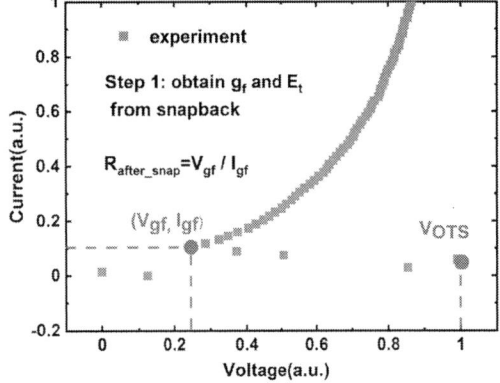

Fig. 4. Snapback characteristics of experimental PCM are used in step one. All the data are normalized for a general description.

The two parameters g_f and E_t are calculated.

$$g_f = \frac{1}{R_{after\_snap}}, \quad E_t = \frac{V_{OTS}}{r_a} \quad (5)$$

Step 2: Obtain E_a and R_{th}

Current pulse with fixed width $t_p = 2us$ and increasing amplitudes is applied to cell whose initial state is completely amorphous state to get the R-I curve in Fig 5. It can be observed that from experiment data, the current of start point of SET process is denoted as I_1, which is corresponding to the cell temperature of T_g. While the current of start point of RESET one is labelled as I_2, which is corresponding to cell temperature of T_m. T_g and T_m are material properties obtained from material characterizations. According to the thermal circuit, equation (6) is easily solved.

$$I_1(R_c + \frac{1}{g_f}\exp(\frac{E_a}{k}(\frac{2}{T_g+T_0} - \frac{1}{T_0}))) \cdot R_{th} + T_0 = T_g$$

$$I_2(R_c + \frac{1}{g_f}\exp(\frac{E_a}{k}(\frac{2}{T_m+T_0} - \frac{1}{T_0}))) \cdot R_{th} + T_0 = T_m$$

(6)

in which k is Boltzmann constant.

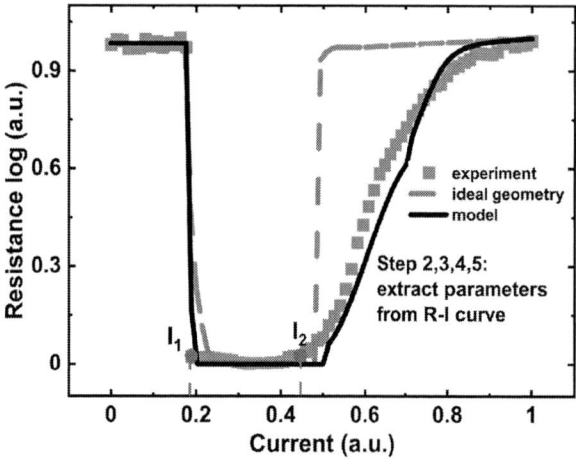

Fig. 5. R-I characteristics of experimental PCM are used in steps two to five. The assumption with an ideal mushroom geometry leads to an abrupt change when the active region size increases beyond the BEC. The improved model agree well to the experimental results.

Step 3: Obtain the parameter set of fully RESET

It is found that the resistance only increases slightly in log scale after r_a goes beyond $W_{be}/2$, hence the state with r_a beyond $W_{be}/2$ is defined as a fully RESET state. Ideally, conformal mapping method is adopted to calculate the resistance of mushroom-type PCM only for $r_a > W_{be}/2$. Meanwhile, the experimental data shows that the RESET is completed more gradually. The possible reasons include a non-ideal mushroom active region as reported in literature. Partial RESET region (with r_a smaller than $W_{be}/2$) is taken into consideration. Equations (7)-(9) are used to correct the conformal mapping calculations. The resistance of fully RESET R_2 is expressed in (7):

$$R_1 = R_{set} + \frac{(\alpha_{f\_rst1} \cdot (\alpha_{f\_rst2} \cdot r_a + \alpha_{f\_rst3})^{\frac{1}{3}} - R_{set})}{1 + \exp(-\frac{r_a - W_{be}/2}{S_{f\_rst}})}$$

(7)

in which, α_{f_rst1}, α_{f_rst2}, α_{f_rst3} are parameters of fully RESET. S_{f_rst} is the parameter that determined the steepness of transition.

Step 4: Obtain parameter set of partial RESET

For partial RESET, the mapping relationship between resistance and r_a is described in (8):

$$R_2 = R_{set} + \frac{(R_{p\_rst} - R_{set})}{1 + \exp(-\frac{(r_a - W_{be}/2)}{S_{p\_rst}})}$$

(8)

in which, R_{p_rst} stands for the resistance after transition. and S_{p_rst} is a parameter that determined the steepness of transition from low resistance partial RESET.

Step 5: Obtain parameter set for transition

A smooth transition from partial to fully RESET is provided with a fermi-like function in (9):

$$R_{act3d} = R_2 + \frac{R_1 - R_2}{1 + \exp(-\frac{(r_a - W_{be}/2 - r_{shift})}{S_{rst}})}$$

(9)

in which R_{act3d} is the resistance of active region. S_{rst} is a fitting parameter, r_{shift} is a s a parameter that determined the steepness of transition between R_1 and R_2.

Step 6: Obtain k_{gr} and C_{th} from R-SET width curve

In the measurements, current source with a fixed amplitudes I_1 and varying SET width is applied to reflect the crystallization dynamics. Before applying each SET pulse, the PCM cell is initialized to the amorphous state. Resistance of cell drops with the extension of SET pulse width, but the rate of decline first increases and then decreases to zero. C_{th} and k_{gr} are extracted according to the key timing. At the same time, the two parameters also affect the R-I curve in the quenching period of the RESET. As such, global optimizations are needed in the next step 7.

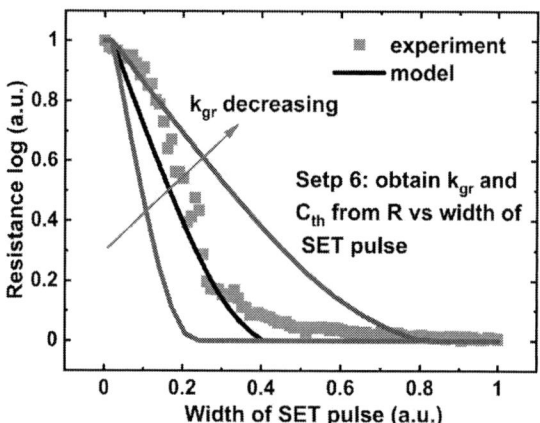

Fig. 6. Resistance of cell of experimental PCM and the model after applying current pulse with fixed amplitude I_1 and increasing pulse width.

Step 7: Global optimization

Global optimizations are performed with initial values of all model parameters from previous steps. Mean square errors (MSE) of the model against experimental data of the R-I curve in Fig.5 and the crystallization dynamic curve in Fig.6 are minimized in the process. The related parameters in the steps 3-6 are involved in this step. When the MSE is below an error tolerance, the global optimization stops and the extraction flow is finished. General optimization algorithm applies in this step.

The model results in Fig.5 and Fig.6 corresponds to the extracted parameter set when the above global optimization is finished.

IV. SIMULATION DEMONSTRATION WITH EXTRACTED PARAMETERS

To verify the extracted parameters, they are used in PCM circuit simulations with the model implementations in simulators. Fig. 7 shows the simulation details of a PCM cell. Current source is applied to the cell, and a SET-READ-RESET-READ process is designed. SET is applied first, snapback occurs in Fig. 7(b), meanwhile, filament begin to appear. Crystallization is proceeding with SET pulse. After a SET pulse, resistance fell into low resistance in Fig. 7(f). Then, a RESET pulse is applied, temperature exceeds T_m in Fig. 7(c), the PCM cell goes into its high resistance state. Functional verifications and convergences of PCM circuits validate the model and extracted parameters.

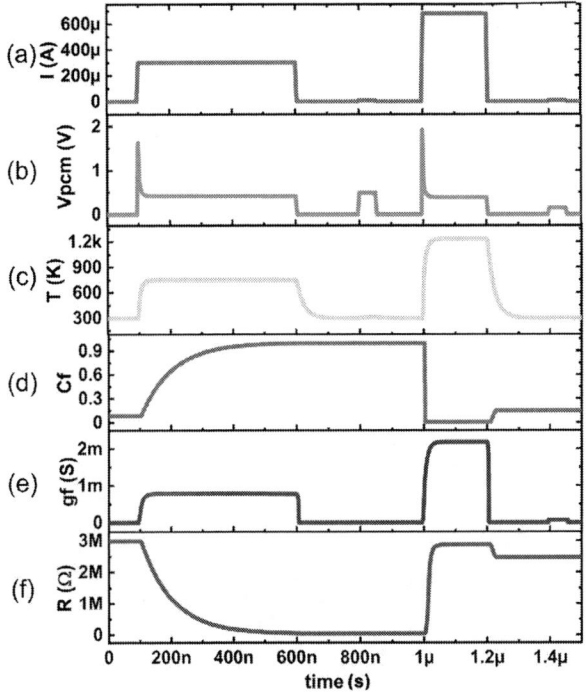

Fig. 7 Waveform of model simulation. (a) Current of the electrical pulse, represents SET-READ-RESET-READ operations. (b) Voltage drop on the cell. (c) Highest temperature of cell.(d) Crystalline fraction(C_f). (e) Filament conductance. (f) Total resistance of cell.

V. CONCLUSION

An experimental data verified PCM model is reported in this work. The core model is comprised of four key modules for the general phase change physics from Ovinic threshold switching to memory switching. With further extensions from the perspective of non-ideal geometries, the model is then verified with experimental PCM prepared with 40nm CMOS process. The extraction flow include seven steps for all the model parameters, and high accuracy has been achieved between the model and experimental data of R-I curves and crystallization kinetics. The model is implemented in circuit simulators with Verilog-A and its functions have been shown with successful simulation convergence.

ACKNOWLEDGMENT

This work is supported in part by the National Natural Science Foundation of China (62074006, 91964204), in part by the Major Scientific Instruments and Equipment Development (61927901), the Shenzhen Science and Technology Project (GXWD20201231165807007-20200827114656001), Strategic Priority Research Program of the Chinese Academy of Sciences (XDB44010200)，Science and Technology Council of Shanghai (19JC1416801), the Shanghai Research and Innovation Functional Program (17DZ2260900), and in part by the 111 Project (B18001).

REFERENCES

[1] T. Kim and S. Lee, "Evolution of Phase-Change Memory for the Storage-Class Memory and Beyond," IEEE Trans. Electron Devices, vol. 67, no. 4, pp. 1394-1406, Apr. 2020.

[2] N. B. Gong, "Multi level cell (MLC) in 3D crosspoint phase change memory array," Science China-Information Sciences, vol. 64, no. 6, pp. 2, Jun. 2021.

[3] W. Zhang, R. Mazzarello, and E. Ma, "Phase-change materials in electronics and photonics," MRS Bull., vol. 44, no. 9, pp. 686-690, Sept. 2019.

[4] A. Redaelli et al., "Improving Ge-rich GST ePCM reliability through BEOL engineering," in ESSDERC 2021 IEEE 51st European Solid-State Device Research Conference, 13-22 Sept. 2021 2021, pp. 231-234.

[5] T. Kim et al., "High-performance, cost-effective 2z nm two-deck cross-point memory integrated by self-align scheme for 128 Gb SCM," 2018 IEEE International Electron Devices Meeting, New York, 2018.

[6] M. Le Gallo and A. Sebastian, "An overview of phase-change memory device physics," Journal of Physics D-Applied Physics, vol. 53, no. 21, pp. 27, May. 2020.

[7] F. L. Ding et al., "A review of compact modeling for phase change memory," Journal of Semiconductors, vol. 43, no. 2, pp. 14, Feb. 2022.

[8] Z. Chen, H. Tong, W. Cai, L. Wang, and X. Miao, "Modeling and Simulations of the Integrated Device of Phase Change Memory and Ovonic Threshold Switch Selector With a Confined Structure," IEEE Trans. Electron Devices, vol. 68, no. 4, pp. 1616-1621, 2021.

[9] R. A. Cobley and C. D. Wright, "Parameterized SPICE model for a phase-change RAM device," IEEE Trans. Electron Devices, vol. 53, no. 1, pp. 112-118, Jan. 2006.

[10] X. Q. Wei et al., "HSPICE macromodel of PCRAM for binary and multilevel storage," IEEE Trans. Electron Devices, vol. 53, no. 1, pp. 56-62, Jan. 2006.

[11] N. Xu et al., "Multi-Domain Compact Modeling for GeSbTe-based Memory and Selector Devices and Simulation for Large-scale 3-D Cross-Point Memory Arrays," 2016 IEEE International Electron Devices Meeting, New York, 2016.

[12] K. Sonoda, A. Sakai, M. Moniwa, K. Ishikawa, O. Tsuchiya, and Y. Inoue, "A compact model of phase-change memory based on rate equations of crystallization and amorphization," IEEE Trans. Electron Devices, vol. 55, no. 7, pp. 1672-1681, Jul. 2008.

[13] C. Pigot et al., "Comprehensive Phase-Change Memory Compact Model for Circuit Simulation," IEEE Trans. Electron Devices, vol. 65, no. 10, pp. 4282-4289, Oct. 2018.

[14] H. F. Hu et al., "A Compact Phase Change Memory Model With Dynamic State Variables," IEEE Trans. Electron Devices, vol. 67, no. 1, pp. 133-139, Jan. 2020.

[15] X. H. Chen et al., "A SPICE Model of Phase Change Memory for Neuromorphic Circuits," IEEE Access, vol. 8, pp. 95278-95287, 2020.

[16] X. H. Chen et al., "A Robust and Efficient Compact Model for Phase-Change Memory Circuit Simulations," IEEE Trans. Electron Devices, vol. 68, no. 9, pp. 4404-4410, 2021.

[17] X. Li et al., "Enhancing the Performance of Phase Change Memory for Embedded Applications," Physica Status Solidi-Rapid Research Letters, vol. 13, no. 4, pp. 13, Apr. 2019.

A novel Approach to measure and model Plasma Noise in Avalanche Diodes

Dr. Elmar Gondro
(IFAG ATV PTP TD EDA MUC)
Infineon Technologies
Neubiberg, Germany
elmar.gondro@infineon.com

Dr. Joost Willemen
(IFAG PSS RFS D2 MS CON)
Infineon Technologies
Oberhaching, Germany
joost.willemen@infineon.com

Peter Bauer
(IFAG ATV PTP TD M DS)
Infineon Technologies
Neubiberg, Germany
peter.bauer2@infineon.com

Abstract—**High voltage Zener diodes showing an avalanche breakdown suffer from a statistical random noise effect. This paper describes a new method to measure this effect on a state-of-the-art automotive technology with highest accuracy. Furthermore, a simulation model has been developed reflecting this behavior and showing its impact on circuit performance.**

Index Terms—**diode, avalanche, RF measurement, model**

I. INTRODUCTION

As device dimensions (as well as supply voltages) are getting smaller and smaller, noise is gaining more and more importance.

A breakdown diode is characterized by a certain voltage level V_{bd} where the current suddenly increases. Zener diodes with a higher breakdown voltage (approx. $V_{bd} > 6\,\text{V}$) show an avalanche behavior which varies statistically over time: lightly doped junctions cause a wider depletion region, where electrons are accelerated by the electric field, and collisions with bound electrons produce free electron-hole pairs. This effect—also known as "microplasma crackling", "telegraph noise switching" or "popcorn noise"—is worst at cold environment where phonon scattering is less prominent.

Fig. 1 illustrates this avalanche breakdown process in a reverse-biased pn-junction [1]:

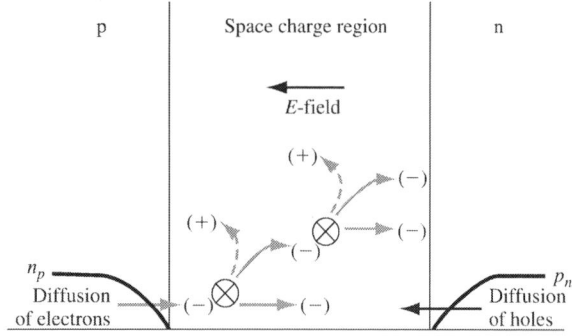

Fig. 1. Avalanche breakdown process in a reverse-biased pn-junction.

At real Zener breakdown (approx. $V_{bd} < 6\,\text{V}$) this effect disappears and — due to thinner junction width — Zener tunneling with positive temperature coefficient dominates.

Zener effect dominates for approx. $V_{bd} < 6\,\text{V}$, avalanche effect dominates for approx. $V_{bd} > 6\,\text{V}$, but having a negative temperature coefficient. The avalanche effect causes current ripples within the breakdown region which are higher (1) for higher breakdown voltage, (2) lower temperature, and (3) cleaner processes.

Nevertheless, for usage as reference diode circuit designers prefer the region near the limit between those effects in order to obtain a temperature-stable breakdown voltage, thus making an accurate noise description more important.

II. MEASUREMENTS

A. Old approach with external parasitics

Fig. 2 shows the typical voltage ripples up to $1.5\,\text{V}$ in amplitude occurring in a circuit using diodes in breakdown:

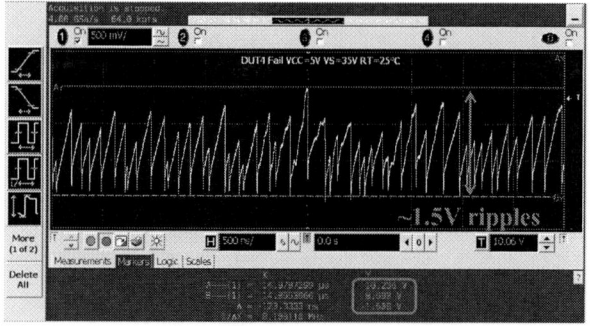

Fig. 2. Typical voltage ripples caused by avalanche diodes.

The saw-tooth-like signal shape is caused by the RC load of the circuit as well as by the measurement setup (parasitic capacitance of measurement cables). This emphasizes the necessity for a low parasitics setup for accurate measurements.

B. New integrated approach

As noise switching occurs mainly in the high frequency regime within time scales in the nanoseconds, a sophisticated RF measurement setup is proposed (fig. 3):

here the parasitics are minimized by using an integrated poly resistor R in series to the diode D to limit the on-state current. This current is measured in series over the $50\,\Omega$ entry of the

978-1-6654-8498-5/22 $31.00 © 2022 IEEE

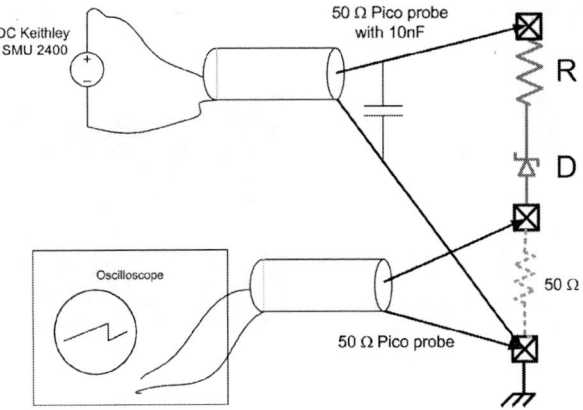

Fig. 3. RF capable $50\,\Omega$ measurement setup.

oscilloscope with a voltage at the anode side varying from $0\,\mathrm{V}$ in non-conductive state to $V_{out} = 50\,\Omega/(50\,\Omega + R_{ser}) \cdot V_{in}$ in conductive state.

The setup is completely driven as a $50\,\Omega$ system using RF probes and enabling current measurement via a $6\,\mathrm{GHz}$ oscilloscope down to a nanosecond scale. Up to time windows of $1\,\mathrm{s}$ continuous measurement with adapted sample rate was applied. For observation of longer-lasting time windows with very rare events ($>1\,\mathrm{s}$) we used the single event trigger mode of the oscilloscope.

For obtaining statistical data (> 100 switching events) we used an extremely high number of samples ($32\,\mathrm{mio}$) to gain $T_{meas} >> t_{on}+t_{off}$. The sampling rate depends on the selected time window. This setup allows measurements like in fig. 4 showing the quasi-rectangular switching between conducting and non-conductive states of the diode.

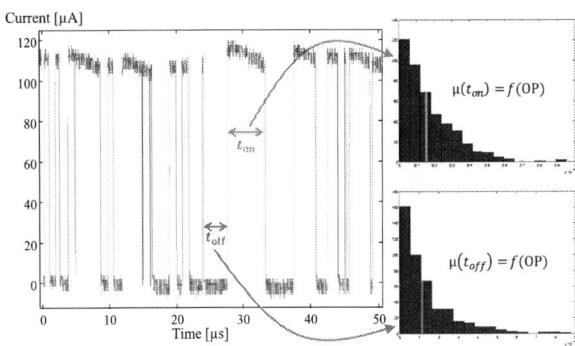

Fig. 4. Statistical switching behavior over time and its distributions.

III. PARAMETER EXTRACTION

By extracting these time slots in on-state as well as in off-state we obtained the distributions shown on the right side of fig. 4. The most important outcome of our work were the dependencies of those on- and off-times on the operating point:

A. On-state Time

The average duration of current conducting state (on-state) strongly depends on current level: the higher the diode current, the longer the avalanche lasts (fig. 5).

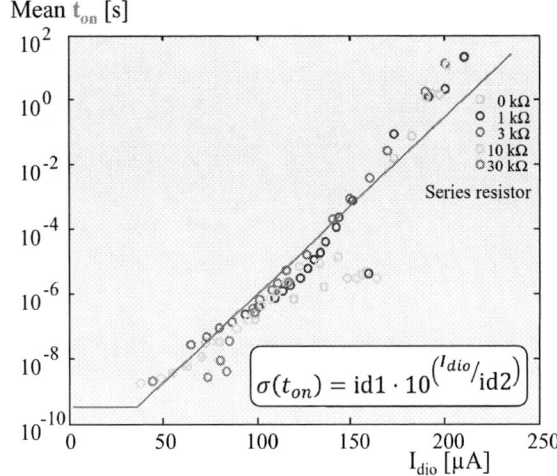

$$\sigma(t_{on}) = \mathrm{id}1 \cdot 10^{\left(I_{dio}/\mathrm{id}2\right)}$$

Fig. 5. Average statistical on-time.

B. Off-state Time

Contrary to the on-state, the average duration of non-conducting state (off-state) decreases exponentially with the voltage across the junction (fig. 6).

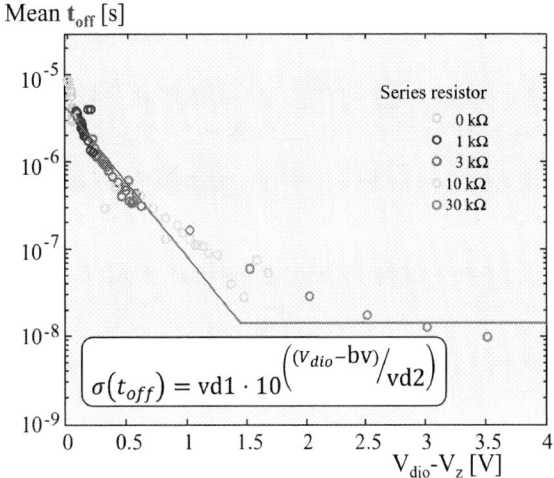

$$\sigma(t_{off}) = \mathrm{vd}1 \cdot 10^{\left((V_{dio}-\mathrm{bv})/\mathrm{vd}2\right)}$$

Fig. 6. Average statistical off-time.

C. Type of Distribution

For the ease of use in a simulator we considered the distributions on the right side of fig. 4 to have an exponential decrease in time (comparable to decay processes); the quantile plot of the voltage drop justifies this approach within limits of ± 3 sigma (fig. 7).

978-1-6654-8498-5/22 $31.00 © 2022 IEEE   313

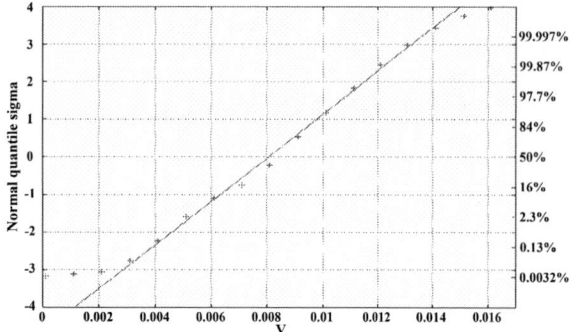

Fig. 7. Quantile plot of the voltage drop.

and time scale. Due to its random nature the absolute switching distances between individual samples must differ.

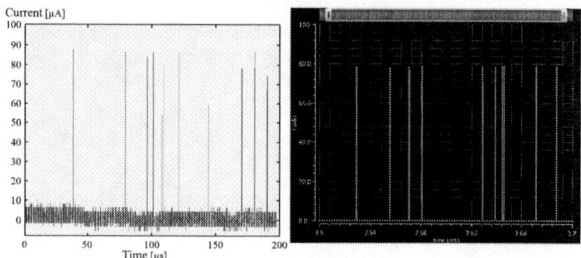

Fig. 8. Comparison of measurements with simulation.

IV. MODELING

An increasing demand on behavioral compact models describing the stochasticity of distinct circuit elements (like magnetic tunnel junctions) has been reported ([2] and [3]).

A. Randomized Switching

To model this diode noise behavior in the transient simulation regime a switching element in series to the diode has been implemented in VerilogA harboring a voltage and current dependent random approach. The principle algorithm sequence is the following:

1) Switch diode to conducting state
2) Wait until operating point is settled
3) Calculate average on-time (mean(t_{on})) as function of diode current
4) "Dice" a random on-time t_{on} with a sigma proportional to mean(t_{on})

On the other hand, after switching off a mean t_{off} time is calculated as function of the voltage overshoot to produce and apply a random t_{off}.

Furthermore, the AC and DCOP simulation behavior should not be influenced, i.e. simulation has to start in on-state (switch closed @ $t = 0$).

Special attention has to be paid to circuits with (nearly) ideal current sources: applied to the switched diode in off-state they charge parasitic capacitances and may cause crucial voltage overshoots! To avoid this the time step control monitoring the operating point within VerilogA has to be adapted accordingly.

B. Transition Smoothing

The combination of statistical and normal transient simulation with different time scales raises an additional challenge: in order to obtain convergence during transient analysis, it is absolutely mandatory to smoothen the switching signal by a parabolic or sinusoidal transition slope to achieve continuously differentiable signals.

C. Comparison of Measurements with Simulation

Fig. 8 shows a comparison between measurements and simulation: a good agreement is achieved in terms of amplitudes

D. Implications on Compact Models of typical Design Kits

Compact models for circuit simulations are usually derived from measurements using SMUs[1]. Those measurement units perform an averaging over time.

As consequence voltage spikes in the low current region (A) are not visible, but influence the measured value via the averaging process. Vice versa, complete turn-offs of the diode in high current operating points (B) are not taken into account (fig. 9).

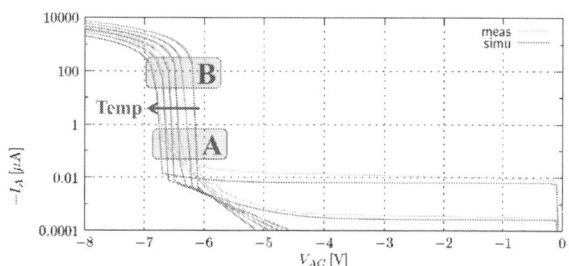

Fig. 9. Typical diode breakdown vs. voltage.

An estimation how long the diode will remain in on-state depending on its DC current can be seen in fig. 10 derived by evaluation of the normalized on-time $t_{on}/(t_{on} + t_{off})$.

E. Avalanche Analogon

The observation that the on-time depends on the current (fig. 5), whereas the off-time depends on the voltage (fig. 6), might seem astonishing. Therefore, a comparison with a nature event like a snow avalanche may enhance clarity (tab. I).

TABLE I
COMPARISON OF DIODE AVALANCHE WITH SNOW AVALANCHE

	Diode Avalanche	Snow Avalanche
Triggered by	$V_{dio} > V_{breakdown}$	slope angle $> \approx 30°$
Triggering[a] raises with	potential difference	steepness of slope
Duration[b] raises with	current density	amount of material

[a]Probability(off→on). [b]Probability(on→off).

[1]source measure unit

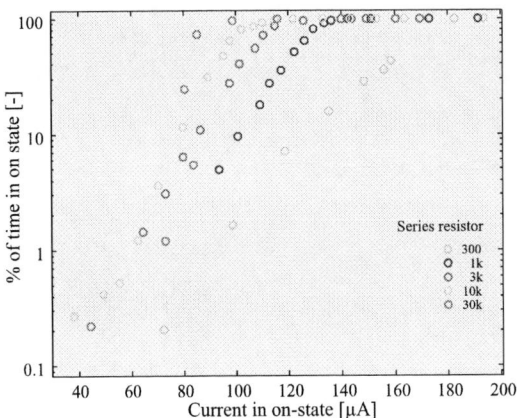

Fig. 10. Statistical occurrence of on-state vs. on-state diode current.

V. CIRCUIT SIMULATION

Fig. 11 shows a part of a gate driver circuit detecting an over-voltage (OV) level by a comparison of a resistive divider versus a Zener voltage.

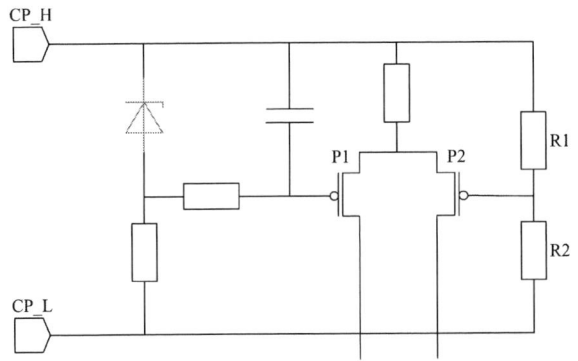

Fig. 11. Overvoltage detection circuit with avalanche diode.

Measurement on FIB[2] structures revealed a voltage increase of approx. 1.5 V on the operating point of the MOSFET gate.

Normal simulations were not able to reflect this, whereas the new noise model continuously charging capacitances could explain the effect (fig. 12 and 13):

The gate voltage V_{P2} (blue) of P2 follows the input voltage $V_{in} = V(\text{CP\_H}) - V(\text{CP\_L})$ (red) according to the ratio of R_1/R_2, whereas the gate voltage V_{P1} (green) of P1 follows the input signal with a difference of the Zener voltage. At the intersection of V_{P1} and V_{P2} the OV detection (orange) is triggered in the digital part (not shown here).

The plasma noise of the reference Zener diode causes an increase of the OV level of $1.7\,\text{V} = 18.6\,\text{V} - 16.9\,\text{V}$ showing good agreement to the measurements mentioned above.

[2]Focused ion beam

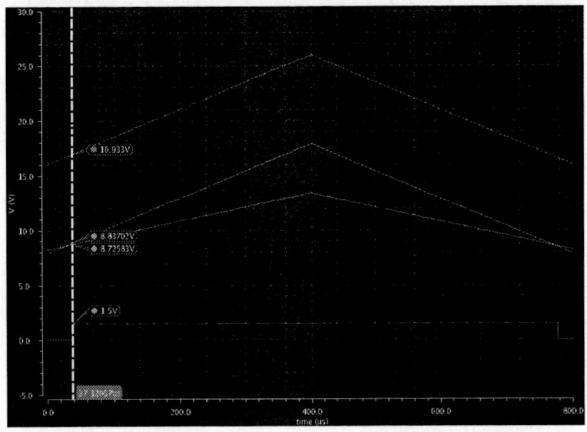

Fig. 12. Simulation of overvoltage detection with old diode model.

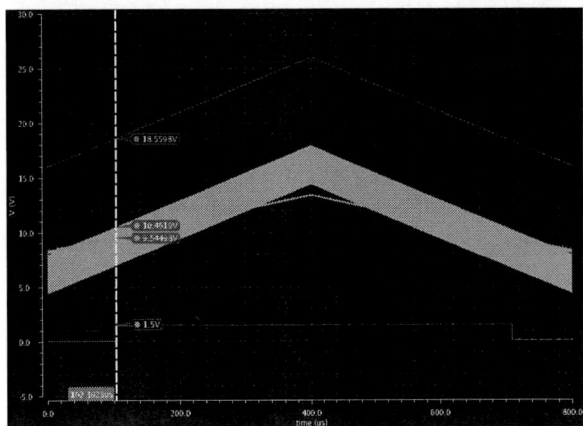

Fig. 13. Simulation of overvoltage detection with new avalanche diode model.

VI. CONCLUSION

Zener diodes driven in their avalanche breakdown regime show a certain kind of instability: The observed on-time and off-time of the voltage ripples vary statistically according to the voltage overshoot and breakdown current, respectively. Investigations showed that simulating this effect can be crucial for certain circuit topologies like voltage references.

Therefore, a novel VerilogA model has been developed and demonstrated advantages in circuit simulation on an automotive gate driver application.

REFERENCES

[1] A. D. Neamen: Semiconductor Physics and Devices. Fourth Edition. McGraw-Hill, 2012, p. 259.
[2] E. Becle, P. Talatchian, G. Prenat, L. Anghel, and I.-L. Prejbeanu: Fast Behavioral VerilogA Compact Model for Stochastic MTJ, European Solid-State Device Research Conference (ESSDERC) 2021.
[3] M. Karami, C. Niclass, and E. Charbon: Random Telegraph Signal in Single-Photon Avalanche Diodes, International Image Sensor Workshop 2009, p. 1-4.

Self-consistent Automated Parameter Extraction of RRAM Physics-Based Compact Model

Tommaso Zanotti, Paolo Pavan, Francesco Maria Puglisi

DIEF, University of Modena and Reggio Emilia, Via P. Vivarelli 10/1, 41125 Modena (MO), Italy

corresponding author phone: (+39) 059-2056320 email: tommaso.zanotti@unimore.it

Abstract— RRAM physics-based compact models are an essential tool for studying and designing novel RRAM-based circuits. Ideally, the compact model parameters should be easily calibrated on different RRAM technologies to enable the adoption of device-circuit co-optimization strategies. Still, most models in the literature lack a simple and self-consistent parameter extraction procedure. In this work, we devise a self-consistent automated parameter extraction procedure for the UniMORE RRAM compact model. The proposed procedure requires the execution of a few experiments that are commonly performed during device characterization. The procedure is validated on data collected experimentally on a TiN/Ti/HfOx/TiN RRAM technology, and on data from three RRAM technologies from the literature. The results show that the proposed automated parameter extraction procedure enables correctly calibrating the model parameters on all four considered RRAM technologies, enabling the simulation of the device characteristic in different operating conditions using a single set of parameters, and the implementation of device-circuit co-optimization strategies.

Keywords—RRAM, Compact modeling, Parameter extraction.

I. INTRODUCTION

Resistive Random access memory (RRAM) technologies [1] are considered one of the most promising emerging nonvolatile memory technologies thanks to their simple fabrication, relatively high performance, and possible exploitation for the development of highly energy efficient non von Neumann computing architectures, that are needed to enable the transition to a more distributed computing infrastructure. In fact, several works have demonstrated that RRAM devices enable the fabrication of in-memory computing hardware accelerators for deep learning applications [2], logic operations [3], but also the implementation of neuromorphic computing circuits which, by taking inspiration from the brain, promise to achieve extremely high energy efficiency performance [4]. Nevertheless, RRAM devices are characterized by intrinsic stochastic effects, which result in several device nonidealities (e.g., cycle-to-cycle variability, random telegraph noise, conduction nonlinearity), which can affect the circuit reliability and increase the design complexity [4]. Also, the use of different materials for the fabrication of RRAM devices leads to devices with different characteristics, demanding the adoption of device-circuit co-optimization strategies. Thus, accurate RRAM device compact models that can be easily calibrated on different RRAM technologies are essential to enable the adoption of appropriate design and circuit optimization strategies. Although several RRAM compact models have been proposed in the literature [5], an accurate RRAM compact model paired with a simple yet rigorous and self-consistent parameter extraction procedure is still missing. Specifically, although some parameter extraction procedures were developed for general purpose RRAM compact models [6], [7], the lack of a direct link between the device physics and the model parameters makes

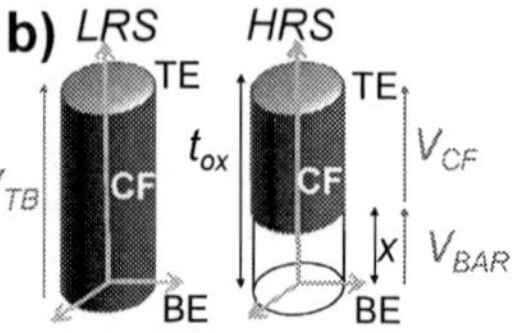

Fig. 1 – a) UniMORE RRAM physics-based compact model equations [9]. The equations model the conductive filament (CF) and the dielectric barrier components, and the field- and temperature- assisted defect generation and recombination which result in the change of the dielectric barrier thickness (x). b) Sketch representation of the modeled RRAM device both in low resistive state (LRS) and high resistive state (HRS).

calibrated general purpose compact models reliable only for a specific experimental condition, thus limiting their accuracy when simulating circuits where devices undergo different stimulations (e.g., neuromorphic circuits). Conversely, physics-based RRAM compact models enable simulating the device behavior in different operating conditions using a single parameter set for excellent dependability. However, very few parameter extraction procedures have been proposed in the literature [8] and commonly require a skilled user and do not cover all the possible operating conditions.

In this work, we develop a simple automated and self-consistent parameter extraction procedure for the UniMORE RRAM physics-based compact model [9]. The proposed procedure only requires the execution of few experiments that are commonly perform during device characterization and is validated on a *TiN/Ti/HfOx/TiN* RRAM technology from SEMATECH. Also, a simplified version of the parameter extraction procedure which requires fewer data is discussed and validated on three RRAM technologies from the literature [10]–[12].

II. UNIMORE RRAM PHYSICS-BASED COMPACT MODEL

The UniMORE RRAM physics-based compact model [9] considers the physical mechanism occurring in RRAMs introducing appropriate approximations to enable a compact implementation. Specifically, as shown in Fig. 1, the model considers the existence of a conductive filament (CF) and of a dielectric barrier, that together determine the total device resistance. A set of differential equations, see Fig. 1a, model the field-assisted and temperature accelerated oxygen ions and vacancies generation and recombination during set and reset, and the internal device temperature of the CF and barrier components, leading to high accuracy also when simulating fast voltage pulses. Also, the model includes the

978-1-6654-8498-5/22 $31.00 © 2022 IEEE

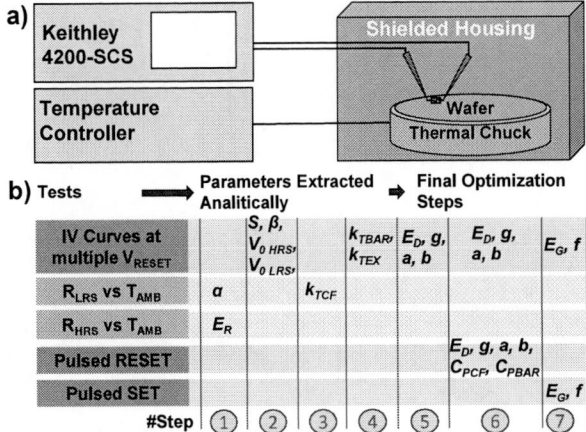

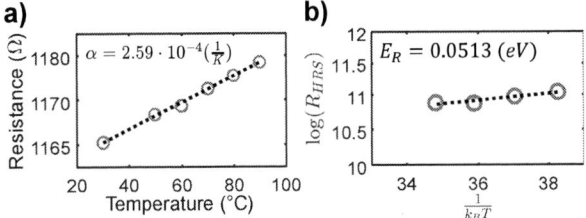

Fig. 2 – a) Sketch of the experimental setup. RRAM devices were characterized using a Keithley 4200-SCS semiconductor parameter analyzer and a temperature-controlled thermal chuck. b) Flow chart of the proposed parameter extraction procedure. The set of experiments and the sequence of steps required to extract all the parameters are reported. The extracted parameters are aligned with the corresponding experimental data used for their calculation. R_{LRS} and R_{HRS} indicate the device resistance when in LRS and HRS, respectively.

Fig. 3 – Experimentally measured data (red circles) and fitting lines (dotted black lines), from which the parameters a) α, and b) E_R.

most relevant device non-idealities such as device-to-device and cycle-to-cycle variability, multilevel Random Telegraph Noise, self-heating, and thermal effects, making it a complete tool for studying the circuit reliability and performance of novel RRAM-based circuits [13].

III. EXPERIMENTAL SETUP

To develop and validate the performance of the proposed parameter extraction procedure, the set of measurements listed in Fig. 2b is required. These measurements are usually performed during standard device characterization, specifically the measurement of the IV characteristic at various reset voltages (i.e., V_{RESET}), the temperature dependence of the device resistance (i.e., R_{LRS} vs T_{AMB}, R_{HRS} vs T_{AMB}, see Fig. 2b), and the pulsed characteristic. In this work, we use the experimental setup sketched in Fig. 2a, which includes a Keithley 4200-SCS semiconductor parameter analyzer, a temperature-controlled thermal chuck, and a shielded enclosure, to collect data on *TiN/Ti/HfOx/TiN* devices from SEMATECH. The full IV switching curve was measured at three different reset voltages (i.e., -0.9 V, -1 V, and -1.25 V), using a current compliance of 500 µA, and a slew rate of 0.0911 V/s. The use of a very slow slew rate is essential to enable a simple parameter extraction, as discussed in the Section IV. The device resistance both in HRS and LRS was measured at increasing temperatures (i.e., from 30 °C to 50 °C and from 30 °C to 90 °C, respectively) using 100 mV read voltage pulses. Also, the IV relation for a device in LRS was measured between 0 and 0.6 V at

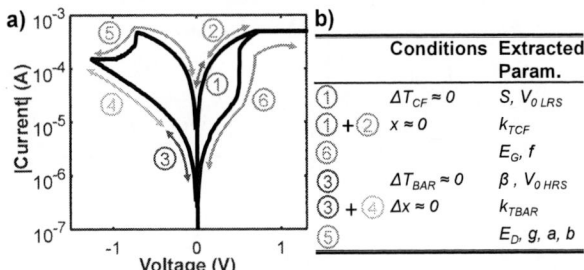

Fig. 4 – a) RRAM IV curve divided into segments. In each segment the approximations shown in b) are assumed to extract the corresponding parameters.

increasing temperature (i.e., from 30 °C to 50 °C), using a slew rate of 1 mV/s. Furthermore, the reset device pulsed characteristic was measured for different reset voltages (i.e., -1.1 V, and -1.3 V) and pulse widths (i.e., 10 µs, 1 µs, 100 ns) by first programming the device in LRS then delivering 200 voltage pulses to the device and reading the device resistance after each reset pulse with a 1 ms long, 50 mV read pulse. Finally, the pulsed set was characterized by first resetting the device (i.e., V_{RESET} = -1.25 V), then using set voltage pulses (i.e., V_{SET} = 1 V, and T_{PULSE} = 1 µs) and employing a 680 Ω resistor as a current limiter. Every measurement was repeated at least 50 times to reduce the effect of cycle-to-cycle variations.

IV. PARAMETER EXTRACTION PROCEDURES

A. Complete parameter extraction procedure

The complete parameter extraction procedure consists of 7 steps, in which parameters are either extracted analytically or through appropriate optimization algorithms, see Fig. 2b. In the first step, the measured relations of R_{LRS} and R_{HRS} with temperature are used to extract the CF resistivity thermal coefficient (i.e., α) and the trap assisted tunneling (TAT) activation energy (i.e., E_R), see Fig. 3a,b. As shown in Fig. 4, most of the other parameters can be extracted from different portions of the quasi-static IV curves, where appropriate approximations of the equations in Fig. 1 are used to enable a simple computation. Specifically, in segment 1 (i.e., the red segment in Fig. 4a) small voltages are applied to a device that is in low resistive state (LRS), thus both the barrier thickness and variations of the CF temperature are assumed to be zero. By knowing the oxide thickness (i.e., t_{OX}) of the specific RRAM technology, and by taking the value of the CF resistivity (i.e., ρ) from the literature (i.e., between 400 Ω nm and 10^4 Ω nm for HfOx based devices [13]) the nominal value of the CF cross-section (i.e, S) is determined by performing a linear regression on the experimental data, as shown in Fig. 5a. Similar to the approach adopted in other RRAM compact models, the nonlinear current conduction associated with TAT for a device in high resistive state (HRS), is modeled using the Simmons' tunneling barrier model [5], which employs the hyperbolic sine function (i.e., *sinh*) and a nonlinearity parameter (i.e., V_{0HRS}). Also, possible imperfections in the metallic-like CF can introduce nonlinearity in the conduction in LRS, which are also modeled using the *sinh* function and a nonlinearity parameter V_{0LRS}, that is commonly higher than V_{0HRS}. Using segments 3 and 1 (see Fig. 4) of the measured IV characteristic, where temperature variations are negligible for a device in HRS and LRS, respectively, both V_{0HRS} and V_{0LRS} are estimated with a nonlinear solver (e.g., MATLAB® *fsolve* function), as

978-1-6654-8498-5/22 $31.00 © 2022 IEEE

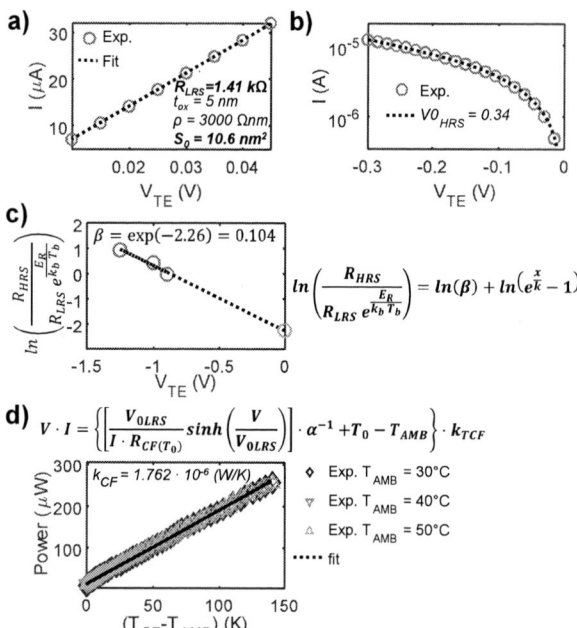

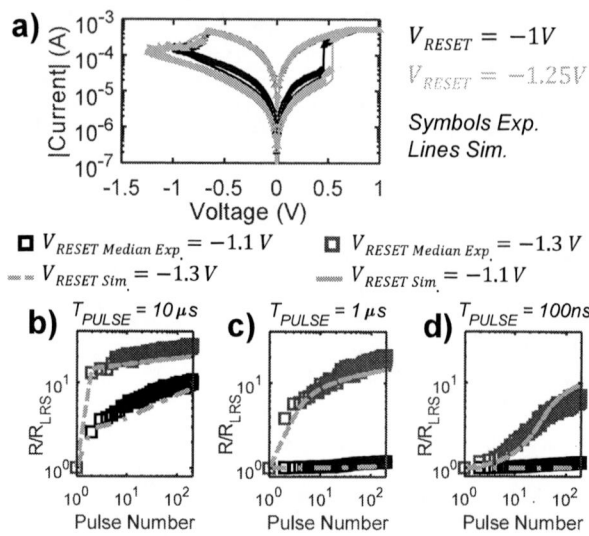

Fig. 5 – Experimentally measured data (red circles) and fitting lines (dotted black lines), from which the parameters a) S_0, b) V_{0HRS}, c) β, and d) k_{TCF} are extracted.

Fig. 6 – Experimental (symbols) and simulated (lines) of a) the quasi-static DC IV characteristic, and b), c), and d) the response to trains of reset pulses considering different reset voltages (V_{RESET}) and pulsed widths (T_{PULSE}).

shown in Fig. 5b for V_{0HRS}. Segment 3 of the IV at multiple V_{RESET} is also used to estimate β (i.e., the proportionality constant modeling the relation between R_{BAR} and the barrier thickness, see Fig. 1a). In this region of the IV the dielectric barrier is formed and stable, the effect of self-heating is negligible, and $R_{HRS} \approx R_{BAR}$. Thus, by rearranging the equation of R_{BAR} (see Fig. 1a), and using the measured R_{HRS} at different V_{RESET}, β is determined as the point at which the regression line in Fig. 5c intersects the y-axis at $V_{TE}=0$. In the third step of the parameter extraction procedure the CF thermal conductance (i.e., k_{TCF}) is computed using segments 1 and 2 of the experimental IV characteristic at different temperature, measured with a very slow slew-rate. In these segments the device is in LRS (i.e., $x \approx 0$), and thanks to the quasi-static approximation enabled by the very slow slew-rate, the CF temperature reaches equilibrium for each applied voltage (i.e., $dT_{CF}/dt \approx 0$). Thus, the equation reported in Fig. 5d is obtained by substitution and k_{TCF} is computed by linear regression, as the slope of the line fitting the dissipated power over the difference between the CF and the ambient temperatures (i.e., T_{CF} and T_{AMB}, respectively), as shown in Fig. 5d. After solving for k_{TCF}, it is possible to estimate the thermal conductivity of the dielectric barrier (i.e., k_{TBAR}) using the experimental segments 3 and 4 of the IV, preferably at various V_{RESET}. In these segments the quasi-static approximation is valid, the effect of self-heating is significant, and the dielectric barrier is fixed. Thanks to the quasi-static approximation, and by considering the term modeling the heat exchange between the barrier and the CF (i.e., k_{TEX}) negligible as a first order approximation, the equations of the model can be rearranged to build a nonlinear system of equations where the unknowns (i.e., T_{CF}, T_{BAR}, V_{CF}, V_{BAR}, and k_{TBAR}) are solved using a nonlinear solver (e.g., MATLAB *fsolve* function) and an optimization algorithm (e.g., particle swarm algorithm). The nonlinear system solver determines T_{CF}, T_{BAR}, V_{CF}, and V_{BAR} that are used to compute the cost function for the optimization

algorithm which searches for the optimal value for k_{TBAR}. However, for some RRAM technologies a more accurate calibration is achieved when also considering k_{TEX}. To extract k_{TEX}, the same system of equations used to extract k_{TBAR} with the inclusion of k_{TEX} is solved but optimizing at the same time k_{TCF}, k_{TBAR}, and k_{TEX}, using as a constraint the value of k_{TCF} extracted in the previous step. The remaining parameters model the device set/reset processes (i.e., E_D, g, a, b, E_G, and f) and the dynamic temperature variations (i.e., C_{PCF}, and C_{PBAR}). These parameters are used in differential equations, thus, to simplify the complex optimization problem we split the problem into three steps to progressively reach the optimal solution. First, an initial guess for the parameters influencing the reset process (i.e., E_D, g, a, and b) is obtained by adopting an optimization algorithm (i.e., particle swarm) on experimental data where the quasi-static approximation is valid, i.e., the portion of the IV characteristic measured at different V_{RESET} (i.e., segments 3, 4, and 5 in Fig. 4). The Mean Absolute Percentage Error (MAPE) between the measured and simulated current is used as the cost function for the optimization algorithm. However, since the cost function presents numerous local minima, the solution must be restricted to values that are physically plausible. For instance, a valid range for the oxygen ions diffusion activation energy (i.e., E_D) is between 1 and 2 eV [14]. In the second step, the parameters C_{PCF}, and C_{PBAR} are also optimized, using as starting point the previously estimated E_D, g, a, and b. To estimate the thermal capacitance for the CF and the barrier, the experimental pulsed response to trains of reset pulses are used in the optimization algorithm. Thus, the cost function considers two terms, the MAPE on the IV and on pulsed characteristic, respectively. To form the complete cost function, these two error terms and their absolute difference are summed together, avoiding overfitting only one of the device characteristics. Finally, the parameters E_G, and f which influence the set operation are calibrated on the experimental IV set curve (segment 6) and on the pulsed set characteristics. As a result of the parameter extraction procedure, both the quasi-static IV characteristic and the pulsed response of the device for different pulse amplitudes

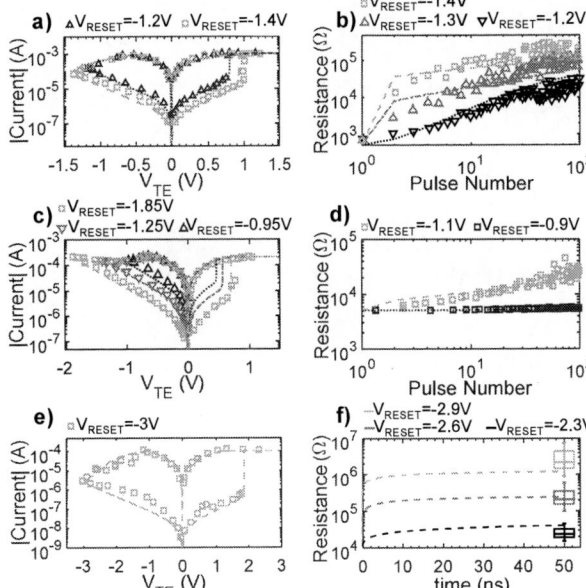

Fig. 7 – Compact model calibrated on three RRAM technologies from the literature. a), c), e) Experimental (symbols) and simulated (lines) quasi-static DC IV curve at different reset voltages (V_{RESET}). b), d), f) Experimental (symbols) simulated (lines) response to fast reset pulses for different reset voltages (V_{RESET}). a), b) data from [10]. c), d) Data from [11]. e), f) Data from [12].

and widths are well reproduced by the compact model using a single set of parameters, as shown in Fig. 6.

B. Simplified parameter extraction procedure

Although data from several RRAM technologies are available in the literature, the set of experiments that are commonly reported does not include all the data that are required for applying the complete parameter extraction procedure described in Section IV.A. Still, the parameter extraction procedure can be adapted to enable the calibration of the model parameters also when only the data for the IV and the pulsed characteristics, possibly at multiple V_{RESET}, are available – that is a typical case for data in the literature. Thanks to the physics-based modeling approach, a solution for missing data is to extract some parameters from the literature or material repositories. Specifically, parameters ρ, α, and E_R, can be taken from the literature (e.g., $E_R = 0.12$ eV for HfO$_x$-based RRAM technologies [15]). The remaining parameters can be extracted following the same steps from 2 through 7 shown in Fig. 2b. Specifically, S, V_{0HRS}, V_{0LRS}, and β are estimated using the available IV characteristic, preferably at multiple V_{RESET}. Also, the CF thermal conductance (i.e., k_{TCF}) can be estimated from the IV characteristic, with only small possible accuracy drops due to the availability of data for a single ambient temperature (i.e., T_{AMB}). Finally, the remaining parameters (i.e., k_{TBAR}, k_{TEX}, E_D, g, a, b, E_G, f, C_{PCF}, and C_{PBAR}) are extracted following the same optimization steps from 5 to 7, see Fig. 2b. The proposed simplified parameter extraction procedure was validated on three RRAM technologies from the literature. Specifically, the data for a Pt/TiO$_x$/HfO$_x$/TiO$_x$/HfO$_x$/TiN RRAM [10], a TiN/HfO$_2$/Ti/TiN RRAM [12], and a TiN/HfO$_x$/AlO$_x$/Pt RRAM [11], were used. As shown in Fig. 7, by using the developed parameter extraction procedure also in this case the compact model is able to correctly

reproduce the device characteristic in different operating conditions using a single set of parameters.

V. CONCLUSIONS

In this work, we developed and validated on four RRAM technologies an automated and self-consistent parameter extraction procedure for the UniMORE physics-based RRAM compact model. The proposed parameter extraction procedure requires the execution of few experiments that are commonly performed during the device characterization step and enables to quickly calibrate the model on new RRAM technologies, assisting the implementation of device-circuit co-optimization strategies that are essential for the development of novel RRAM-based circuits.

REFERENCES

[1] H.-S. P. Wong et al., "Metal–Oxide RRAM," *Proceedings of the IEEE*, vol. 100, no. 6, pp. 1951–1970, Jun. 2012.

[2] I. Chakraborty et al., "Resistive Crossbars as Approximate Hardware Building Blocks for Machine Learning: Opportunities and Challenges," *Proc. IEEE*, vol. 108, no. 12, pp. 2276–2310, Dec. 2020.

[3] J. Borghetti, G. S. Snider, P. J. Kuekes, J. J. Yang, D. R. Stewart, and R. S. Williams, "'Memristive' switches enable 'stateful' logic operations via material implication," *Nature*, vol. 464, no. 7290, pp. 873–876, Apr. 2010.

[4] I. Chakraborty, A. Jaiswal, A. K. Saha, S. K. Gupta, and K. Roy, "Pathways to efficient neuromorphic computing with non-volatile memory technologies," *Applied Physics Reviews*, vol. 7, no. 2, p. 021308, Jun. 2020.

[5] D. Panda, P. P. Sahu, and T. Y. Tseng, "A Collective Study on Modeling and Simulation of Resistive Random Access Memory," *Nanoscale Res Lett*, vol. 13, no. 1, p. 8, Jan. 2018.

[6] C. Yakopcic, T. M. Taha, D. J. Mountain, T. Salter, M. J. Marinella, and M. McLean, "Memristor Model Optimization Based on Parameter Extraction From Device Characterization Data," *IEEE Transactions on Computer-Aided Design of Integrated Circuits and Systems*, vol. 39, no. 5, pp. 1084–1095, May 2020.

[7] I. Messaris et al., "A TiO2 ReRAM parameter extraction method," in *2017 IEEE International Symposium on Circuits and Systems (ISCAS)*, May 2017, pp. 1–4.

[8] P. Huang et al., "Parameters extraction on HfOX based RRAM," in *2014 44th European Solid State Device Research Conference (ESSDERC)*, Sep. 2014, pp. 250–253.

[9] F. M. Puglisi, T. Zanotti, and P. Pavan, "Unimore Resistive Random Access Memory (RRAM) Verilog-A Model," *nanoHUB*, 2019.

[10] S. Yu, B. Gao, Z. Fang, H. Yu, J. Kang, and H.-P. Wong, "A neuromorphic visual system using RRAM synaptic devices with Sub-pJ energy and tolerance to variability: Experimental characterization and large-scale modeling," in *2012 International Electron Devices Meeting*, Dec. 2012, p. 10.4.1-10.4.4.

[11] S. Yu, Y. Wu, Y. Chai, J. Provine, and H.-S. P. Wong, "Characterization of switching parameters and multilevel capability in HfOx/AlOx bi-layer RRAM devices," in *Proceedings of 2011 International Symposium on VLSI Technology, Systems and Applications*, Hsinchu, Taiwan, Apr. 2011, pp. 1–2.

[12] J. Woo et al., "Improved Synaptic Behavior Under Identical Pulses Using AlOx/HfO2 Bilayer RRAM Array for Neuromorphic Systems," *IEEE Electron Device Letters*, vol. 37, no. 8, pp. 994–997, Aug. 2016.

[13] E. Hildebrandt, J. Kurian, M. M. Müller, T. Schroeder, H.-J. Kleebe, and L. Alff, "Controlled oxygen vacancy induced p-type conductivity in HfO2−x thin films," *Appl. Phys. Lett.*, vol. 99, no. 11, p. 112902, Sep. 2011.

[14] N. Capron, P. Broqvist, and A. Pasquarello, "Migration of oxygen vacancy in HfO2 and across the HfO2/SiO2 interface: A first-principles investigation," *Appl. Phys. Lett.*, vol. 91, no. 19, p. 192905, Nov. 2007.

[15] F. M. Puglisi, A. Qafa, and P. Pavan, "Temperature Impact on the Reset Operation in HfO2 RRAM," *IEEE Electron Device Letters*, vol. 36, no. 3, pp. 244–246, Mar. 2015.

InP DHBT test structure optimization towards 110 GHz characterization

Nil Davy[1], Marina Deng[2], Virginie Nodjiadjim[1], Chhandak Mukherjee[2], Muriel Riet[1], Colin Mismer[1], Jérémie Renaudier[3], Cristell Maneux[2]

[1] III-V Lab, joint lab between Nokia Bell Labs, Thales and CEA Leti, Palaiseau, France.
[2] IMS Laboratory, University of Bordeaux, CNRS UMR5218, Bordeaux INP, Talence, France.
[3] Nokia Bell Labs, Nozay, France

Abstract—In this paper, three different designs of test structures are explored in order to accurately characterize InP DHBTs up to 110 GHz. In particular, a new design, optimized for high frequency measurements while keeping high device density, has been proposed. De-embedding test structures are analyzed and InP DHBT RF figures of merit are extracted for the three designs. Extraction of the maximum oscillation frequency, f_{MAX}, confirms the relevance of optimized test structures as well as good performances of the new design.

Keywords—Characterization, Millimeter-wave, Double heterojunction bipolar transistor (DHBT), InP/InGaAs, Indium Phosphide (InP)

I. INTRODUCTION

Due to the increase in the volume of data exchanged, developing high-speed communication systems has become crucial, be it for optical or wireless communications. For this purpose, developing transistor technologies that combine high speed and high breakdown voltage has become mandatory. InP double heterojunction bipolar transistors (InP DHBT) have these characteristics. Both Type-I [1] and Type-II [2] DHBTs have already demonstrated f_{MAX} above 1 THz while having breakdown voltages of 4.1 V and 5.4 V, respectively.

To benefit from these performances, accurate characterization of these transistors is crucial in order to confirm the high cutoff frequency and to validate the associated compact model used for integrated circuit (IC) design. Transistor characterization below 110 GHz often involves off-wafer calibration and open-short de-embedding [1]–[3]. It has been shown that measurements beyond 110 GHz are particularly difficult [4]. To perform InP DHBT characterization at higher frequencies, on-wafer thru-reflect-line (TRL) calibration has been introduced along with optimized test structures [5]–[7]. This method has demonstrated excellent results. However, it requires more area on the wafer to accommodate additional calibration structures and one needs to change the inter-probe distance during the calibration step.

However, most of these previous works do not take into account the "industrial" point of view of characterizing transistors and rather invest of developing techniques that are area consuming. With that in mind, in this paper, we introduce a new industrially compatible design. This new design contains optimized test structures as well as a high density of transistors. The rest of this paper is organized as follows: section II details the transistor technology and the three RF test structures design; section III presents the comparison of the three test structure designs on deembedding structures characterization and InP DHBT RF figures of merit extraction.

II. RF TEST STRUCTURE DESIGN OPTIMIZATION

A. Technology description

The InP DHBT structure is grown on a 3-inch semi-insulating substrate using solid source molecular beam epitaxy (SSMBE). The intrinsic emitter is 40-nm thick InP. The highly C-doped ($8 \cdot 10^{19}$ cm^{-3}) 28-nm InGaAs base is compositionally graded (from In$_{0.47}$Ga$_{0.53}$As on the emitter side to In$_{0.53}$Ga$_{0.47}$As on the collector side) in order to reduce the base transit time. The 130-nm thick composite collector is composed of a non-intentionally doped InGaAs spacer, a 20-nm thick highly doped layer and a lightly doped collector region. The transistors are processed using a wet-etch self-aligned triple mesa technology as described in [8]. This technology, which also includes three levels of gold metallization for interconnections, is compatible with circuit design [9], [10]. Among the characterized devices, different transistor geometries were considered: emitter width varied from 0.4-μm to 0.7-μm while the emitter length varied from 5-μm to 10-μm. Fig. 1 shows a scanning electron microscopy view of a 0.4×7 μm² InP DHBT.

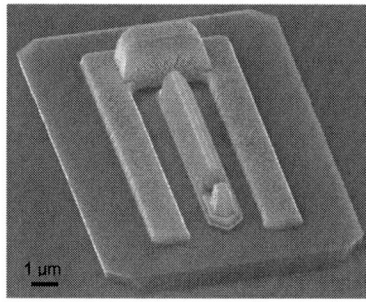

Fig. 1 . Scanning electron microscopy view of a 0.4×7 μm² InP DHBT before interconnection level

The performances of the 0.4×5 μm² InP DHBT under test have been presented in [11]. The common emitter breakdown voltage BV$_{CE0}$ is greater than 4.5 V and the static current gain β is around 30. The transition frequency, f_T, and the maximum oscillation frequency, f_{MAX}, are 380 GHz and 605 GHz, respectively.

B. Layout and floor-plan optimization of RF test structures

In this study an optimized design of RF test structures for industrial application is proposed. In order to analyze its performances, three different RF test structures are compared. The first design is used in an "industrial" context

978-1-6654-8498-5/22 $31.00 © 2022 IEEE

while the second has been designed to prove the feasibility of measurement up to 500 GHz. The third one is the new optimized design proposed in this work

1) High density pads (design 1)

This design (see Fig. 2) is the pre-defined design in standard III-V Lab process. It is designed with the purpose of maximizing the number of test structures within the chip area with different geometries of transistors. These high-density constraints are imposed due to the fact that a large number of transistors is required for process yield and variability analysis. The RF pads are compatible with 100-µm pitch probes. The density of transistor targeted with this design is 20 transistors per chip.

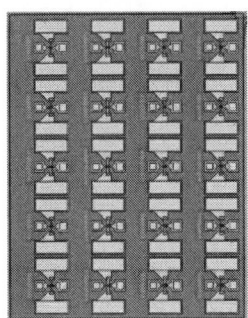

Fig. 2. Photograph of a die with high density pads

2) High quality pads (design 2)

As presented in [5], this design (see Fig. 3) combines two optimization strategies to improve the on-wafer measurement accuracy up to 500 GHz. First, a continuous ground plane was implemented in order to minimize the probe-to-substrate coupling, as well as the coupling between neighbouring structures. By connecting all the ground pads together and creating a signal pad shielding, this continuous ground plane also contributes to the improvement of the RF signal propagation along the access line. Second, on-wafer TRL calibration method was used up to 500 GHz, requiring the design of TRL standards on the same InP substrate as the devices to characterize. In addition, the RF pads were compatible with the 50-100 µm probing pitch required by the RF probes in the different frequency bands up to 500 GHz. The density of this design is 8 transistors per chip (Fig. 3 shows a chip, which has twice the area of the one in Fig. 2).

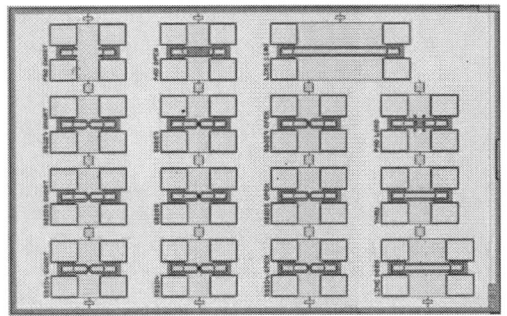

Fig. 3. Photograph of a die with high quality pads

3) Optimized pads for industrial process fabrication (design 3)

This third design (see Fig. 4) re-uses the continuous ground plane method used in the design 2 as it reduces the probe to substrate interaction for high frequency measurements. Moreover, a checkerboard configuration, as introduced in [12], was implemented. It allows the reduction of the distance between the neighbouring structures that are above and below. Thus, this configuration allows to optimize the density of test structures while maintaining a reduced coupling between neighbouring test structures. These pads are designed for 100-µm pitch probes. The density is 14 transistors per chip.

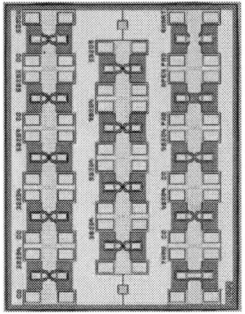

Fig. 4. Photograph of a die with optimized pads

III. RESULTS

To analyze the performances of these three test structure designs, the measurements on a 0.4×5 µm² InP DHBT and on its associated de-embedding test structures were performed. The S-parameters measurements were carried out from 1 GHz to 110 GHz with an Anritsu Vectorstar network analyzer using an off-wafer Short-Open-Load-Through (SOLT) calibration and 100-µm pitch Picoprobe RF probes.

A. De-embedding test structures characterization up to 110 GHz

The accurate extraction of transistor parameters relies on a precise de-embedding of parasitic capacitances and inductances that surround the device. In order to verify the validity of the de-embedding, open and short structures are associated to electrical equivalent circuits (see Fig. 5). To verify that open capacitances and short inductances are not frequency-dependent, their values were extracted using Y and Z parameters (see Fig. 7.).

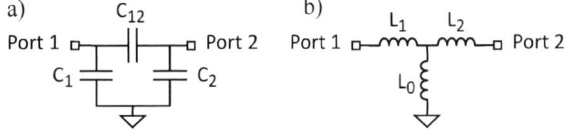

Fig. 5. Equivalent electrical circuit of (a) open and (b) short structures

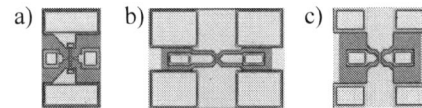

Fig. 6. Photograph of the open de-embedding structure (a) design 1 (b) design 2 (c) design 3

978-1-6654-8498-5/22 $31.00 ©2022 IEEE

As observed from Fig. 7, the measured open capacitances and short inductances of the three different test structure designs are quite constant over the frequency range. There is no resonance over the 110 GHz band. Design 1 is strongly asymmetric; it explains why C_1 and C_2 (as well as L_1 and L_2) show different values. The value of C_1 and C_2 can be correlated to the metal that surrounds the signal pad, which is why capacitances of design 2 (which is designed for 50-μm probe pitch) are twice the capacitances of design 1. Moreover, for the same design, a difference between L_1 and L_2 is visible. It is caused by the different RF probe tip positioning on the pads at port 1 and port 2, resulting in an unequal distance from the center of the test structure. The sensitivity to the probe tips positioning is one of the main drawbacks of probe-tip calibration methods, such as the SOLT calibration used in this work.

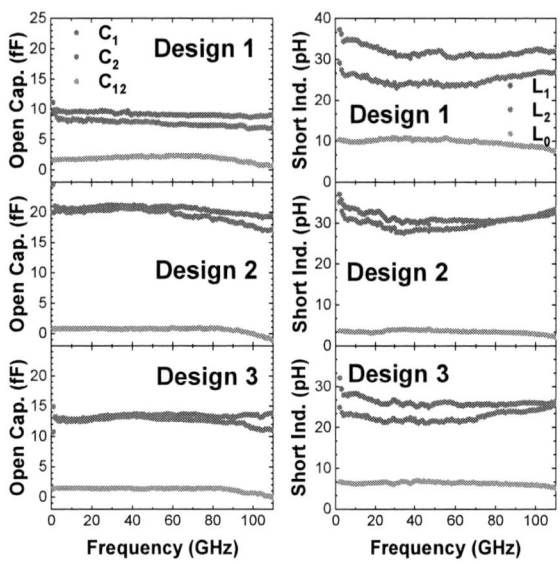

Fig. 7. Measured open capacitances and short inductances of the 3 test structure designs

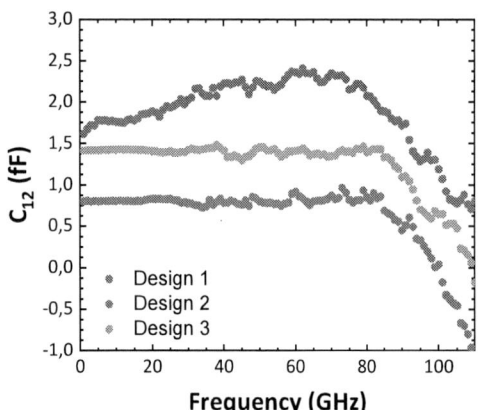

Fig. 8. Extracted port-1 to port-2 open capacitance C_{12} for the three test structure designs

Fig. 8 shows the C_{12} capacitances of the three test structure designs. It appears that in all three cases, the capacitance value starts to drop after 80 GHz. This behavior is certainly due to the signature of the Picoprobe RF probes, since it has already been observed both in EM simulation and measurement in [13], where Picoprobe RF probes were also used.

The capacitance in design 1 is not constant (up to 80 GHz), this indicates that coupling (with substrate or adjacent structures) still occurs. The probe-to-probe distance is different in the three test structure designs which explains the difference between the C_{12} capacitances values.

B. Extraction InP DHBT RF figures of merit up to 110 GHz

The InP DHBT were measured at $V_{CE} = 1.6$ V and different collector current densities J_C (from 0.25 mA/μm² to 9 mA/μm²). This chosen value of V_{CE} is a trade-off to combine high values of both f_{MAX} and f_T. Open-short de-embedding is performed to extract the performances of the device.

The extraction results of f_T are shown in Fig. 9. The extraction of f_T up to 110 GHz gives pretty similar results between the three designs.

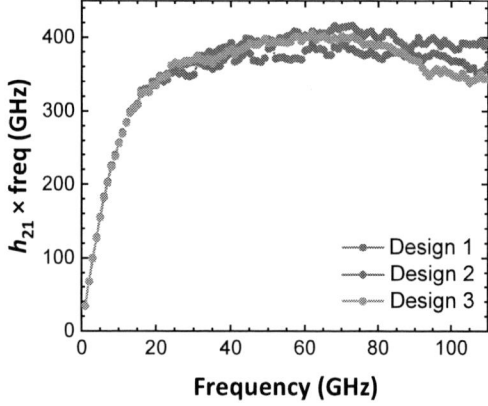

Fig. 9. Determination of f_T with gain-bandwidth product of a 0.4 × 5 μm² DHBT at $V_{CE} = 1.6$ V and $J_C \approx 6$ mA/μm

On the other hand, extraction of the maximum oscillation frequency f_{MAX} is more challenging and is sensitive to the design (see Fig. 10). Design 1 leads to an inaccurate extraction of f_{MAX}, certainly due to a combined effect of crosstalk between ports and additional coupling effects for using this non-optimized RF test structure design. The industrial design 3 produces results comparable to the optimized design 2 for the extraction of f_{MAX} up to 70 GHz. This confirms the satisfying tradeoff made in design 3 between optimization of RF test structures design and high device density. The drop of the three curves after 70 GHz is attributed to RF probes (and their coupling with structures). This effect is described in [14] where probe effects appear after 40 GHz. In our case, these effects appear at higher frequencies, which is why it is still possible to extract f_{MAX} by fitting a single pole function on the measurements up to 60 GHz (as shown on Fig. 10).

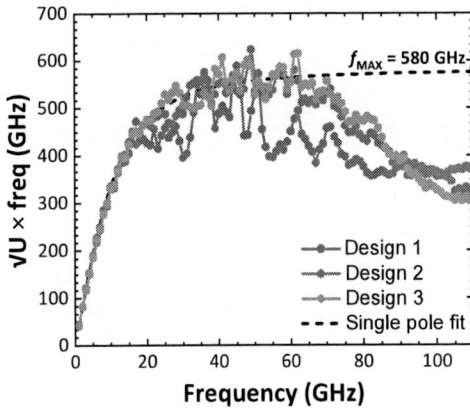

Fig. 10. Determination of f_{MAX} with gain-bandwidth product of a $0.4 \times 5 \ \mu m^2$ DHBT at $V_{CE} = 1.6$ V and $J_C \approx 6 \ mA/\mu m$

IV. CONCLUSIONS

Three different designs of test structures were explored in order to accurately extract RF figures of merit of InP DHBTs. De-embedding test structures were characterized and transistors RF figures of merit were extracted. The need of optimized test structures for accurate f_{MAX} extraction has been demonstrated. The new design, which maintains a high-density of transistors (1.75 times of that of design 2) while containing optimized test structures, shows similar performances as the design optimized for 500 GHz measurements (in the 110 GHz range). This work opens new perspectives to high frequency measurements beyond 110 GHz as well as accurate characterization with other RF probes.

ACKNOWLEDGMENT

This work was supported by ANR-FNS through ULTIMATE project (ANR-16-CE93-0007) and by the European Commission through the Photonics Public Private Partnership Initiatives under Grant H2020-ICT-2019-2 (Twilight Project).

REFERENCES

[1] J. C. Rode *et al.*, 'Indium Phosphide Heterobipolar Transistor Technology Beyond 1-THz Bandwidth', *IEEE Trans. Electron Devices*, vol. 62, no. 9, pp. 2779–2785, Sep. 2015, doi: 10.1109/TED.2015.2455231.

[2] A. M. Arabhavi *et al.*, 'InP/GaAsSb Double Heterojunction Bipolar Transistor Emitter-Fin Technology With f_{MAX} = 1.2 THz', *IEEE Trans. Electron Devices*, pp. 1–8, 2022, doi: 10.1109/TED.2021.3138379.

[3] H. Xu, B. Wu, E. W. Iverson, T. S. Low, and M. Feng, '0.5 THz Performance of a Type-II DHBT With a Doping-Graded and Constant-Composition GaAsSb Base', *IEEE Electron Device Lett.*, vol. 35, no. 1, pp. 24–26, Jan. 2014, doi: 10.1109/LED.2013.2290299.

[4] K. Yau, E. Dacquay, I. Sarkas, and S. P. Voinigescu, 'Device and IC Characterization Above 100 GHz', *IEEE Microw. Mag.*, vol. 13, no. 1, pp. 30–54, Jan. 2012, doi: 10.1109/MMM.2011.2173869.

[5] M. Deng *et al.*, 'InP DHBT Characterization up to 500 GHz and Compact Model Validation Towards THz Circuit Design', in *2021 IEEE BiCMOS and Compound Semiconductor Integrated Circuits and Technology Symposium (BCICTS)*, Dec. 2021, pp. 1–4. doi: 10.1109/BCICTS50416.2021.9682466.

[6] H. Lu, W. Cheng, O. Li, Y. Kong, and T. Chen, 'Measurement and Modeling Techniques for InP-Based HBT Devices to 220GHz', in *2018 International Conference on Microwave and Millimeter Wave Technology (ICMMT)*, May 2018, pp. 1–3. doi: 10.1109/ICMMT.2018.8563540.

[7] D. F. Williams *et al.*, 'Calibrations for Millimeter-Wave Silicon Transistor Characterization', *IEEE Trans. Microw. Theory Tech.*, vol. 62, no. 3, pp. 658–668, Mar. 2014, doi: 10.1109/TMTT.2014.2300839.

[8] V. Nodjiadjim *et al.*, '0.7- μ m InP DHBT Technology With 400-GHz f_T and f_{MAX} and 4.5-V BVCE0 for High Speed and High Frequency Integrated Circuits', *IEEE J. Electron Devices Soc.*, vol. 7, pp. 748–752, 2019, doi: 10.1109/JEDS.2019.2928271.

[9] A. Konczykowska, J.-Y. Dupuy, F. Jorge, M. Riet, V. Nodjiadjim, and H. Mardoyan, 'Extreme Speed Power-DAC: Leveraging InP DHBT for Ultimate Capacity Single-Carrier Optical Transmissions', *J. Light. Technol.*, vol. 36, no. 2, pp. 401–407, Jan. 2018, doi: 10.1109/JLT.2017.2760507.

[10] R. Hersent *et al.*, '160-GSa/s-and-Beyond 108-GHz-Bandwidth Over-2-V $_{ppd}$ Output-Swing 0.5-μm InP DHBT 2:1 AMUX-Driver for Next-Generation Optical Communications', *IEEE Microw. Wirel. Compon. Lett.*, pp. 1–4, 2022, doi: 10.1109/LMWC.2022.3161706.

[11] N. Davy *et al.*, '0.4-μm InP/InGaAs DHBT with a 380-GHz f_T, > 600-GHz f_{max} and BV_{CE0} > 4.5 V', in *2021 IEEE BiCMOS and Compound Semiconductor Integrated Circuits and Technology Symposium (BCICTS)*, Dec. 2021, pp. 1–4. doi: 10.1109/BCICTS50416.2021.9687209.

[12] M. Cabbia, C. Yadav, M. Deng, S. Fregonese, M. De Matos, and T. Zimmer, 'Silicon Test Structures Design for Sub-THz and THz Measurements', *IEEE Trans. Electron Devices*, vol. 67, no. 12, pp. 5639–5645, Dec. 2020, doi: 10.1109/TED.2020.3031575.

[13] M. Deng *et al.*, 'Design of On-Wafer TRL Calibration Kit for InP Technologies Characterization up to 500 GHz', *IEEE Trans. Electron Devices*, vol. 67, no. 12, pp. 5441–5447, Dec. 2020, doi: 10.1109/TED.2020.3033834.

[14] S. Fregonese, M. De Matos, M. Deng, D. Céli, N. Derrier, and T. Zimmer, 'Importance of Probe Choice for Extracting Figures of Merit of Advanced mmW Transistors', *IEEE Trans. Electron Devices*, pp. 1–8, 2021, doi: 10.1109/TED.2021.3118671.

Trap Behavior of Metamorphic HEMTs

with Pulsed IV and $1/f$ Noise Measurement

Ki-Yong Shin
Department of Electrical, Electronic
and Computer Engineering,
University of Ulsan
Ulsan 44610, South Korea
Email: rldyd3140@ulsan.ac.kr

Ju-Won Shin
Department of Electrical, Electronic
and Computer Engineering,
University of Ulsan
Ulsan 44610, South Korea
Email: swinwins@ulsan.ac.kr

Surajit Chakraborty
Department of Electrical, Electronic
and Computer Engineering,
University of Ulsan
Ulsan 44610, South Korea
Email: surajit5103@ulsan.ac.kr

Walid Amir
Department of Electrical, Electronic
and Computer Engineering,
University of Ulsan
Ulsan 44610, South Korea
Email: walid37@ulsan.ac.kr

Chan-Soo Shin
Korea Advanced Nano Fab Center,
Suwon-si 17550, South Korea
Email: chansoo.shin@kanc.re.kr

Tae-Woo Kim
Department of Electrical, Electronic
and Computer Engineering,
University of Ulsan
Ulsan 44610, South Korea
Email: twkim78@ulsan.ac.kr

Abstract— **This work investigates the impact of the metamorphic buffer on the reliability of InAlAs/InGaAs metamorphic high-electron mobility transistors (MHEMT) on GaAs substrate. The pulsed IV measurement technique was employed to study the effect of transient charging in buffer layer of both devices. In pulse width (PW) ranging from 50 ~ 1000 μs, a greater rapid drain current degradation was shown in MHEMT due to the trapping effect in a metamorphic buffer. The dominant trapping location was estimated via low-frequency ($1/f$) noise characteristic. The extracted frequency exponent indicates that the dominant trapping location is in the bulk distant from the InGaAs channel in MHEMT. The value of extracted bulk trap density using the carrier-number fluctuation (CNF) model was 1.89×10^{17} eV^{-1}·cm^{-3} for LM-HEMTs and 2.31×10^{18} eV^{-1}·cm^{-3} for MHEMT.**

Keywords—Metamorphic Buffer, MHEMT, Lattice mismatch, Trap, Pulsed IV, I_D degradation, Low-frequency ($1/f$) noise measurement, Bulk trap density.

I. INTRODUCTION

High-electron mobility transistors (HEMTs) with InGaAs channels showed great performance suitable for next-generation THz communication and low-noise amplifiers due to the high electron mobility characteristics [1]. Based on these high electron mobility characteristics, InP-based pseudomorphic HEMTs (pHEMTs) showed the over 700GHz cut-off frequency (f_T) and over THz maximum oscillation frequency (f_{max}) and recorded the highest value among all FETs [2]–[4]. In spite of the impressive performance of InP-based InGaAs HEMTs, there is limitation of substrate with InP substrate which has the maximum growth size is 3 or 4 inches. For the commercial purpose, MHEMT has been investigated by utilizing a GaAs substrate with a bigger substrate size of up to 6 inches and fewer brittle wafers at a lower cost than InP substrate. The large wafer size is highly desirable for the semiconductor industry to reduce the production cost. In particular, the DC and RF performance comparable to pHEMTs were shown in MHEMT using a graded metamorphic buffer [5]–[7].

For commercial applications including low noise mm-wave amplifiers, the reliability of the device was one of the great concerns. In particular, MHEMTs have many concerns

due to lattic mismatch between channel and substrate that can be degraded on reliabilies. Various studies except lifetime and bias stressing tests mainly have been done on reliability analysis according to dislocation in metamorphic buffer. The metamorphic buffer between the substrate and active layer of HEMT structure is lattice-mismatched, which results in a high dislocation density and rough surface morphology [8]. As a matter of fact, it will be potential problems when device is working. Therefore, it is necessary to verify reliability problems that are expected to occur due to traps caused by dislocation.

In this work, we investigated the reliability issues due to lattice-mismatched metamorphic buffer in MHEMT. And it was compared with InP-based LM-HEMTs having the same standard HEMTs structure. For comparison, the same fabrication process was performed for both devices. In two devices with comparable DC and RF performance, we used pulsed IV measurement to observe the rapid degradation in drain current due to the transient charging effect according to the buffer difference. We also used $1/f$ noise measurements to figure out where the dominant trap was. Along with these characteristics, bulk trap density (N_t) was extracted for quantitative comparison.

The sample was grown on 3-inch InP and 6-inch GaAs substrates by molecular beam epitaxy (MBE). From top to bottom, the epitaxial layer consists of a 40 nm heavily doped multi-layer cap (In$_{0.7}$Ga$_{0.3}$As, In$_{0.53}$Ga$_{0.47}$As, In$_{0.52}$Al$_{0.48}$As), 4 nm InP etch-stopper, 8 nm In$_{0.52}$Al$_{0.48}$As barrier, Si δ-doping, 3 nm In$_{0.52}$Al$_{0.48}$As spacer, 10 nm In$_{0.53}$Ga$_{0.47}$As channel, 100 nm In$_{0.52}$Al$_{0.48}$As buffer. LM-HEMTs were grown on InP substrates with these structures, and MEHMT was grown on GaAs substrates with 300 nm graded metamorphic buffer. Figure 1(a) shows a schematic cross section of InAlAs/InGaAs LM-HEMTs and MHEMT. In order to inspect high dislocation density and rough surface morphology in metamorphic buffer, transmission electron microscopy (TEM) inspection was carried out. Figures 1(b) and 1(c) show TEM images of LM-HEMTs and MHEMT. In MHEMT, it can be confirmed that the high dislocation density and rough surface morphology are exhibited due to the lattice mismatch between the GaAs substrate and the

978-1-6654-8498-5/22 $31.00 © 2022 IEEE

metamorphic buffer. On the other hand, a neat interface is formed between the lattice-matched InP substrate and the $In_{0.52}Al_{0.48}As$ buffer in LM-HEMTs.

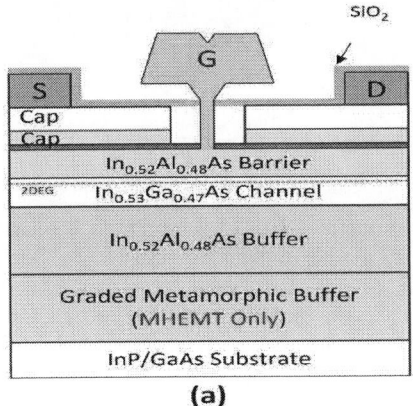

(a)

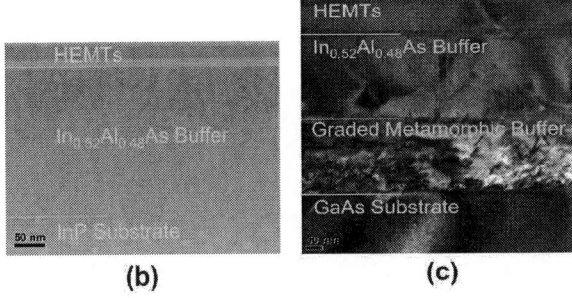

(b)   **(c)**

Fig 1: (a) Schematic cross section of InAlAs/InGaAs HEMTs with two different substrates. TEM image of schematic cross section showing : (b) InP-based LM-HEMTs; (c) GaAs-based MHEMT.

Device fabrication consists of mesa isolation to isolate the devices by wet chemical etching, non-alloyed ohmic contact with a Mo/Ti/Pt/Au/Ni (10/10/30/350/5 nm), SiO_2 passivation by sputter, sub-200 nm T-gate formation by electron-beam lithography. And SiO_2 passivation layer was etched by reactive ion etching. Before gate metallization, a heavily doped multi-layer cap was etched by the citric acid-based etchant. Finally, gate metal was deposited with a Pt/Ti/Pt/Au (8/20/20/350 nm) metal stack. The gate width (W_g), source to gate distance (L_{SG}), and drain to source distance (L_{DG}) of the fabricated devices were 50, 1, and 1 µm, respectively.

II. RESULTS AND DISCUSSION

We have compared the DC and RF characteristics of both fabricated devices with the Keysight B1500A Semiconductor Parameter Analyzer. Figure 2(a) shows I_D and I_G characteristics of L_g = 150 nm LM-HEMTs and MHEMT at V_{DS} = 0.05, 0.5 V. MHEMT had a high off-state current, so the subthreshold swing was 65.10 mV/decade in LM-HEMTs and 148.81 mV/decade in MHEMT. Despite the g_m of MHEMT was 1.16 mS/µm, which was higher than LM-HEMTs at V_{DS} = 0.5 V. Figure 2(b) plot the measured short-circuit current gain ($|h_{21}|^2$) for both devices with L_g = 150 nm at V_{DS} = 0.8 V and near the peak g_m conditions. All the RF characterization was carried out from 1 to 45 GHz using HP

8510C Network Analyzer with on-wafer calibration. The de-embedding of the parasitic pad components from the measured S-parameter data was performed using on-wafer open and short patterns. By extrapolating, the obtained f_T shows similar RF performance with 226 GHz at LH-HEMTs and 235 GHz at MHEMT. MHEMT with a graded metamorphic buffer structure that was lattice-mismatched with a GaAs substrate has a high off-state current but comparable device performance including DC and RF performance.

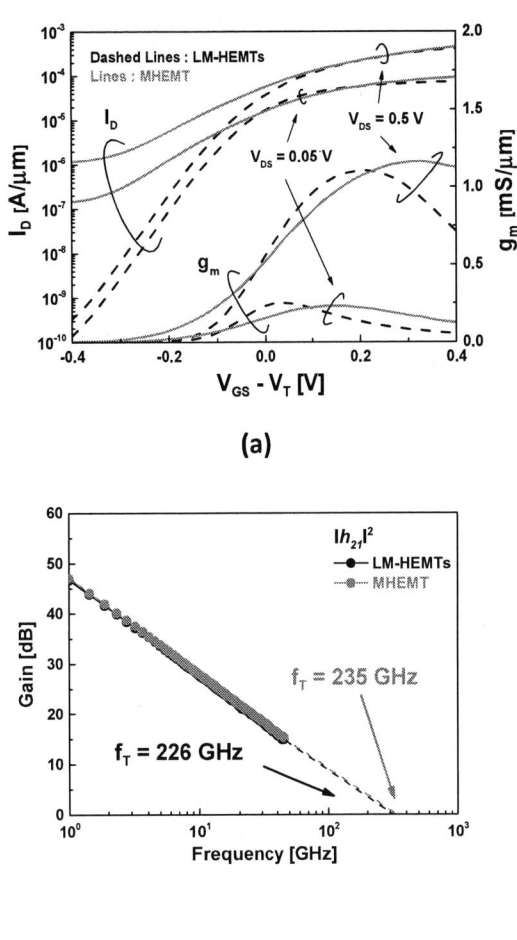

(a)

(b)

Fig 2: DC and RF Characteristic of LM-HEMTs and MHEMT with L_g = 150 nm. (a) Drain current and transconductance(g_m) characteristics at V_{DS} = 0.05, 0.5 V. (b) RF performances at V_{DS} = 0.8 V and near the peak g_m conditions.

To check the transient charging effect in two devices, we used the pulsed IV measurement with Keysight B1500A semiconductor parameter analyzer and B1530A waveform generator module [9]–[11]. The pulsed IV measurements were performed by inputting a trapezoidal-shaped V_G pulse to the gate and then measuring the drain current. In order to sufficiently minimize the trapping effect while the trapezoidal V_G pulse is applied, we used a rise time (t_r) and a fall time (t_f) of 50 ns. This is because, when t_r and t_f become longer, the trapping effect already occurs while the pulse is applied [9]. The transient charging effect was checked under various PW time conditions (50, 100, 500, 1000 µs) with holding the t_r and t_f constant. Figure 3 is the pulsed IV result

978-1-6654-8498-5/22 $31.00 © 2022 IEEE

of two devices, which can be expressed in pulse time domain. The charge trapping effect causes the degradation of drain current in various PW conditions. When PW = 1000 μs, it can be seen that when 8.76 μA was reduced for LM-HEMTs, 46.79 μA was reduced for MHEMT, which is more than five times. This indicates that more electrons are trapped in the defect area of the device during the same time, which leads to greater current degradation.

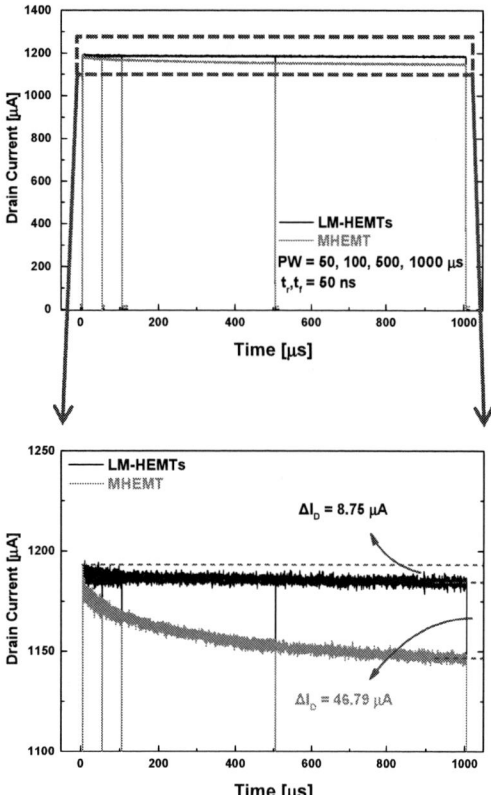

Fig 3: Pulsed IV measurement result of two devices with various PW conditions holding the t_r, t_f constant at $V_{DS} = 0.5$ V.

To confirm that the dominant trapping location caused by the device defect site happened in response to the channel carrier fluctuation, the $1/f$ noise measurement was conducted by the use of an SRS-570 low-noise current amplifier and HP 35670A signal analyzer, covering frequency up to 10^4 Hz at $V_{DS} = 0.05$ V. From the normalized drain current noise (S_{Id}/I_D^2) where $I_D = 1$ A/um The S_{Id}/I_D^2 characteristics according to frequency can be described by the power law equation of $1/f^\gamma$, which was shown along with the $1/f^\gamma$ function in the frequency range [12], [13]. At this time, the frequency exponent (γ) can be extracted, and the extracted γ values are generally close to 1, which means that the dominant traps involved in 1/f noise are constant for energy and depth [12], [13]. The corresponding γ is 1.13 for LM-HEMTs and 0.93 for MHEMT. The γ value for the LH-HEMTs above 1 indicates that the dominant trapping location is close to the InGaAs channel interface. On the other hand, the dominant trapping location of the MHEMT with a γ value lower than 1 is in the bulk traps, which are more distant from the InGaAs interface. Furthermore, N_t was extracted for quantitative

comparison based on the carrier number fluctuation (CNF) model [14]–[16] :

$$\frac{S_{Id}}{I_D^2} = \left(\frac{g_m}{I_D}\right)^2 S_{Vfb} \quad (1)$$

with

$$S_{Vfb} = \frac{q^2 N_t kT \lambda}{WLC_{ins}^2 f} \quad (2)$$

S_{Vfb} denotes the power spectral density of flat band voltage fluctuation, k is the Boltzmann constant, T is the absolute temperature, WL is the channel area, C_{ins} is the insulator capacitance, f is the frequency and N_t is the bulk trap density. λ is the tunneling attenuation distance, which is expressed as $\lambda = [4\pi(2m^*\Phi_B)^{1/2}/h]^{-1}$, where Φ_B is the barrier height and m^* is the effective mass. The terms S_{Id}/I_D^2 and $(g_m/I_D)^2$ fluctuate in comparable ranges with the drain current or gate voltage, according to the CNF model.

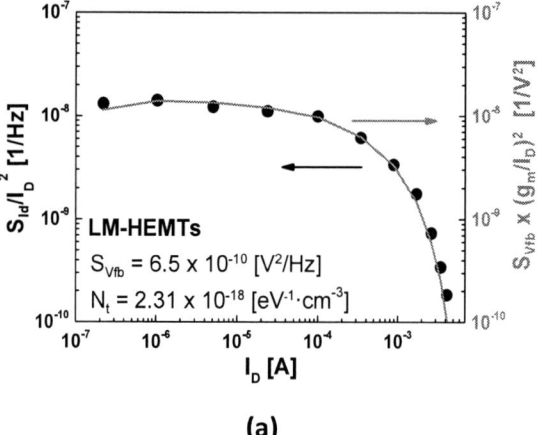

(a)

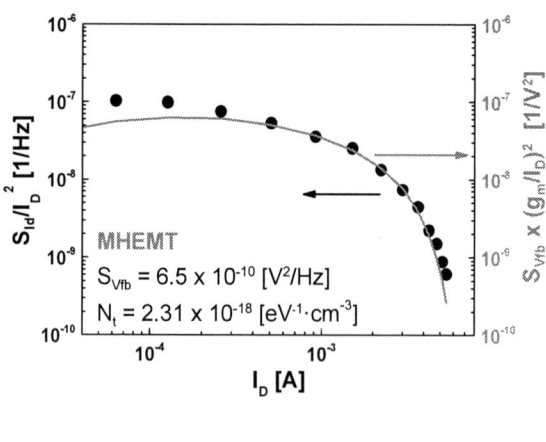

(b)

Fig 4: S_{Id}/I_D^2 (black dot) and $(g_m/I_D)^2$ (red line) for (a) LM-HEMTs and (b) MHEMT at $V_{DS} = 0.05$ V at $L_g = 150$ nm.

Figures 4(a) and 4(b) illustrate that with changing I_D both S_{Id}/I_D^2 and $(g_m/I_D)^2$ fluctuate similarly over several decades at a frequency of 10 Hz and $L_g = 150$ nm in two devices. The extracted values of S_{Vfb} are 1.2×10^{11} V²·Hz⁻¹ for LM-HEMTs and 6.5×10^{10} V²·Hz⁻¹ for LM-HEMTs, respectively. The corresponding N_t is 1.89×10^{17} eV⁻¹·cm⁻³ for LM-HEMTs and 2.31×10^{18} eV⁻¹·cm⁻³ for MHEMT.

III. CONCLUSION

In summary, we investigated the impact of the metamorphic buffer on the reliability of InAlAs/InGaAs MHEMT grown on GaAs substrates and LM-HEMTs grown on InP substrates. MHEMT showed high off-state current, but still showed comparable DC and RF characteristics. However, through pulsed IV measurement, it was confirmed that the rapid drain current degradation due to the transient charging effect, which is more than five times greater in MHEMT. The dominant trapping location of LM-HEMTs was close to the $In_{0.53}Ga_{0.47}As$ channel, whereas the dominant trapping location of MHEMT exists in the bulk. The N_t value was calculated using the CNF model via $1/f$ noise measurement was 1.89×10^{17} eV⁻¹·cm⁻³ for LM-HEMTs and 2.31×10^{18} eV⁻¹·cm⁻³ for MHEMT. We experimentally confirmed the mechanism of degradation caused by the metamorphic buffer layer. This means that MHEMT is structurally unstable, and it is necessary to optimize and verify the reliability of the metamorphic buffer layer.

ACKNOWLEDGMENT

This work was supported by a National Research Foundation of Korea (NRF) grant funded by the Korean government (MSIP; Ministry of Science, ICT and Future Planning, NRF-2019R1A2C1009816 and NRF-2019M3F5A1A01076973), and by the Civil-Military Technology Cooperation Program (No. 19-CM-BD-05).

REFERENCES

[1] J. A. del Alamo, "Nanometre-scale electronics with III-V compound semiconductors," *Nature*, vol. 479, no. 7373, pp. 317–323, Nov. 2011, doi: 10.1038/NATURE10677.

[2] H. B. Jo *et al.*, "Lg= 19 nm In0.8Ga0.2As composite-channel HEMTs with fT= 738 GHz and fmax= 492 GHz," *Technical Digest - International Electron Devices Meeting, IEDM*, vol. 2020-December, pp. 8.4.1-8.4.4, Dec. 2020, doi: 10.1109/IEDM13553.2020.9372070.

[3] R. Lai *et al.*, "Sub 50 nm InP HEMT device with Fmax greater than 1 THz," *Technical Digest - International Electron Devices Meeting, IEDM*, pp. 609–611, 2007, doi: 10.1109/IEDM.2007.4419013.

[4] D. H. Kim, J. A. del Alamo, P. Chen, W. Ha, M. Urteaga, and B. Brar, "50-nm E-mode in0.7Ga0.3As PHEMTs on 100-mm InP substrate with fmax > 1 THz," *Technical Digest - International Electron Devices Meeting, IEDM*, 2010, doi: 10.1109/IEDM.2010.5703453.

[5] D. H. Kim, B. Brar, and J. A. del Alamo, "f T = 688 GHz and f max = 800 GHz in L g = 40 nm In 0.7Ga 0.3As MHEMTs with g m-max > 2.7 mS/μm," *Technical Digest - International Electron Devices Meeting, IEDM*, 2011, doi: 10.1109/IEDM.2011.6131548.

[6] A. Leuther *et al.*, "20 nm Metamorphic HEMT technology for terahertz monolithic integrated circuits," *European Microwave Week 2014: "Connecting the Future", EuMW 2014 - Conference Proceedings; EuMIC 2014: 9th European Microwave Integrated Circuits Conference*, pp. 84–87, Dec. 2014, doi: 10.1109/EUMIC.2014.6997797.

[7] D. v. Lavrukhin *et al.*, "MHEMT with a power-gain cut-off frequency of f max = 0.63 THz on the basis of a In0.42Al0.58As/In0.42Ga0.58As/In0.42Al0.58As/GaAs nanoheterostructure," *Semiconductors 2014 48:1*, vol. 48, no. 1, pp. 69–72, Jan. 2014, doi: 10.1134/S1063782614010187.

[8] D. Lubyshev *et al.*, "Strain relaxation and dislocation filtering in metamorphic high electron mobility transistor structures grown on GaAs substrates," *Journal of Vacuum Science & Technology B: Microelectronics and Nanometer Structures Processing, Measurement, and Phenomena*, vol. 19, no. 4, p. 1510, Aug. 2001, doi: 10.1116/1.1376384.

[9] C. D. Young, Y. Zhao, D. Heh, R. Choi, B. H. Lee, and G. Bersuker, "Pulsed Id-Vg methodology and its application to electron-trapping characterization and defect density profiling," *IEEE Transactions on Electron Devices*, vol. 56, no. 6, pp. 1322–1329, 2009, doi: 10.1109/TED.2009.2019384.

[10] H. M. Kwon, D. H. Kim, and T. W. Kim, "Impact of fast and slow transient charging effect on reliability instability in In0.7Ga0.3As quantum-well MOSFETs with high-κ dielectrics," *Japanese Journal of Applied Physics*, vol. 59, no. 11, p. 110903, Oct. 2020, doi: 10.35848/1347-4065/ABBFE5.

[11] C. D. Young, D. Heh, A. Neugroschel, R. Choi, B. H. Lee, and G. Bersuker, "Electrical characterization and analysis techniques for the high-κ era," *Microelectronics Reliability*, vol. 47, no. 4–5, pp. 479–488, Apr. 2007, doi: 10.1016/J.MICROREL.2007.01.053.

[12] T. Boutchacha, G. Ghibaudo, G. Guégan, and M. Haond, "Low frequency noise characterization of 0.25 μm Si CMOS transistors," *Journal of Non-Crystalline Solids*, vol. 216, pp. 192–197, 1997, doi: 10.1016/S0022-3093(97)00212-3.

[13] I. S. Han *et al.*, "Effect of nitrogen concentration on low-frequency noise and negative bias temperature instability of p-channel metal-oxide-semiconductor field-effect transistors with nitrided gate oxide," *Japanese Journal of Applied Physics*, vol. 50, no. 10 PART 2, p. 10PB03, Oct. 2011, doi: 10.1143/JJAP.50.10PB03/XML.

[14] G. Ghibaudo, O. Roux, C. Nguyen-Duc, F. Balestra, and J. Brini, "Improved Analysis of Low Frequency Noise in Field-Effect MOS Transistors," *physica status solidi (a)*, vol. 124, no. 2, pp. 571–581, Apr. 1991, doi: 10.1002/PSSA.2211240225.

[15] P. Viktorovitch *et al.*, "Low frequency noise sources in InAlAs/InGaAs MODFET's," *IEEE Transactions on Electron Devices*, vol. 43, no. 12, pp. 2085–2100, 1996, doi: 10.1109/16.544379.

[16] W. Amir *et al.*, "A quantitative approach for trap analysis between Al0.25Ga0.75N and GaN in high electron mobility transistors," *Scientific Reports*, vol. 11, no. 1, Dec. 2021, doi: 10.1038/S41598-021-01768-4.

Analysis and Optimization of an Analog MOSFET with a Slit Well at Channel Center Towards Higher Output Resistance

Hiroki Fujii[*‡], Jaehyun Yoo[*], Dawon Jeong[*], Seongsik Min[†], Myoungsoo Kim[†], Uihui Kwon[*], and Dae Sin Kim[*]

[*]*Innovation Center /* [†]*Foundry Division, Samsung Electronics Co., Ltd., Hwaseong-si, Gyeonggi-do, Korea*
[‡]E-mail: h.fujii@samsung.com

Abstract— **This paper reports on a novel approach to improve and optimize an output resistance (Rout) which is critical to a long-channel analog MOSFET. The Rout degraded by halo doping can be overcompensated by the slit well inserted along the channel center, reaching the target value of 10Mohm*µm at the channel length of 0.5µm. This improvement is brought by the pinch-off generation at the channel center which makes the drain-side half channel act as a buffer layer for the source-side half channel potential against the drain voltage. The increased fitting parameters can be precisely regressed and optimized by the machine-learning based TCAD scheme, maximizing the overall electrical performance including the Rout.**

Keywords— Output resistance (Rout), Long-channel, Analog, Halo, Slit well, Drain-induced threshold voltage shift (DITS), Machine-learning, TCAD

I. INTRODUCTION

Halo doping is generally applied in digital CMOS devices to reduce the short channel effect (SCE) which becomes prominent with design scaling. However, the halo doping is not preferable for long-channel analog MOSFETs since it degrades the output resistance (Rout) in saturation mode [1, 2]. The mechanism is that the pinch-off occurs in the drain-side halo region due to the relatively weak vertical field, and the pinch-off point moves towards the source within the halo region while the low-doped center region still operates in a linear mode [2]. Hence, the increase in net voltage applied to the center region enlarges the drain current even after the pinch-off occurs. The lower Rout reduces the various analog performance including the gain of amplifier and the precision of current mirror circuit [3]. As a practical method to solve this issue, we propose a structure using a slit well at channel center with a dedicated mask step. Then, we investigate the improvement effect, and also present a machine-learning TCAD optimization scheme to handle the increased fitting parameter number.

II. DEVICE SETUP

A 1.2V-class NMOSFET from 0.06µm CMOS technology is used for evaluation. The Rout target is 10Mohm*µm at the gate length (Lg) of 0.5µm. It is monitored at Id/Wg= 1µA/µm at half Vdd in saturation region. The halo dose decrease or removal improves Rout as expected (Fig. 1), however, the Rout value is still behind the target. Furthermore, the SCE reduces the linear threshold voltage (Vtlin) considerably (Fig. 2a), and the breakdown voltage (BVdss) goes below the target line (Fig. 2b) due to SD punch-through. As a countermeasure, a slit well is inserted along the channel center in a long-channel analog MOSFET with Lg≥ 0.5µm (Fig. 3).

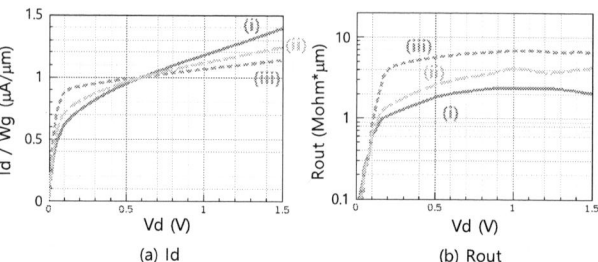

(a) Id   (b) Rout

Fig. 1. Halo effect on output curves at Lg= 0.5µm. The normalized halo dose values in (i), (ii), and (iii) are 1 (nominal), 0.5, and 0, respectively. To avoid velocity saturation for analog use, Vg is suppressed so that the current can pass the point of (Vd, Id/Wg) = (0.6V, 1µA/µm) where the Rout is calculated as d(Vd)/d(Id).

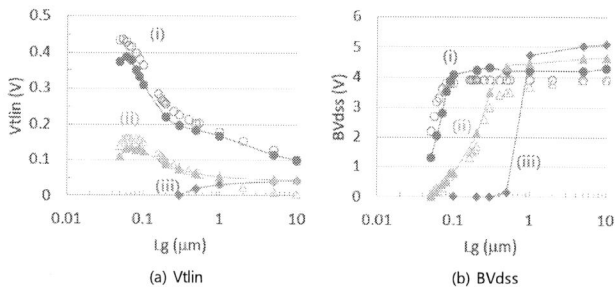

(a) Vtlin   (b) BVdss

Fig. 2. Halo effect on lowering curves of Vtlin at Vd= 0.1V (a) and BVdss (b). The halo doses in (i), (ii), and (iii) are the same as in Fig. 1. The SCE is enhanced as halo dose decreases. Simulation data (filled) are calibrated with measurement data (open). The BVdss must be 2V or more.

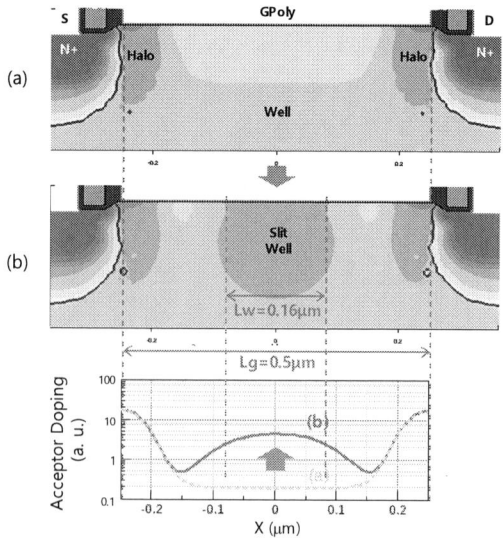

Fig. 3. Structure comparison between conventional (a) and proposed (b) at Lg= 0.5µm. The slit well is inserted along the channel center in (b) by adding a p-type dopant implant. The 1D acceptor profile along the Si surface is plotted in the bottom figure.

III. Effect and Analysis of Slit Well at Center

The Rout has a peak with respect to the slit width (Lw) at constant Lg (Fig. 4a). It surpasses 10Mohm*μm when both Lw/Lg ratio and slit well dose are set in proper ranges. In these ranges, Lw and its dosage should be set so that both Vtlin and BVdss can be within the target scopes (Fig. 4b).

Rout is found to be strongly related to the drain-induced threshold voltage shift (DITS) which is calculated as difference of Vtlin from Vtsat (Fig. 5). The power law relation is maintained regardless of halo and slit well dosages. The comparison of channel potential explains this relation clearly. In the conventional structure, the channel potential continues to increase until Vd reaches 1.2V (Fig. 6a) since the low-doped center channel region can be considered as a source resistance due to its continuous linear operation even though the halo region is already saturated [1, 2].

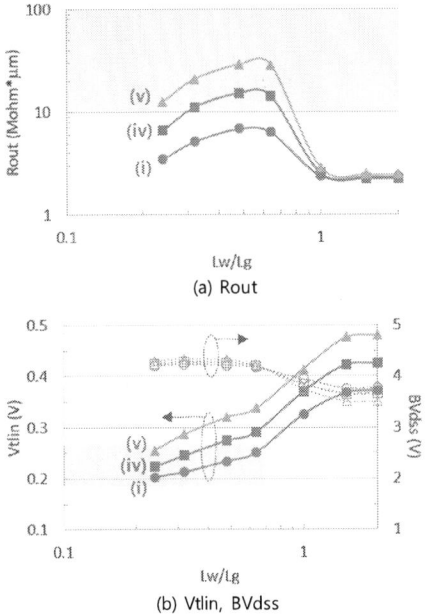

(a) Rout

(b) Vtlin, BVdss

Fig. 4. Simulated Rout, Vtlin, and BVdss as functions of slit well width ratio (Lw/Lg) at Lg= 0.5μm. The (i) is the same as (i) in Fig. 1. The slit well implant doses in (iv) and (v) are respectively 1.4 and 1.8 times higher than (i). The filled and open marks in (b) represent Vtlin and BVdss, respectively. The shaded areas stand for the target scopes. The BVdss must be 2V or more.

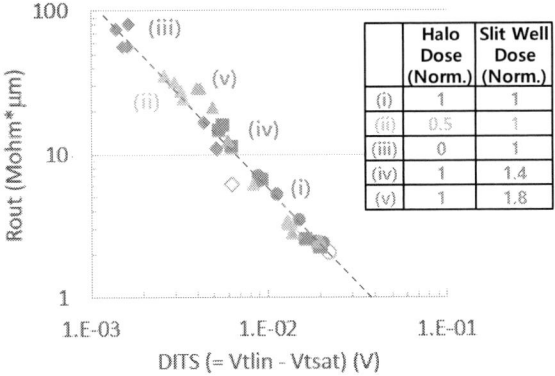

	Halo Dose (Norm.)	Slit Well Dose (Norm.)
(i)	1	1
(ii)	0.5	1
(iii)	0	1
(iv)	1	1.4
(v)	1	1.8

Fig. 5. Relation between Rout and DITS at Lg= 0.5μm. The Vd values for Vtlin and Vtsat are 0.1V and 1.2V, respectively. The normalized halo and slit well implant doses from (i) to (v) are in the inset table. The open and filled marks stand for measured conventional structure and simulated proposed structure, respectively. In the simulated data, slit width (Lw) is skewed from 0.12μm to 10μm. All plots are converged on one log-log line.

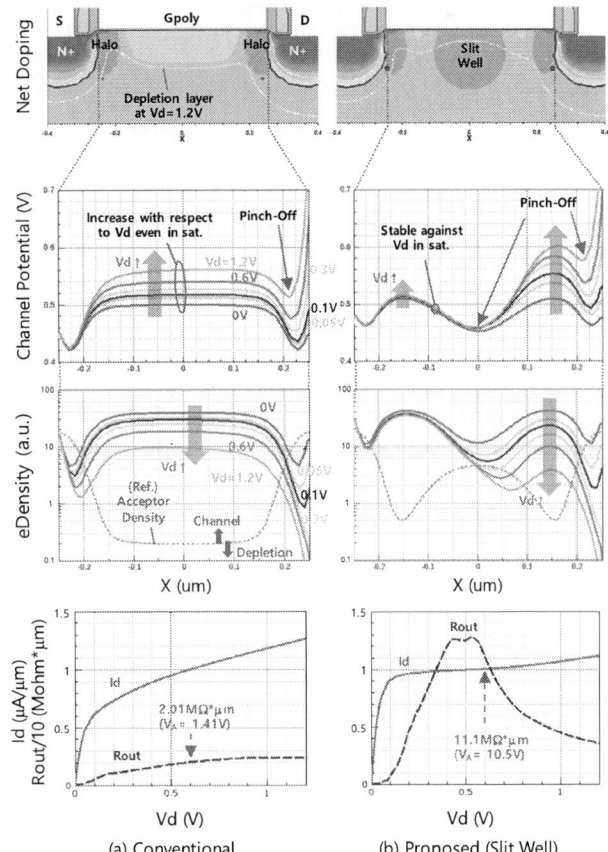

(a) Conventional

(b) Proposed (Slit Well)

Fig. 6. Net doping profiles of conventional (a) and proposed (b) structures with Lg of 0.5μm are compared in the first row. The corresponding channel potential and electron density along the surface follow in the 2nd and 3rd rows of figures, respectively. The slit well width (Lw) in (b) is 0.16μm. Halo implant dose is set at the value of (iv) shown in Figs 4 and 5. Vg is set so that the current can pass the point of (Vd, Id/Wg) = (0.6V, 1μA/μm) as shown in the bottom row, that is, 0.224V and 0.298V in (a) and (b), respectively. As for the channel potential, six curves are drawn at Vd= 0, 0.05, 0.1, 0.3, 0.6, and 1.2V. The Rout curves are also shown in the bottom figures with the values of Rout and early voltage (V$_A$) at Vd= 0.6V.

On the other hand, in the proposed structure, it saturates in the source-side half of channel beyond 0.1V of Vd thanks to the generation of pinch-off point at channel center (Fig. 6b). The drain-side half of channel acts as a buffer layer against Vd to shield this pinch-off point since the channel potential increase in the drain-side half of channel is more suppressed than Vd itself. The almost invariable channel potential in the source-side channel reduces the DITS. And also, the Rout increases by stable pinch-off point.

The slit well is also effective even when the halo is removed, which is actually not able to be applied for products due to SD punch-through at the minimum Lg of 0.06μm for digital applications. In this case, Rout considerably improves thanks to the saturation of channel potential even in the drain-side channel (Fig. 7b). The center pinch-off point becomes immovable against Vd.

The misalignment effect of slit well should be considered since it is not self-aligned with gate poly. In the expected range of maximum deviation, the shifts of all the Rout, Vtlin, and Idsat are confirmed to be within the acceptable ranges at the minimum Lg of 0.5μm for analog applications (Fig. 8).

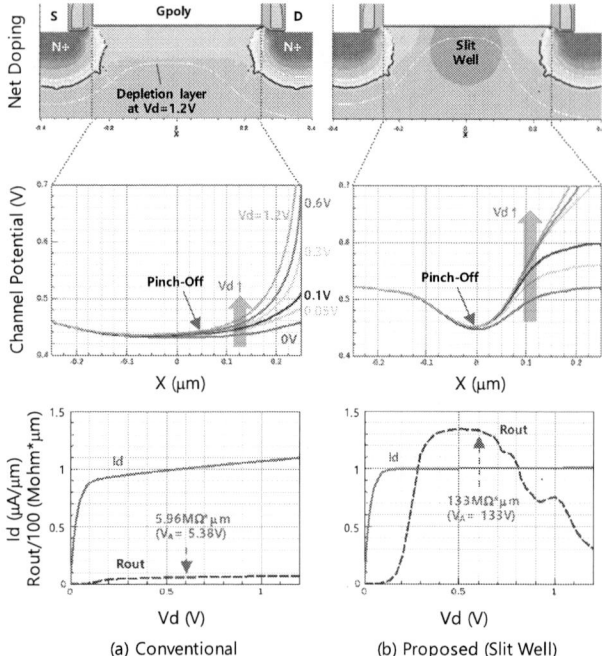

followed by the TCAD simulation of each condition. Then, the regression between input and output parameters (Table 1) has been done by our original machine-learning based auto regressor called GOAT (Generic Optimization Algorithm Tank) [6] which produces the excellent values of coefficients of determination (R^2) compared with conventional linear or quadratic equation model (Table 2 and Fig. 9).

Table. 1 The input and output parameter list for the GOAT procedure. The practical input parameter ranges are shown as normalized values except Lw/Lg ratio and Lx which are actual values. Lg is fixed at 0.5μm. Lw and Lx respectively represent slit well width and its misalignment volume as described before. Lx is included in regression for misalignment check only. The Vd is set at 1.2V for Idsat and Isubmax (maximum Isub). For Idoff, it is 10% higher (1.32V).

Input					GOAT regression	Output
Process	Parameter	Unit	Min.	Max.		Parameter
Slit Well	Lw/Lg	-	0.05	0.5		Rout
	Lx	um	-0.05	0.05		Vtlin
	Dose			5		Vtsat
	Energy			4		BVds
Halo	Dose	normalized	1	10		Idsat
	Energy			2		Idoff
PW Vth Adjust	Dose			15		Isubmax

Table 2. R^2 result of regression. Logarithmic values are used for Rout, Idoff, and Isubmax. GOAT shows the highest scores compared with other models.

	Rout	Vtlin	Vtsat	BVdss	Idsat	Idoff	Isubmax
Linear	0.803	0.792	0.795	0.907	0.898	0.814	0.848
Quadratic	0.975	0.975	0.977	0.964	0.990	0.980	0.974
GOAT	0.993	1.000	1.000	0.988	0.999	1.000	0.992

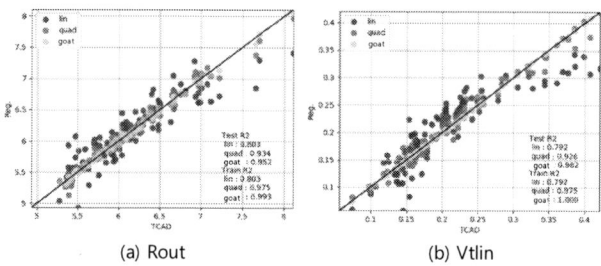

(a) Rout   (b) Vtlin

Fig. 9. Regression diagrams of Rout (a) and Vtlin (b) obtained by GOAT scheme. Logarithmic values are used for Rout. Blue, red, and green marks represent linear, quadratic, and GOAT regressor, respectively. GOAT shows outstanding matching results thanks to the in-house machine-learning algorithm [6].

(a) Conventional   (b) Proposed (Slit Well)

Fig. 7. Net doping and the corresponding channel potential along the surface are compared between conventional (a) and proposed (b) structures in the top and middle rows of figures, respectively. The Lg is 0.5μm. The slit well width (Lw) in (b) is 0.16μm. Halo is not implanted this time, which is actually not allowed due to SD punch-through at minimum Lg. Vg is set so that the current can pass the point of (Vd, Id/Wg) = (0.6V, 1μA/μm) as shown in the output curve in the bottom row, that is, 0.0672V and 0.286V in (a) and (b), respectively. The Rout curves are also shown in the bottom figures with the values of Rout and early voltage (V_A) at Vd= 0.6V.

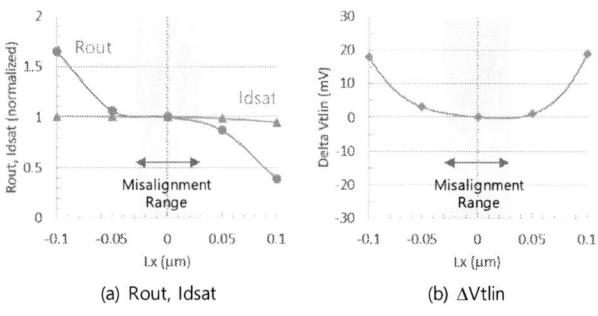

(a) Rout, Idsat   (b) ΔVtlin

Fig. 8. Slit well shift (Lx) dependences of Rout, Idsat (a), and ΔVtlin (b) at (Lg, Lw)= (0.5, 0.16) μm obtained by simulation. The Lx is regarded as mask misalignment from the channel center. The halo and slit well doses are default values which is shown as (i) in the inset table of Fig. 5. The shaded areas stand for the target scopes.

IV. TCAD OPTIMIZATION WITH MACHINE LEARNING

The insertion of slit well makes the manual optimization more exhaustive since the parameter number is increased, and also the optimum parameter values of slit well such as width, implant energy, and dosage vary depending on halo implant condition. Moreover, the SCE in digital MOSFETs which has no slit well should be simultaneously suppressed so that the minimum Lg can maintain 0.06μm with the same halo doping.

Our in-house automatic optimization tool [4] greatly relaxes this issue. The 100 conditions of 7 input parameter values (Table 1) are set by Latin Hypercube Sampling [5],

Sensitivity results obtained by GOAT scheme show that both slit well and halo are influential to the electrical characteristics within the assigned ranges (Fig. 10), which indicates the importance of usage of automatic optimization scheme in this slit well structure. It also shows that the sensitivity with Lx is quite small as indicated in the previous analysis (Fig. 8). The input parameter values optimized and extracted by genetic algorithm (GA) [7] successfully produce sufficient electrical characteristics without any critical Lx-induced variations for analog MOSFETs (Table 3). The Rout is improved in the entire Lg region above 0.5μm (Fig. 11a). The minimum Lg is also confirmed to maintain 0.06μm for digital MOSFETs (Fig. 11b). Vtlin increase by slit well slightly reduce Idsat, which is limited in the acceptance range (Fig. 11 (c)). No risk increase of hot carrier injection is expected since Isubmax is suppressed (Fig. 11 (d)).

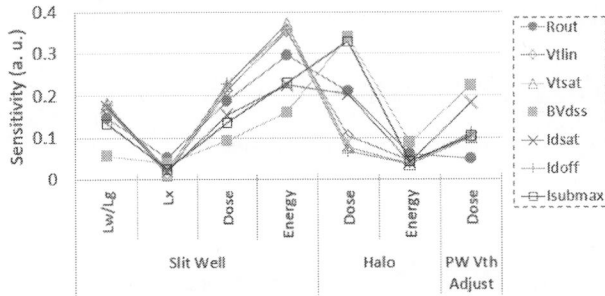

Fig. 10. Sensitivity results of electrical characteristics against several input parameters. The misalignment factor (Lx) is confirmed to be less influential compared with the other key input parameters in the practical parameter ranges described in Table 1.

Table 3. The deviation of electrical characteristics at Lg= 0.5μm obtained by GOAT scheme. The deviation is caused by the expected maximum mask misalignment (Lx) of slit well which is +/-0.03μm.

Lx	Rout	Vtlin	Vtsat	BVds	Idsat	Idoff	Isubmax
um	order	mV	mV	%	%	order	order
-0.03	-0.08	+/-0	+2	+0.7	+0.3%	-0.02	-0.02
0.03	+0.03	-1	+/-0	+0.2	-0.6%	+0.01	-0.02

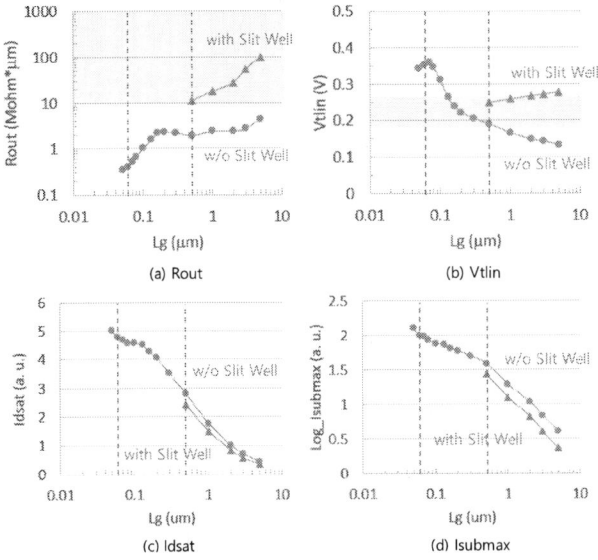

(a) Rout

(b) Vtlin

(c) Idsat

(d) Isubmax

Fig. 11. Lg dependences of Rout (a), Vtlin (b), Idsat (c), and Isubmax (d) depending on the slit well existence after the GOAT procedure. The NMOSFETs without and with slit well are prepared for digital and analog applications, respectively. Each minimal Lg is 0.06μm and 0.5μm in order. The shadowed area stands for the target scope. For the slit well, the Lw/Lg ratio is kept constant with respect to Lg. BVdss (not shown) is confirmed to be above the target level of 2V at each minimal Lg.

V. CONCLUSION

We have shown that the slit well inserted along the channel center improves Rout by creating pinch-off point there. The simulation analysis has revealed the mechanism that the drain-side of the channel acts as a buffer layer against Vd, which stabilizes the center pinch-off point. It is also confirmed that the slit well contributes the Rout improvement regardless of halo dosage. The optimization complicated by the addition of slit well related parameters is resolved by the machine-learning based TCAD scheme. The proposed approach will greatly enhance the flexibility of effective analog MOSFET design.

REFERENCES

[1] K. M. Cao, W. Liu, X. Jin, *et al.*, "Modeling of Pocket Implanted MOSFETs for Anomalous Analog Behavior," *in IEDM Tech. Dig.*, pp. 171-174, 1999.

[2] S. Mudanai, W.-K. Shih, R. Rios, *et al.*, "Analytical Modeling of Output Conductance in Long-Channel Halo-Doped MOSFETs," *IEEE Trans. Electron Devices*, Vol. 53, No. 9, pp. 2091-2097, Sep. 2006.

[3] W. Sansen, "Analog Design Essentials," *Springer*, Chapter 3, 2006.

[4] J. Kim, J. Yoo, J. Jung, *et al.*, "Novel Optimization Method Using machine-Learning for Device and Process Competitiveness of BCD Process," *in Proc. SISPAD*, pp. 343-346, 2020.

[5] B. Tang, "Orthogonal Array-Based Latin Hypercubes," *Journal of the American Statistical Association*, 88:424, pp. 1392-1397, 1993.

[6] J. Ko, J. Lee, S. Kim, I. Huh, and C. Jeong, "Method and Apparatus for Selecting Model of Machine Learning Based on Meta-Learning," US2020042896A1, Feb. 2020.

[7] C. M. Fonseca and P. J. Fleming, "Genetic Algorithms for Multiobjective Optimization: Formulation, Discussion, and Generalization," *in Proc. the Fifth International Conference*, pp. 416-423, 1993.

978-1-6654-8498-5/22 $31.00 © 2022 IEEE

Highly robust and reliable power amplifiers in 22FDX and 45RFSOI technologies

A. Bossuet[1], A. Divay[1], B. Martineau[1], C. Dehos[1], B. Blampey[1] and Y. Morandini[2]

[1]CEA, Grenoble, France
[2]SOITEC, Bernin, France
{[1]alice.bossuet}@cea.fr

Abstract—**This paper reports the VSWR (Voltage Standing Wave Ratio) ruggedness and aging measurements of two power amplifiers designed in 22FDX and 45RFSOI technologies. The PA is one of the most critical function in a front-end module as it is operating at high power and directly impacted by mismatch load. De-rating of 20% is applied on the supply voltage to ensure the operation and durability of the power amplifiers for a 10 years lifetime. The measured and modelled degradation versus time are presented and show excellent results on both technologies.**

Keywords—Reliability, Aging, power amplifier, 5G, 22FDX, 45RFSOI

I. INTRODUCTION

Reliability and aging are a main concern for power amplifiers, which is the module in the RF transceiver that delivers the highest output power. This implies voltages across transistors that could be more than two-times higher than the recommended foundry Vdd_{max} when used at maximum power operation. Important design margins must be considered to keep the transistor in a safe area to prevent breakdown and quick degradation over time. In beamforming 5G transceiver, the required output power per transceiver depends on the number of elements for a given EIRP (Effective Isotropic Radiated Power) [1]. For large antenna array, a III-V front end module is not required, but advanced silicon technologies are preferred for PAs (40nm node or below) in order to be integrated into the RF transceiver (SoC) and to provide enough gain and PAE at mmW frequencies. The SOI (Silicon On Insulator) technologies are good candidates to answer these needs for low cost, good gain (f_{MAX}) and power handling capabilities. In that context, this work proposes to evaluate the ruggedness and aging of the 45RFSOI and the 22FDX technologies from GlobalFoundries.

II. POWER AMPLIFIERS IN 22FDX AND 45RFSOI

Implementing >200 mW power PA in sub-micron CMOS technology at 40 GHz is a challenging task due to the low transistor breakdown voltage as well as losses of passive silicon components. There are currently several topologies proposed by the designer community to generate high output power. In typical beamforming applications, the output power levels of multiple low to medium power PAs are combined to generate a total high output power at a low supply voltage. However, the effectiveness of power combining techniques is limited due to the losses linked to the substrate and the low thickness of the metals of CMOS technology inducing high series resistances. A combination of 4 to 8 outputs is a maximum. One of the most efficient topology is the transformer-based combiner due to its small size and its ability to perform the function of combiner and impedance transformer simultaneously, reducing the additional losses generated by a matching network (Fig. 1.a). Another possibility is to stack transistors in a cascode

configuration, with 2 or more transistors, to use a larger supply voltage and reduce the risks of breakdown (BD) at high mismatches. Therefore, a voltage combination is performed instead of a power/current combination. The use of thick oxide transistor is a good way to increase the BV but should be limited to the stacked transistor due to its limited performance in mmW. On the other hand, increasing the number of stacks decreases the overall efficiency showing a structural limit to the use of thin oxide transistors in a stack greater than 3 as designed Fig. 1.b [2]. In this paper, two architectures are evaluated, using the voltage combination in 22FDX and the stacking technique in 45RFSOI and presented Fig. 1.

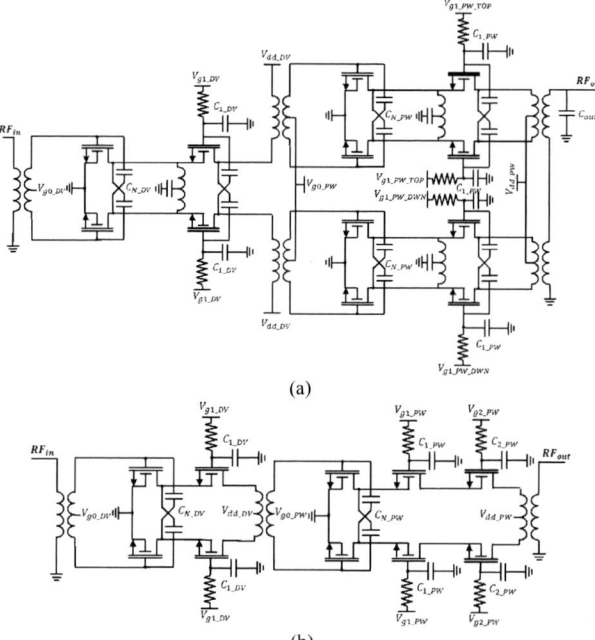

(a)

(b)

Fig. 1 : Power amplifiers schematics in 22FDX (a) and in 45RFSOI (b)

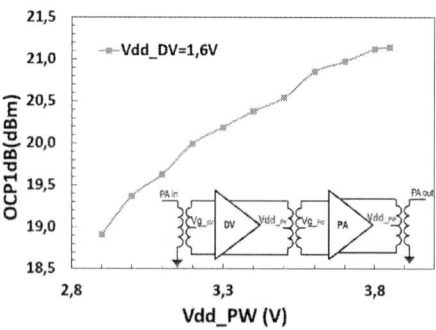

Fig. 2 : Measured OCP1dB versus V_{DD_PW} at 39 GHz of the 45RFSOI power amplifier for different V_{DD_DV}

Any devices biased at nominal technology voltage is exposed to reliability issues in large signal operation, making them sensitive to Hot Carrier Injection (HCI) and Time Dependent Dielectric Breakdown (TDDB). It appears that

978-1-6654-8498-5/22 $31.00 © 2022 IEEE

most designs are biased at the upper limit to reach higher output power as show Fig. 2. In our case, increasing V_{dd_PW} by 1V (2nd stage PA45RFSOI Fig. 1.a) improves the measured OCP1dB by 3dBm, but the supply voltage is not compliant in the context of industrial use.

In these designs, a de-rating of 20% is applied on the supply voltage to ensure the operation and durability of the power amplifiers for a 10 years lifetime as described TABLE 1.

TABLE 1 : MAXIMUM Vdd VOLTAGE FOR 22FDX AND 45RFSOI TECHNOLOGY AT 125°C

FET	L (nm)	Vdd (V) Max @125 °C	Vdd (V) with a de-rating of 20%
SLVT 22FDX	20	0,8	**0,64**
EGSLVT 22FDX	150	1,8	**1,44**
FB Regular Vt nfet, pfet 45RFSOI	40	1,1	**0,88**
FB 1.5V FETs tonfet, topfet 45RFSOI	112	1,65	**1,32**

III. AGING AND VSWR RUGGEDNESS METHODOLOGY

A. Ageing methodology

The ageing deals with the degradation of a device that occurs because of exposure to high voltages over time. This degradation is induced by two physical mechanisms: TDDB and HCI. A drain current drop I_{dsat} of more than 10% (~0,5dBm RF output drop) is chosen as a critical degradation since the current trend line is exponential after reaching this point. The RF stress sequence is summarized on Fig. 3. An initial power sweep is performed at t_0 and the source power is searched in order to get the appropriate P_{out_stress} and P_{out_BO}. The power is then set to the stress level desired and left as is, while the power levels, currents and voltages are monitored periodically. At specific moments following a logarithmic sampling, the input power is lowered to monitor the back-off performance before returning to stress level. At the end of the RF stress, another power sweep is performed. The RF and DC feed are cut off, a new probe contact and another power sweep are performed in order to verify potential probe contact issues during the stress, due to pad oxidation. The total stress time is limited to 9h, which corresponds to a compromise between total stress time and potential contact issues.

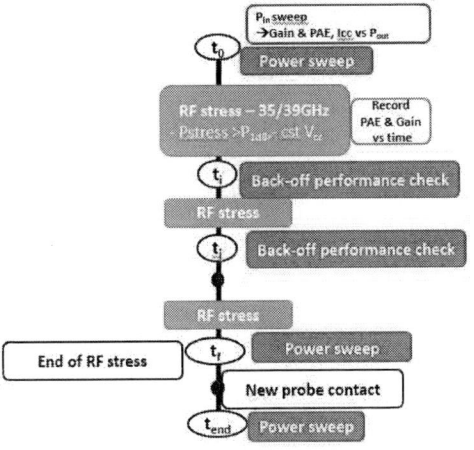

Fig. 3 : RF Stress algorithm

In order to evaluate the power amplifier ageing through a modulated mission profile, an output power-based acceleration model is used. The power dynamic under modulation is presented on Fig. 4. Most of the PA operating time is spent at ~8-9dB back-off or even less while power peaks up to P1dB represent only a very short cumulated time. Considering a 10 years lifetime, always ON for simplicity purposes, the amplifier will spend approximately 9 years at back-off (Peak to Average power ratio or PAPR = 0dB), 100 hours at P1dB-3dB and 13 hours at P1dB.

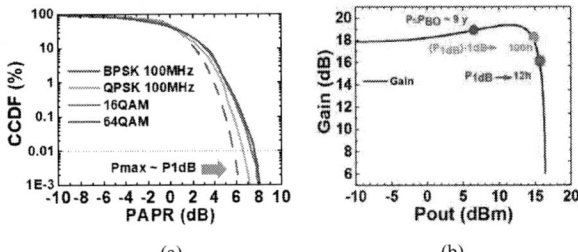

(a)   (b)

Fig. 4 : CCDF of a power amplifier for various modulation profiles (a) Gain, PAE vs Pout of the 45RFSOI power amplifier, with time passed @PBO, P1dB -3dB & P1dB for a 10 years 100% ON mission profile under OFDM modulation (b)

The large signal performance drift is measured for continuous wave (CW) stresses at different output power, chosen above P1dB for acceleration purposes. The parameter drift ($Gain_{BO}$ and/or P_{stress}) versus time is then modelled using a saturated power law (eq.1), which is a common model used in reliability [3]. This model depends on a saturation parameter *sat*, a scaling factor *A*, an output power acceleration factor γ and the time to the power *n*. Three different stress for each PA is a minimum requirement in order to model accurately the output power acceleration and is described next part. Eq.1 represent the degradation of a parameter under RF stress, following a saturated power law.

$$\Delta Parameter(dB\ or\%) = \frac{1}{\dfrac{1}{sat} + \dfrac{1}{10^A . P_{out}^{\gamma} . t^n}} \quad (1)$$

B. VSWR ruggedness methodology

In the product lifetime in 5G telecommunication, antenna impedance mismatches may occur, inducing high VSWR on the PA. The transistors will function at higher Vd & Id compared to the matched condition, running the risk of inducing higher HCI degradation or even TDDB. The measurement verification for the breakdown voltage (BVDS) and the circuit ruggedness to VSWR are then essential. In FDSOI the FETs are not sensitive to parasitic bipolar activation as the channel is fully depleted, compare to the PDSOI technology. The Vds limit is defined for $V_{G=0V}$ and V_{dgmax}. Based on gate length and a temperature of 25°C, Vds limit in 22FDX is 3,25V for slvt FET and 4,6V for the egslvt FET. In 45RFSOI the Vds limit is 1,8V for the adnfet and 2,35V for the tonFET. Simulation are performed on the PA in 45RFSOI for a VSWR of 4:1, the worst-case voltage across transistors are on the 2nd stage Fig. 1.a, with V_{ds3max}=3.5V for the upper thick oxide transistor, and V_{ds2max}=2V / V_{ds1max}=1.6 V for the two adnfet thin oxide common base and common source transistor respectively. Compare to the V_{dsmax} of thick and thin oxide transistors of 2.35 V and 1.8 V, the voltages seen by the transistors are

49% and 11% higher than the recommended voltage for the tonFET and adnFET respectively.

Two reliability measurements are performed on the power amplifiers based on [4] to see if any break down occurs: a soft stress test corresponding to a short 27 min of stress loading VSWR 4:1 and a hard test corresponding to 15 hours of stress, same load.

IV. MEASUREMENTS RESULTS

A. Measurements Ageing CW results

The two different PA architectures (22FDX PA and 45RFSOI PA) were stressed at three different power levels and on a fresh die each time.

- Pout=P1dB at 25°C for 9h, Vdd nominal

- Pout=P1dB+**1dB** at 25°C for 9h, Vdd nominal

- Pout=P1dB+**2dB** at 25°C for 9h, Vdd nominal

From these measurements the gain at back-off, the P_{stress}, the currents I_{DV}, I_{PW} and I_{gates} vs time are extracted. Fig. 5 describes a power sweep before/after stress and with a new contact afterwards. The new contact is used in order to de-correlate a potential contact issue on the RF probes and may retrieve some performances. If a large drift is initially observed and is then recovered with a new probe contact, the data are not considered and another die is stressed for the same condition, or with a lower stress power.

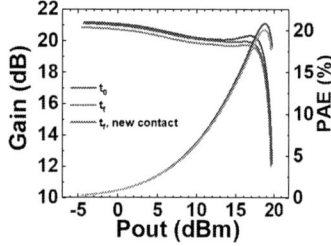

Fig. 5 : Gain, PAE vs Pout before, after stress, and with a new contact (45RFSOI PA)

The measured and modelled degradation versus time are presented on Fig. 6 and Fig. 7 for back-off Gain and Pout. The G_{BO} drift is relatively difficult to model because the drift window is very small and some of the measurements that are not consistent with the others (lower power level on Fig. 6.a or middle level on Fig. 6.b). The limitation in test time and dies to stress is not allowing many different conditions and dies for statistics. However, the modeling is successful on the P_{out} vs time degradation (output power at P_{stress}, in compression). A general P_{out} drift model is extracted, with a saturation value close to 0.5 dB for both power amplifiers.

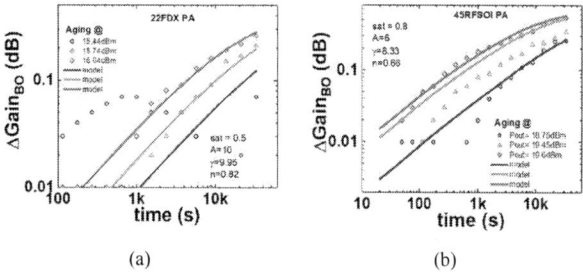

Fig. 6 : Back-off gain versus time degradation and modeling for both PAs

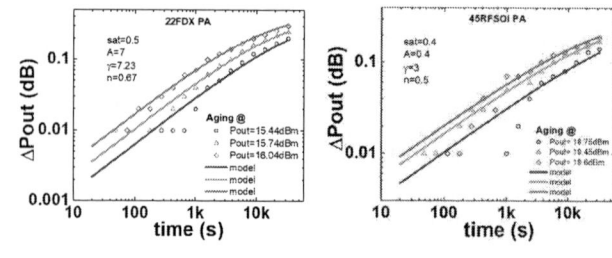

(a)   (b)

Fig. 7 : Pout (in saturation) versus time degradation and modeling for both PAs

It is then possible to estimate the Pout degradation for 9 years at back-off, 100 hours at P1dB-3dB and 13h at P1dB separately. These estimations are presented in TABLE 2.

TABLE 2 : POUT DEGRADATION ESTIMATED FOR DIFFERENT OUTPUT POWER LEVELS FOR THE TWO POWER AMPLIFIERS

	22FDX PA P_{out} drift	**45RFSOI PA P_{out} drift**
9 years @ 9dB BO	0	0.02dB
100h @ P_{1dB}-3dB	-0.17dB	-0.16dB
13h @ P_{1dB}	-0.2dB	-0.13dB

A graphical representation of these CW-like RF stress modeling is presented on Fig. 8. For the 45RFSOI power amplifier, the output power acceleration factor is relatively low (γ=3), which explains why the degradation simulated for 13h @P1dB is comparable compared to 100h with 3dB less output power. The 22FDX PA seems to be more accelerated with power compared to the 45RFSOI PA. This may be explained by the technology node (22nm vs 45nm) which implies more HCI acceleration with equivalent voltage but is also very design dependent.

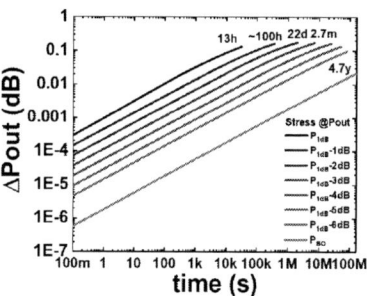

Fig. 8 : Estimated Pout degradation for PA 45RFSOI, with different power levels and for time interval corresponding to CCDF under 64QAM mission profile.

In order to obtain the real estimated degradation under modulation, a proper sum of the degradation along a typical 5G modulation profile must be conducted.

B. Ageing simulation on modulated data

The non-linear characteristics of both 22FDX and 45RFSOI PAs are measured on wafer using Vector Network Analyzer. From these data, AM/AM and AM/PM polynomial model are extracted by curve fitting. A 6th order model is enough to fit the characteristics with low discrepancies. Then a 5GNR-FR2 waveform feeds those PA nonlinear models to get the signal distribution by simulation. The waveform corresponds to a 1ms frame with numerology 2, modulated with OFDM 64QAM and sampled at 245.76MHz. Then the waveform is up-converted to 39.32GHz (Fs/4) in the FR2 band. The mmW signal is adjusted in amplitude to be set to 9dB input power back off from P1dB. Due to PA compression, the signal spectrum exhibits regrowth, and the power distribution is truncated close to the saturation. The modulated signal power at PA output (Pout) is recorded and feeds the ageing model.

Using the extracted Pout vs time samples and the quasi-static approximation, it is then possible to estimate the degradation for each PA under the 5G modulation profile. First, the degradation rate is calculated for each power in the sample eq. (2), using the CW aging model generated previously (parameters A, γ and n).

$$Degradation\ rate(t) = (10^A . P_{out}(t)^\gamma)^{\frac{1}{n}} \qquad (2)$$

Then, the degradation for the total samples is calculated using the sum of degradation rates multiplied by the time sampling. Finally, the estimated degradation under 5G modulation profile is calculated following eq. (3) for N x 100μs samples, where *sat* and *n* are respectively the saturation and time dependence parameters in the CW aging model and *D* is the total degradation for one 100μs sample.

$$\Delta P_{out}(t) = \cfrac{1}{\cfrac{1}{sat} + \cfrac{1}{(D.N)^n}} \qquad (3)$$

The estimated PA degradation under various modulation schemes are presented on Fig. 9 for a total ON time of 10 years. In practice, the actual mission profile of a PA would be around ~50% of that time in order to switch to the receiver (Rx) path. Most of the degradation is visible during the first year of operation, afterwards the Pout drift saturates around 0.5 dB. The degradation rate is highly dependent on the back-off level : the higher the back-off, the lower the degradation rate observed for HCI degradation, as the PA spends less time at high output powers, where most of the degradation occurs. However, this model does not take into account TDDB lifetime even if the same behavior might be expected, as high power is consistent with large voltages seen by the transistors. For all modulation schemes, the absolute level of degradation remains quite low and is consistent with the reliability margins taken in the design phase.

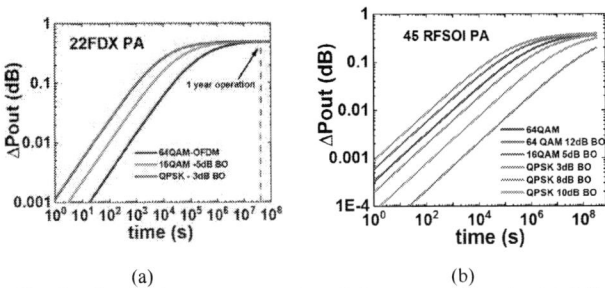

(a)   (b)

Fig. 9 : Saturated output power degradation versus time simulated for various modulation schemes for both PAs

The saturation parameter accuracy is crucial for a reliable degradation estimation in the long term. CW stress tests gives evidence of this saturation but its absolute value may be subject to discussion as the test time is limited by probe contact resistance degradation.

C. Ruggedness measurements results

Circuit ruggedness are tested using VSWR conditions from 4:1 to a maximum of 7:1 which is the limitation of the tuner capability at 40 GHz. The different time stress on both power amplifiers are shown Fig. 10. The test performed on the 45RFSOI power amplifier is done as follow: the input power is first fixed at IP1dB, the gamma_mag is set to VSWR 4:1 and the gamma phase is swept 3 times from 0 to 350 degree by 10 deg. A 15 sec waiting time is set at every point leading to a total stress of 27 min. Same operation is performed for hard test with gamma phase swept 100 times leading to a total stress of 15h. No breakdown is observed even after 15h of 4:1 VSWR applied. The output power of the 22FDX power amplifier was measured during 60 min VSWR 4:1, also here no breakdown was observed. However, mismatch load on the power amplifiers leads to a degradation over time. The 22FDX amplifier shows an output power degradation of 0.3 dB after 27 min of 4:1 VSWR stress and 0,36 dB after 1 hour. On the 45RFSOI power amplifier, at t0 + 27h VSWR 4:1 an output power degradation of 0,2 dB is observed, and at t0 + 15h VSWR 4:1, a degradation of 0,8 dB.

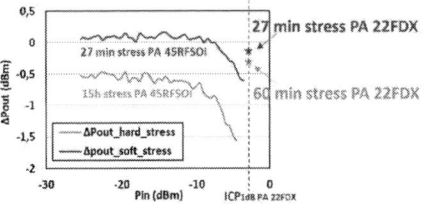

Fig. 10 : pre and post 27 min and 15h VSWR 4:1 stress test: ΔPout (dB) vs. Pin (dBm) for the PA in 45RFSOI(a) and the PA in 22FDX(b)

These measurements show an excellent robustness above expectation up to 7:1 VSWR.

V. CONCLUSION

Two different power amplifiers in recent node technologies <40 nm in 22FDX and 45RFSOI are tested for reliability and ageing for 10 years use. They are stressed in CW operation and their degradation are modelled using a saturated power law. The 20% de-rating on V_{DD} allows a low stress window as the PA are designed for reliability purposes. The results demonstrate very good reliability and ageing performances. The saturated Pout degradation can be estimated in CW mode. A 5G modulation scheme has been simulated on these circuits and fed to the reliability model. The estimated output power degradation has been calculated for a 10 years operation with this modulated profile, showing a maximum drift of 0.5dB in the worst case, which is very acceptable for the application.

REFERENCES

[1] A. Valdes-Garcia et al., "A Fully Integrated 16-Element Phased-Array Transmitter in SiGe BiCMOS for 60-GHz Communications," in IEEE Journal of Solid-State Circuits, vol. 45, no. 12, pp. 2757-2773, Dec. 2010.

[2] B. Martineau and D. Belot, "Si and SOI CMOS technologies for millimeter wave wireless applications," 2020 IEEE International Electron Devices Meeting (IEDM), 2020, pp. 17.3.1-17.3.4.

[3] S.W. Sun, et. al. in IEEE Electron Device Letters, vol. 11, no. 7, pp. 297-299, July 1990

[4] Shafiullah Syed et al., "A highly rugged 19 dBm 28GHz PA using novel PAFET device in 45RFSOI technology achieving peak efficiency above 48%", in IEEE/MTT-S International Microwave Symposium (IMS), 4-6 Aug. 2020.

Joint Modeling of Multi-Domain Ferroelectric and Distributed Channel towards Unveiling the Asymmetric Abrupt DC Current Jump in Ferroelectric FET

Simon Thomann[1], Kai Ni[2], and Hussam Amrouch[1]

[1]University of Stuttgart, Germany; [2]Rochester Institute of Technology, USA

Email: kai.ni@rit.edu, {thomansn, amrouch}@iti.uni-stuttgart.de

Abstract—We have developed a modeling framework to explain the experimentally-observed asymmetric abrupt current jump in DC I_D-V_G sweeps in ferroelectric FET (FeFET). We demonstrate that: i) with a model that combines the TCAD model for distributed channel and a custom-built physics-based model for ferroelectric (FE) polarization (P_{FE}) switching dynamics, the abrupt current jump during reverse sweep can be reproduced through modeling for the first time; ii) abrupt current jump exists in both forward and reverse sweep, with the forward sweep jump below the instrument noise level and typically not observed; iii) the abrupt jump is due to the significant amount of P_{FE} switching, not due to the formation or rupture of conductive channel path, i.e. percolation.

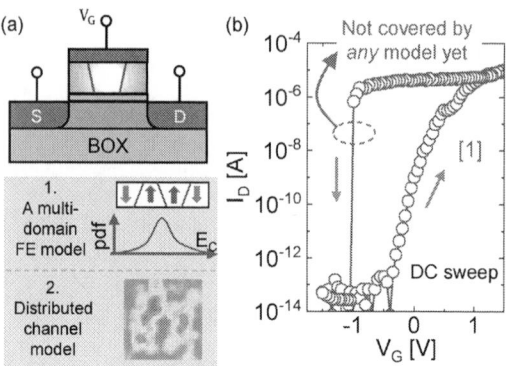

Fig. 1: (a) Practical device structure has a multi-domain FE layer and distributed channel. (b) Measured I_D-V_G data exhibits abrupt jumps which are not covered by *any* available model.

I. INTRODUCTION

The ever-increasing interest in ferroelectric (FE) HfO$_2$-based FeFET devices demands a comprehensive model that not only reproduces measurements but also provides insights into the underlying physics. In particular, abrupt current jumps during reverse DC I_D-V_G sweeps are frequently observed in FDSOI FeFET [1], as shown in Fig. 1(b), or in back-end-of-line FeFET [2]. Interestingly, this abrupt jump is typically missing for the forward sweep. This abrupt jump has been previously thought to be the evidence of steep-slope transistors and is now generally considered to be caused by the dynamical polarization switching. However, understanding what is happening inside the device is not available. One explanation is due to the single/few domain switching that is dominating the overall response of the FeFET. To model that, it is necessary to capture the realistic multi-domain features and nucleation-limited switching dynamics in HfO$_2$ thin film. Another possibility is the channel percolation effect [3]–[6], where an abrupt formation or dis-rupture of the channel occurs due to the random spatial fluctuation of the polarization. Modeling such a phenomenon inevitably requires capturing the distributed channel. Therefore, describing the abrupt jump in the DC I_D-V_G would necessitate the most general description of the FeFET device, i.e., multi-domain FE thin film and the distributed channel, as shown in Fig. 1(a), which are not captured by any model reported so far. Achieving this goal is the key focus of this work.

As summarized in Fig. 2(a), existing FeFET models can be categorized based on the model description of the ferroelectric layer and the semiconductor channel. To model the channel, a lumped model, e.g., the BSIM compact model, or a distributed model, e.g., TCAD model, can be applied. The most accurate approach to capture the ferroelectric behaviors is to adopt a multi-domain model, yet the single domain model or analytical Preisach model is frequently used. For the analytical Preisach model, because it is possible to get a polarization state with an arbitrary value between the saturation polarization, it suggests that it contains an infinite number of domains. These models have limited applicability due to the various assumptions made. So far, the most general FeFET model remains to be developed, which incorporates the multi-domain ferroelectric model with realistic nucleation-limited switching dynamics and the TCAD models to capture the distributed channel, close to realistic devices. This work fills in the space and demonstrates that such a comprehensive model is essential to clarify unexplained behaviors and provide insights into the device's operation.

II. PROPOSED MODELING FRAMEWORK

We employ, for the first time, a feedback loop between TCAD simulations and a custom-built MATLAB model for multi-domain FE thin film [7], as shown in Fig. 2(b). Given an applied electric field and the previous P_{FE} configuration, the MATLAB model simulates the nucleation-limited polarization switching dynamics and calculates the polarization charge, P_{FE}. TCAD receives the P_{FE} output from MATLAB, simulates the FE electric field, and sends it to MATLAB to finish the loop. Fig. 3 shows a more detailed implementation, which is capable of processing arbitrary input voltage waveform. First, the FE domains are initialized in TCAD; then, given the applied biases, the electric field vector in the FE layer is calculated

978-1-6654-8498-5/22 $31.00 © 2022 IEEE

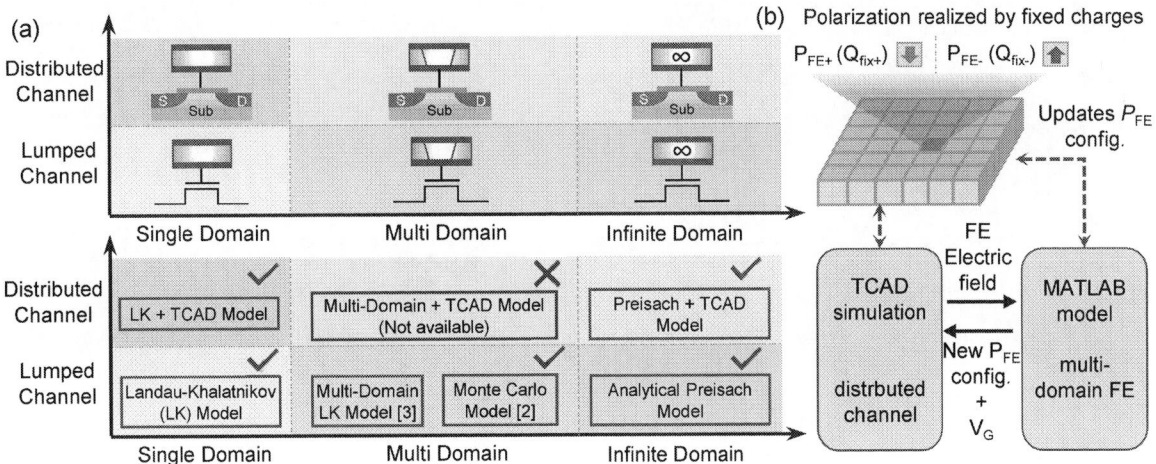

Fig. 2: (a) Summary of available models organized by domain and channel abstraction leaves an empty spot. (b) Our proposed model in which a feedback loop between TCAD and MATLAB is realized.

by TCAD, which is sent back to the MATLAB model for P_{FE} switching. The FE model is a Monte-Carlo model that assumes that each domain is independent, and its switching probability accumulates over time, from which the P_{FE} configuration (S) for each domain can be updated [7]. Finally, the time progresses, the updated S along with the next V_G value are fed back into TCAD, closing the loop and repeated until the input waveform ends.

To have a predictive modeling, Fig. 4(a) shows a 14 nm node FDSOI logic transistor built in TCAD, which is calibrated for electrostatics and carrier transportation with the reported transistor I_D-V_G characteristics obtained from measurement data [8], as shown in Fig. 4(b). To build the FeFET model, the high-κ dielectric in the logic transistor is then replaced with a 10 nm thick $Hf_{0.5}Zr_{0.5}O_2$ FE layer, as shown in Fig. 4(d). The FE parameters of MATLAB model are calibrated with the measured Q_{FE}-V_{FE} hysteresis loops (Fig. 4(c)) and the switching dynamics of P_{FE} (Fig. 4(e)). To facilitate the TCAD simulations, the P_{FE+}/P_{FE-} charge (Fig. 2(b)) is emulated with fixed charges (Q_{fix+}/Q_{fix-}) at the FE interfaces [5].

III. MODELING RESULTS

Fig. 5 shows the simulation results during the DC V_G sweep. The triangular V_G sweep waveform similar as that used in DC I_D-V_G measurements [1] is shown in Fig. 5(a). The simulated I_D-V_G characteristics very well reproduce what is observed in experiments (Fig. 5(b)). The forward sweep has a larger subthreshold swing, while the reverse sweep has an abrupt current jump. A few insights can be gained from this result. First, abrupt jumping does exist in the forward sweep. However, it happens at a very low current level ($< 10^{-14}$ A), around the measurement noise floor, thus typically not observed. Second, the abrupt current jump aligns well with the P_{FE} switching, as shown in Fig. 5(c). Therefore, it suggests that the P_{FE} switching is the cause of the abrupt jump in the DC I_D-V_G curves. In the forward direction, the P_{FE} switching happens when transitioning from the accumulation region to the depletion

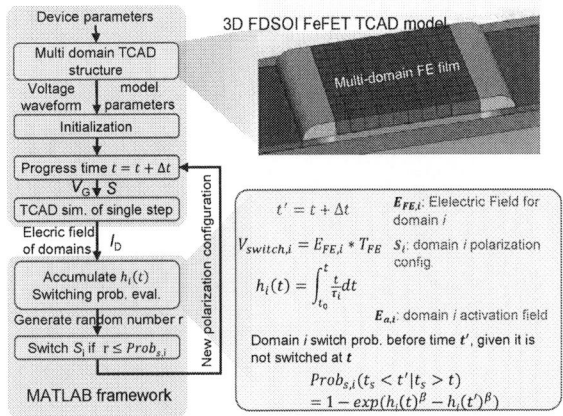

Fig. 3: Flow of the model using custom-built MATLAB FE Monte-Carlo model together with TCAD simulations. The resulting electric field from TCAD enables the FE model to calculate accurate switching probabilities and update P_{FE}, which can be fed back to TCAD to simulate the next step.

region and the P_{FE} switching in the reverse sweep happens when transitioning from the inversion region to the depletion region. Therefore, abrupt jumping is typically observed in the reverse sweep, where the current goes from high to low. Fig. 5(d) shows the I_D-V_G characteristics for different voltage sweep rate. The hysteresis window increases with the increase of the sweep rates. Due to faster V_G sweeping, less time is left for the P_{FE} to switch, and hence the larger the applied voltage needs to be to switch the P_{FE}, as expected [7].

Third, the abrupt jump is not due to sudden rupture or formation of a conduction path, i.e., not channel percolation. Fig. 6(a) shows the P_{FE} configuration and channel electron density during the reverse V_G sweep. Right before the current abrupt jump at $V_G = -0.96$ V, there is a small percentage of P_{FE} getting switched (i.e., red blocks). Right after the jump at

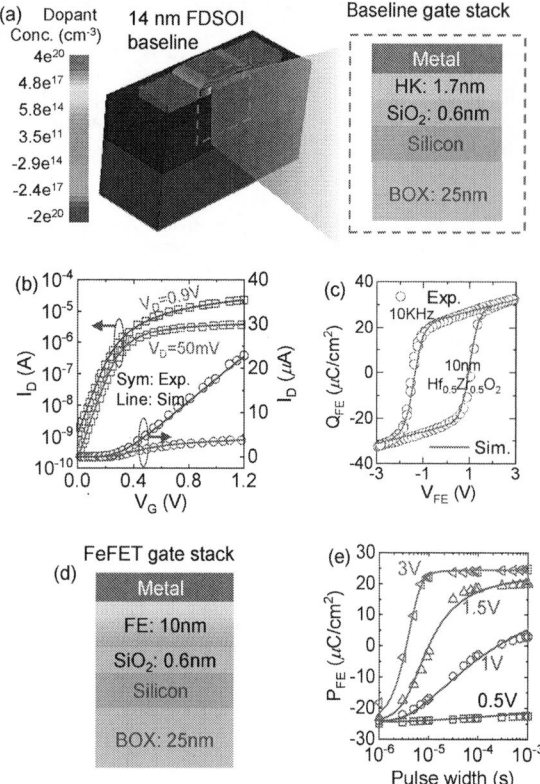

Fig. 4: (a) The baseline 14 nm FDSOI logic transistor built in TCAD and (b) its calibration against measurements. (c) Replacing HK with a 10 nm FE layer to realize FeFET. FE Monte-Carlo model calibration with 10 nm $Hf_{0.5}Zr_{0.5}O_2$ MFM capacitor using (d) Q_{FE}-V_{FE} hysteresis loops and (e) P_{FE} switching dynamics.

$V_G = -1.12$ V, a significant number of domains get switched, and the channel is depleted of electrons, causing the abrupt current jump. But from the P_{FE} configuration, conduction paths formed by connecting the P_{FE+} blocks are still available after the abrupt current jump at $V_G = -1.12$ V and $V_G = -1.28$ V (i.e., shown as the black arrows on the figure), suggesting that the abrupt current jump is not due to the conduction path rupture, but rather due to the significant P_{FE} switching.

Similarly, during the forward V_G sweep, the current jump is also independent of the percolation path formation, as shown in Fig. 6(b). Comparing the P_{FE} configuration right before and after the current jump, no conductive path is formed even after the current jump. These results suggest that the abrupt current jump is due to the significant P_{FE} switching rather than the percolation path formation. The significant P_{FE} switching can result from a dominating domain of significant size or simultaneous switching of multiple domains of a similar coercive field.

The developed model is general that it can reveal internal behaviors of the ferroelectric along with the interaction with the semiconductor channel for various interesting applications.

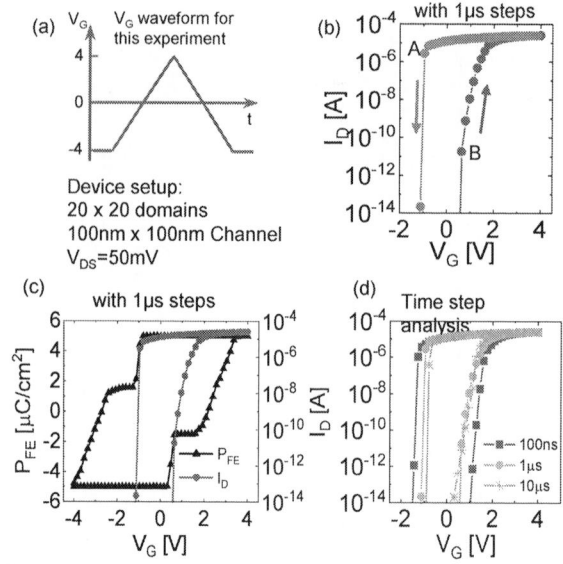

Fig. 5: Prediction results of our proposed model. (a) Input waveform with variable time steps and device setup. (b) I_D-V_G characteristics exhibiting the same abrupt current jumps as in measurements. (c) Current jumps are caused by abrupt P_{FE} switching. (d) I_D-V_G for different sweep rates.

One promising application of FeFET is the synaptic weight cell for neural network accelerator [9], [10], leveraging the partial polarization switching of ferroelectric. The developed model is also applied to study the synaptic behaviors of FeFET under identical or growing amplitude pulse trains, as shown in Fig. 7(a). For the identical pulse train shown in Fig. 7(b), the conductance tuning is mainly dominated by the first few pulses, and the remaining pulses can not switch the P_{FE} efficiently; thus, the linearity of conductance tuning is poor. In contrast, the growing amplitude pulse train induces a more linear modulation of conductance, as shown in Fig. 7(c). Fig. 7(d) shows the P_{FE} evolution with the applied pulses, again showing the growing amplitude pulse is more suitable for synapse application.

IV. CONCLUSION

We have developed a complex and general modeling framework of FeFET that can simulate the device behavior given any input voltage waveforms. With the model describing the distributed channel and multi-domain ferroelectric switching dynamics, we are able to explain the previously unexplained abrupt current jump during DC reverse sweep. Such a framework provides significant insights into the device operation and can help guide the FeFET development.

ACKNOWLEDGEMENT

This work was supported partially by the U.S. Department of Energy, Office of Science, Office of Basic Energy Sciences Energy Frontier Research Centers program under Award Number DE-SC0021118 and partially by the SRC and GRC Program under Contract 2020-LM-2999. Authors thank Om Prakash from KIT for the device calibration and his support.

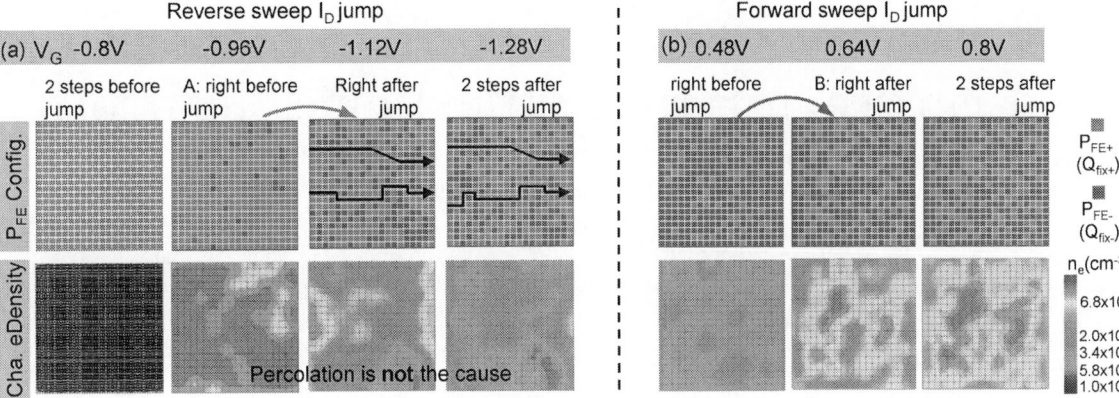

Fig. 6: Evolution of P_{FE} and channel electron density during V_G sweep. P_{FE} and channel electron density progression of (a) low-V_{TH} and (b) high-V_{TH} current jump which are not caused by percolation.

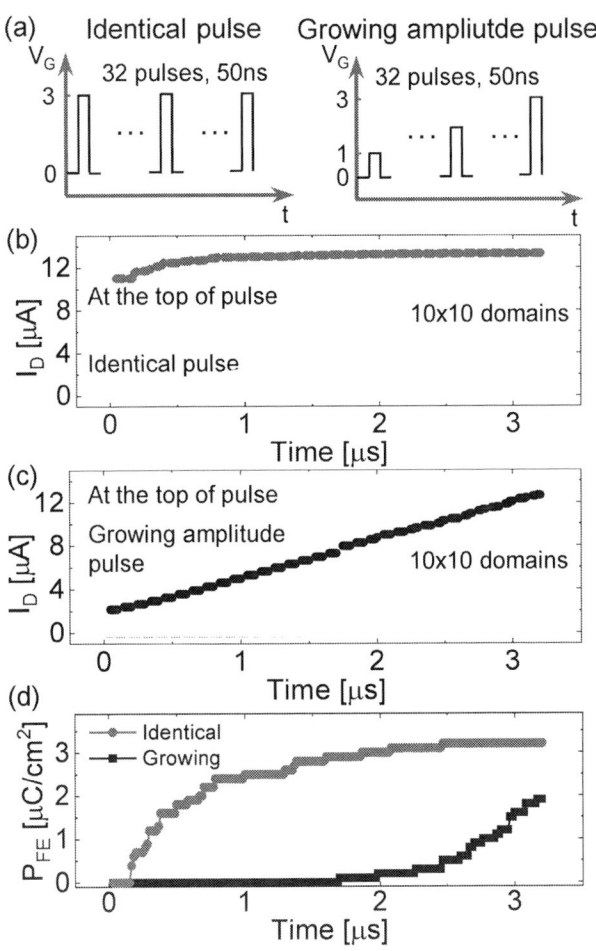

Fig. 7: Model application for FeFET-based synapses. (a) Applied pulses for training. (b) and (c) show the increase in current for identical and growing amplitude pulses, respectively. (d) demonstrates the switched polarization over time.

REFERENCES

[1] H. Mulaosmanovic, S. Dünkel, J. Müller, M. Trentzsch, S. Beyer, E. T. Breyer, T. Mikolajick, and S. Slesazeck, "Impact of read operation on the performance of hfo 2-based ferroelectric fets," *IEEE Electron Device Letters*, vol. 41, no. 9, pp. 1420–1423, 2020.

[2] S. Dutta, H. Ye, A. A. Khandker, S. G. Kirtania, A. Khanna, K. Ni, and S. Datta, "Logic compatible high-performance ferroelectric transistor memory," *IEEE Electron Device Letters*, vol. 43, no. 3, pp. 382–385, 2022.

[3] Y. Xiang, M. G. Bardon, B. Kaczer, M. N. K. Alam, L.-Å. Ragnarsson, K. Kaczmarek, B. Parvais, G. Groeseneken, and J. Van Houdt, "Compact modeling of multidomain ferroelectric fets: Charge trapping, channel percolation, and nucleation-growth domain dynamics," *IEEE Transactions on Electron Devices*, vol. 68, no. 4, pp. 2107–2115, 2021.

[4] F. Müller, M. Lederer, R. Olivo, T. Ali, R. Hoffmann, H. Mulaosmanovic, S. Beyer, S. Dünkel, J. Müller, S. Müller *et al.*, "Current percolation path impacting switching behavior of ferroelectric fets," in *2021 International Symposium on VLSI Technology, Systems and Applications (VLSI-TSA)*. IEEE, 2021, pp. 1–2.

[5] K. Ni, S. Thomann, O. Prakash, Z. Zhao, S. Deng, and H. Amrouch, "On the channel percolation in ferroelectric fet towards proper analog states engineering," in *2021 IEEE International Electron Devices Meeting (IEDM)*. IEEE, 2021, pp. 15–3.

[6] K. Ni, O. Prakash, S. Thomann, Z. Zhao, S. Deng, and H. Amrouch, "Suppressing channel percolation in ferroelectric fet for reliable neuromorphic applications," in *2022 IEEE International Reliability Physics Symposium (IRPS)*. IEEE, 2022, pp. 1–8.

[7] S. Deng, G. Yin, W. Chakraborty, S. Dutta, S. Datta, X. Li, and K. Ni, "A comprehensive model for ferroelectric fet capturing the key behaviors: Scalability, variation, stochasticity, and accumulation," in *2020 IEEE Symposium on VLSI Technology*. IEEE, 2020, pp. 1–2.

[8] Q. Liu, M. Vinet, J. Gimbert, N. Loubet, R. Wacquez, L. Grenouillet, Y. Le Tiec, A. Khakifirooz, T. Nagumo, K. Cheng *et al.*, "High performance utbb fdsoi devices featuring 20nm gate length for 14nm node and beyond," in *2013 IEEE International Electron Devices Meeting*. IEEE, 2013, pp. 9–2.

[9] M. Jerry, P.-Y. Chen, J. Zhang, P. Sharma, K. Ni, S. Yu, and S. Datta, "Ferroelectric fet analog synapse for acceleration of deep neural network training," in *2017 IEEE International Electron Devices Meeting (IEDM)*. IEEE, 2017, pp. 6–2.

[10] M.-L. Wei, M. Yayla, S.-Y. Ho, J.-J. Chen, C.-L. Yang, and H. Amrouch, "Binarized snns: Efficient and error-resilient spiking neural networks through binarization," in *2021 IEEE/ACM International Conference On Computer Aided Design (ICCAD)*. IEEE, 2021, pp. 1–9.

Polarization switching and AC small–signal capacitance in Ferroelectric Tunnel Junctions

M. Segatto[1,†], M. Massarotto[1,†], S. Lancaster[2], Q. T. Duong[2], A. Affanni[1], R. Fontanini[1], F. Driussi[1], D. Lizzit[1], T. Mikolajick[2,3], S. Slesazeck[2], D. Esseni[1]

[1]DPIA, University of Udine, Udine, Italy; [2]NaMLab gGmbH, Dresden, Germany;
[3]Chair of Nanoelectronics, IHM, TU–Dresden, Germany. [†]These authors have equally contributed to the work.

Abstract—**We here report a joint experimental and simulation analysis for large signal *P-V* and AC small–signal *C-V* curves in ferroelectric tunnel junctions. The attempt to reproduce both experimental data sets with the same model and material parameters challenges our understanding of the underlying physics, but it also helps develop a sound background for the device design.**

I. INTRODUCTION

Memories and memristors based on ferroelectric $Hf(Zr)O_2$ have recently emerged as competitive options for conventional and novel neuromorphic hardware [1]. In Ferroelectric Tunnel Junctions (FTJs) the tunnelling barrier is altered by the ferroelectric polarization state [2], whereas in ferroelectric field–effect transistors (FeFETs) the polarization affects the threshold voltage and the read current of the FeFET [3], [4]. The polarization P at the interface between the ferroelectric and a thin dielectric layer governs the operation of both devices.

Experimental characterization and sound modelling are both paramount for an optimal design of ferroelectric devices. The multi–domain Landau, Ginzburg, Devonshire (LGD) theory is well credited for the ferroelectric dynamics, and it has been used for negative capacitance effects [5]–[9], as well as for the operation of FTJs [10] and of FeFETs [11]. The most appropriate thermodynamic potential in the presence of free charges in the dielectric stack has been recently revisited in [12].

This paper presents an investigation of the AC small–signal *C–V* curves (SSCV) in metal–ferroelectric-dielectric–metal (MFDM) FTJs (Fig. 1), whereby measurements are obtained with a purposely developed experimental setup, and simulations with a rigorous linearization of the LGD model. An instructive insight is reported by comparing simulations and experiments for both the large–signal *P–V* curves (LSPV) and the SSCV response.

II. DEVICE FABRICATION

The MFDM FTJ structure consists of $\approx 10\,nm$ $Hf_{0.5}Zr_{0.5}O_2$ (HZO) and $\approx 2\,nm$ Al_2O_3 deposited via ALD on top of a W (30 nm)/TiN (10 nm) electrode. The 10 nm TiN top electrode was also deposited via sputtering under ultra–high vacuum. The HZO was crystallized by annealing at $500\,°C$ for 20 s. Finally, capacitor structures

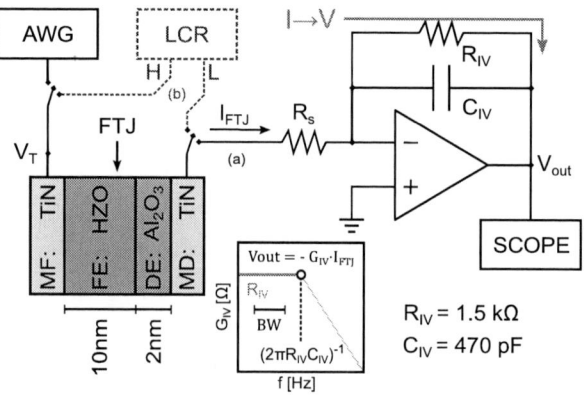

Figure 1: Sketch of the MFDM FTJs of this work and of the experimental setup. The setup consists of two distinct parts: (a) Virtual–grounded I→V converter (R_{IV}=1.5 kΩ, C_{IV}=470 pF) to measure the switching current I_{FTJ} through an oscilloscope. The inset shows the bandwidth (BW) of the amplifier; (b) An LCR meter to measure the AC small–signal capacitance.

were formed by depositing 10 nm Ti/25 nm Pt through a shadow mask and etching the TiN layer. For both electrodes, UHV sputtering ensures a low resistivity of around $3 \cdot 10^{-6}\,\Omega\,m$ so that the voltage is dropped mainly over the active bilayer. The Al_2O_3 tunneling layer in series with the ferroelectric increases the coercive voltage V_c of the stack to $\approx \pm 2\,V$. More switching properties of these devices have been reported elsewhere [13].

III. EXPERIMENTAL SETUP AND RESULTS

Triangular pulses with an amplitude of several Volts are typically used to measure LSPV curves, while an AC small-signal is used for SSCV measurements. It has been argued that the irreversible polarization switching dominates LSPV measurements, while it gives a negligible contribution to SSCV curves [14], [15]. In order to directly inspect the current response to the AC small-voltage in an FTJ, we developed the experimental setup of Fig. 1. An arbitrary waveform generator (AWG, Agilent 33250A) supplies V_T at the MF metal electrode, while the current I_{FTJ} is measured at the virtual–grounded MD metal contact through an I→V converter [13]. The OPAMP (TI TL082CP) feedback loop defines the trans–impedance of the amplifier ($V_{out}=-R_{IV}I_{FTJ}$) inside its

978-1-6654-8498-5/22 $31.00 © 2022 IEEE

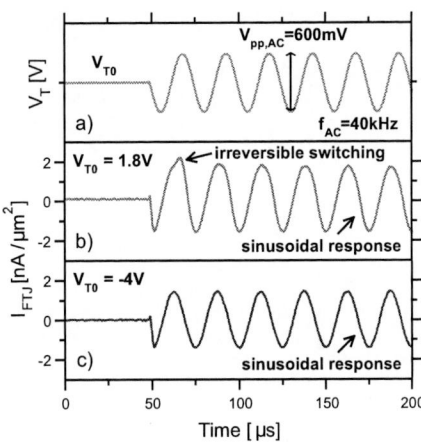

Figure 2: Emulation of an AC measurement by using the I→V converter. (a) Sinusoidal V_T waveform superimposed to a DC bias V_{T0}. (b) Measured I_{FTJ} for V_{T0}=1.8 V which is close to the positive coercive voltage of the FTJ (see Fig. 4a). A deviation of the I_{FTJ} from the sinusoidal waveform is observed during the first positive half-period, which we ascribe to irreversible polarization switching. (c) I_{FTJ} for V_{T0}=−4 V showing a sinusoidal-only response.

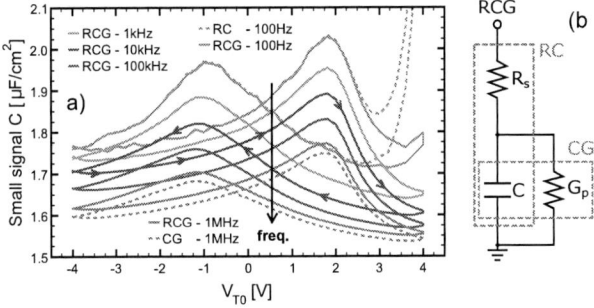

Figure 3: (a) SSCV curves measured with the LCR meter in the 100 Hz–1 MHz range and according to the different equivalent circuits sketched in (b), namely the RC series (green, dashed), the CG parallel (red, dashed) and the hybrid series–parallel RCG (solid line) circuit.

bandwidth $BW=(2\pi R_{IV} C_{IV})^{-1}$. The V_{out} is finally monitored through an oscilloscope (Tektronix TDS520B). This versatile setup allows us to measure the P–V curves using triangular pulses, as well as the AC small–signal response to a sinusoidal input. Moreover, we also measured SSCV curves by using an LCR meter (HP 4284A).

As shown in Fig. 2, we used the AWG to emulate the V_T waveform used in LCR based measurements, and recorded the current for different DC bias values V_{T0}. Figure 2 shows that an evidence of irreversible switching is observed only during the first positive AC semi–period at V_{T0}=1.8 V, which is close to the positive coercive voltage (see Fig. 4a). The I_{FTJ} is instead clearly sinusoidal in the following periods, despite the relatively large 300 mV amplitude of the AC signal. The analysis in Fig. 2 excludes that a non–linear I_{FTJ} response to the AC V_T waveform can affect the SSCV curves detected by an LCR meter.

Figure 3a shows the hysteretic butterfly–shaped C–V

curves measured with an LCR meter at different AC frequencies f ranging from 100 Hz to 1 MHz. The reasonably bias–independent capacitance measured at $V_{T0}\approx\pm4$ V is quite close to the estimated series capacitance C_s=1.44 µF cm^{-2} due to the sole linear dielectric response of the FTJ stack (for ε_{FE}=30, ε_{DE}=9). Since the frequency dependence in Fig. 3a may be affected by the capacitance extraction method, we compared the results for different equivalent circuits (see Fig. 3b). Solid lines in Fig. 3a were obtained with a hybrid series–parallel RCG circuit, whereby the R_s at each V_{T0} was estimated as the real part of the impedance at 1 MHz (assuming that at 1 MHz the influence of leakage through the FTJ is negligible). As expected, the series RC model (green, dashed) deviates from the RCG results at small f and large V_{T0} because it misrepresents the leakage. The accuracy of the parallel CG scheme (red, dashed), instead, degrades at high frequency because it cannot capture the influence of R_s. The experimental data compared with simulations have always been extracted with the RCG model.

IV. SIMULATION METHODS AND RESULTS

The SSCV curves in ferroelectric materials and devices have been investigated to a lesser extent compared to the LSPV counterpart. In particular, it is unclear if the same model and material parameters explaining a given set of P–V curves can reproduce equally well the corresponding SSCV curves. There is, however, a consensus that LSPVs measure the irreversible polarization switching, while the SSCVs mainly probe a reversible component, typically interpreted as a domain wall motion (DWM) [15]. In this latter respect, it has been argued that the DWM effects are adequately represented in the LGD equations [16], [17], thus even in their small–signal linearization.

Our modeling methodology solves the multi-domain LGD equations, that for n_D domains read [7], [8]

$$\frac{\partial P}{\partial t} = \frac{1}{t_F\,\rho}\left[-\left(2\alpha_i P_i + 4\beta_i P_i^3 + 6\gamma_i P_i^5\right)t_F + \right.$$
$$-\frac{t_F\,k}{d\,w}\sum_n\left(P_i-P_n\right)+$$
$$\left.-\frac{1}{2}\sum_{j=1}^{n_D}\left(\frac{1}{C_{i,j}}+\frac{1}{C_{j,i}}\right)\left(P_j+Q_{Tj}\right)+\frac{C_D}{C_0}V_T\right] \quad (1)$$

where P_i, Q_{Ti} are the polarization and trapped charge in domain i, and the sum over n is restricted to the domains sharing a domain wall with domain i. Moreover, α_i, β_i, γ_i are the anisotropy constants, ρ is a switching resistivity, d is the side of square domains, k and w are respectively the domain wall coupling coefficient and wall width, and $C_{i,j}$ are capacitive couplings between domains [7].

The ferroelectric dynamics is self-consistently solved with first order dynamic equations for traps at the FE-DE interface. Traps are assumed to exchange charge only with the MD electrode (in virtue of the much thicker ferroelectric layer) and the dynamic equations read [7]:

$$\frac{\partial n_{tr,i}(E_T)}{\partial t} = c_n [N_T - n_{tr,i}] - e_n n_{tr,i} \qquad (2)$$

where $n_{tr,i}(E_T)$ is the density of trapped electrons at energy E_T in domain i, while N_T is the corresponding trap density. The $c_n(E_T)$ and $e_n(E_T)$ denote the capture and emission rates with $e_n = e_{n0} F_0[(E_{f,MD} - E_T)/K_B T]$, where e_{n0} is a bias independent rate, $F_0(\eta)$ is the Fermi-Dirac function and $E_{f,MD}$ the Fermi level at the MD electrode.[1] The c_n is linked to e_n by $c_n = e_n \exp[(E_{f,MD} - E_{Tr})/K_B T]$, which ensures that the steady-state occupation of traps deriving from Eq. 2 is in thermodynamic equilibrium with $E_{f,MD}$. A set of Eqs. 2 is solved for both donor and acceptor type traps, and the overall trapped charge Q_{Ti} in each domain is $Q_{Ti} = Q_{Ti,acc} + Q_{Ti,don}$.

Figure 4a reports the experimental $P-V$ curves corresponding to triangular pulses at a frequency $f=1\,\text{kHz}$, and Fig. 4b shows also an effective large-signal capacitance curve (LSCV), that was extracted by dividing the current during the triangular pulse by the slope of the voltage ramp, namely as $\text{LSCV} = I_{FTJ}/(dV_T/dt)$. Figure 4b confirms that the measured LSCV curves are much larger than the SSCV counterparts (see Fig. 3a), due to the irreversible switching component.

Figures 4a and 4b also report the simulated LSPV and LSCV curves. In simulations we used nominal values of anisotropy constants α, β and γ equal to respectively $-3.8 \cdot 10^8$ m/F, $-3.2 \cdot 10^{10}$ m⁵/(FC²) and $7.9 \cdot 10^{11}$ m⁹/(FC⁴), and then introduced a domain to domain fluctuations of α, β, γ corresponding to a standard deviation $\sigma_{Ec} = 30\%$ of the coercive field (normalized to mean value). Simulations assume a fully ferroelectric HZO film (i.e. 100% orthorhombic phase), unless otherwise stated. The switching resistivity was set to $\rho = 110\,\Omega\,\text{m}$ [18], while the domain wall coupling k was set to zero, if not otherwise stated, by following recent first principle calculations for HfO₂ [19]. Figure 4a shows that simulations neglecting any trapping at the FE-DE interface result in much narrower and more tilted curves compared to experiments. We have already emphasized this behavior in [20], and discussed the links to the previous literature [21]. Only a fairly large density of traps at the FE-DE interface can reconcile simulations with experiments, and our trap densities are consistent with values extracted in [22], [23]. Even for the LSCV curves the simulations neglecting trapping show a large discrepancy with experiments, whereas the agreement improves drastically by including traps.

As for the simulation of the AC small–signal response, we note that Eqs. 1 and 2 can be collectively denoted as

$$\frac{\partial Y_h}{\partial t} = F_h(Y_h, V_T(t)) \qquad (3)$$

[1]The Fermi-Dirac occupation function $F_0(\eta)$ is defined as $F_0(\eta) = 1/[1 + \exp(\eta)]$.

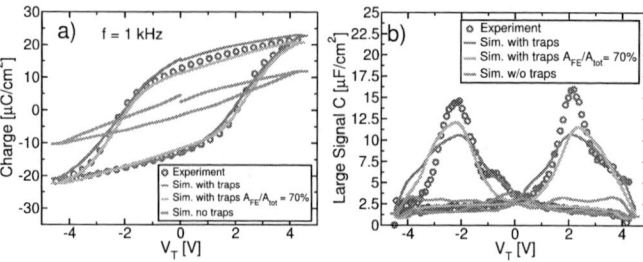

Figure 4: Comparison of measured (symbols) and simulated (lines) LSPV curves (a), and LSCV curves (b), for an Al₂O₃ thickness $t_{DE} = 2\,\text{nm}$. Simulations are shown for no trapped charge (green line), and for a density of acceptor and donor type traps equal to respectively $N_{acc} = 4.7 \cdot 10^{13}\,\text{cm}^{-2}\,\text{eV}^{-1}$ and $N_{acc} = 5.4 \cdot 10^{13}\,\text{cm}^{-2}\,\text{eV}^{-1}$ and over a 2.5 eV energy range. Simulations for a 70% ferroelectric area in the HZO film (i.e. $A_{FE}/A_{tot} = 70\%$) are also reported for $N_{acc} = 5.9 \cdot 10^{13}\,\text{cm}^{-2}\,\text{eV}^{-1}$ and $N_{don} = 9.6 \cdot 10^{13}\,\text{cm}^{-2}\,\text{eV}^{-1}$. The bias-independent trapping emission rate in Eq.2 is $e_{n0} = 2 \cdot 10^4\,\text{s}^{-1}$.

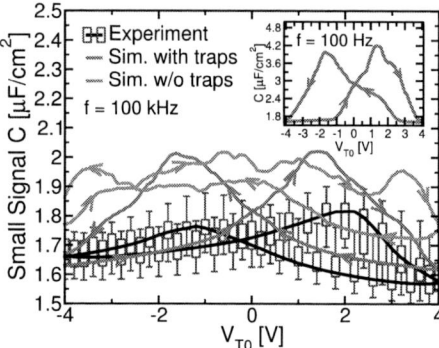

Figure 5: Measured (box plot, blue) and simulated (lines) small–signal capacitance curves at an AC frequency $f=100$ kHz and for the same device as in Fig. 4a. Measurements correspond to ten nominally identical devices. Simulations with and without trapping are displayed. The inset shows simulations at $f=100$ Hz.

where Y_h is a generic unknown (i.e. P_i or $n_{tr,i}(E_T)$), with $h = 1, 2, \cdots N_{PT}$ and N_{PT} being the number of equations. With a standard notation, the AC small–signal version of Eq. 3 at the radial frequency ω can be written as:

$$j\omega \widetilde{Y}_h = \sum_{k=1}^{N_{PT}} J_{h,k} \widetilde{Y}_k + \left. \frac{\partial f_h}{\partial V_T} \right|_{V_{T,0}} \widetilde{V}_T \qquad (4)$$

where $J_{h,k}$ is an entry of the Jacobian matrix $J_{hk} = \partial F_h/\partial Y_k$, while $\widetilde{V}_T$ is the AC external bias. Equation 4 is a linear problem for the unknowns $\widetilde{Y}_h$, from which the AC- small–signal terminal currents and thus the small–signal capacitance can be readily calculated.

Figure 5 compares the experimental SSCV with simulations at an AC frequency $f=100$ kHz. The simulations without traps do not show the capacitance peaks at the coercive V_T voltages observed in experiments. Simulations with traps reproduce quite well the capacitance at large $|V_T|$ values and the coercive voltages, but they overestimate the measured peak capacitance. At lower frequencies the discrepancy with experiments gets worse (see inset), because the simulated capacitance enlarges significantly due to AC response of traps (which is negligible at f=100 kHz), whereas the increase in experiments is comparatively much smaller (see also Fig. 3a). This mismatch may hint that traps are located deeper inside the ferroelectric bulk and have longer time constants compared to simulations.

Figure 6a illustrates the effect of a non negligible domain wall coupling k, which is expected to enhance the capacitance contribution due to domain wall motion. Larger k values emphasize the discrepancy with experiments in the peak capacitance region, besides being in contrast with first principle calculations [19].

We now recall that in our HZO films the fraction of orthorhombic ferroelectric phase can reasonably vary between 50% to 70% [24]. Figures 4a and 4b reveal that simulations with 70% ferroelectric area (i.e. A_{FE}/A_{tot}=70%) can still reproduce well the experimental LSPV and LSCV curves by adjusting the trap densities. Moreover Fig. 6b shows that, by accounting for a non ferroelectric area in the HZO film, the simulated peak capacitance in SSCV curves is reduced (for fixed values of the LGD anisotropy constants), thus improving the agreement with experiments, particularly for the positive V_T values.

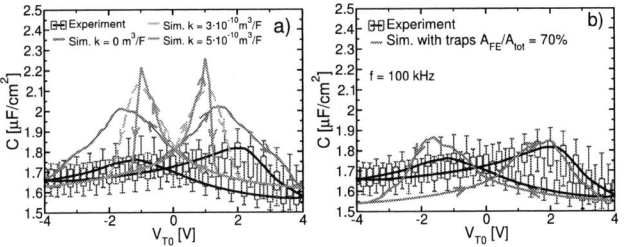

Figure 6: Measured and simulated capacitance as in Fig. 5 but for: (a) different values of the domain wall coupling k; (b) 70% fraction of ferroelectric over total area. The AC frequency is f=100 kHz.

V. CONCLUSIONS

We have reported experimental characterization and numerical modelling for LSPV and SSCV curves of MFDM based FTJs. As already pointed out in [20], our simulations can be reconciled with experiments only by accounting for charge trapping at the FE-DE interface. In our simulations the AC response of spontaneous polarization is not due to domain wall motion, in fact the domain wall coupling k was set to zero, and an increase of k impairs the agreement with experiments. By duly accounting for a fraction of non ferroelectric domains in the HZO film, the agreement with expriments of simulated SSCV curves is improved.

Acknowledgements: This work was supported by the European Union through the BeFerroSynaptic project (GA:871737).

REFERENCES

[1] S. Slesazeck and T. Mikolajick, "Nanoscale resistive switching memory devices: a review", Nanotech., vol. 30, p. 352003, 2019.

[2] B. Max et al., "Direct correlation of ferroelectric properties and memory characteristics in Ferroelectric Tunnel Junctions", IEEE JEDS, vol. 7, pp. 1175-1181, 2019.

[3] M. Jerry et al., "Ferroelectric FET analog synapse for acceleration of deep neural network training" IEEE IEDM, 2017, pp. 6.2.1-6.2.4.

[4] H. Mulaosmanovic et al., "Investigation of accumulative switching in Ferroelectric FETs: Enabling universal modeling of the switching behavior" IEEE TED, vol. 67, p. 5804, 2020.

[5] M. Hoffmann et al., "On the stabilization of ferroelectric negative capacitance in nanoscale devices", Nanoscale, vol. 10, pp. 10891–10899, 2018.

[6] P. Lenarczyk and M. Luisier, "Physical modeling of ferroelectric field-effect transistors in the negative capacitance regime", IEEE SISPAD, pp. 311-314, 2016.

[7] T. Rollo et al., "Stabilization of negative capacitance in ferroelectric capacitors with and without a metal interlayer", Nanoscale, vol. 12,, pp. 6121–6129, 2020.

[8] D. Esseni and R. Fontanini, "Macroscopic and microscopic picture of negative capacitance operation in ferroelectric capacitors", Nanoscale, vol. 13, pp. 9641–9650, 2021.

[9] M. Hoffmann et al., "Intrinsic nature of negative capacitance in multidomain $Hf_{0.5}Zr_{0.5}O_2$-based ferroelectric/dielectric heterostructures", Adv. Funct. Mater., vol. 32, p. 2108494, 2022.

[10] Z. Zhou et al., "Time-dependent Landau-Ginzburg equation-based Ferroelectric Tunnel Junction modeling with dynamic response and multi-domain characteristics", in IEEE EDL, vol. 43, no. 1, pp. 158-161, 2022.

[11] Y. Liu et al., "Investigation of the impact of externally applied out-of-plane stress on Ferroelectric FET," IEEE EDL, vol. 42, no. 2, pp. 264-267, 2021.

[12] J. Bizindavyi et al., "Thermodynamic equilibrium theory revealing increased hysteresis in ferroelectric field-effect transistors with free charge accumulation", Communic. Physics, 4:86, 2021.

[13] M. Massarotto et al., "Versatile experimental setup for FTJ characterization", accepted for presentation at EuroSOI-ULIS 2022.

[14] Y. Qu et al., "Quantitative characterization of interface traps in ferroelectric/dielectric stack using conductance method", IEEE TED, vol. 67, pp. 5315–21, 2020.

[15] S. Deng et al., "Examination of the interplay between polarization switching and charge trapping in Ferroelectric FET", IEEE IEDM, pp. 4.4.1-4.4.4, 2020.

[16] P. Marton et al., "Domain walls of ferroelectric $BaTiO_3$ within the Ginzburg-Landau-Devonshire phenomenological model", Phys. Rev. B, vol. 81, p. 144125, 2010.

[17] A. K. Saha et al., "Phase field modeling of domain dynamics and polarization accumulation in ferroelectric HZO", Applied Physics Letters, vol. 114, p. 202903, 2019.

[18] M. Kobayashi et al., "Experimental study on polarization-limited operation speed of Negative Capacitance FET with ferroelectric HfO_2", IEEE IEDM, pp. 12.3.1-12.3.4, 2016.

[19] H. J. Lee et al., "Scale-free ferroelectricity induced by flat phonon bands in HfO_2", Science, vol. 369, no. 6509, p. 1343–1347, 2020.

[20] R.Fontanini et al., "Polarization switching and interface charges in BEOL compatible Ferroelectric Tunnel Junctions", IEEE ESSDERC, no. 51, pp. 255-258, 2021.

[21] H. W. Park et al., "Polarizing and depolarizing charge injection through a thin dielectric layer in a ferroelectric–dielectric bilayer", Nanoscale, vol. 13, pp. 2556–2572, 2021.

[22] K. Toprasertpong et al., "Direct observation of interface charge behaviors in FeFET by quasi-static Split C-V and Hall techniques: Revealing FeFET operation", IEEE IEDM, pp. 23.7.1-23.7.4, 2019 methods", IEEE VLSI, pp. 1–2, 2020.

[23] S. Zhao et al., "Experimental extraction and simulation of charge trapping during endurance of FeFET with TiN/HfZrO/SiO2/Si (MFIS) gate structure", IEEE TED, vol. 69, no. 3, pp. 1561-1567, 2022.

[24] U. Schroeder et al., "Recent progress for obtaining the ferroelectric phase in hafnium oxide based films: impact of oxygen and zirconium", Jap. Journal of Appl. Phys., vol. 58, p. SL0801, 2019.

Multi-level Operation of FeFETs Memristors: the Crucial Role of Three Dimensional Effects

Daniel Lizzit, Thomas Bernardi and David Esseni

DPIA, University of Udine, via delle Scienze 206, 33100 Udine, Italy, e-mail: daniel.lizzit@uniud.it

Abstract—**This paper investigates and compares through a comprehensive TCAD analysis 2D and 3D simulations for ferroelectric based FETs. We provide clear evidence that the multiple read conductance values experimentally observed in FeFETs stem from source to drain percolation current paths, which are governed by the polarization patterns in the ferroelectric domains. Such a physical picture makes 3D simulations indispensable to capture even the qualitative features of the device behaviour, not to mention the quantitative aspects.**

I. INTRODUCTION

Ferroelectric based FETs (FeFETs) are very promising memories and memristors for neuromorphic computing [1], [2]. Unlike other device concepts, such as resistive RAMs (ReRAM) or Phase Change Memories (PCM) [3], that require a current flow for the switching operation, the physical substrate offered by FeFETs exploits a displacement current to set the ferroelectric polarization, thus holding the promise of low-power consumption memristors [4].

Ferroelectric oxides for CMOS compabile devices are based either on hafnium oxides with Si doping (HSO) or on hafnium zirconium oxides (HZO), both exhibiting remnant polarization values ranging from 15 to 30 μC/cm^2 and coercive fields in the range of 1-2 MV/cm [5]–[8].

It is well known that ferroelectric materials tend to minimize their free energy by creating domains [9], [10], whose size is usually associated to the size of the polycrystalline grains [11]. Indeed, for the film thicknesses of practical interest for nanoscaled electronic devices, namely in the range of 5-10 nm, ferroelectric oxides are polycristalline with a grain size in the nanometer range [11], [12].

In this paper, a physical based TCAD simulation approach is used to account for the multi-domain nature of the ferroelectric material and thus explore a multi-level operation of FeFETs memristors.

II. MODELLING OF THE FeFET

Simulations are carried out by using the TCAD tool Sentaurus-Device package [13], that couples the drift-diffusion equations with a kinetic model for the ferroelectric dynamics based on the phenomenological Landau-Ginzburg-Devonshire (LGD) equation [13].

A sketch of the simulated device is shown in Fig.1, where the polycrystalline nature of the ferroelectric HfO$_2$ is described by defining square ferroelectric domains having a 6 nm size in the y-z plane, namely in the range of experimentally reported values [11], [12]. An essentially uniform polarization inside

each grain is enforced by using an artificially large value of the domain wall coupling parameter [13]. Ferroelectric domains are separated by a thin non-ferroelectric spacer having the same dielectric constant as HSO (*i.e.* ε_{FE}=30 [14]). The electrostatic coupling between the domains is inherently accounted for in simulations. However the model does not assign any energy penalty to the formation of domains with an anti-parallel polarization or, equivalently, to the formation of 180° domain walls. This is consistent with recent *ab-initio* calculations of the domain wall energy in HfO$_2$ based ferroelectrics [15].

The calibration of the Landau's anysotropic constants has been previously carried out by comparison against experimental data for an HSO capacitor [16], leading to α=-5.37×10^8 m/F, β=9.62×10^8 m^5/(FC^2), γ=9.59×10^{10} m^9/(FC^4), corresponding to a remnant polarization P$_R$=20 μC/cm^2 and a coercive field E$_C$=1.1 MV/cm. Our simulations include a domain to domain statistical dispersion of the ferroelectric properties by using a normal distribution for the coercive electric field featuring a mean value E$_C$=1.1 MV/cm and a standard deviation σ_{E_C}=0.3E$_{C,0}$. The resistivity for the ferroelectric switching employed in simulations is ρ=110 Ωm [17], resulting in a time constant τ=$\rho/2|\alpha|\sim$100 ns.

In this work we consider a channel length of 40 nm, corresponding to six ferroelectric domains in the source to drain direction. Furthermore, we denote as 2D simulations those including a single domain in the device width direction (i.e. 6×1 FeFET), and as 3D simulations those featuring more domains in the transverse direction (e.g. 6×3 and 6×6 FeFETs). Between the ferroelectric oxide and the polysilicon channel, which is compatible with a BEOL integration, the simulated FeFET has a 1 nm thick SiO$_2$ interlayer (IL), which is typically observed in conventional growth processes for hafnia-based ferroelectrics on silicon [5], [18]. This layer plays an important role in the modelling of the device in several respects. Indeed, experimental results based on Hall and split-CV measurements suggest that most of the charge compensating for the ferroelectric polarization is located in the IL at the ferroelectric-dielectric interface [19], as it's been also confirmed in more recent experiments and simulations [8]. Moreover, in the absence of traps, the voltage drop across the IL reduces the drop on the ferroeletric and can lead to the so called minor-loop operation, where domains with trajectories located along nonsaturated hysteresis loops tend to back-switch after the write programming phase. Indeed, we have previously shown that, by neglecting altogether the charge trapping, simulations exhibit irreconcilable discrepancies with the experimental results [16]. Based on these premises, in the simulations of this work we employed an interfacial trap density at the FE-IL interface as reported in Fig. 1b. Interfacial traps in this work can exchange carriers with the polysilicon channel through a non-local trapping model that includes both elastic and phonon-mediated tunneling processses across the SiO_2 IL [13]. Tunneling emission and capture rates are described by the trap volume, the Huang-Rhys factor, the phonon energy and the tunneling masses [13], which can be regarded as fitting parameters and for the SiO_2 IL are taken from Ref. [20].

Simulations are performed by using a constant carrier mobility of 10 cm^2/Vs in the polysilicon channel [21]. The V_{DS} in this work is 50 mV, therefore the longitudinal electric fields are such that velocity saturation effects are negligible.

In these simulations, the bottom of the polysilicon channel is connected to ground to effectively provide majority carriers (holes) that allow for a positive to negative polarization switching within tens or hundred of nanoseconds. This simulation setup can still represent a BEOL compatible FeFET with the channel lying on a thick oxide layer and that is electrically contacted from the top side.

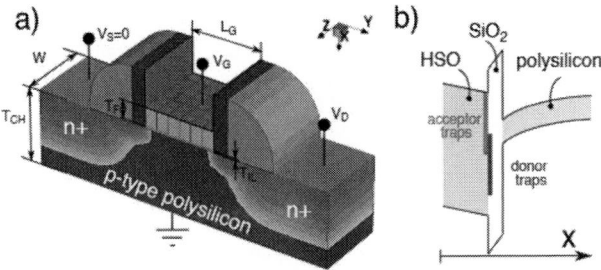

Fig. 1. a) Simulated FeFET with a gate length L_G=40 nm and a gate width W=40 nm, and consisting of 6×6 domains of ferroelectric HSO with a 10 nm thickness. The dielectric spacer between each grain is 5 Å thick. The interfacial SiO_2 layer has a thickness t_{IL}=1 nm. Source and drain regions are n-type doped with a density of 5×10^{20} cm^{-3}, whereas the channel is p-type with a doping of 1×10^{18} cm^{-3} and a thickness t_{CH}= 40 nm. b) Sketch of the bands profile in the device gate stack and of the acceptor and donor type traps at the FE/IL interface, having a uniform density of respectively $D_{it;acc}$=0.8×10^{14} $cm^{-2}eV^{-1}$ and $D_{it;don}$=1.0×$10^{14}cm^{-2}eV^{-1}$.

III. SIMULATION RESULTS

To investigate the maximum shift of the threshold voltage (V_{th}), or equivalently the memory window (MW) of simulated FeFETs, we first used a positive and negative gate voltage with sufficient amplitude and duration to drive the device in the low resistance (LRS) and high resistance state (HRS), where all the domains have respectively positive or negative polarization[1]. Figure 2a shows the applied gate voltage scheme, consisting of a write pulse followed by a V_{GS} ramp for the readout (with a read voltage small enough to leave the polarization pattern unperturbed). The I_{DS} versus V_{GS} characteristics are shown

[1]The polarization is defined positive when it points to the transistor channel, namely along the x direction in Fig.1.

in Fig. 2b for a 6x1, a 6x3 and a 6x6 domains FeFET. In the three devices the MW is essentially the same and equal to approximately 1.7 V. Hence, the uniform polarization state enforced by the write pulse makes the device behaviour insensitive to the number of domains in the width direction. Such a simulated MW is in good agreement with the experimental values for the HSO based FeFETs reported in Refs. [11], [22].

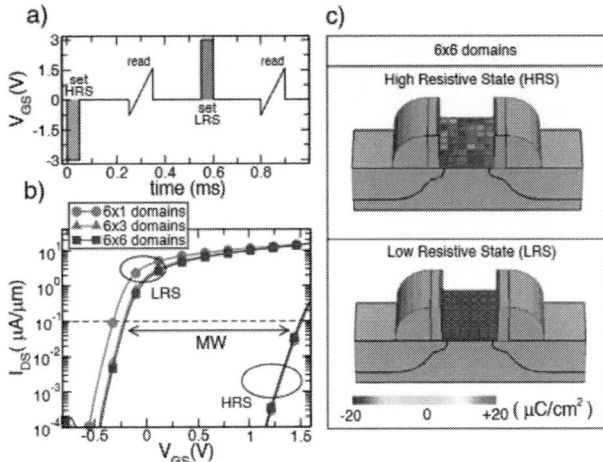

Fig. 2. a) Pulsed waveforms for the write operation have duration t_P=50 μs and are followed by a gate sweep from -1 to 1.6V to perform the readout. b) I_{DS} vs. V_{GS} curves at V_{DS}= 0.05 V for the HRS and LRS in a FeFET with 6 domains along the transport direction and 1, 3 or 6 domains along the device width. The horizontal dashed line corresponding to 100 nA/μm is used to determine the MW width. c) Domain polarization patterns for the 6x6 domains FeFET obtained at the maximum read voltage amplitude V_{GS}=1.6V.

In order to explore with simulations the possibility of setting multiple V_{th} values, we applied a series of increasing number of constant amplitude pulses at the gate terminal followed by a read sweep, as illustrated by the waveforms in Fig. 3. The idea behind this pulse scheme is to exploit the accumulative switching behaviour, where multiple pulses separated by a time delay, t_{delay}, can induce a progressive polarization switching involving very few and possibly even a single domain [1], [6].

Figure 4 shows the polarization trajectories of each domain (dashed, coloured lines) for the 6×1 FeFET. The negative to positive switching of polarization in one or a few domains occurs almost at each group of V_{GS} pulses, whereas the read operation does not perturb the stored polarization.

Figure 5 shows the I_{DS} versus V_{GS} curves obtained during the V_{GS} ramp used for the readout operation (see Fig.3). The FeFET with a single domain along the width direction (i.e. the 6×1 FeFET shown in Fig. 5a) exhibits a digital behaviour. This is because any domain with negative polarization precludes the formation of a source to drain conduction path. Therefore the FeFET is basically in the LRS when all domains have a positive polarization, and it is in the HRS for all other polarization patterns. On the other hand, FeFETs having 3 and 6 domains along the device width exhibit several intermediate V_{th} values or, equivalently, intermediate current levels in the readout mode. In order to illustrate the physics behind this

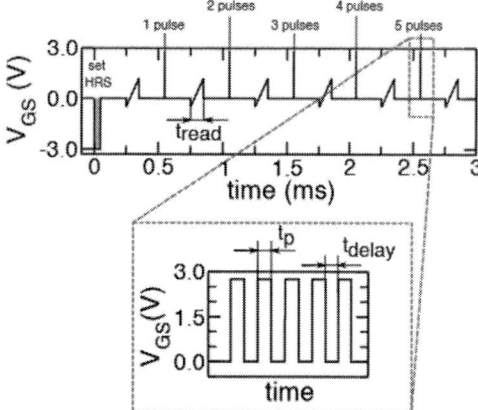

Fig. 3. Pulse scheme for potentiation simulations. After a pre-set signal with amplitude of -3 V, a series of pulses with duration t_p=200 ns separated by t_{delay}=200 ns and with amplitude of 2.75 V are applied to the gate contact. Between each series of pulses a read operation is performed by sweeping the V_{GS} in the range -0.5 to 1.2 V in 100 μs.

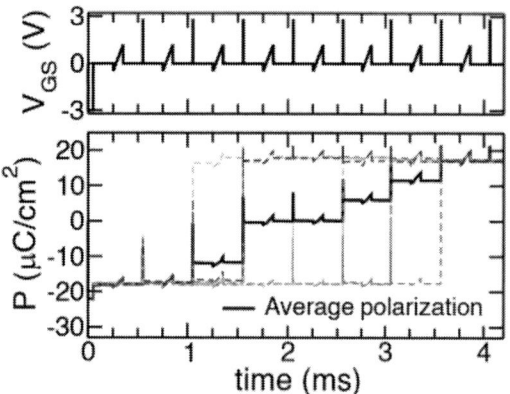

Fig. 4. Top: V_{GS} waveforms with write pulses separated by a read sweep. Bottom: polarization trajectories for the 6×1 FeFET of each domain (dashed line) and average value (solid line). All domains have positive polarization after 28 pulses.

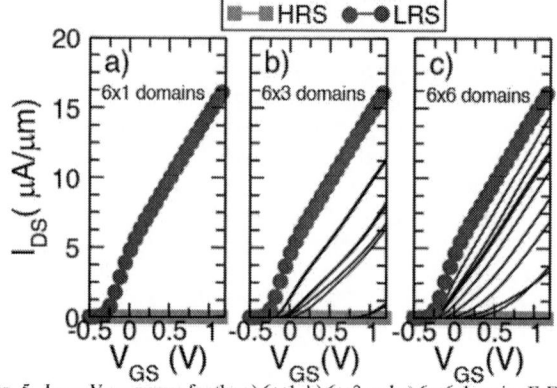

Fig. 5. I_{DS}- V_{GS} curves for the a) 6×1, b) 6×3 and c) 6×6 domains FeFETs obtained after series of gate pulses with same amplitude and duration as shown in Fig. 3. Simulations corresponding to V_{DS}= 0.05 V.

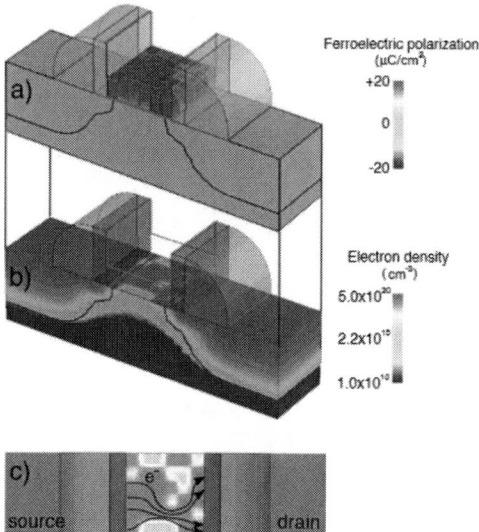

Fig. 6. Simulation results for the 6×6 FeFET. a) ferroelectric polarization, b) electron density obtained for the maximum reading voltage of V_{GS}= 1.2 V. c) sketch of the percolation paths from source to drain.

multi level behaviour, Fig. 6a shows the polarization pattern for the 6×6 domains FeFET after 15 potentiation pulses. From the electron density map distribution in Fig. 6b-c it can be observed the formation of high conductance paths from source to drain. The number of these conductance paths enlarges by increasing the number of potentiation pulses, which results in the multi-level operation observed in Fig. 5b-c.

The read current corresponding to different read V_{GS} values is illustrated in Fig. 7 versus the number of potentiation pulses and for the different simulated structures. Consistently with the results shown in Fig. 5, only a single LRS is observed for the 6×1 FeFET for any read V_{GS}, which is reached after 28 potentiation pulses that set all the ferroelectric domains to a positive polarization. The 6×3 FeFET exhibits instead a multi-level behaviour, where discrete read I_{DS} values are observed by increasing the number of pulses (and for a fixed read V_{GS}), corresponding to the activation of new percolation

current paths or the reinforcement of existing ones. Hence the read current in the 6×3 FeFET shows some discrete values which are modulated by the potentiation pulses. Finally, Fig. 7 shows that the 6×6 FeFET offers an almost continuous set of read I_{DS} values, in virtue of the much larger number of possible source to drain percolation current paths. Consistently with the results in Fig. 5, the three FeFETs have essentially the same read I_{DS} per unit width in the LRS.

IV. DISCUSSION AND CONCLUSIONS

This work has presented a comprehensive TCAD study of the multi-level operation in FeFETs memristors, focusing in particular on the difference between 2D and 3D simulations. We found that, if an accumulative switching behavior can be induced in the HSO film, then the modulation of the

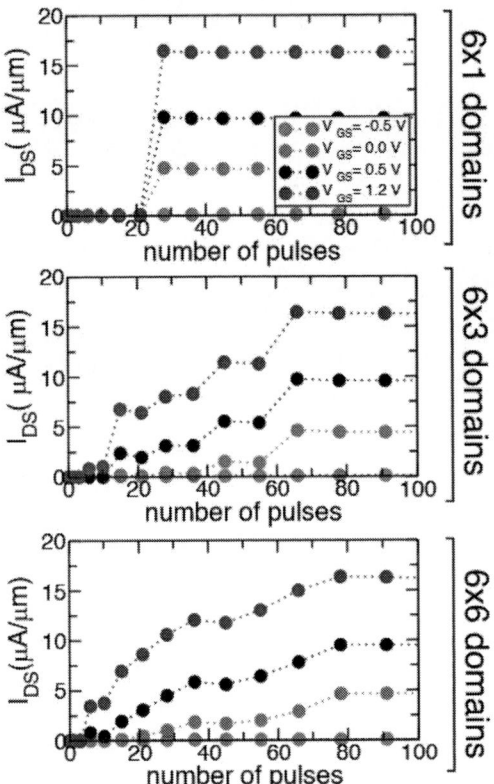

Fig. 7. FeFET simulated channel current versus number of pulses for the 6×1, 6×3 and 6×6 FeFETs, and for different V_{GS} values. Results obtained for V_{DS}=0.05 V.

percolation current paths in the device offers a powerful pathway for a multi-level memristor operation.

Despite the encouraging results in Fig. 7, it should be noticed that the displayed behaviour depends critically on a number of simulation parameters, including the density of traps and their emission and capture rates, as well as the features of the V_{GS} waveforms including the delay times between the groups of potentiation pulses. In our simulations, for instance, the accumulative switching behavior is linked to the trapping dynamics, and it is understood that the properties of traps cannot be regarded as a design knob. Nevertheless, trap densities can be partly engineered in ALD deposited Al_2O_3 films by tuning the deposition parameters, such as temperature, oxidant precursor type and dosing time [23].

Although the material and device engineering for FeFETs memristors is still a challenging path, our simulations predict good potentials for this class of ferroelectric devices.

Acknowledgments This work was supported by the European Union through the BeFerroSynaptic project (GA:871737).

REFERENCES

[1] E. Covi, H. Mulaosmanovic, B. Max, S. Slesazeck and T. Mikolajick, "Ferroelectric-based synapses and neurons for neuromorphic computing", *Neuromorphic Computing and Engineering*, vol. 2, no. 1, 2022.

[2] S. Slesazeck and T. Mikolajick, "Nanoscale resistive switching memory devices: a review", *Nanotechnology*, vol. 30, no. 35, 2019.

[3] F. Zahoor, T. Z. Azni Zulkifli and F. A. Khanday, "Resistive Random Access Memory (RRAM): an Overview of Materials, Switching Mechanism, Performance, Multilevel Cell (mlc) Storage, Modeling, and Applications", *Nanoscale Res. Lett.*, vol. 15, no. 90, 2020.

[4] S. Majumdar "Back-End CMOS Compatible and Flexible Ferroelectric Memories for Neuromorphic Computing and Adaptive Sensing", *Adv. Intell. Syst.*, pp. 2100175, 2021.

[5] M. Trentzsch, S. Flachowsky, R. Richter, J. Paul, B. Reimer, D. Utess *et al.*, "A 28nm HKMG super low power embedded NVM technology based on ferroelectric FETs", *Proceed. of IEDM*, pp. 11.5.1-11.5.4, 2016.

[6] M. Jerry, P.-Y. Chen, J. Zhang, P. Sharma, K. Ni, S. Yu and S. Datta, "Ferroelectric FET analog synapse for acceleration of deep neural network training", *Proceed. of IEDM*, pp. 6.2.1-6.2.4, 2017.

[7] B. S. Kim, S. D. Hyun, T. Moon, K. D. Kim, Y. H. Lee, H. W. Park *et al.*, "A Comparative Study on the Ferroelectric Performances in Atomic Layer Deposited $Hf_{0.5}Zr_{0.5}O_2$ Thin Films Using Tetrakis(ethylmethylamino) and Tetrakis(dimethylamino) Precursors", Nanoscale Res. Lett., vol. 15, no. 72, 2020.

[8] S. Zhao, F. Tian, H. Xu, J. Xiang, T. Li, J. Chai, "Experimental Extraction and Simulation of Charge Trapping During Endurance of FeFET With TiN/HfZrO/SiO2/Si (MFIS) Gate Structure", IEEE Transactions on Electron Devices, vol. 69, no. 3, 2022.

[9] H.W. Park, M. Oh, C.S. Hwang, "Negative Capacitance from the Inhomogenous Stray Field in a Ferroelectric–Dielectric Structure", *Adv. Funct. Mater.*, pp. 2200389, 2022.

[10] H.W. Park, J. Roh,Y.B. Lee, C.S. Hwang, "Modeling of Negative Capacitance in Ferroelectric Thin Films", *Adv. Mater.*, vol. 31, pp. 1805266, 2019.

[11] H. Mulaosmanovic, J. Ocker, S. Müller, U. Schroeder, J. Müller, P. Polakowski *et al.*,"Switching Kinetics in Nanoscale Hafnium Oxide Based Ferroelectric Field-Effect Transistors", *ACS Applied Materials & Interfaces*, vol. 9, no. 4, 2017.

[12] D.H. Lee, Lee Y., K. Yang, J. Y. Park, S. H. Kim, P. R. S. Reddy *et al.*, "Domains and domain dynamics in fluorite-structured ferroelectrics", *Applied Physics Reviews*, vol. 8, no. 2, pp. 021312, 2021.

[13] Synopsys Inc., Sentaurus Device User Guide, Q-2019.12.

[14] J. Muller, P. Polakowski, S. Mueller and T. Mikolajick, "Ferroelectric Hafnium Oxide Based Materials and Devices: Assessment of Current Status and Future Prospects", *ECS J. Solid State Sci. Technol.*, vol. 4, no. 30, 2015.

[15] H.J. Lee, M. Lee, K. Lee, J. Jo, H. Yang, Y. Kim *et al.*, "Scale-free ferroelectricity induced by flat phonon bands in HfO_2", *Science*, vol. 369 , no. 6509, 2020.

[16] D. Lizzit and D. Esseni, "Operation and Design of Ferroelectric FETs for a BEOL Compatible Device Implementation", *Proceed. of ESSDERC*, pp. 215-218, 2021.

[17] M. Kobayashi, N. Ueyama, K. Jang and T. Hiramoto, "Experimental study on polarization-limited operation speed of negative capacitance FET with ferroelectric HfO_2", *Proceed. of IEDM*, pp. 12.3.1-12.3.4, 2016.

[18] H. Mulaosmanovic, S. Dünkel, D. Kleimaier, A. el Kacimi, S. Beyer, E.T. Breyer *et al.*, "Effect of the Si Doping Content in HfO_2 Film on the Key Performance Metrics of Ferroelectric FETs", *IEEE Transactions on Electron Devices*, vol. 68, no. 9, 2021

[19] K. Toprasertpong, M. Takenaka and S. Takagi, "Direct Observation of Interface Charge Behaviors in FeFET by Quasi-Static Split C-V and Hall Techniques: Revealing FeFET Operation", *Proceed. of IEDM*, pp. 23.7.1-23.7.4, 2019.

[20] L. Vandelli, A. Padovani, L. Larcher, R.G. Southwick, W. B. Knowlton and G.Bersuker, "A Physical Model of the Temperature Dependence of the Current Through SiO2/HfO2 stacks", *IEEE Transactions on Electron Devices*, vol. 58, no. 9, 2011.

[21] N. Lifshitz, and S . Luryi, "Enhanced Polysilicon Channel Mobility in Thin Film Transistors", *IEEE Trans. on Electron Devices*, vol. 15, n. 8, 1994.

[22] Dünkel, S. Trentzsch, M. Richter, R. Moll, P. Fuchs, C. Gehring, O. Majer, M. , "A FeFET based super-low-power ultra-fast embedded NVM technology for 22nm FDSOI and beyond", *Proceed. of IEDM*, pp. 19.7.1-19.7.4, 2017.

[23] M. M. Rahman, K.-Y. Shin, and T.-W. Kim, "Characterization of electrical traps formed in Al_2O_3 under various ALD conditions," *Materials*, vol. 13, no. 24, 2020.

978-1-6654-8498-5/22 $31.00 © 2022 IEEE

Spin Torques in ULTRA-Scaled MRAM Devices

Simone Fiorentini[1,2,*], Mario Bendra[1,2], Johannes Ender[1,2], Roberto L. de Orio[1,2], Wolfgang Goes[3],
Siegfried Selberherr[2], and Viktor Sverdlov[1,2]

[1]*Christian Doppler Laboratory for Nonvolatile Magnetoresistive Memory and Logic at the*
[2]*Institute for Microelectronics, TU Wien, Gußhausstraße 27–29/E360, 1040 Vienna, Austria*
[3]*Silvaco Europe Ltd., Cambridge, United Kingdom*
[*]*fiorentini@iue.tuwien.ac.at*

Abstract—**We present an accurate extension of the coupled spin and charge transport model used to compute the spin-transfer torques in nanovalves to enable the anlysis of ultra-scaled magnetic tunnel junctions. The charge current behavior is reproduced with low conductivity locally dependent on the angle between the magnetization vectors, while for the spin current proper boundary conditions at the tunnel barrier interfaces are introduced. We demonstrate that the experimentally measured voltage and angle dependencies of the torques acting on the free layer are accurately reproduced, and that using our extended approach is key to accurately capture the interplay of the Slonczewski and Zhang-Li torque contributions acting on a textured magnetization. The described approach is successfully employed for switching simulation of recently proposed ultra-scaled MRAM cells.**

Index Terms—**Spin and charge drift-diffusion, spin-transfer torque, magnetic tunnel junctions, STT-MRAM**

I. INTRODUCTION

Emerging nonvolatile spin-transfer torque (STT) magnetoresistive random access memory (MRAM) offers high speed and endurance and is attractive for stand-alone [1], embedded automotive [2], MCU and IoT [3] applications, frame buffer memory [4], and slow SRAM [5]. The core of an STT-MRAM cell is the magnetic tunnel junction (MTJ), which in modern devices is usually composed of a CoFeB magnetic reference layer (RL) and a free layer (FL) separated by an MgO tunnel barrier (TB). Introducing additional MgO spacer layers in the FL allows to boost the interface anisotropy, while elongating the FL and reducing its diameter allows for the shape anisotropy to also contribute to the total perpendicular magnetic anisotropy [6]. For the development of accurate simulation tools which could aid the design of ultra-scaled MRAM cells, like the one reported in Fig. 1, it is paramount to generalize the traditional [7] approach, applicable only to thin FLs, to take into account normal metal buffers, MgO barriers between the CoFeB layers, as well as the barrier

Fig. 1: Example of an elongated ultra-scaled MRAM cell. RL indicates the reference layer, FL the free layer, TB is the tunnel barrier, and NM are nonmagnetic contacts.

between the RL and FL, and the torques acting on the textured magnetization which can appear in elongated ultra-scaled FLs.

II. MODEL

In micromagnetic simulations, the magnetization dynamics is described by the Landau-Lifshitz-Gilbert equation

$$\frac{\partial \mathbf{m}}{\partial t} = -\gamma \mu_0 \mathbf{m} \times \mathbf{H}_{\text{eff}} + \alpha \mathbf{m} \times \frac{\partial \mathbf{m}}{\partial t} + \frac{1}{M_S} \mathbf{T_S}, \quad (1)$$

where γ is the gyromagnetic ratio, μ_0 is the magnetic permeability, α is the Gilbert damping constant, M_S is the saturation magnetization, $\mathbf{H}_{\text{eff}}$ is an effective magnetic field containing the contribution of external field, exchange interaction, and demagnetizing field, and $\mathbf{T_S}$ is the STT term. We implemented the equation in a Finite Element (FE) solver using the Open Source library MFEM [8]. The contribution of the demagnetizing field is evaluated only on the disconnected magnetic domain by using a hybrid approach combining the boundary element method and the FE method [9]. A complete description of the torque term, which allows to include all physical phenomena responsible for proper ultra-scaled MRAM operation, can be obtained by computing the non-equilibrium spin accumulation through the transport equations [10], [11]

$$\overline{\mathbf{J_S}} = -\frac{\mu_B}{e} \beta_\sigma \mathbf{m} \otimes \left(\mathbf{J_C} - \beta_D D_e \frac{e}{\mu_B} \left[(\nabla \mathbf{S})^T \mathbf{m} \right] \right) - D_e \nabla \mathbf{S}, \quad (2a)$$

$$-\nabla \cdot \overline{\mathbf{J_S}} - D_e \frac{\mathbf{S}}{\lambda_{sf}^2} - \mathbf{T_S} = \mathbf{0}, \quad (2b)$$

$$\mathbf{T_S} = -\frac{D_e}{\lambda_J^2} \mathbf{m} \times \mathbf{S} - \frac{D_e}{\lambda_\varphi^2} \mathbf{m} \times (\mathbf{m} \times \mathbf{S}), \quad (2c)$$

where μ_B is the Bohr magneton, e is the electron charge, β_σ and β_D are polarization parameters, D_e is the electron diffusion coefficient, λ_{sf} is the spin-flip length, λ_J is the spin exchange length, and λ_φ is the spin dephasing length. $\mathbf{J_C}$ is the charge current density, while $\overline{\mathbf{J_S}}$ is the spin polarization current density tensor, where the components $J_{S,ij}$ indicate the flow of the i-th component of spin polarization in the j-th direction.

As the transport approach described by (2) is suitable to compute the torques in spin metal valves and, therefore, only accounts for semi-classical transport properties, it must be supplemented with appropriate conditions for the TB to properly

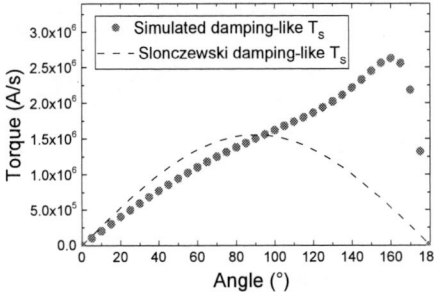

Fig. 2: Angular dependence of the torque acting on a semi-infinite FL based on the DD (dotted line) and on the Slonczewski expression [7] (dashed line).

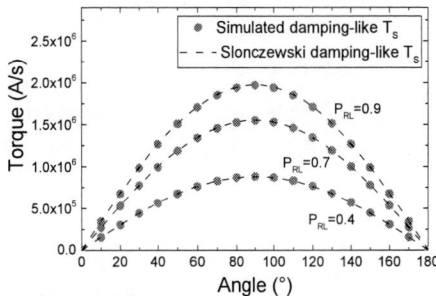

Fig. 3: Angular dependence of the torque with the spin-current boundary conditions, for semi-infinite FL and RL.

describe the dependence of the torque on the tunneling process across the MTJ.

Model extension to include MTJ properties

In order to include the tunneling magnetoresistance effect, we modeled the TB as a poor conductor with a local conductivity dependent on the relative orientation of the magnetization [13]. While it is then possible to use (2) to mimic the torque magnitude expected in an MTJ by tuning the tunnel barrier parameters, some of the torque properties are not reproduced. In Fig. 2, the angular dependence of the torque acting on a semi-infinite FL is compared with the Slonczewski expression [7]. The results show a clear deviation from the expected ones. Therefore, the spin current transport approach (2) must be modified. The traditional FE approach to the spin transport equations enforces the spin currents and the spin accumulations to be continuous through all the interfaces [10], [11]. In order to include the tunneling spin current in the model, we take the diffusion coefficient of the TB to be low, proportionally to the conductivity, and apply the following expression as a boundary condition for both the RL|TB and TB|FL interface [14], [15]:

$$\mathbf{J_S^{TB}} = -\frac{\mu_B}{e} \frac{\mathbf{J_C^{TB}} \cdot \mathbf{n}}{1 + P_{RL} P_{FL} \mathbf{m_{RL}} \cdot \mathbf{m_{FL}}} [\alpha P_{RL} \mathbf{m_{RL}}$$
$$+\alpha P_{FL} \mathbf{m_{FL}} + 1/2 \left(P_{RL} P_{RL}^\eta - P_{FL} P_{FL}^\eta \right) \mathbf{m_{RL}} \times \mathbf{m_{FL}}]$$
$$(3)$$

Here, $\mathbf{J_C^{TB}}$ is the electric current density at the interface, $\mathbf{n}$ is the interface normal, and $\mathbf{m_{RL(FL)}}$ is the unit magnetization vector of the RL(FL) at the interface. $P_{RL(FL)}$ is the in-plane Slonczewski polarization parameter [7], $P_{RL(FL)}^\eta$ is the out-of-plane polarization parameter [14], and α describes the influence of the interface spin-mixing conductance on the transmitted in-plane spin current. Doing this, we fix the spin current to the value prescribed by (3), when $\mathbf{J_C}$ flows through the TB. This is the key to describe the spin currents and the spin accumulations in the RL and FL of an MTJ. Employing our approach gives the opportunity to describe the spin and charge transport coupled to the magnetization in arbitrary stacks of MTJs and metallic spin valves in a unified manner,

and allows to compute a full three-dimensional solution in the presence of non-uniform magnetization configurations.

III. Results and Discussion

A. MTJ torques

The proposed approach is applied to the computation of the torque in a structure with semi-infinite ferromagnetic leads separated by a 1 nm thick tunnel junction. For these simulations, we set $\lambda_\varphi = 0.4$ nm, $\lambda_J = 1$ nm and $\lambda_{sf} = 10$ nm in (2). The short value of the dephasing length is employed to guarantee the fast absorption of the transverse components of the spin accumulation near the interface, as expected in the presence of strong ferromagnets [7]. The angular dependence of the resulting damping-like torque is shown in Fig. 3. The typical sinusoidal dependence [7], [14] of the torque acting on the FL in an MTJ is now reproduced exactly, for various values of the RL|TB interface spin polarization. The structure is biased by a fixed voltage, so that the torque is independent of the TB|FL polarization.

The implementation discussed produces a linear dependence of the torques on the bias voltage. Fabricated MRAM devices usually exhibit a clear non-linearity in the observed voltage dependence of both the torques and the TMR [12]. We include the voltage dependence in the polarization parameters P_{RL} and P_{FL} as an on-demand feature. It can be postulated after [16] as

$$P_{RL}(V) = \frac{1}{1 + P_0 \exp(V/V_0)}, \quad P_{FL}(V) = P_{RL}(-V) \quad (4)$$

where P_0 can be extracted from the TMR at zero bias, and V_0 from the high bias behavior. A comparison of both TMR and torque results with experimental ones [12] is reported in Fig. 4, showing a good agreement.

B. Torques in elongated ultra-scaled devices

When in the presence of elongated FLs, the switching of the whole layer at the same time is not guaranteed, a domain wall (DW) can be generated, with its propagation through the FL causing an increase in the critical current required for switching. In this case, the additional spin torques created by the presence of magnetization gradients in the bulk of the

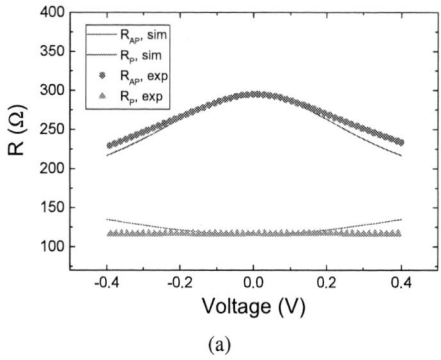
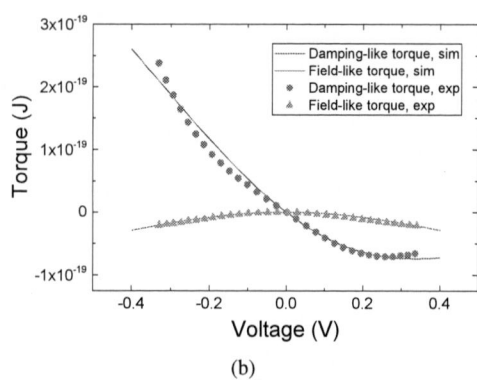

(a)

(b)

Fig. 4: Dependence of both TMR (a) and torques (b) on the bias voltage, compared with experimental data [12]. The results were obtained for $P_{RL}(0) = P_{FL}(0) = 0.66$, $P_{RL}^{\eta} = P_{FL}^{\eta} = 0.11$, $\alpha = 0.36$, $V_0 = 0.65\,\mathrm{V}$, and σ_0 extracted from the anti-parallel resistance $R_{AP} = 294\,\Omega$

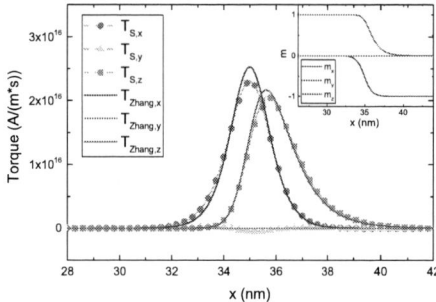

Fig. 5: Comparison of the spin torque $\mathbf{T_S}$ to the Zhang-Li torque $\mathbf{T_{ZL}}$ for a magnetization texture shorter than λ_{sf}, for $\lambda_J = 1\,\mathrm{nm}$ and $\lambda_\varphi = 0.4\,\mathrm{nm}$. The magnetization texture is shown in the inset.

ferromagnetic layers must be taken into account. These torques are modeled by the Zhang and Li (ZL) [17] equation. We generalized the ZL torques to include λ_φ using the expression

$$\mathbf{T_{ZL}} = -\frac{\mu_B}{e} \frac{\beta}{1+(\epsilon+\epsilon')^2} \left((1+\epsilon'(\epsilon+\epsilon')) \, \mathbf{m} \times \right.$$
$$\left. [\mathbf{m} \times (\mathbf{J_C} \cdot \nabla)\,\mathbf{m}] - \epsilon \, \mathbf{m} \times (\mathbf{J_C} \cdot \nabla)\,\mathbf{m} \right), \quad (5)$$

where $\epsilon = (\lambda_J/\lambda_{sf})^2$ and $\epsilon' = (\lambda_J/\lambda_\varphi)^2$. This expression can be derived from the spin accumulation equation by taking $\nabla \mathbf{S} = 0$, and is strictly valid only, when the change of magnetization in space happens over length scales longer than λ_{sf}. Fig. 5 shows that the presence of a short spin dephasing length, $\lambda_\varphi = 0.4\,\mathrm{nm}$, guarantees the fast absorption of the transverse spin, and keeps a good agreement between $\mathbf{T_S}$ and $\mathbf{T_{ZL}}$ for the magnetization texture in the inset.

In MRAM cells with elongated FLs, both the MTJ and ZL torque contribution can act at the same time, in the presence of magnetization textures in the bulk of the layer. We compute the torque in an MTJ structure with a $5\,\mathrm{nm}$ RL, $0.9\,\mathrm{nm}$ TB, and an elongated FL of $15\,\mathrm{nm}$ with a magnetization profile in the FL like the one in the inset of Fig. 5. The magnetization

in the RL is pointing towards the x-direction. The torque $\mathbf{T_S}$ acting in the FL for this magnetization profile is shown in Fig. 6a. Both the interface contribution from the MTJ and the bulk ZL contribution are present. In Fig. 6b we show a close-up of the bulk portion of $\mathbf{T_S}$, compared with the ZL torque $\mathbf{T_{ZL}}$. The comparison reveals a substantial difference between the two, even in the presence of a short spin dephasing length. Our modeling approach clearly demonstrates that in the presence of an MTJ the Slonczewski and ZL torques are not independent, as the presence of the TB also generates a spin accumulation component parallel to the magnetization, whose decay is dictated by λ_{sf}. This component interacts with the magnetization texture, modifying the ZL torque contribution. A unified treatment of the torque is thus necessary to accurately describe the switching process in ultra-scaled MRAM cells.

Finally, Fig. 7 demonstrate the magnetization behavior for anti-parallel to parallel switching in an ultra-scaled MRAM cell (cf. Fig. 1) with a diameter of $2.3\,\mathrm{nm}$ studied experimentally [6] under $-1\,\mathrm{V}$ bias voltage. The cell has a composite FL made of two $5\,\mathrm{nm}$ long parts separated by an MgO layer. The spin-current polarization allows for the first part of the FL to be reverted first, followed shortly after by the second one thanks to both, the STT contribution and the strong magnetostatic coupling between the two sections. The whole FL switches successfully, in agreement with the experimental result [6].

Conclusion

We presented an approach to accurately describe the torques and the magnetization dynamics in ultra-scaled MRAM cells consisting of several elongated pieces of ferromagnets separated by multiple tunnel barriers. We showed how the fully three-dimensional spin and charge transport equations can be supplied with appropriate conditions at the tunneling barrier to reproduce the angular and voltage dependence of the torque expected in MTJs. We demonstrated that the Slonczewski and Zhang and Li torques are not additive and must be derived from the spin accumulation to account for their interplay and

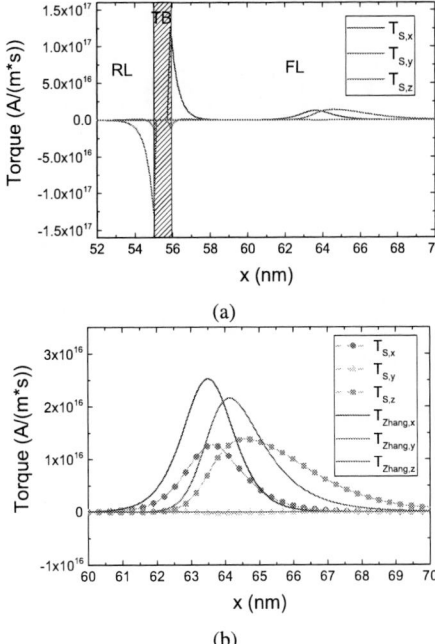

(a)

(b)

Fig. 6: (a) Torques computed for an MRAM cell with elongated RL and FL and a magnetization profile in the FL like the one in the inset of Fig. 5. The magnetization in the RL is in x-direction. (b) Close-up of the spin torque $\mathbf{T_S}$ compared to the Zhang-Li torque $\mathbf{T_{ZL}}$.

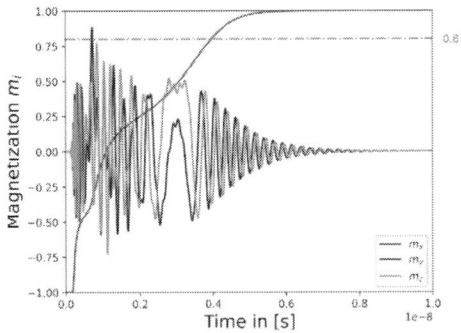

Fig. 7: Switching under -1 V bias of an MRAM cell made of 5 nm RL, 0.9 nm TB, and composite FL separated from the contact by a 0.9 nm MgO layer. The FL consists of two ferromagnetic parts of 5 nm separated by a 0.9 nm MgO layer.

correctly describe the torques on textured magnetization in elongated FLs. Finally, we applied our method to analyze the switching of an MRAM cell with elongated and composite FL. The obtained results validate the use of the proposed simulation approach as support for the design of advanced ultra-scaled MRAM cells.

ACKNOWLEDGMENT

The financial support by the Austrian Federal Ministry for Digital and Economic Affair, the National Foundation for Research, Technology and Development and the Christian Doppler Research Association is gratefully acknowledged.

REFERENCES

[1] S. Aggarwal, H. Almasi, M. DeHerrera, B. Hughes, S. Ikegawa et al., "Demonstration of a reliable 1 Gb standalone spin-transfer torque MRAM for industrial applications," in Proc. IEDM Conf., 2019, pp. 2.1.1–2.1.4.

[2] V. B. Naik, K. Yamane, T. Lee, J. Kwon, R. Chao et al., "Jedec-qualified highly reliable 22nm fd-soi embedded mram for low-power industrial-grade, and extended performance towards automotive-grade-1 applications," in Proc. IEDM Conf., 2020, pp. 11.3.1–11.3.4.

[3] Y.-C. Shih, C.-F. Lee, Y.-A. Chang, P.-H. Lee, H.-J. Lin et al., "A reflow-capable, embedded 8mb stt-mram macro with 9ns read access time in 16nm finfet logic cmos process," in Proc. IEDM Conf., 2020, pp. 11.4.1–11.4.4.

[4] S. H. Han, J. M. Lee, H. M. Shin, J. H. Lee, K. S. Suh et al., "28-nm 0.08 mm^2/Mb embedded mram for frame buffer memory," in Proc. IEDM Conf., 2020, pp. 11.2.1–11.2.4.

[5] J. G. Alzate, U. Arslan, P. Bai, J. Brockman, Y. J. Chen et al., "2 Mb array-level demonstration of STT-MRAM process and performance towards L4 cache applications," in Proc. IEDM Conf., 2019, pp. 2.4.1–2.4.4.

[6] B. Jinnai, J. Igarashi, K. Watanabe, T. Funatsu, H. Sato et al., "High-performance shape-anisotropy magnetic tunnel junctions down to 2.3 nm," in Proc. IEDM Conf., 2020, pp. 24.6.1–24.6.4.

[7] J. C. Slonczewski, "Currents, torques, and polarization factors in magnetic tunnel junctions," Phys. Rev. B, vol. 71, p. 024411, 2005. [Online]. Available: https://link.aps.org/doi/10.1103/PhysRevB.71.024411

[8] R. Anderson, J. Andrej, A. Barker, J. Bramwell, J.-S. Camier et al., "MFEM: A modular finite element library," Comp. & Math. with Appl., 2020.

[9] J. Ender, M. Mohamedou, S. Fiorentini, R. Orio, S. Selberherr et al., "Efficient demagnetizing field calculation for disconnected complex geometries in stt-mram cells," in Proc. SISPAD Conf., 2020, pp. 213–216.

[10] C. Abert, M. Ruggeri, F. Bruckner, C. Vogler, G. Hrkac et al., "A three-dimensional spin-diffusion model for micromagnetics," Sci. Rep., vol. 5, no. 1, p. 14855, 2015. [Online]. Available: https://doi.org/10.1038/srep14855

[11] S. Lepadatu, "Unified treatment of spin torques using a coupled magnetisation dynamics and three-dimensional spin current solver," Sci. Rep., vol. 7, no. 1, p. 12937, 2017. [Online]. Available: https://doi.org/10.1038/s41598-017-13181-x

[12] H. Kubota, A. Fukushima, K. Yakushiji, T. Nagahama, S. Yuasa et al., "Quantitative measurement of voltage dependence of spin-transfer torque in mgo-based magnetic tunnel junctions," Nat. Phys., vol. 4, no. 1, pp. 37–41, Jan 2008. [Online]. Available: https://doi.org/10.1038/nphys784

[13] S. Fiorentini, J. Ender, S. Selberherr, R. de Orio, W. Goes, and V. Sverdlov, "Coupled spin and charge drift-diffusion approach applied to magnetic tunnel junctions," Sol.-St. El., vol. 186, p. 108103, 2021. [Online]. Available: https://www.sciencedirect.com/science/article/pii/S0038110121001489

[14] M. Chshiev, A. Manchon, A. Kalitsov, N. Ryzhanova, A. Vedyayev et al., "Analytical description of ballistic spin currents and torques in magnetic tunnel junctions," Phys. Rev. B, vol. 92, p. 104422, 2015.

[15] K. Y. Camsari, S. Ganguly, D. Datta, and S. Datta, "Physics-based factorization of magnetic tunnel junctions for modeling and circuit simulation," in Proc. IEDM Conf., 2014, pp. 35.6.1–35.6.4.

[16] M. M. Torunbalci, P. Upadhyaya, S. A. Bhave, and K. Y. Camsari, "Modular compact modeling of mtj devices," IEEE Trans. on El. Dev., vol. 65, no. 10, pp. 4628–4634, 2018.

[17] S. Zhang and Z. Li, "Roles of nonequilibrium conduction electrons on the magnetization dynamics of ferromagnets," Phys. Rev. Lett., vol. 93, p. 127204, Sep 2004. [Online]. Available: https://link.aps.org/doi/10.1103/PhysRevLett.93.127204

The Environmental Footprint of IC Production: Meta-Analysis and Historical Trends

Thibault Pirson[†], Thibault Delhaye[†], Alex Pip[‡], Grégoire Le Brun[†], Jean-Pierre Raskin[†], David Bol[†]

[†]ICTEAM Institute - [‡]IMMC Institute, Université catholique de Louvain, Belgium

Email: {thibault.pirson, thibault.delhaye, alex.pip, gregoire.lebrun, jean-pierre.raskin, david.bol}@uclouvain.be

Abstract—Given the ubiquity of electronic devices and the increasing technical challenges to continue CMOS technology downscaling following Moore's Law, it becomes critical to better quantify the environmental impacts of IC production. In this context, this paper compares 22 key relevant sources, from scientific literature, commercial state-of-the-art databases, industry roadmaps, and foundries CSR reports, over the period 2010-2020. Three environmental indicators, i.e., carbon footprint, primary energy, and water consumption, are considered and normalized per silicon area. Despite a significant variability between the sources, we observe a clear increase of the environmental footprint with technology downscaling. We then use historical data from foundries to analyze the long-term evolution of environmental impacts normalized to the silicon area over 40 years. We highlight the arduous challenge of further decreasing environmental indicators per cm^2 below these values. As the global IC production volumes keep on increasing, it raises serious concerns regarding the sustainability of the IC production sub-sector and calls for rethinking the road ahead with sobriety.

Index Terms—life-cycle assessment, sustainability, carbon footprint, integrated circuits, semiconductors, sobriety.

I. INTRODUCTION

THE worldwide production of integrated circuits (ICs) is continuously growing, supported both by the increasing pervasiveness of electronic devices in our societies and the very low re-usability of these components. Even though the size of electronic components is usually very small, their production is known to be very intensive in terms of resources and energy due to the complex processes involved [1]. As these electronic devices are the building blocks of information and communication technologies (ICTs) which support current digital products and services, it becomes critical to better evaluate their environmental impacts. This is even more important given the high investments in the ICT sector [2] and the need for environmentally-aware roadmaps and policies. Although there is no consensus regarding the future trend of the ICT environmental footprint, its carbon footprint was evaluated at about 1000–2000 MtCO2-eq in 2020, or equivalently 2%–4% of the world greenhouse gas (GHG) emissions [3]. A significant part of this footprint is attributed to the production of ICT equipments, i.e., 281-543 MtCO2-eq [3]. More specifically, the production of IC is known to be responsible for an important share of cradle-to-gate impacts [4]–[6], therefore justifying special attention.

This work was supported by the FEDER and the Wallonia within the Wallonie-2020.EU program. It was also supported by the FRS-FNRS through a FRIA grant.

To support the evaluation of environmental impacts over multiple indicators, a standardized methodology known as life-cycle assessment (LCA) [7] is generally used. However, unlike the environmental impacts from the use phase, the impacts related to the production are more difficult to measure directly and usually out of reach for people outside of semiconductor foundries [4], [8]. This relates to a limitation pointed out for years in the field of semiconductors [4], [6], [8], [9], i.e., the lack of up-to-date data due to confidentiality barriers as well as the complexity and quick evolution of manufacturing processes. This hinders a proper understanding of the environmental footprint of the IC production sub-sector.

To tackle this issue, in this work we gather 22 relevant sources from different literature categories to provide a clear overview of the data available. In addition, we identify the gaps and similarities between the sources. To the best of our knowledge, this is the first study of this kind, going well beyond previous scientific literature and covering multiple environmental indicators. The paper is structured as follows. Section II outlines the methodology while Section III presents and interprets the results, which are then put in perspective with historical data. Finally, Section IV highlights the main conclusions.

II. METHODOLOGY

A. Four Different Literature Categories

In order to cover a wide range of data while avoiding field-related biases, four categories of sources are considered, i.e., foundry reports, semiconductor industry roadmaps, scientific literature, and commercial state-of-the-art databases.

Foundries must report annually their environmental performance in corporate social responsibility (CSR) reports which ensure a systematic source of information. Four *pure-play* foundries are considered, i.e., TSMC, UMC, SMIC, and GF, together with one integrated device manufacturer, i.e., STMicroelectronics. The data for Samsung and Intel are either too aggregated or not available, which prevents us from including them in this study. Roadmaps provide targets rather than actual data from on-site measurements. A group of industry experts from Europe, Japan, Tawain, United States and Korea published the International Technology Roadmap for Semiconductors (ITRS), including guidelines for *environment, safety and health (ESH)*. Apart from the ITRS roadmaps, no other official document with explicit target for the environmental impacts of semiconductors was found. Scientific literature is an important source of information as several peer-reviewed

978-1-6654-8498-5/22 $31.00 © 2022 IEEE

studies have been carried out for years. Let us mention that we also included three relevant non peer-reviewed reports in this literature category. Finally, commercial databases provide specific LCA data and usually benefit from maintenance. Three commercial state-of-the-art databases are considered, i.e., GaBi, EIME, and Ecoinvent.

B. Functional Unit and Scope

In this study, we focus on the environmental impacts of IC production over three indicators, i.e., primary energy demand (PED), global warming potential (GWP), and water consumption. Even though these indicators do not provide a comprehensive picture of all environmental impacts [6], they already cover important aspects. In addition, they are the ones usually reported in the different literature categories, which makes the analysis possible. The core results of this study are based on data published over the time period 2010-2020.

Choosing the most appropriate functional unit and metrics to characterize the environmental impacts of IC production still remains an open question, especially because semiconductor manufacturing is a fast-evolving sector [1], [9]. In this study, we choose to normalize all environmental indicators by the silicon die area in cm^2, as done in [9] and [10]. This is also supported by the presence of area in the common power-performance-area-cost (PPAC) targets.

A well-known challenge for LCA practitioners is the concordance of scopes when gathering data from different sources. Consequently, a transparent methodology is critical to ensure reliable conclusions in the identification of gaps and similarities. To tackle this challenge, we define 10 criteria which are then used to compare all sources included in this study, i.e., (1) technology type and inclusion of front-end/back-end; (2) CMOS technology node; (3) wafer size; inclusion of (4) production yields and of (5) infrastructure and upstream materials; (6) geographical location; (7) covered environmental indicators; (8) process details and transparent normalization; (9) assessment method; and (10) primary data and dependencies.

III. RESULTS

A. Transversal Analysis of the Sources

Table I shows the classification of all sources with respect to the 10 selected criteria. This clearly shows the existing mismatch between the scope and LCA approach of the considered sources, revealing the complexity to implement a proper harmonization across all sources. We decided not to perform such an harmonization as sufficient level of information was rarely disclosed. Nevertheless, two clusters can be identified: scientific literature and databases on the one side and foundry reports and industry roadmaps on the other side. While the first cluster allows for a node-wise analysis as illustrated in Fig. 1, the second one provides node-aggregated temporal trends as shown in Fig. 2. These two clusters cannot be directly compared to each other, but they both bring useful insights to better understand the environmental impacts of IC production from different yet complementary perspectives, as explained in the following sections.

B. The Environmental Impacts of Technology Downscaling

There is an obvious inter-sources variability within the scientific literature and databases cluster as shown in Fig. 1, mainly due to the mismatch of scopes. Consequently, comparing the absolute values is of limited interest and relevance. However, focusing on the general trend depicted in Fig. 1 reveals a more compelling conclusion: the environmental impacts per area are clearly increasing with technology downscaling below 0.13 μm. This is observed both in the scientific literature and in databases, despite the scope mismatch. It seems also that this increase is accelerating for advanced nodes.

C. The Impact of Pursuing Advanced Scaling for Foundries

We found no public data or evidence disclosed by foundries to allow a consistent estimation of the environmental impacts with respect to the technology node. Consequently, we analyzed the evolution of aggregated environmental impacts per area with respect to time instead of the technology node, as illustrated in Fig. 2. This highlights two opposing trends: the impacts per area either (slightly) decrease over the period 2010-2020 for most of the foundries while they increase for TSMC. More importantly, these trends seem to be foundry-dependent but consistent for all environmental indicators. Aside from the geographical location, the maturity of their process, and the output production volume, an important difference is the mix of technology nodes in the foundry production volume. Indeed, TSMC is the only industry in this study engaged in the production of technology nodes below 10 nm, which could well explain why it demonstrates a net increase in the impacts per area. This is not due to a lack of effort from the foundry to pursue efficiency but rather to the introduction of more demanding processes such as extreme ultra-violet (EUV) lithography for scaling purpose [10]. Other foundries keep demonstrating slight overall impacts reduction per cm^2, suggesting that a reduction of the impacts per area is possible, provided that advanced scaling is not pursued.

The comparison between the foundry impacts and the roadmap targets shows that the industry is still facing huge challenges to reach these targets. The high impact of EUV is also explicitly pointed out in ITRS roadmaps and illustrated in Fig. 2(a).

D. Long-Term Perspectives from Historical Data

Although LCA is generally used either to evaluate, compare, and optimize the impacts of a specific device or service, it can also be useful to analyze trends over time. In this context, we gathered historical data reported for the period 1980-2010 and we used the same 10 criteria to support the analysis and the identification of scope mismatch, as depicted in Table I. This aims at providing long-term perspectives on the evolution of environmental impacts per cm^2, while it does not intervene with the core results. For most of the historical data, impact values were disclosed in the scientific literature, although they were directly coming from the foundries.

TABLE I
CLASSIFICATION OF ALL SOURCES CONSIDERED IN THIS STUDY ACCORDING TO THE 10 CRITERIA DEFINED IN SECTION II.

Criteria	(1)			(2)	(3)	(4)	(5)		(6)	(7)	(8)		(9)		(10)
	Tech. type	Front-end	Back-end	Tech. node (nm)	Wafer size (mm)	Yield	Infra.	Upstr. materials	Geogr. location	Indic. coverage	Process details	Transp. norm.	Assess. method	Prim. data	Depend-encies
Foundry reports															
TSMC, CSR & 20-F (2010-2020)	mix	✓	✗	mix	mix	⊖	✓	✗†	GLO	Full	✗	✗	↓	✓	✗
UMC, CSR & 20-F (2011-2020)	mix	✓	✗	mix	mix	⊖	✓	✗†	GLO	Full	✗	✗	↓	✓	✗
SMIC, CSR (2011-2020)	mix	✓	✗	mix	mix	⊖	✓	✗†	GLO	Full	✗	✗	↓	✓	✗
STmicro, CSR & 20-F (2011-2020)	mix	✓	✗	mix	mix	⊖	✓	✗†	GLO	Full	✗	✗	↓	✓	✗
GF, CSR & F-1 (2011-2020)	mix	✓	✗	mix	mix	⊖	✓	✗†	GLO	Full	✗	✗	↓	✓	✗
Industry roadmap															
ITRS ESH (2015)	⊖	✓	✗†	⊖	200;300	✗	✓	✗	⊖	PED;W	✗	✓	⊖	⊖	✗†
Scientific literature															
Boyd, PhD (2012)	L;M	✓	✗†	350→32	200;300	✓	✓	✓	USA	Full	✓	✓†	⇃⇂	✓	✓
Schimdt, Int J LCA (2012)	L;M	✓	✗	⊖	300	✓	✓	✗	GLO	GWP	✓	✓	⇃⇂	✗	✓
Prakash, ProBas report* (2013)	M	✓	✗	45	300	✓	✗	✗	⊖	PED;W	✓	✓	⊖	✗	✓
Bol, FTFC (2013)	L	✓	✓†	130;65	⊖	⊖	⊖	⊖	⊖	GWP	✗	✗	↑	✗	✓
Jones, ICCAD (2013)	mix	✓†	✓	45	⊖	⊖	⊖	⊖	⊖	PED;GWP	✗	✓	↓	✗	✗
Teehan, PhD (2014)	⊖	✓†	✗	⊖	⊖	⊖	⊖	✓	⊖	GWP;PED	✗	✗	⇃⇂	✗	✗
Andrae, Challenges (2014)	mix	✓	✗	⊖	⊖	⊖	⊖	⊖	⊖	GWP	✗	✓†	↓	✗	✗
Proske, FF2 report* (2016)	L;M	✓	✓	32	300	✓	⊖	✗†	CN	GWP	✓	✓	⇃⇂†	✗	✓
Ercan, ICT4S (2016)	L;M	✓	✗	⊖	⊖	✓	✓	✓	⊖	GWP;PED	✗	✓	⊖	✗	✗†
Andrae, Challenges (2017)	L	✓†	⊖	⊖	⊖	⊖	⊖	⊖	GLO	GWP	✗	✓	↑†	✗	✓
Bardon, IEDM (2020)	L	✓	✗	28→3	300	✗†	✓	✗†	GLO	Full	✗	✓	↑	✗	✓
Proske, FF3 report* (2020)	L;M	✓	✗†	60;45;32	300	✓†	✓	✗†	CN	GWP	✓	✓	⇃⇂†	✗	✓
Das, SEGAN (2020)	M	✓†	⊖	57	⊖	⊖	⊖	⊖	⊖	PED	✗	✓	⊖	⊖	✓
Commercial databases															
Ecoinvent v3.5 (2018)	L	✓	✓	⊖	200	✓	✓	✗	GLO	Full	✗	✗	⇃⇂	✗	✓
EIME, Bureau Veritas (2019)	M	✓	✗	130→7	300	✗	✓	✓	TWN	Full	✗	✗	↑	✓	✓
GaBi, Sphera (2020)	L;M	✓	✓	250→14	⊖	✓	✓	✓	GLO	Full	✗	✗	⇃⇂	✓	✓
Historical data															
Williams, EST (2002|1996-1998)	mix	✓	⊖	⊖	150	⊖	✓†	✗†	GLO	PED°;W	✗	✓	↓	✗	✗
Williams, EST (2002)	M	✓	⊖	⊖	200	✗	⊖	✓	GLO	PED°	✗	✓	↑	✗	✗
Plepys, PhD (2004|1983-2003)	mix	✓	✗†	mix	150;300	⊖	⊖	✗†	GLO	PED°;W;GWP°	✗	✓	⇃⇂	✗	✗
Krishnan, EST (2008)	L	✓	✗	130	300	✓	✓	✓	USA	Full	✗	✓	⇃⇂	✗	✗
Deng, JCP (2011|1995-2006)	mix	✓	✗†	⊖	⊖	✗	✓†	✗†	GLO	PED	✗	✓	↓	✗	✗
Ercan, ICT4S (2016|1995)	mix	✓	✗†	⊖	⊖	⊖	✓†	✓†	GLO	GWP;PED°	✗	✓	⊖	⊖	✗

L : logic   M : memory   ↑ : bottom-up   ↓ : top-down   ⇃⇂ : hybrid   ⊖ : not available   † : implicit, deduced by the authors   * : not peer-reviewed (scientific literature only)   ✓ : included   ✗ : excluded
° : converted from the electrical consumption in kWh (PED: assuming a primary energy factor (PEF) of 2.5 | GWP : assuming an average carbon intensity of 0.475 kgCO2eq/kWh)

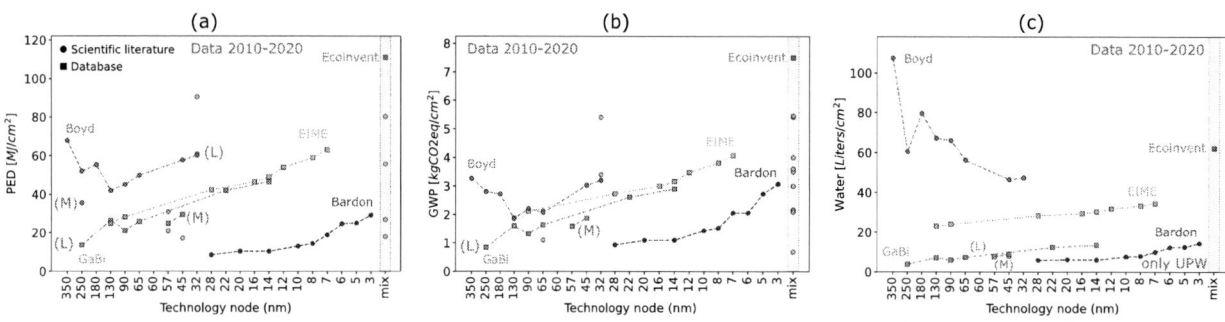

Fig. 1. Node-wise trends based on data from the scientific literature and the databases. This shows the environmental impacts per cm^2 with respect to (a) PED, (b) GWP, and (c) water consumption.

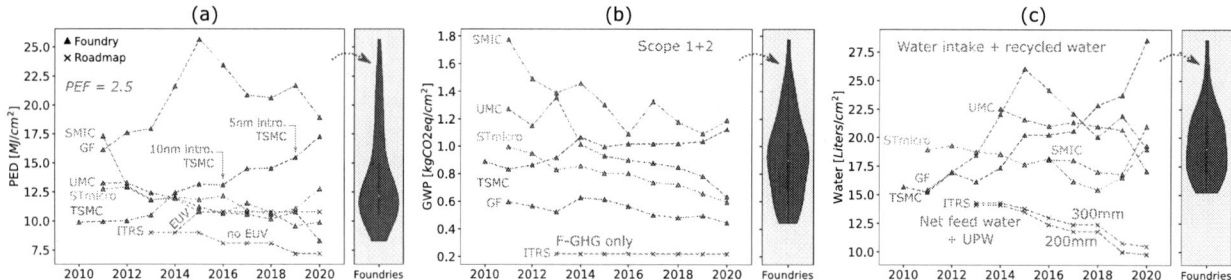

Fig. 2. Node-aggregated temporal trends based on data from foundry reports and industry roadmaps. This shows the environmental impacts per cm^2 with respect to (a) PED, (b) GWP, and (c) water consumption. The left sub-figure exhibits the data for each source while the right sub-figure reflects the aggregated data for all the foundries considered in this study. Assumptions made by the authors are shown in *italic*.

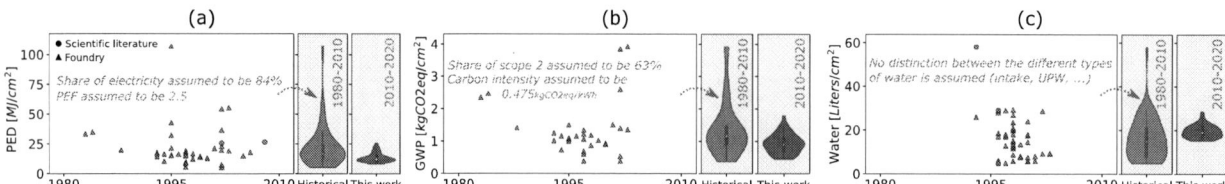

Fig. 3. Long-term perspectives between the results presented in this work (2010-2020) and historical data (1980-2010). The comparison per cm^2 is done only for data from foundries, with respect to (a) PED, (b) GWP, and (c) water consumption. Assumptions made by the authors are shown in *italic*.

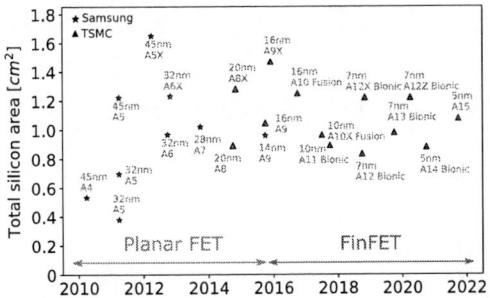

Fig. 4. Evolution of the average chip area of Apple's application processors.

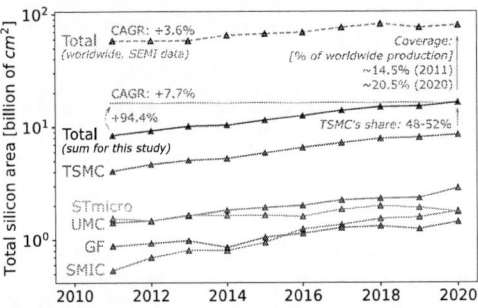

Fig. 5. IC production volumes for the foundries considered in this study and for the worldwide production.

The comparison of the data from the foundries over the period 1980-2010 and 2010-2020 reveals an important finding: environmental indicators normalized per cm^2 did not significantly decrease compared to historical values from 1980-2010, as shown in Fig. 3. Undeniably, the number of transistors fitted on the same area of silicon has tremendously increased since then, as captured by Moore's Law. Nevertheless, the computational demand and the constant increase of functionality also surged in the meantime [1], [9], [11]. The increase in functionality is at the heart of Moore's Law, which explains the almost constant die size despite technology downscaling. When analyzing the evolution of application processor chips from Apple over the last 10 years, Fig. 4 shows that the average die area is roughly constant even though technology node shifted from 45 nm to 5 nm. The environmental consequences of this increase of functionality were already pointed out more than 10 years ago [1]. They can be seen as a rebound effect (Jevons' paradox) of technology downscaling which improves the environmental indicators per transistor [1], [11]–[13].

Focusing on GWP, the Paris Agreement sets an explicit reduction target of global GHG emissions at a rate of 7.6%/year to limit climate change, assuming a start in 2020 [14]. This would therefore require a reduction of the GWP/cm^2 close to 11-14%/year if we assume that the total IC production keeps increasing at the same rate as last decade, as shown in Fig. 5. Therefore, conflicting trends are to be expected if the IC production sub-sector pursues aggressive downscaling with increasing production volumes while committing to rapid reduction in GHG emissions. Nevertheless, advanced nodes could be game changing if used to provide constant functionality and performance with less area, rather than more functionalities with constant area.

IV. CONCLUSION

It appears very unlikely that optimization of the environmental impacts per cm^2 will be sufficient to quickly reduce the global absolute environmental footprint of the IC production sub-sector, especially for advanced nodes. Moreover, the trend towards more functionalities observed over the last decades reveals a clear rebound effect at the hardware level, absorbing or backfiring efficiency improvements captured by Moore's Law. This is likely to continue as long as higher performance is available, which attests the limits of technology downscaling in being an effective solution for sustainability.

Therefore, we conclude that sobriety must be integrated into technological innovation and technology usage, which calls for rethinking our socio-economic models and innovation roadmaps currently fostering growth in the IC production. Nevertheless, an important question holds: could downscaling remain economically viable if the global IC production volumes flatten or decrease in the long term?

ACKNOWLEDGMENTS

The authors would like to thank Marie Garcia Bardon, Etienne Lees-Perasso, Constantin Herrmann, Stephan Benecke, and Noel Ullrich for the valuable discussions and constructive inputs as well as ECS group members for their proofreading.

REFERENCES

[1] L. Deng *et al.*, "Measures and trends in energy use of semiconductor manufacturing," in *2008 IEEE International Symposium on Electronics and the Environment*, pp. 1–6, IEEE, 2008.

[2] IDC, "Global ICT Spending : Forecast 2020 – 2023," 2018.

[3] C. Freitag *et al.*, "The real climate and transformative impact of ICT: A critique of estimates, trends, and regulations," *Patterns*, vol. 2, no. 9, p. 100340, 2021.

[4] P.-V. C. Louis-Philippe *et al.*, "Sources of variation in life cycle assessments of smartphones and tablet computers," *Environmental Impact Assessment Review*, vol. 84, p. 106416, 2020.

[5] T. Pirson and D. Bol, "Assessing the embodied carbon footprint of IoT edge devices with a bottom-up life-cycle approach," *Journal of Cleaner Production*, vol. 322, p. 128966, 2021.

[6] Y. Arushanyan *et al.*, "Lessons learned–Review of LCAs for ICT products and services," *Computers in industry*, vol. 65, no. 2, pp. 211–234, 2014.

[7] I. O. for Standardization, "ISO 14044:2006," 2006.

[8] A. Plepys, "The environmental impacts of electronics: going beyond the walls of semiconductor fabs," in *IEEE International Symposium on Electronics and the Environment, 2004. Conference Record. 2004*, pp. 159–165, IEEE, 2004.

[9] S. B. Boyd, *Life-cycle assessment of semiconductors*. Springer Science & Business Media, 2012.

[10] M. G. Bardon *et al.*, "DTCO including sustainability: Power-performance-area-cost-environmental score (PPACE) analysis for logic technologies," in *2020 IEEE International Electron Devices Meeting (IEDM)*, pp. 41–4, IEEE, 2020.

[11] D. Bol *et al.*, "Moore's Law and ICT Innovation in the Anthropocene," in *Proceedings of the IEEE Design and Test in Europe Conference, Grenoble, France*, pp. 1–5, 2021.

[12] A. Plepys, "Substituting computers for services-potential to reduce ICT's environmental footprint," in *International Congress and Exhibition on Electronics Goes Green 2004+*, pp. 217–222, Fraunhofer IRB Verlag, 2004.

[13] C. Gossart, "Rebound effects and ICT: a review of the literature," *ICT innovations for sustainability*, pp. 435–448, 2015.

[14] V. Masson-Delmotte *et al.*, "Global warming of 1.5°C," *IPCC Special Report*, 2018.

Experimental fabrication of an ESF3 floating gate flash cell in an FD-SOI process

Nicki Mika
Module One LLC & Co. KG
GlobalFoundries Dresden
Dresden, Germany
Nicki.Mika@gf.com

Thomas Melde
Module One LLC & Co. KG
GlobalFoundries Dresden
Dresden, Germany
Thomas.Melde@gf.com

Stefan Dünkel
Module One LLC & Co. KG
GlobalFoundries Dresden
Dresden, Germany
Stefan.Duenkel@gf.com

Michael Otto
Module One LLC & Co. KG
GlobalFoundries Dresden
Dresden, Germany
Michael.Otto2@gf.com

Francois Weisbuch
Module One LLC & Co. KG
GlobalFoundries Dresden
Dresden, Germany
Francois.Weisbuch@gf.com

Peter Krottenthaler
Module One LLC & Co. KG
GlobalFoundries Dresden
Dresden, Germany
Peter.Krottenthaler@gf.com

Thomas Mikolajick
Chair of Nanoelectrics
Technical University of Dresden
Dresden, Germany
thomas.mikolajick@tu-dresden.de

Abstract—**As minimum feature sizes in semiconductor processes are continuously decreasing, the parallel implementation of embedded non-volatile memory cells becomes increasingly complex for technology nodes below 28 nm. In this publication it is demonstrated the first ESF3 fabrication in an advanced SOI process using the buried oxide as a tunnel oxide. This enables a small cell area of 0.086 μm² with only few additional process steps and promising electrical data.**

Index Terms—**floating gate, embedded flash, NVM, MTPM, ESF3, SOI**

I. INTRODUCTION

The demand for embedded non-volatile memory is increasing and the dominant node of 40nm is expected to lose market share towards the 28nm node [1]. Due to the scaling of dimensions as well as operation voltages, it becomes increasingly challenging to meet the requirements for a floating gate flash cell. Therefore, emerging memory concepts have a big attraction as alternatives. So far STT-MRAM [2] and RRAM [3], which are processed in the back-end-of-line (BEoL) and FeFETs [4] which is processed in the front-end-of-line (FEoL), are considered as possible replacements to floating gate flash cells. All the emerging technologies rely on sophisticated material compositions that require special process integration and are not fully matured yet [1].

The fully depleted silicon-on-insulator (FD-SOI) technology offers a unique opportunity to re-evaluate constrains from bulk technologies as already shown in the possibility of shrinking single-poly floating gate (FG) flash cells [5]. The ESF3 structure is currently dominating nodes down to 28nm through its robust design but requires at least 7 additional masks [6]. The SOI technology with thin buried oxides (BOX) below 30 nm simplifies the fabrication of this architecture in the established logic flow, using just one gate deposition.

II. CELL DESCRIPTION

A typical vertical floating gate memory stack consists of 5 layers: bulk – oxide – floating gate – oxide – control gate or in terms of electrical properties: conducting – insulating – conducting – insulating – conducting. These electrical properties are identical to the ones of the FD-SOI logic stack as shown in Fig. 1. On a raw SOI wafer, the first 3 layers are already preprocessed. Therefore, only one gate oxide and one gate material need to be processed in order to resemble a floating gate stack.

The fabricated cell is shown in Fig. 2. The select gate (SG) acts as the word line (WL) connection in such it enables the column selection within an array, by opening or closing the channel of the memory cell. One of the key features of this approach is to use the BOX of the FD-SOI wafer as the tunneling oxide, instead of forming at tunnel oxide during memory device fabrication. The BOX isolates the channel from the SOI that acts as a floating gate. In the floating SOI layer, electrons are injected and stored, or removed. This changes the threshold voltage of the memory device as is discussed in Section IV. All three gates are isolated from the SOI by the gate oxide.

Like a standard ESF3 cell, the programming is performed through the source side injection (SSI). There, the control gate (CG) is biased around 11 V. The resulting electrical field sets the channel beneath the tunnel oxide in strong inversion. Together with a slightly opened SG and a source line (SL) usually biased around 4.5 V, a strong lateral and strong vertical field are present at the same time. This leads to hot carrier generation which are redirected through the BOX into the SOI. The memory cell gets programmed.

The erase is done by use of Fowler-Nordheim tunneling through the erase gate (EG). For read condition the bit line (BL) is biased, and the SL is set to ground. For write, erase and read operations, only positive voltages are necessary. In the fabricated macro, each cell has only an area of roughly 0.086 μm².

978-1-6654-8498-5/22 $31.00 © 2022 IEEE

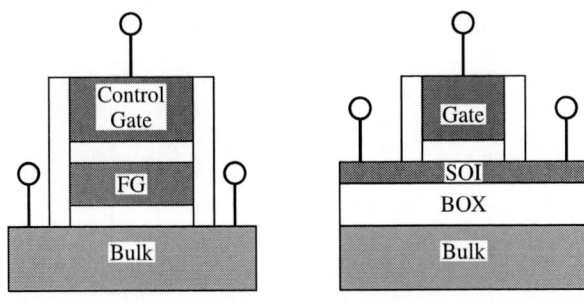

Fig. 1. Schematics of a typical floating gate cell stack (left) and an SOI transistor (right). Both architectures share the same sequence of conducting and insulating layers.

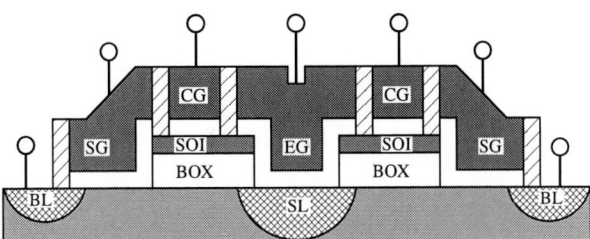

Fig. 2. Realized cell structure. The select gates (SG) are located on the bit line (BL) sites and act as word line selectors through opening or closing the channels. The control gate (CG) is located on top of the SOIs which act as the charge trapping layers. Similar to other ESF3 approaches, the erase gate (EG) located in the middle, is shared by both cells and warps slightly over the charge trapping layers. It withdraws the stored electrons from the SOI via FN tunneling. The source line (SL) is buried below the EG and connected through cuts in the erase gate line.

III. MANUFACTURING

The processing does not follow the flash first principle, but the memory is fabricated at the same time as the logic. At first the FG is formed during the bulk etch. This process step usually exists in an SOI processes, to remove SOI and BOX to apply back bias to the SOI. The source line formation is done by a standard N-well logic implant. This will have implications on the program efficiency and will be discussed in Section IV. In our experiment, the gate formation follows the same flow as the logic, but an extra gate oxide thickness of 85 Å was processed, to withstand the applied voltages and achieve sufficient retention. The formation of all three gates was done with a single high-κ and poly deposition, where the patterning was achieved by a cut mask. At this time, also a hole in the erase gate is etched to connect the SL to the first metal layer. Finally, source and drain of the memory cells are formed at the same time as for the logic. In total, the cell needs 3 additional DUV mask, which is a reduction of at least 4 masks (2 ArF immersion & 2 DUV), compared to the cell proposed in [6] and can be processed simultaneously with the logic. This reduces process time and costs.

IV. ELECTRICAL CHARACTERIZATION

The electrical characterization has been performed on small NOR arrays. Figure 3 shows one example cell in the erased

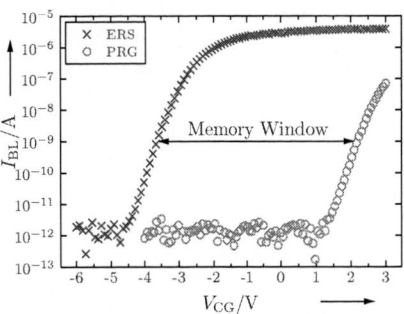

Fig. 3. Transfer characteristic of programmed (PRG) and erased (ERS) state. The bit line current is shown as a function of the control gate voltage, while select gate voltage is set to 1.8 V. The shift between both curves marks the memory window.

and programmed state. The bias conditions shown in Table I were applied for program, erase and read operations. While the select gate is complete open, the bit line current I_{BL} is measured as the function of control gate voltage V_{CG} for both states. There, a threshold voltage shift of roughly 6 V is shown between both states, which marks the memory window (MW).

TABLE I
PROPOSED BIAS CONDITIONS FOR PROGRAM, ERASE, AND READ OF
MEMORY CELLS IN THE MINI-ARRAY.

Terminal	Bias conditions		
	Program	Erase	Read
Bit-Line (BL)	0 V	0 V	0.8 V
Source-Line (SL)	4.5 V	0 V	0 V
Select-Gate (SG)	0.8 V	0 V	1.8 V
Control-Gate (CG)	13 V	0 V	0 V
Erase-Gate (EG)	4.5 V	11.5 V	0 V
Pulse duration	20 µs	10 ms	–

This is sufficient for the targeted applications. The used bias conditions are similar to the ones in other ESF3 cells [6]–[8] with the exception of the control gate voltage being roughly 2 V higher, which will be a topic for further development iterations of the cell. As discussed in Section III. The formation of the buried source line was done by using the logic N-well implant. This doping concentration is too low for a high conducting common source, resulting in a high resistance. The influence on the program efficiency can be seen in Fig. 4. Here data of six programmed cells are depicted, where three groups are shown. The numbers next to the transfer curves indicates the position of the cell along the source line. The smaller the number, the closer the cell is to the contact, leading to an improved program efficiency. A separate high dose implant will fix this dependency on the position along the SL.

In Figure 5 the program efficiency is shown for varying program pulse heights V_{CG} and width T_{PRG}. The threshold voltages V_{th} were determined by the maximum transconductance $\max(g_m)$ method. The figure shows a clear linear correlation between V_{CG} and reached threshold voltage V_{th}. Furthermore, it shows the opportunity to reduce the necessary voltage in

978-1-6654-8498-5/22 $31.00 © 2022 IEEE   357

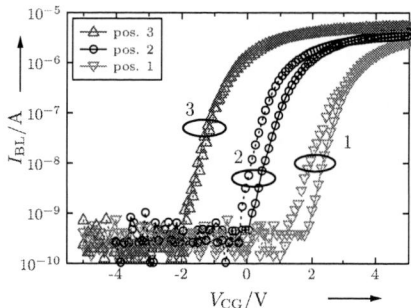

Fig. 4. Program efficiency depending on the position along the source line as a result of the used SL doping concentration. Pos. 1 is the one closest to the source contact. Pos. 3 has the longest distance to this contact.

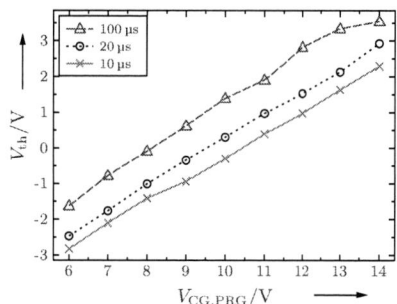

Fig. 5. Programmed V_{th} of the cell as a function of the programming pulse voltage for three different pulse times. The cell was erased before every repeated program pulse. V_{th} was extracted using the $\max(g_m)$ criteria. An onset of program saturation is visible for extracted threshold values around 4 V.

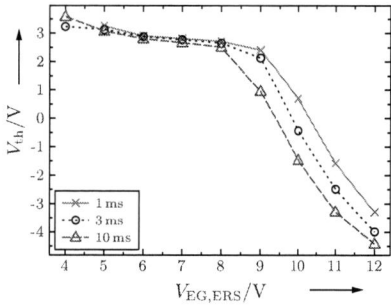

Fig. 6. Erase V_{th} of the cell as a function of the erase pulse voltage for three different pulse times. An increase of erase efficiency for erase pulses above 8 V is indicated by the change of the slope. An onset of erase saturation is visible for extracted threshold values around -4.5 V. The cell was programmed before every repeated erase pulse. V_{th} was extracted using the $\max(g_m)$ criteria.

exchange for a longer pulse time.

The erase performance is shown in 6. Here, also voltage and pulse time were varied. As it can be seen, voltages below 8 V only lead to a small V_{th} shift. This changes towards higher voltages and the erase efficiency starts to increase.

Besides the memory window, two figures of merit are of in-

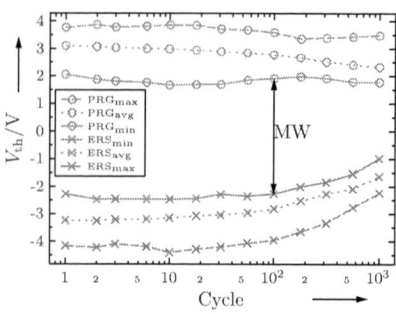

Fig. 7. Endurance of 32 cells. Determination of V_{th} via $\max(g_m)$ criteria. The minimum and maximum reached values within the states indicate the deviation of the mean. The memory window is spanned between the weakest values of both states. The programmed states remains stable over the complete shown number of cycles, where the erase state narrows down the memory window after 100 times of re-erasing.

terest for multiple-time-programmable (MPT) memory blocks. Endurance is the first parameter, stating how often the cell can be re-written between the different states maintaining a sufficient memory window. Fig. 7 shows the results for the presented cell. During the measurement, the pulse conditions remained unchanged. The solid lines represent the weakest erased and programmed states within this array, which marks the memory window for a specific cycling step. The array starts with a MW of roughly 4.4 V. After 100 and 1000 cycles remain 4.2 V and 2.8 V, respectively. In general, the behavior of a nearly constant programmed state and a weaker erased state over cycling is typical for floating gate flash [9]. The narrowing of MW on chips is typically compensated by an increase of used erase voltages known as trimming. Finally, the average and maximum values were plotted, to indicate the V_{th} distribution between different cells.

The second parameter of interest is the retention, which indicates, how long data can be stored on the cell at a given temperature. Typical industrial practice is to pre-conditions the cells in either programmed or erase state and measure repetitively the threshold voltage after several bake steps. The chosen temperature marks in most of the cases the upper temperature limit for an application. Through the high temperatures, the electrons get more energy and the possibility for tunneling through the barriers increases. This is measured in shift of the threshold voltage over total bake time. Subsequently, the threshold voltage shift is extrapolated towards 10 years. The Fig. 8 proves that for all three cycling variants, the arrays remain a memory windows at a bake temperature of 105 °C. For a fresh macro the MW remains 2 V and for an array cycled 1000 times, which starts with a lower MW, 1 V. Going down towards industrial temperature ranges of 85 °C the remaining MW is expected to further increase.

V. SUMMARY/OUTLOOK

A new opportunity for the fabrication of embedded non-volatile memories in SOI technologies has been demonstrated. This design is smaller than the one, we proposed in [5] and

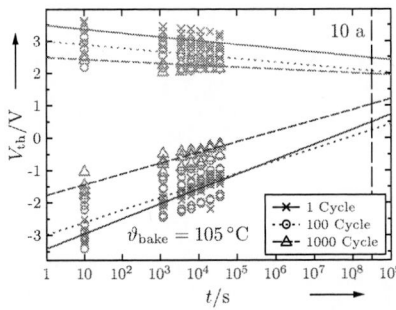

Fig. 8. Retention data with bake temperature of $105\,^{\circ}\mathrm{C}$ for 8 dies and three different numbers of previous cycles. The memory window narrows towards longer total bake times for each pre-cycled variation. The extrapolation towards 10 years show a remaining window, which is crucial for storing data.

uses only positive voltages, which eases the circuit development and requires less area. This cell design was applied on a 22nm process and in theory also works for smaller nodes. By using the buried oxide as tunneling layer and the SOI as floating gate, the cell could be processed simultaneously with the logic. This save costs, which is especially crucial for Internet of things (IoT) devices. Although the gate stack contains high-κ for all gates, the shown endurance and retention data prove suitability for MTPs. Further improvements can be done on chip level by error correction code (ECC) and trimming.

Many applications only need to store a key or some calibration data. There a 256 kbit up to 1024 kbit memory block that is re-writable ten or hundred times, fulfills those needs. This is feasible with the demonstrated cell.

REFERENCES

[1] R. Strenz, "Review and Outlook on Embedded NVM Technologies – From Evolution to Revolution," in *2020 IEEE International Memory Workshop (IMW)*, 2020, pp. 1–4.

[2] V. B. Naik, K. Lee, K. Yamane, R. Chao, J. Kwon, N. Thiyagarajah, N. L. Chung, S. H. Jang, B. Behin-Aein, J. H. Lim, T. Y. Lee, W. P. Neo, H. Dixit, S. K, L. C. Goh, T. Ling, J. Hwang, D. Zeng, J. W. Ting, E. H. Toh, L. Zhang, R. Low, N. Balasankaran, L. Y. Zhang, K. W. Gan, L. Y. Hau, J. Mueller, B. Pfefferling, O. Kallensee, S. L. Tan, C. S. Seet, Y. S. You, S. T. Woo, E. Quek, S. Y. Siah, and J. Pellerin, "Manufacturable 22nm FD-SOI Embedded MRAM Technology for Industrial-grade MCU and IOT Applications," in *2019 IEEE International Electron Devices Meeting (IEDM)*, 2019, pp. 2.3.1–2.3.4.

[3] J.-H. Yoon, M. Chang, W.-S. Khwa, Y.-D. Chih, M.-F. Chang, and A. Raychowdhury, "A 40-nm, 64-Kb, 56.67 TOPS/W Voltage-Sensing Computing-In-Memory/Digital RRAM Macro Supporting Iterative Write With Verification and Online Read-Disturb Detection," *IEEE Journal of Solid-State Circuits*, vol. 57, no. 1, pp. 68–79, 2022.

[4] S. Beyer, S. Dünkel, M. Trentzsch, J. Müller, A. Hellmich, D. Utess, J. Paul, D. Kleimaier, J. Pellerin, S. Müller, J. Ocker, A. Benoist, H. Zhou, M. Mennenga, M. Schuster, F. Tassan, M. Noack, A. Pourkeramati, F. Müller, M. Lederer, T. Ali, R. Hoffmann, T. Kämpfe, K. Seidel, H. Mulaosmanovic, E. T. Breyer, T. Mikolajick, and S. Slesazeck, "FeFET: A versatile CMOS compatible device with game-changing potential," in *2020 IEEE International Memory Workshop (IMW)*, 2020, pp. 1–4.

[5] T. Melde, M. Trentzsch, S. Duenkel, R. Richter, M. Otto, H. Giesler, F. Weisbuch, N. Weddeler, and S. Beyer, "Novel embedded single poly floating gate flash demonstrated in 22nm FDSOI technology," in *2021 IEEE International Memory Workshop (IMW)*, 2021, pp. 1–4.

[6] R. Richter, M. Trentzsch, S. Dünkel, J. Müller, P. Moll, B. Bayha, K. Mothes, A. Henke, M. Mazur, J. Paul, P. Krottenthaler, J. Poth, S. Jansen, R. Hüselitz, H. Kim, A. Zaka, T. Herrmann, E. M. Bazizi, S. Beyer, P. Ghazavi, H. Om'mani, S. Lemke, Y. Tkachev, F. Zhou, J. Kim, X. Liu, V. Tiwari, and N. Do, "A cost-efficient 28nm split-gate eFLASH memory featuring a HKMG hybrid bit cell and HV device," in *2018 IEEE International Electron Devices Meeting (IEDM)*, Dec 2018, pp. 18.5.1–18.5.4.

[7] D. Shum, L. Q. Luo, Y. Kong, F. Deng, X. Qu, Z. Teo, J. Q. Liu, F. Zhang, X. Cai, K. Tan, K. Lim, P. Khoo, P. Yeo, B. Nguyen, S. Jung, S. Siah, K. Pey, K. Shubhakar, C. Wang, J. Xing, G. Liu, Y. Diao, G. Lin, F. Luo, L. Tee, V. Markov, S. Lemke, P. Ghazavi, N. Do, V. Tiwari, and X. Liu, "40nm Embedded Self-Aligned Split-Gate Flash Technology for High-Density Automotive Microcontrollers," in *2017 IEEE International Memory Workshop (IMW)*, 2017, pp. 1–4.

[8] I. Mazzetta and F. Irrera, "Influence of Common Source and Word Line Electrodes on Program Operation in SuperFlash Memory," *Electronics*, vol. 10, no. 3, 2021. [Online]. Available: https://www.mdpi.com/2079-9292/10/3/337

[9] G. Torrente, J. Coignus, A. Vernhet, J.-L. Ogier, D. Roy, and G. Ghibaudo, "Physically-based evaluation of aging contributions in HC/FN-programmed 40nm NOR Flash technology," *Microelectronics Reliability*, vol. 79, pp. 281–287, 2017. [Online]. Available: https://www.sciencedirect.com/science/article/pii/S0026271417301841

Gap in pagination due to withheld paper.

Pages 360-363

GeSn Vertical Gate-all-around Nanowire n-type MOSFETs

Yannik Junk
Peter-Grünberg-Institut (PGI-9)
Forschungszentrum Jülich
Jülich, Germany
y.junk@fz-juelich.de

Omar Concepción Diaz
Peter-Grünberg-Institut (PGI-9)
Forschungszentrum Jülich
Jülich, Germany
o.diaz@fz-juelich.de

Detlev Grützmacher
Peter-Grünberg-Institut (PGI-9)
Forschungszentrum Jülich
Jülich, Germany
d.gruetzmacher@fz-juelich.de

Marvin Frauenrath
CEA-LETI
MINATEC Campus and University of
Grenoble Alps
Grenoble, France
marvin.frauenrath@cea.fr

Jin-Hee Bae
Peter-Grünberg-Institut (PGI-9)
Forschungszentrum Jülich
Jülich, Germany
j.bae@fz-juelich.de

Dan Buca
Peter-Grünberg-Institut (PGI-9)
Forschungszentrum Jülich
Jülich, Germany
d.m.buca@fz-juelich.de

Yi Han
Peter-Grünberg-Institut (PGI-9)
Forschungszentrum Jülich
Jülich, Germany
y.han@fz-juelich.de

Jean-Michel Hartmann
CEA-LETI
MINATEC Campus and University of
Grenoble Alps
Grenoble, France
jean-michel.hartmann@cea.fr

Qing-Tai Zhao
Peter-Grünberg-Institut (PGI-9)
Forschungszentrum Jülich
Jülich, Germany
q.zhao@fz-juelich.de

Abstract—**Vertical GeSn gate-all-around (GAA) nanowire nMOSFETs fabricated using a top-down approach are presented. The devices are benchmarked with similar Ge and Ge/GeSn/Ge heterostructure devices to underline the great potential of GeSn for future nMOS devices. Device measurements are performed in the temperature range from 12 K to room temperature (RT, 300 K). At RT the all-GeSn n-MOSFETs show a subthreshold swing (SS) of ~120 mV/dec that decreases at cryogenic temperatures to a very steep 20mV/dec. The abrupt transition from subthreshold to on-state shows the suitability of GeSn alloys for cryogenic CMOS applications.**

Keywords—**Germanium-tin (GeSn) alloys, gate-all-around nanowires, MOSFET, mobility, cryogenic electronics,**

I. INTRODUCTION

Germanium-tin (GeSn) alloys have received a lot of attention in recent years, due to their outstanding properties [1-4]. Pure Ge has already been considered a viable alternative to Si CMOS, stemming from its increased hole carrier mobility compared to Si [5]. Alloying Ge with Sn results in the formation of an alloy which further enhances this advantage of Ge and additionally increases the electron mobility [6]. This is related to the most interesting property of GeSn: At Sn contents above ~8%, the bandgap turns from indirect to direct [7]. A direct bandgap, here at the Γ-point, enables a very low electron mass leading to more high-mobility electrons contributing to carrier transport, as opposed to the higher-mass-lower-mobility L-valley electrons, resulting in a strong increase of electron mobility. Moreover, being a Si-group element, GeSn retains the compatibility to conventional Si-CMOS platforms which makes it a strong contender for future nanoelectronic and optoelectronic devices.

The continuous scaling of transistors led to increasing demands regarding electrostatic gate control ability.

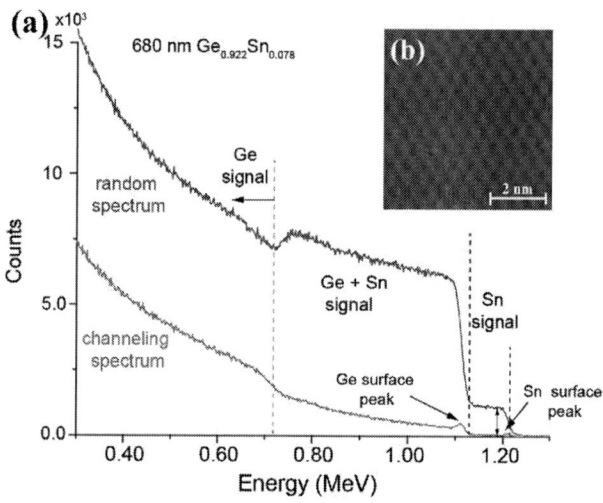

Fig. 1. (a) RBS/channeling measurement of the grown wafer. A Sn-content of ~7.8% is extracted from the data. (b) High resolution TEM image of the lattice demonstrating the high crystal quality.

Therefore, alternative device designs have been implemented in the past, like e. g. FinFETs [8, 9]. Recently, gate-all-around (GAA) nanowires have come into focus for further downsizing of devices. The architecture, that employs a gate that wraps around the nanowire, offers enhanced electrostatic control over the channel [10] and is therefore regarded as the path towards the ultimate CMOS device [11]. While GAA nanowire FETs can be fabricated in a horizontal or vertical configuration, the vertical design exhibits several advantages over the horizontal configuration, like potentially reduced power consumption [12], the use of epitaxial heterostructures with sharp interfaces and a decoupling of the gate length from the device footprint [13].

In this work we present vertical GAA GeSn nanowire n-MOSFETs fabricated with Si compatible top-down process.

978-1-6654-8498-5/22 $31.00 © 2022 IEEE

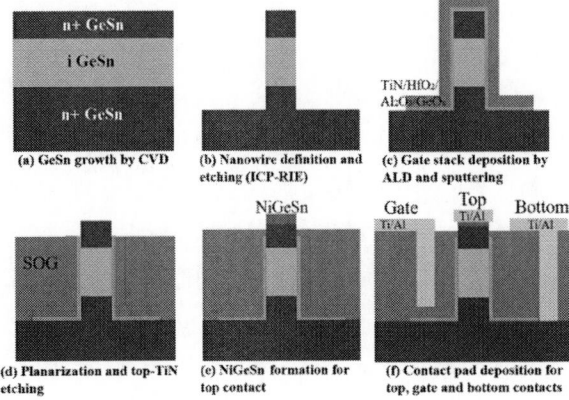

Fig. 2. Schematic showing the key steps of the device fabrication flow.

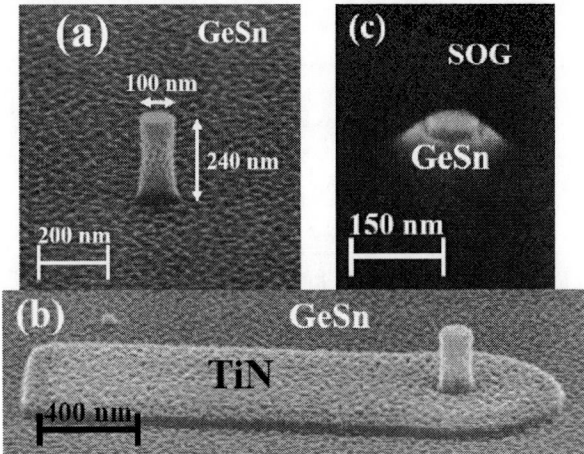

Fig.3. SEM images of the device at different stages of fabrication. (a) fabricated NW with a diameter of 100 nm, (b) After gate patterning, (c) After etchback for top contact deposition.

The characterization of the device shows an SS of 120 mV/dec at 300K, but reduced to 20 mV/dec at 12K. Most interestingly, the average subthreshold swing measured from the off state voltage to the threshold voltage is also very small, providing high potential for cryogenic CMOS applications.

II. EXPERIMENTAL

The GeSn layers were grown by chemical vapor deposition (CVD) on a Ge virtual substrate (VS). First, a thick GeSn buffer layer was grown to introduce strain relaxation. It offers a larger lattice for low strain epitaxy of an n^+-GeSn/i-GeSn/n^+-GeSn stack with thicknesses of 80 nm, 100 nm and 200 nm, respectively. The Sn content was kept constant at 7.8%, as confirmed by RBS measurements (Fig. 1 (a)). The channeling spectrum shows a very low yield of Sn, demonstrating very high single crystalline quality of the GeSn layer which is further confirmed by a TEM image of the GeSn layer in Fig. 1 (b) with no visible interface or doping induced dislocations. Electrochemical CV (ECV) measurements indicate a phosphorus (P) doping concentration of about 7×10^{19} cm^{-3} in the top and bottom layers (not shown).

Using a top-down fabrication process, as schematically shown in Fig. 2, vertical nanowire GAA nMOSFETs were fabricated. Nanowires with diameters between 50 nm, and 100 nm were defined using electron beam lithography and anisotropically etched in an ICP-RIE reactor using a Cl_2/Ar-plasma. Exemplarily, Fig. 3 (a) shows a nanowire with a diameter of 100 nm. The noticeable surface roughness is most likely caused by Sn-containing etch products with a low volatility, since the roughness is significantly lower on pure Ge samples. Using atomic layer deposition (ALD), 1 nm of Al_2O_3 was deposited, followed by a post-oxidation step in an O_2-plasma to form a layer of $GeSnO_x$ under the Al_2O_3. Such a process step has been proved to reduce the interface states density D_{it} [14-18], increasing the the gate control of the finished device. The gate stack dielectric was continued by a 7 nm thick HfO_2 layer deposited by ALD. The final equivalent oxide thickness (EOT) of the gate stack was ~2.9 nm. The metal gate of a 40 nm TiN layer deposited by RF sputtering was patterned (Fig. 3 (b)). Next, the sample was planarized using spin-on glass (SOG) with subsequent curing in a furnace in N_2-ambient. The TiN at the top of the nanowires was then removed by isotropic dry etching to expose the nanowire top. After a second planarization step (Fig. 3 (c)), Ni was deposited on top of the nanowires by sputtering, followed by a forming-gas annealing step to form a NiGeSn alloy. NiGe(Sn)-alloys have been shown to offer a lower contact resistivity on n-type Ge(Sn) compared to plain metal contacts without alloying [19-21]. Finally, contact windows for gate and bottom were etched and a Ti/Al stack was deposited for contact pads.

III. RESULTS AND DISCUSSION

Fig. 4 shows the transfer characteristics measured at room temperature for a vertical all-GeSn GAA nanowire n-FET with a NW diameter of 60 nm. The device was measured in two configurations: top NW contact as the drain (TCD), and top NW contact as the source (TCS). The subthreshold swing (SS) is ~120 mV/dec in TCD configuration and ~221 mV/dec in TCS configuration. This large discrepancy can be explained by the larger contact resistance at the top of the NWs due to the smaller contact size, which decreases the effective gate-source voltage. The reduced effective gate-source voltage is further proved by the curve shift to the right for the TCS configuration. The output characteristics of these devices (not shown here) demonstrate that the device exhibits Schottky characteristics. To improve this, higher doping in top and bottom layer of the structure should be introduced and the contact annealing process should be optimized with respect to annealing temperature and time.

Fig. 5 shows the transfer characteristics of the n-MOSFET measured at temperatures of 12.1 K, 60 K and 120 K in TCD configuration. Both ON and OFF-currents decrease with decreasing temperature. At 12.1 K the strong decrease of the OFF-current leads to a large improvement of the I_{ON}/I_{OFF}-current ratio to $>10^4$ for V_d=0.1 V and $>10^5$ for V_d=0.3 V, being only limited by the threshold current of the measurement equipment. The SS decreases with temperature (Fig. 6), following a linear trend, as it is predicted by the relation $SS \propto kT$. At room temperature the SS is around 120 mV/dec and reaches ~20 mV/dec at 12.1 K. The transition from the subthreshold to the threshold becomes very abrupt at cryogenic temperatures. This is extremely beneficial for device application in quantum computing at

978-1-6654-8498-5/22 $31.00 © 2022 IEEE

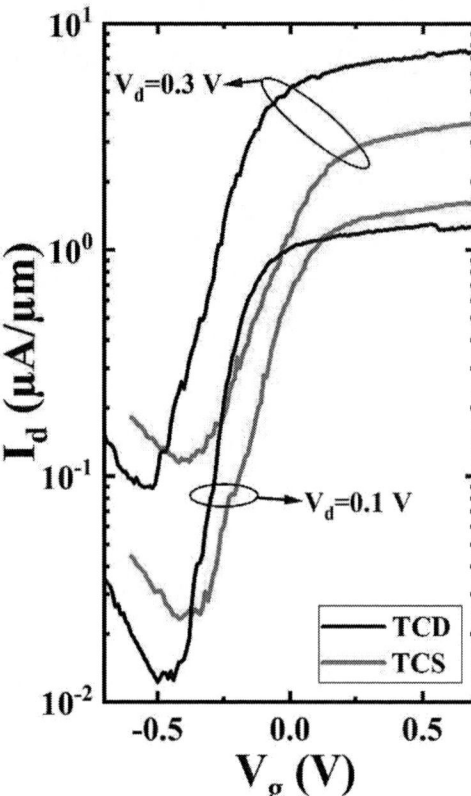

Fig. 4. Transfer characteristics of a device with a diameter of 60 nm. Measurement with top as drain (TCD) and top as source (TCS). The drain voltages were 0.1 V and 0.3 V.

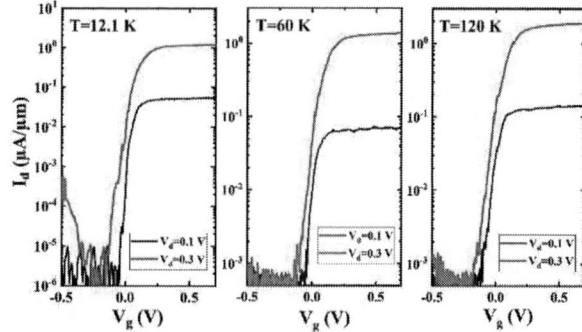

Fig. 5. Transfer characteristics of a GAA NW nFET measured at different temperatures. (a) 12.1 K, (b) 60 K, (c), 120 K.

resistance contacts on n-type GeSn, since a high doping level helps to overcome the Fermi level pinning which is strong in Ge-based materials. Furthermore, the device performance could be enhanced further by increasing the Sn content in the channel. A higher Sn content is associated with a higher carrier mobility, as the low-mass-high-mobility Γ-valley electrons contribute more and more strongly to the carrier transport. On the other hand, it can be an advantage to use a wider-bandgap material for source and drain, in order to minimize the GIDL and therefore the off-current, which enhances the I_{on}/I_{off} ratio. SiGeSn alloys, where a larger Si content widens the bandgap, could be a solution, which is under processing in our lab.

cryogenic temperatures since it decreases the voltage change necessary to lower the power consumption.

At low temperatures higher currents would be expected because of the increase of the carrier mobility due to less phonon scattering. The decrease of the ON-current at low temperatures shown in Fig.5 is explained by the non-optimized Ohmic characteristics of the contacts due to the relatively low doping in the source/drain GeSn layers. At low temperature, the partial freeze-out of dopants leads to a higher barrier for electron tunneling and less thermal energy for electrons to overcome the Schottky barrier, eventually leading to an increase of resistance and a decrease in current.

In comparison to Ge devices and devices based on a Ge/GeSn/Ge heterostructure (GeSn channel only), that were presented in a previous work [22], the all-GeSn device outperforms the Ge/GeSn/Ge devices with respect to the SS for a similar nanowire diameter: all-GeSn: ~120 mV/dec at 60 nm diameter; Ge/GeSn/Ge: ~136 mV/dec at 65 nm diameter. The all-Ge devices showed poor performance regardless of the nanowire diameter: ~136 mV/dec even at 25 nm NW diameter. These results demonstrate again the improvement that can be achieved by implementing GeSn as the channel material instead of Ge for n-type devices. It can be expected that decreasing the diameter of the GeSn devices can further improve the device performance.

An advantage that GeSn has over Ge is that the achievable *in-situ* doping concentrations are significantly higher. This is beneficial for the fabrication of low

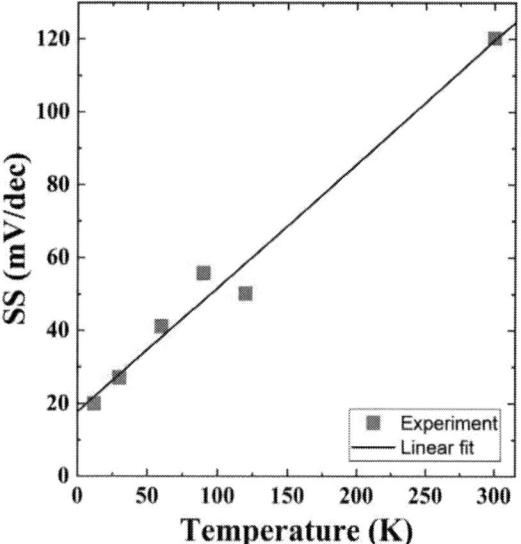

Fig. 6. SS of the device from Fig. 4 as a function of the temperature, showing a linear decrease with temperature.

IV. CONCLUSION

Vertical gate-all-around nanowire nFETs with GeSn alloy have been fabricated and characterized for the first time. The devices hold a strong potential for future CMOS technologies, illustrated by the lower subthreshold swing than the Ge counterpart. They appear to be especially suitable for cryogenic CMOS due to their low SS and extremely abrupt subthreshold/on-state transition. Further work will focus on optimization of the contact resistances to increase the on-current and on the improvement of the gate stack interface to further improve the SS. Furthermore, the influence of Sn-content on vertical gate-all-around device performance will be addressed.

ACKNOWLEDGMENT

This work is supported by the German BMBF project "SiGeSn NanoFETs".

REFERENCES

[1] S. Wirths, D. Buca , and S. Mantl, "Si-Ge-Sn alloysw: From growth to applications," Prog. Cryst. Growth Charact. Mater., vol. 62, pp. 1-39, 2016

[2] S. Zaima, O. Nakatsuka, N. Taoka, M. Kurosawa, W. Takeuchi, and M. Sakashita, "Growth and applications of GeSn-related group-IV semiconductor materials," Sci. Technol. Adv. Mater., vol. 16, no. 4, 043502, 2015

[3] J. Zheng, Z. Liu, C. Xue, C. Li, Y. Zuo, B. Cheng, and Q. Wang, "Recent progress in GeSn growth and GeSn-based photonic devices," J. Semicond., vol. 39, 061006, 2018

[4] R. Geiger, T. Zabel, and H. Sigg, „Group IV direct band gap photonics: methods, challenges, and opportunities," Front. Mater., vol. 2, art. 52, 2015

[5] R. Pillarisetty, "Academic and industry research progress in germanium nanodevices", Nature, vol. 479, pp. 324-328, 2011

[6] J. D. Sau and M. L. Cohen, "Possibility of increased mobility in Ge-Sn alloy system", Phys. Rev. B, vol. 75, 045208, 2007

[7] K. L. Low, Y. Yang, G. Han, W. Fan, and Y.-C. Yeo, "Electronic band structure and effective mass parameters of $Ge_{1-x}Sn_x$ alloys", J. Appl. Phys., vol. 112, 103715, 2012

[8] C. Auth et al. "A 22nm High Performance and Low-Power CMOS Technology Featuring Fully-Depleted Tri-Gate Transistors, Self-Aligned Contacts and High Density MIM Capacitors", 2012 Symposium of VLSI Technology, pp. 131-132

[9] C.-H. Jan et al. "A 22nm SoC Platform Technology Featuring 3-D Tri-Gate and High-k/Metal Gate, Optimized for Ultra Low Power, High Performance and High Density SoC Applications", IEEE International Electron Devices Meeting 2012, pp. 44-47

[10] C. P. Auth and J. D. Plummer, "Scaling Theory for Cylindrical, Fully-Depleted, Surrounding-Gate MOSFETs", IEEE Electron Device Lett., vol. 18, no. 2, pp. 74-76, 1997

[11] K. J. Kuhn, "Considerations for Ultimate CMOS Scaling", IEEE Trans. Electron Devices, vol. 59, pp. 1813-1828, 2012

[12] A. Veloso et al., "Vertical Nanowire FET Integration and Device Aspects", ECS Trans. Vol. 72, no. 4, pp. 31-42, 2016

[13] D. Yakimets et al., "Lateral versus Vertical Gate-all-around FETs for Beyond 7nm Technologies", 72nd Device Research Conference Digest, pp.133-134, 2014

[14] R. Zhang, X. Yu, M. Takenaka, and S. Takagi, "Physical Origins of High Normal Field Mobility Degradation in Ge p- and n-MOSFETs With GeO_x/Ge MOS Interfaces Fabricated by Plasma Postoxidation", IEEE Trans. Electron Devices, vol. 61, no. 7, pp. 2316-2323, 2014

[15] R. Zhang, P.-C. Huang, J.-C. Lin, M. Takenaka, and S. Takagi, "Physical Mechanism Determining Ge p- and n-MOSFETs Mobility in High N_s Region and Mobility Improvement by Atomically Flat GeO_x/Ge Interfaces", IEEE International Electron Devices Meeting 2012, pp. 371-374

[16] R. Zhang, J.-C. Lin, X. Yu, M. Takenaka, and S. Takagi, "Impact of Plasma Postoxidation Temperature on the Electrical Properties of Al_2O_3/GeO_x/Ge pMOSFETs and nMOSFETs", IEEE Trans. Electron Devices, vol. 61, no. 2, pp 416-422, 2014

[17] R. Zhang, P.-C. Huang J.-C. Lin, N. Takoa, M. Takenaka, and S. Takagi, "High-Mobility Ge p- and n-MOSFETs With 0.7-nm EOT Using HfO_2/Al_2O_3/GeO_x/Ge Gate Stacks Fabricated by Plasma Postoxidation", IEEE Trans. Electron Devices, vol. 60, no. 3, pp. 927-934 2013

[18] R. Zhang, X. Tang, X. Yu, J. Li, and Y. Zhao, "Aggressive EOT Scaling of Ge pMOSFETs With HfO_2/AlO_x/GeO_x Gate-Stacks Fabricated by Ozone Postoxidation", IEEE Electron Device Lett., vol. 37, no. 7, pp. 831-834, 2016

[19] S. Gaudet, C. Detavernier, A. J. Kellock, P. Desjardins, and C. Lavoie, „Thin film reaction of transition metals with germanium", J. Vac. Sci. Technol. A, vol 24, pp. 474-485, 2006

[20] K. Gallacher, P. Velha, D. J. Paul, I. MacLaren, M. Myronov, and D. R. Leadley, "Ohmic contacts to n-type germanium with low specific contact resistivity", App. Phys. Lett., vol. 100, 022113, 2012

[21] S. Wirths, R. Troitsch, G. Mussler, P. Zaumseil, J. M. Hartmann, T. Schroeder, S. Mantl, and D. Buca, „Ni(SiGeSn) Metal Contact Formation on Low Bandgap Strained (Si)Ge(Sn) Semiconductors", ECS Trans., vol. 64, no. 6, pp. 107-112, 2014

[22] M. Liu, "Ge(Sn)-Based Vertical Gate-all-around Nanowire MOSFETs and Inverters for Low Power Logic", Dissertation, RWTH Aachen, Germany, 2021.

978-1-6654-8498-5/22 $31.00 © 2022 IEEE

Defects Motion as the Key Source of Random Telegraph Noise Instability in Hafnium Oxide

Sara Vecchi, Paolo Pavan, Francesco Maria Puglisi

Dipartimento di Ingegneria "Enzo Ferrari", Via P. Vivarelli 10/1, 41125 – Modena (MO) - Italy
Università degli Studi di Modena e Reggio Emilia
Corresponding author email: sara.vecchi@unimore.it   phone: +39-059-2056320

Abstract – **Besides standard two- and multi-level Random Telegraph Noise (RTN), more complex cases of RTN are commonly reported which show peculiar current signal instabilities. The physical origin of such phenomena is typically traced back to the presence of metastable defects states, the Coulomb interaction between traps, and the possible interaction of hydrogen species with oxide defects. However, the effect of the motion of atomic species on RTN phenomena has never been brought to the picture, even though such a mechanism is extremely relevant for oxygen ions in HfO₂, e.g., it guarantees resistive switching in HfO₂ RRAM. In this paper, we demonstrate that complex RTN signals observed in experiments naturally emerge when considering the combination of the Coulomb field due to the trapped charge at defects together with their field-assisted motion. Strikingly, we demonstrate that multilevel RTN signals with high instability and complex time evolution, which are conventionally though to be caused by an intricate many-bodies problem involving several defects, can in fact result by the activity of one single defect drifting within the oxide.**

Keywords – MIM, HfO₂, RTN, TAT, Drift-Diffusion.

I. INTRODUCTION

Random Telegraph Noise (RTN) is well known to be one of the most prominent reliability concerns in electron devices. However, despite being intensively studied, some of its features (e.g., anomalous, temporary, coupled RTN, instabilities) still hinder its full physical understanding. The switching between two (or more) current levels typical of RTN is commonly attributed to charge trapping/de-trapping into/from defects [1], affecting the performance of nanoscale devices, e.g., FinFETs and Resistive Random Access Memory (RRAM) [2-5]. Many reports focus on the characterization of simple and stable two- or multi-level RTN signals, which can be exploited to retrieve information about the defects. However, complex and unstable RTN signals have been reported and tentatively associated to possible metastable states of defects that assist charge transport [1,2,6], Coulomb interactions between traps [2-4], and possible interaction with hydrogen species [6]. Yet, the possible role of defect motion has never been considered, even though it is known to rule over RRAMs physical operations [5,7], and is therefore expected to play a meaningful role. In this work, we implement for the first time Monte-Carlo physics-based simulations of HfO₂-based stacks that include trapping and defects motion together. We show that complex RTN signals (which are commonly recorded in Hf-based devices [8]) can be easily promoted by the motion of just one single defect, which

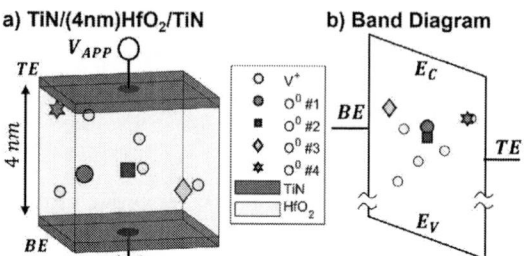

Fig. 1 – (a) MIM structure employed for tMC and kMC simulations and (b) its band diagram. Defects and materials parameters are taken from [2,9] and references therein.

can dramatically change the electrostatic scenario within the oxide over time causing RTN instabilities.

II. DEVICES AND SIMULATIONS

To evaluate the effect of defects motion (i.e., drift and diffusion) on RTN, we performed Monte-Carlo simulations on a TiN/(4nm)HfO₂/TiN cell with an area of 25 nm² (Fig. 1a). Oxygen vacancies (V^+, $N_T = 10^{20}$ cm⁻³) and oxygen ions (O^0) are included since they are the main observed defects in HfO₂ [9-11], both uniformly distributed in space and energy, as depicted in the relative band diagram (Fig. 1b). In MIM cells, the overall leakage current can be the result of many conduction mechanisms [9,12,13] (Fig. 2a) depending on the oxide thickness, applied voltage conditions, traps density and temperature profile. In our stack, we verified that in the explored range 0.45-0.75 V (that spans V_{DD} values forecasted by the IEEE IRDS roadmap for actual and next-gen scaled devices [14]) the leakage current is dominated by trap-assisted tunneling at V^+s (Fig. 2b). In fact, the latter are the defects most involved in charge transport [2,3] and are commonly called *fast* traps since they quickly capture and emit charge carriers (especially compared to O^0s), promoting also the typical white noise (Fig. 3). Even though (*slow*) O^0s do not contribute to leakage current, trapped charge at such defects can locally perturb the potential within the oxide and in turn modulate the overall leakage current promoting RTN. Besides the electrostatic interaction between fast (V^+) and slow (O^0) defects, a possible explanation for RTN signals is the presence of possible metastable states for V^+ [1]. Though here we do not consider this phenomenon to model RTN, the effects of defects motion are expected to be similar also for defects with metastable states. Transient (tMC) and kinetic Monte-Carlo

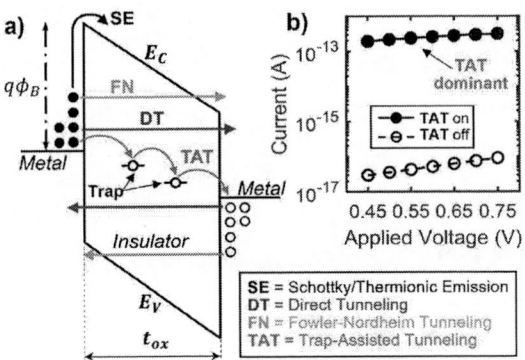

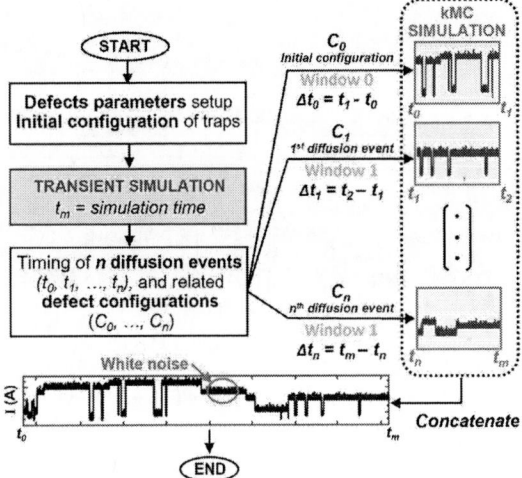

Fig. 2 – (a) Representation of the main conduction mechanisms in MIM cells [9,12,13]. (b) Leakage through the cell w/ all conduction mechanisms included (filled symbols) and w/o TAT (open symbols).

(kMC) 3D simulations were run using Ginestra® [15] to simulate RTN traces including for the first time the drift-diffusion mechanisms for defect species (Fig. 3). The main defects and material parameters used in simulations are shown in Tab. I, and are fully consistent with DFT values [16,17] that allow reproducing in detail the switching characteristics of RRAM devices and their retention [5]. We consider Schottky and thermionic emission, direct (WKB approximation), trap-assisted (TAT - including trap-to-trap contribution), and Fowler-Nordheim tunneling, as well as the trapped charge term in the Poisson equation [18]. To simulate defects drift-diffusion we employed tMC simulations, with a simulation time t_m = 200 s. In tMC simulations the local field and temperature profiles are used to evaluate the steady-state occupancy of all defects and, therefore, the trapped charge contribution to the overall electric field distribution. However, defect motion kinetics is included in a fully MC fashion. In HfO$_2$, O ions have been shown both experimentally and by simulations and ab-initio studies [16,17] to move in the lattice driven by electric field and temperature (V$^+$s motion is negligible due to the much higher diffusion barrier [16]) and this mechanism is crucial in explaining resistive switching in RRAMs [5]. O^{0}s move via interstitial jumps [17] and the rate of the process (thermally- and field-activated) [5] is:

$$R_D(x,y,z) = v_0 \cdot exp\left(-\frac{E_{A,D} - k_D \cdot F_{EFF}(x,y,z)}{k_B \cdot T(x,y,z)}\right) \quad (1)$$

F_{EFF} is the electric field at the defect location, and the remaining parameters and their relative values are specified in Tab. I. At each instant of time, R_D is used to probabilistically determine if a defect motion event is to happen. We progressively labeled each instant of time at which any defect motion is detected as $t_1, ..., t_n$. Therefore, we identified n time intervals (windows), each characterized by given duration ($\Delta t_1, ..., \Delta t_n$) and by a specific spatial configuration of defects ($C_1, ..., C_n$). To fully capture the stochastic nature of trapping and detrapping at defects (and of RTN), each window is again simulated separately with a dedicated kMC simulation in which the kinetics of defect motion is not included (as it is not

Fig. 3 – Flowchart of the proposed simulation approach combining tMC and kMC simulations to include defect motion effect on RTN.

Tab. I – Main defects and materials parameters used in simulations taken from [5,16,17] and reference therein.

Symbol	Descriptions	V$^+$	O^0	HfO$_2$
E_{REL}	Relaxation energy	1.19 eV	3 ± 0.3 eV	-
E_{TH}	Thermal ionization energy	2 ± 0.5 eV	2 ± 0.2 eV	-
σ	Capture cross section	10-14 cm^2	3·10^{-16} cm^2	-
$E_{A,D}$	Activation energy	1.5 eV	0.8 eV	-
k_D	Field acceleration factor	4 eA	9 eA	-
v_0	Frequency prefactor	5·10^{13} Hz	1·10^{13} Hz	-
k	Relative dielectric constant	-	-	21
k_{TH}	Thermal conductivity	-	-	5·10^{-3} W/cmK
E_G	Band gap	-	-	5.8 eV

needed, since within each window the relative defects configuration persists without any new diffusion event) while the MC engine captures the full stochastic kinetics of charge trapping and de-trapping for precise RTN simulation. Finally, all kMC windows traces are concatenated resulting in the complete RTN including charge kinetics and defects motion.

III. RESULTS

In Fig. 4 we shown the results of simulations at 0.6 V. First, we performed a kMC simulation of the MIM depicted in Fig. 1a, that includes four O^{0}s and a handful of V$^+$s, in which defects are held in their initial places without the possibility of moving. In this case (Fig. 4a), we observed a 4-level RTN which corresponds to the modulation of the TAT (driven by V$^+$s, as depicted in Fig. 2b) induced by charge capture and emission at O^0 #1 and #2, that stochastically change their occupancy as represented in the inset of Fig. 4a (yellow band). The contribution of O^0#3 and #4 is not appreciable within our observation window (1 ms - 1000 s), meaning that their average τ_c and/or τ_e are either smaller than our sampling time (1 ms) or larger than our simulation time (1000 s) [18]. Nevertheless, the RTN we observe naturally includes some inter-defect coupling, since O^0 #1 and #2 electrostatically affect each other [18]. For instance, the average value of τ_c

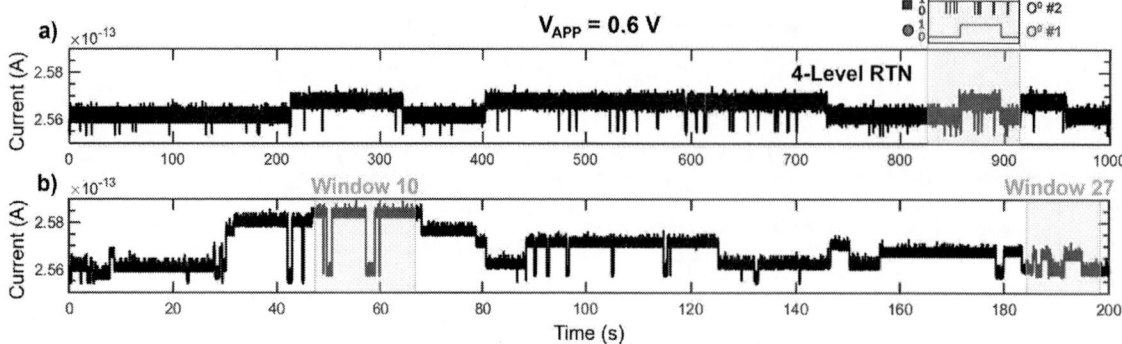

Fig. 4 – RTN current signals without (a) and with (b) considering defects motion. (a) results from a 1ks long kMC simulation. (b) is obtained using the approach in Fig. 3. (b) Yellow bands highlight time intervals in which the 10th (window 10) and the 27th (window 27) defect spatial configurations persist (defect motion happens at the edge between adjacent windows).

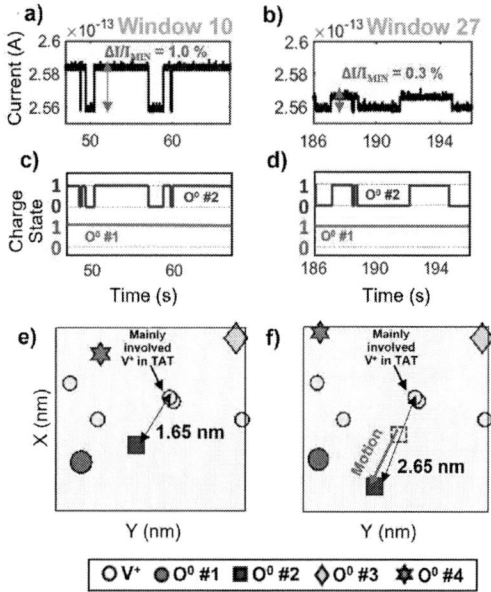

Fig. 5 – RTN signals which characterize window 10 (a) and window 27 (b), along with their respective ions charge states (c) and (d). XY vistas of the defects spatial arrangement following the 10th (e) and the 27th (f) diffusion event (i.e., window 10 and window 27).

and/or τ_e of O^0 #2 is different when O^0 #1 is neutral or charged, being such defects spatially close to each other. Also, in the literature, a handful of papers [2,3,19,20] put in the spotlight the relevance of the Coulomb interactions between traps to explain complicated RTN signals. However, the simulated RTN (Fig. 4a) shows no instabilities since, although defects can interact with each other and modulate TAT, the number of (all) possible defects charge configurations is limited [18] and cannot change over time since defects cannot change their spatial location. Expectedly, when defects motion is included (Fig. 4b) the scenario gets more complicated. In fact, RTN properties (amplitude and dwell times) strongly depend on the mutual distance between traps, which can change over time when considering the defects motion. As such, the

interdependencies between defects will change in a complex fashion according to the spatial traps' configuration evolution. To the best of the authors' knowledge, defects motion has never been considered to explain RTN instabilities, even if defects are known to move within the oxide, and their key role in resistive switching is well established [5].

IV. DISCUSSION: THE DRAMATIC IMPACT OF IONS DRIFT

Conventionally, the complex multilevel RTN trace of Fig. 4b is reasonably associated to the overlapped action of many slow (O^0, in the case of HfO_2) defects, mathematically represented by multiple Markov Chains [21], since the current trace strongly changes its properties (e.g., ΔI) over time passing through many distinct current levels. From an accurate analysis of such RTN signal, we have identified 28 O^0 diffusion events occurring in the whole simulation time (t_m = 200 s), each determining a certain configuration of defects persisting for a given window of time and associated with a specific local RTN current behavior. In Fig. 5, we report the RTN signals observed in window 10 and 27 (highlighted by yellow bands in Fig. 4b), in which the change in the RTN properties is evident, e.g., the relative amplitudes ($\Delta I/I_{MIN}$, Fig. 5a,b), as well as the capture/emission times, which characterize the RTN traces are very different. The comparison of the $\Delta I/I_{MIN}$ values may point to the fact that different defects are involved in these two current windows, i.e., the 2-level RTN characterizing window 10 would be promoted by one defect which is different from the one causing the 2-level RTN of window 27. Alternatively, the signal of window 27 may show different properties as compared to that in window 10 due to defect's interaction with atomic hydrogen [6]. Nevertheless, for both windows, the time-domain analysis of the charge states of O^0 #1 and #2 (Fig. 5c,d) reveals that both current traces, even if very different to each other, are caused by the same O^0 (#2), while #1 is always charged. In fact, throughout the whole simulation time (1 ms - 200 s) O^0 #1 rarely moves over time, and when it moves it negligibly modulates the current, in turn without promoting well-appreciable RTN signals. From the evaluation of defects spatial arrangements at window 10 (Fig. 5e) and at window 27 (Fig. 5f), we can observe that when O^0 #2 is closer to the V^+

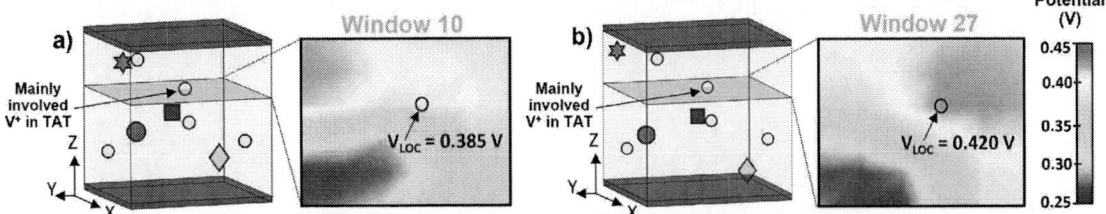

Fig. 6 – 2D maps of the potential along the horizontal planes which cut the V^+ driving the largest fraction of the overall TAT current during windows 10 (a) and 27 (b). Both maps represent the distribution of the potential in the oxide when O^0 #1 and #2 are charged. The local potential at the V^+ highlighted in the figure increases by ≈ 10 % if O^0 #2 is further away (i.e., in window 27 as compared to window 10, see also Fig. 5).

that is responsible for most of the overall charge transport, Fig 5e, the $\Delta I/I_{MIN}$ is higher compared to the case in which O^0 #2 is farther away (Fig. 5f). When O^0 #2 traps an electron it perturbs the local potential (V_{LOC}) at V^+ modulating the current pathway. Obviously, the closer O^0 #2 is to such V^+, the stronger the perturbation at V^+ gets when O^0 #2 traps an electron. To convey the idea, in Fig. 6 we depicted the 2D maps of the potential along the horizontal planes passing through the V^+, for window 10 (Fig. 6a) and 27 (Fig. 6b), when O^0 #1 and #2 are both charged (charge state = 1). As depicted, when O^0 #2 is close to such V^+ (Fig. 5e, window 10), V_{LOC} at V^+ is lower compared to the case in which the defects are more distant from one another (Fig. 6b), where V_{LOC} increases by about 10 %. Nevertheless, even if the increment of V_{LOC} at such V^+ is mild, it can cause very high perturbations of the RTN signal. The remarkable result of this study is that the highly complex RTN trace in Fig. 4b, commonly observed in experiments and difficult to analyze, as it continuously changes its properties over time showing high instability, is not due to the concurrent action of many slow defects as typically thought. Instead, *it can be elegantly explained by trapping/de-trapping at an individual slow defect (O^0 #2) that moves over time in the dielectric under the action of the local field, in turn perturbed by trapped charge at defects themselves, naturally resulting in a complex electrodynamic scenario and, thus, a complex RTN.* This underlines the importance of defects motion to fully explain complex RTN signals often observed in experiments.

V. CONCLUSIONS

In this work, we analyzed the often-overlooked impact of defects motion in HfO$_2$ on RTN. Since the defects move within the oxide, the (already complicated) electrostatic scenario changes over time promoting RTN signal instabilities. However, we shown that the often observed highly complex multilevel RTNs can originate from one single slow defect moving over time within the oxide, changing the local potential at V^+s and causing instabilities in the current and related RTN.

ACKNOWLEDGMENT

The authors acknowledge Applied Materials Italy for their support with Ginestra® device simulation software [15].

REFERENCES

[1] T. Grasser, "Stochastic charge trapping in oxides: From random telegraph noise to bias temperature instabilities," *Microelectron. Reliab.*, vol.52, no.1, pp. 39–70, 2012.
[2] F. M. Puglisi, L. Larcher, A. Padovani, and P. Pavan, "A complete statistical investigation of RTN in HfO2-based RRAM in high resistive state," *IEEE Trans. Electron Devices*, vol. 62, no. 8, pp. 2606–2613, 2015.
[3] F. M. Puglisi, L. Larcher, A. Padovani, and P. Pavan, "Anomalous random telegraph noise and temporary phenomena in resistive random access memory," *Solid. State. Electron.*, vol. 125, pp. 204–213, 2016.
[4] F. M. Puglisi, P. Pavan, L. Vandelli, A. Padovani, M. Bertocchi, and L. Larcher, "A microscopic physical description of RTN current fluctuations in HfOx RRAM," *IEEE Int. Reliab. Phys. Symp. Proc.*, vol. 2015-May, pp. 5B51-5B56, 2015.
[5] A. Padovani, L. Larcher, O. Pirrotta, L. Vandelli, and G. Bersuker, "Microscopic modeling of HfOx RRAM operations: From forming to switching," *IEEE Trans. Electron Devices*, vol. 62, no. 6, pp. 1998–2006, 2015.
[6] Y. Wimmer, A. M. El-Sayed, W. Gös, T. Grasser, and A. L. Shluger, "Role of hydrogen in volatile behaviour of defects in SiO2-based electronic devices," *Proc. R. Soc. A Math. Phys. Eng. Sci.*, vol. 472, no. 2190, 2016.
[7] D. Ielmini, "Modeling the Universal Set / Reset Characteristics of Filament Growth," *IEEE Trans. Electron Devices*, vol. 58, no. 12, pp. 1–9, 2011.
[8] A. Ranjan, N. Raghavan, K. Shubhakar, S. J. O'Shen, K. L. Pey, "Random Telegraph Noise Nano-spectroscopy in High-k Dielectrics Using Scanning Probe Microscopy Techniques," in *Noise in Nanoscale Semiconductor Devices*, 2020, pp. 417–440.
[9] L. Vandelli, A. Padovani, L. Larcher, R. G. Southwick, W. B. Knowlton, and G. Bersuker, "A physical model of the temperature dependence of the current through SiO2/HfO2 stacks," *IEEE Trans. Electron Devices*, vol. 58, no. 9, pp. 2878–2887, 2011.
[10] G. Bersuker *et al.*, "Metal oxide resistive memory switching mechanism based on conductive filament properties," *J. Appl. Phys.*, vol. 110, no.12, 2011.
[11] A. Padovani, L. Larcher, G. Bersuker, and P. Pavan, "Charge transport and degradation in HfO2 and HfOx dielectrics," *IEEE Electron Device Lett.*, vol. 34, no. 5, pp. 680–682, 2013.
[12] L. Larcher, "Statistical simulation of leakage currents in MOS and flash memory devices with a new multiphonon trap-assisted tunneling model," *IEEE Trans. Electron Devices*, vol. 50, no. 5, pp. 1246–1253, 2003.
[13] F. C. Chiu, "A review on conduction mechanisms in dielectric films," *Adv. Mater. Sci. Eng.*, vol. 2014, 2014.
[14] IEEE, "IRDS International Roadmap for Devices and Systems," 2021.
[15] "https://www.appliedmaterials.com/products/applied-mdlx-ginestra-simulation-software." .
[16] N. Capron, P. Broqvist, and A. Pasquarello, "Migration of oxygen vacancy in HfO2 and across the HfO 2/SiO2 interface: A first-principles investigation," *Appl. Phys. Lett.*, vol. 91, no. 19, pp. 89–92, 2007.
[17] A. S. Foster, A. L. Shluger, and R. M. Nieminen, "Mechanism of Interstitial Oxygen Diffusion in Hafnia," *Phys. Rev. Lett.*, vol. 89, no. 22, 2002.
[18] S. Vecchi, P. Pavan, and F. M. Puglisi, "The Relevance of Trapped Charge for Leakage and Random Telegraph Noise Phenomena," *IEEE Int. Reliab. Phys. Symp.*, pp. 4–9, 2022.
[19] P. Hao, D. Mao, R. Wang, S. Guo, P. Ren, and R. Huang, "On the frequency dependence of oxide trap coupling in nanoscale MOSFETs: Understanding based on complete 4-state trap model," *2016 13th IEEE Int. Conf. Solid-State Integr. Circuit Technol. ICSICT 2016 - Proc.*, no. 1, 2017.
[20] S. Guo, R. Wang, D. Mao, Y. Wang, and R. Huang, "Anomalous random telegraph noise in nanoscale transistors as direct evidence of two metastable states of oxide traps," *Sci. Rep.*, vol. 7, no. 1, pp. 1–6, 2017.
[21] F.M. Puglisi, A. Padovani, L. Larcher, P. Pavan, "Random Telegraph Noise: Measurement, Data Analysis, and Interpretation", *2017 IEEE 24th International Symp. on the Physical and Failure Anaysis of Integr. Circuits*.

A Novel Temperature Estimation Technique Exploiting Carrier Emission from Buffer Traps

Marcello Cioni[1,†], Nicolò Zagni[1] and Alessandro Chini[1]

1. Dipartimento di Ingegneria "Enzo Ferrari", Via P. Vivarelli 10, 41125 Modena, Italy,
University of Modena and Reggio Emilia (UNIMORE)

†e-mail: marcello.cioni@unimore.it

Abstract—We propose a novel technique for temperature estimation in electron devices based on the mutual correlation between emission time constant from traps (τ) and temperature (T). Arrhenius equation is employed as the physical model relating τ and T. The reference system used to present the technique is AlGaN/GaN high electron mobility transistors (HEMTs) with Fe-doping in the buffer. Drain Current Transients (DCTs) are used for extracting the emission time constant (τ) from Fe traps and non-linear regression through Trust Region Reflective (TRR) optimization algorithm is used to learn the model parameters from data and infer device temperature. Electro-thermal device simulations are employed for validating the proposed technique, showing that this method is able to provide an improved accuracy with respect to conventional electrical techniques (e.g., McAlister method) promoting it as a valid alternative to state of-the-art optical techniques in GaN HEMTs.

Keywords—*Temperature Estimation, Electro-Thermal Simulations, DCTs, GaN HEMTs, Trap Emission.*

I. INTRODUCTION

One of the most pressing issues causing performance degradation in semiconductor devices is the amount of generated heat due to high power dissipation causing self-heating effect (SHE). Specifically, in high power/speed circuits, operation at very high-power densities exacerbates the problem of Self Heating [1]. Temperature estimation in semiconductor devices is therefore of paramount importance to accurately characterize device behavior as well as its safe operating area.

Several techniques (electrical and optical) for estimating device temperature have been proposed so far [2], each of them presenting both advantages and limitations. While electrical techniques [3] offer versatility and ease of implementation – at the expense of accuracy [4] – optical methods [5] generally have better spatial resolution and accuracy [6] but require specific sample preparation and therefore cannot be used directly on transistor structures and/or on packaged devices [7]. These limitations call for different approaches to estimate the device channel temperature.

To this end, we propose a novel electrical technique – which is both as versatile as DC ones and as accurate as state-of-the-art optical methods – based on the indirect temperature estimation through the characterization of charge emission from traps. Here, we focus our analysis on Fe-doped AlGaN/GaN high-electron mobility transistors (HEMTs), a very popular technology for power/RF applications thanks to

TABLE I
PROS & CONS OF TEMPERATURE ESTIMATION TECHNIQUES

	Technique	PROs	CONs
ELECTRICAL METHODS	McAlister	•Simplicity	•Low accuracy
	Gate Resistance Temperature Detector (RTD)	•Accuracy	•Requires External Circuitry
	Trap Emission (This Work)	•Simplicity •Accuracy •Suitable on Wafers and Packaged Devices	•Requires trap emission characterization
OPTICAL METHODS	Infrared Thermography	•Simple Implementation •Short Characterization Time	•Low Resolution •Low Accuracy
	Raman Microscopy	•Accuracy •Spatial Resolution	•Requires Sample Preparation •No sensing under metals

its unmatched performance in terms of both power and speed. Traps in these devices are associated to Fe-dopants intentionally introduced in the GaN buffer for reducing leakage current, whose occupation dynamics is affected by T_{peak}. In fact, these traps are located near the gate edge towards the drain terminal of the device [8], which is the region where temperature peaks [9]. Accordingly, the temperature (T_{Fe}) activating charge emission from Fe-traps is fairly close to the hotspot, suggesting that the thermally activated emission can be used as an indirect measurement of T_{peak} in the device.

To validate this method, we perform two-dimensional electro-thermal device simulations to extract Drain Current Transients (DCTs) [10] at different (i) base-plate temperatures (T_{BP}) and (ii) bias conditions. The data is then used to extract the trap emission time constant (τ) allowing to estimate device temperature from the τ-T relationship described by the Arrhenius equation [11]. Our results indicate that the proposed method can provide an improved accuracy with respect to the conventional McAlister method, while being comparable with the one of state-of-the-art optical techniques. The main advantages and disadvantages of the proposed technique compared to the conventional ones are summarized in Tab. I.

978-1-6654-8498-5/22 $31.00 © 2022 IEEE

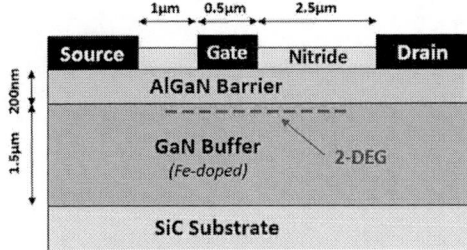

Fig. 1. Schematic cross-section of the simulated devices. 2-DEG stands for two-dimensional electron gas.

II. SIMULATED DEVICE DESCRIPTION

Electro thermal device simulations were performed in DESSIS-ISE. Simulated devices were single heterojunction AlGaN/GaN HEMTs featuring a SiN passivation and a Silicon Carbide (SiC) substrate (see Fig. 1). GaN Buffer was iron (Fe) doped to obtain a semi-insulating layer. Fe trap concentration is constant (10^{18} cm^{-3}) up to 0.6 µm from AlGaN/GaN interface, above which concentration exponentially decays with slope of 1 dec/0.4 µm. The presence of Fe-traps in the buffer layer, introduced by Fe dopants, was simulated by considering a deep-acceptor trap located 0.56 eV below the conduction band edge, consistent with the energy level reported for Fe-traps in GaN [12].

III. PROPOSED METHOD

A. Model definition

Fe-related Buffer traps behave as deep acceptor states located 0.5-0.6 eV below the GaN conduction band edge (E_C) [12] and capture electrons when the device is biased at relatively large drain voltage (V_{DS}) [12]. When this bias is removed, captured electrons are emitted to E_C through a thermally activated emission process characterized by an emission time constant (τ). This parameter can be directly extracted by means of DCTs and strongly depends on T. As such, the τ-T relationship can be used to indirectly extract temperature from DCTs. The mutual correlation between τ and temperature (T) is described by the Arrhenius equation [12]:

$$\tau = \frac{1}{T^2} \times \exp\left(\frac{qE_A}{kT} + C\right) \qquad (1)$$

where k is Boltzmann's constant, q is the elementary electron charge, E_A is the trap activation energy, C is a parameter proportional to the trap cross section (σ), and T (expressed in Kelvin) is the temperature experienced by deep levels. To a first approximation, temperature in the device can be expressed as:

$$T = T_{BP} + R_{TH} \times P_D \qquad (2)$$

where T_{BP} is the base-plate temperature, R_{TH} is the thermal resistance and P_D is the dissipated power. R_{TH} is a temperature-dependent parameter, due to the thermal conductivity dependence on temperature [13]:

$$R_{TH}(T_{BP}) = R_{TH\_300K} \times \left(\frac{T_{BP}}{300\,K}\right)^\alpha. \qquad (3)$$

Thus, by substituting (2), (3) in (1) we obtain a non-linear model for τ with four unknowns (E_A, C, R_{TH_300K}, α). In order to estimate T, these unknowns must be extracted.

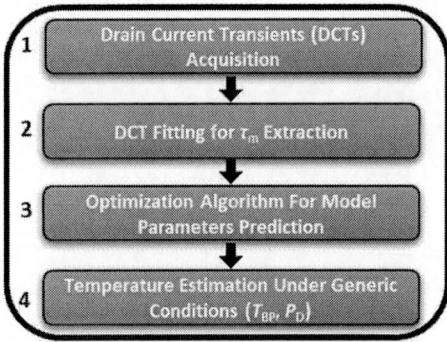

Fig. 2. Schematic representation of the proposed technique for temperature estimation.

B. Procedure description

The procedure describing the proposed method is schematically depicted in Fig. 2 and described in detail as follows.

1) A dataset of emission time constants representing the Fe-traps behavior under different thermal conditions is built by simulating DCTs (see Fig. 3) at several base plate temperatures (T_{BP}) and quiescent bias conditions ($V_{DS,q}$, $I_{D,Q}$) or, equivalently, P_D (since $P_D = V_{DS,q} \times I_{D,Q}$). Moreover, we set the trap filling bias ($V_{DS,fill}$, see Fig. 3) such that comparable DCT amplitudes are obtained in each condition. This is important for consistency, since only DCTs featuring similar trapped charge variation (ΔQ) are used [12].

2) Once acquired, DCTs are fitted to extract the emission time constant (τ_m) in correspondence of the peak in the dI_D/dlog$_{10}t$ signal. Thus, for each DCT, we have one 'observable' (τ_m) and two thermally related features (T_{BP} and P_D).

3) An optimization algorithm (e.g., Trust Region Reflective (TRR) [14]) is applied for non-linear least squares minimization (NLLSQ) of the residuals between τ_m and the τ determined from eqs. (1)-(3) (i.e., τ_p). The least squares problem defines the cost function (CF) which has to be minimized and the optimal parameters are those minimizing this function. Actually, in CF, we consider the logarithm of τ_m and τ_p in order to uniformly weighting both short and long emission times. Thus, CF is defined as:

$$CF = \frac{1}{2}\sum_{i=0}^{n}\left[\log_{10}\left(\frac{\tau_{p,i}}{\tau_{m,i}}\right)\right]^2 \qquad (4)$$

where, $\tau_{p,i}$ is the i-th emission time constant predicted by the model for $T_{BP,i}$ and $P_{D,i}$; $\tau_{m,i}$ is the i-th emission time constant extracted from DCT for the same $T_{BP,i}$ and $P_{D,i}$.

4) After optimization, the extracted parameters can be used in equations (1)-(3) to infer the device temperature for an arbitrary (T_{BP}, P_D) condition.

IV. SIMULATION RESULTS

Emission process from Fe-traps are characterized by means of DCTs [10] simulated with the waveforms shown in Fig. 3. Simulated device is first biased in the on-state to define a quiescent point (V_{GS}; $V_{DS,q}$; $I_{D,q}$) and guaranteeing thermal equilibrium. A short (t_{fill}=1 µs) trap filling pulse (V_{GS}; $V_{DS,fill}$) is then applied to induce electron capture by Fe-related buffer traps. After t_{fill}, the V_{DS} is set back to its

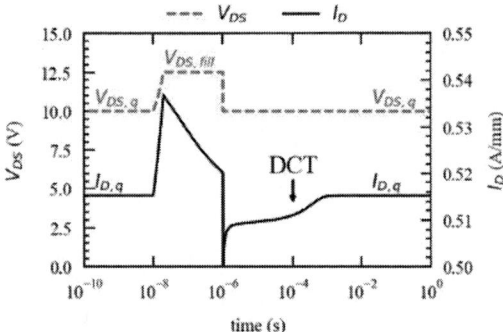

Fig. 3. Typical V_{DS} and I_D during simulation of DCTs characterization in drain lag configuration.

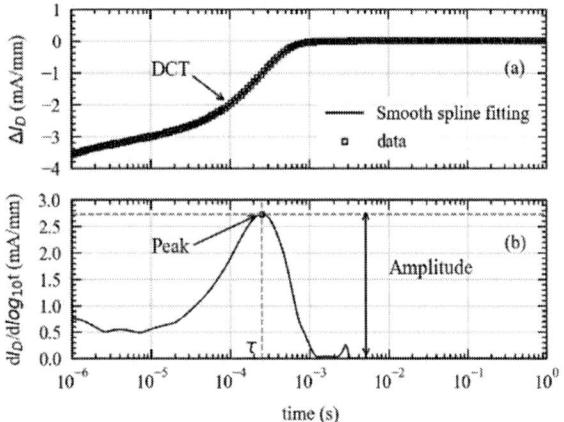

Fig. 4. (a) DCTs simulated at $V_{DS,q}$= 10 V (T_{BP} = 300 K) $I_{D,q}$= 0.5 A/mm and (b) corresponding $dI_D/dlog_{10}t$. Current variation (ΔI_D) is with respect to the steady state current ($I_{D,q}$). τ and amplitude are extracted in correspondence of the peak in the $dI_D/dlog_{10}t$ signal.

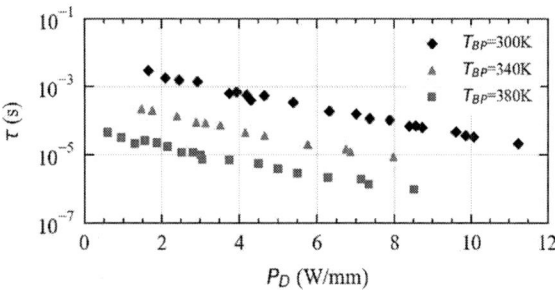

Fig. 5. τ vs P_D extracted from DCTs.

quiescent value and the drain current recovery transient is monitored for several decades. An example of simulated DCT is shown in Fig. 4(a) from which we observe a single and well-defined transient, yielding a single peak in the corresponding $dI_D/dlog_{10}t$ signal (see Fig. 4 (b)). To determine the time constant of the emission process (τ), DCTs are fitted by means of smooth spline functions and τ is extracted in correspondence of the peak in the ($dI_D/dlog_{10}t$) signal [12].

In Fig. 5 we report the emission time constants extracted from simulated DCTs acquired at different T_{BP} and P_D. In Fig. 5 we can see three well-spaced τ vs P_D curves which correspond to the three different T_{BP} levels considered. Moreover, we observe that τ reduces with either increasing T_{BP} or increasing P_D. These dependencies are expected, since both are a direct consequence of thermal dependent emission form Fe-traps [12]. In fact, an increase in T (or P_D) determines

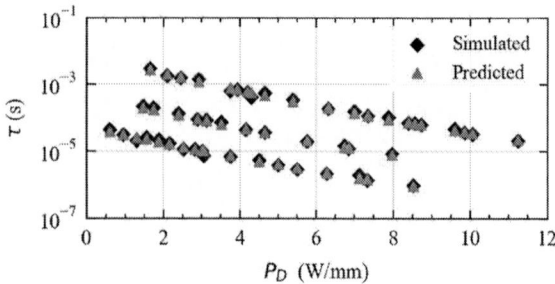

Fig. 6. Comparison between simulated and predicted τ vs P_D.

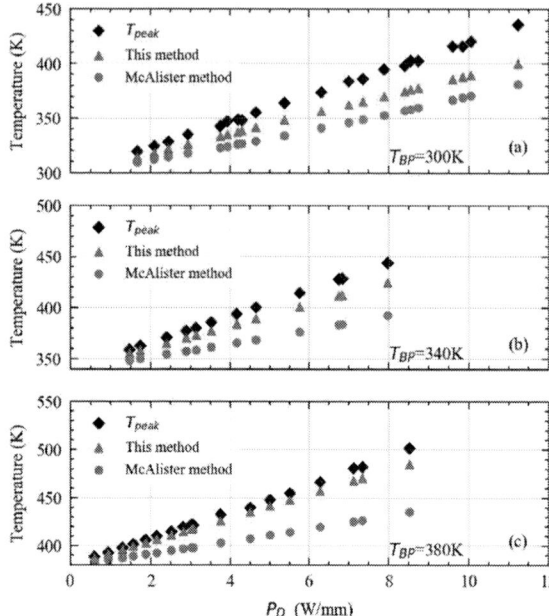

Fig. 7. Comparison between the peak temperature obtained from simulations and the one estimated through the McAlister's method and the proposed technique for three different T_{BP}.

a speed-up in the thermally activated emission process, leading to reduced τ [15].

V. TEMPERATURE ESTIMATION AND DISCUSSION

We thus employ the TRR optimization algorithm (implemented in Python) via a non-linear regression process. Fig. 6 shows the comparison between the emission time constants extracted from DCTs and those predicted by the model after optimization. A reasonable agreement is observed, meaning that the model parameters (minimizing CF) are a good estimate of the actual ones.

The parameters extracted with the TRR algorithm are now employed for temperature estimation. Temperature of the simulated device was also estimated by means of the McAlister method [3]. This way we provide a means of comparison for assessing the accuracy of the proposed technique. Fig. 7 shows the actual peak temperature (T_{peak}) provided by the simulator and the corresponding temperature estimated with the McAlister and the proposed method.

The McAlister method is known to significantly underestimate the peak temperature [4] and this is confirmed by the results obtained in this work. On the other hand, the proposed method allows to estimate a temperature which is closer to the peak one. However, even with the proposed method, we have an underestimation of T_{peak}. This is due to the finite distance between the ionized Fe-traps concentration

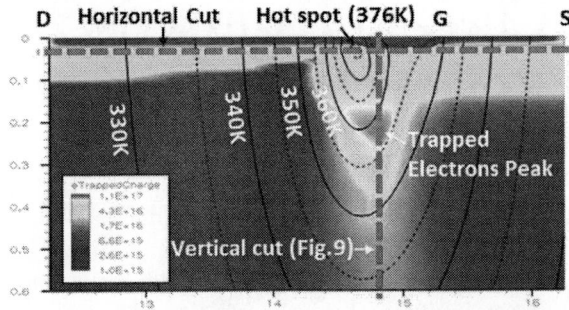

Fig. 8. Contour Plot of temperature and trapped charge concentration in the device (P_D = 6.5 W/mm, T_{BP} = 300 K). We highlight horizontal and vertical cuts along which temperature and trapped charge profiles are evaluated.

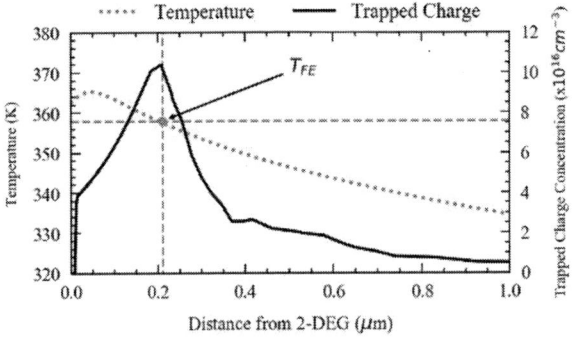

Fig. 9. Temperature and Trapped charge profiles as function of distance from the 2-DEG (P_D = 6.5 W/mm, T_{BP} = 300 K) along vertical cut (see Fig. 8).

peak and the temperature hotspot, which makes T_{Fe} slightly lower than the actual T_{peak}. This can be appreciated by comparing the temperature (T) and trapped charge (N_{FE}^-) distributions in the device's active region (see Fig. 8).

As expected, the largest concentration of trapped charge is located under the gate edge towards the drain terminal of the device [9]. However, the results in Fig. 8 highlight that the peak concentration of ionized Fe-traps is not exactly aligned to the point of maximum temperature, as shown also in Fig. 9 by the temperature and trapped charge concentration profiles along the vertical cut aligned with the peak trapped charge concentration coordinate, see Fig. 8. The misalignment between temperature and ionized trap concentration peaks shown in Fig. 8, is the reason for the error in T_{peak} estimation. In fact, the actual temperature experienced by Fe-traps in the point of peak ionized trap concentration (i.e., T_{Fe}= 358 K) is ~18 K lower than T_{peak} = 376 K (T_{BP} = 300 K, P_D = 6.5 W/mm). This error is consistent with the estimation error observed in Fig. 7, indicating that the proposed technique accurately measures the temperature in correspondence of the Fe-traps location (T_{Fe}). This suggests that, similarly to μ-Raman technique, numerical simulations may be used to correct T_{est} (~T_{FE}) and find a better estimate for T_{peak} [16], further reducing the estimation error.

Nonetheless, as it is, the error associated with the proposed method is relatively low compared to conventional electrical methods, (e.g., McAlister method) while it is comparable to the one of optical methods and Gate resistance temperature detector (RTD) technique [4]. To make this clearer, we reported in Fig. 10 the estimation error ($\varepsilon_E = T_{peak} - T_{est}$) provided by different electrical and optical methods. The results shown in Fig. 10 indicate that the proposed method allows to significantly reduce the estimation error provided by the McAlister method. While the proposed technique

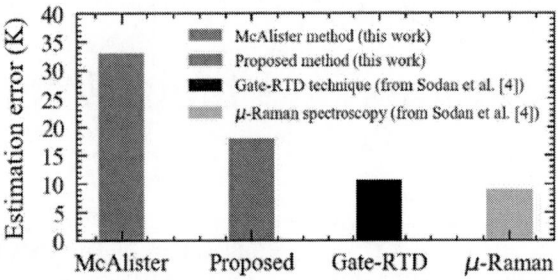

Fig. 10. Estimation error obtained in this work for McAlister's method and Proposed method compared with those reported in literature for Gate-RTD technique and μ-Raman Spectroscopy (P_D = 6.5 W/mm and T_{BP}= 300 K).

shows an accuracy comparable to the one of μ-Raman and Gate-RTD techniques, conversely to these it does not require any special sample preparation; as such, it can be directly employed on both packaged and on-wafer devices.

As final remark, we observe that the proposed technique can be applied to any electron device provided that it is influenced by trap dynamics in such a way that DCTs can be clearly associated to a single carrier emission process, so that Arrhenius equation applies. To this end, specific bias conditions could be required (see [17] for the case of Fe-doped AlGaN/GaN HEMTs), which hence needs to be characterized prior to the application of this technique.

VI. CONCLUSIONS

A novel method for estimating device temperature was proposed, exploiting the thermally activated emission process from electron traps. Results indicate that the proposed method allows to significantly reduce the estimation error provided by conventional McAlister method, while showing an accuracy comparable to the one of μ-Raman and Gate-RTD techniques without the need for specific sample preparation, making it directly employed on both packaged and on-wafer devices.

ACKNOWLEDGMENT

This work was partially supported by the Italian Ministry for University and Research (MIUR) under the PRIN 2017 Project "Empowering GaN-on-SiC and GaN-on-Si technologies for the next challenging millimeter-wave applications".

REFERENCES

[1] A. Darwish et al., IEEE Trans. Microwave Theory Techn., vol. 57, no. 12, 2009.

[2] D. L. Blackburn, IEEE Semi-Therm, San Jose, CA, USA, 2004.

[3] S. McAlister et al., J. Vac. Sci. Technol. A, vol. 24, no. 3, 2006.

[4] V. Sodan et al., IEEE Trans. Electron Devices, vol. 63, no. 6, 2016.

[5] L. Baczkowski et al., IEEE Trans. Electron Devices, vol. 62, no. 12, 2015.

[6] M. Kuball et al., IEEE Electron Device Lett., vol. 23, no. 1, 2002.

[7] A. Sarua et al., IEEE Trans. Electron Devices, vol. 53, no. 10, 2006.

[8] M. Uren et al., IEEE Trans. Electron Devices, vol. 59, no. 12, 2012.

[9] S. Rajasingam et al., IEEE Electron Device Lett., vol. 25, no. 7, 2004.

[10] J. Joh, J. del Alamo, IEEE Trans. Electron Devices, vol. 58, no. 1, 2011.

[11] C. Potier et al., EuMIC, Rome, Italy, 2014.

[12] M. Cioni et al., IEEE Trans. Electron Devices, vol. 68, no. 7, 2021.

[13] J. Paasschens, et al., IEEE BCTM, Montreal, Canada, 2004.

[14] M. Branch et al., SIAM J. Sci. Comput., vol. 21, no. 1, 1999.

[15] H. Choi et al., IEEE IJCNN, Budapest, Hungary, 2004.

[16] R. J T. Simms et al., IEEE Trans. Electron Devices, vol. 55, no. 2, 2008.

[17] N. Zagni et al., WiPDA 2021, Redondo Beach, CA, USA, 2021.

Metastability of Negatively Charged Hydroxyl-E' Centers and their Potential Role in Positive Bias Temperature Instabilities

Christoph Wilhelmer[1], Dominic Waldhoer[2], Markus Jech[2], Al-Moatasem Bellah El-Sayed[2,3],
Lukas Cvitkovich[2], Michael Waltl[1] and Tibor Grasser[2]

[1]Christian Doppler Laboratory for Single-Defect Spectroscopy in Semiconductor Devices at the Institute for Microelectronics,
Gußhausstraße 27–29, 1040 Vienna, Austria
[2]Institute for Microelectronics, Technische Universität Wien,
[3]Nanolayers Research Computing, Ltd., 1 Granville Court, Granville Road, London N12 0HL, United Kingdom
E-mail: [wilhelmer | grasser]@iue.tuwien.ac.at

Abstract—Oxide defects are well known to negatively impact the performance of modern electronics by introducing device reliability degrading phenomena due to their ability to trap charges from the substrate during operation. The hydroxyl-E' center gained considerable attention in the recent past because of the close vicinity of its electron and hole charge transition levels to the band edges of the Si substrate. Here, we employ density functional theory to statistically analyze different hydroxyl-E' center configurations in amorphous SiO$_2$. We identify two negatively charged defect states that are significantly lower in energy than previously discovered configurations and further show that the hydroxyl-E' center is a suitable candidate for a three-state defect model involving electron capturing processes by calculating thermal barriers between different configurations and corresponding charge transition levels. The discovered minimum energy configurations introduce electron trap levels far below the conduction band maximum of Si and SiC substrates, which might explain the experimentally observed mitigated positive temperature bias instability effect in Si/SiO$_2$ systems compared to its negative counterpart.

Index Terms—PBTI, SiO$_2$, Hydroxyl-E', MOSFET, SiC

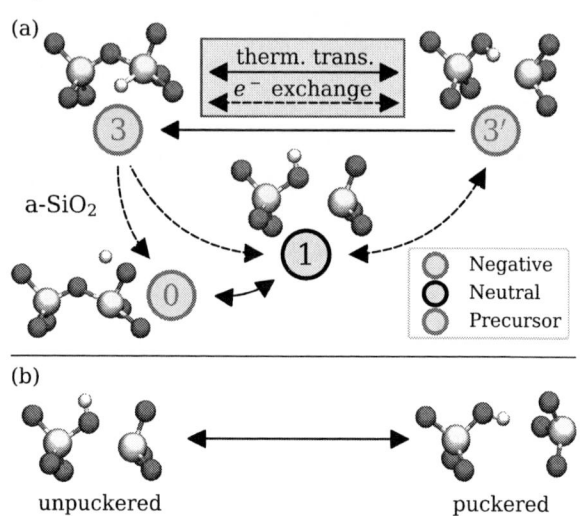

Fig. 1. **Top:** Schematic state diagram of a hydroxyl-E' defect in the neutral (1 black) and negative (3 blue) charge state. Oxygen atoms are colored red, silicon atoms yellow. 0 denotes a precursor configuration with an interstitial H before the defect is formed by breaking a Si-O bond, 3' denotes a metastable negatively charged configuration. Dashed arrows denote a transition involving charge transfer, solid arrows without. **Bottom:** Thermal transition to a puckered configuration that can occur for states 1 and 3'.

I. INTRODUCTION

Due to the ongoing downscaling of electronic devices, the reliability of modern electronics is increasingly governed by single point defects. In metal oxide semiconductor field effect transistors (MOSFETs), defects in the oxide and the oxide-substrate interface are suspected to negatively influence the device performance by trapping charges from the bulk during operation, causing parasitic effects such as bias temperature instability (BTI) [1] or random telegraph noise (RTN) [2]. The amorphous nature of the oxide makes both theoretical and experimental investigations highly challenging, as defect parameters are stochastically distributed over a broad range and can thus only be analyzed in a statistical manner. Furthermore, single defects can exist in different configurations, causing phenomena like anomalous RTN [3] or defect volatility [4] by modulating the observable capture and emission time constants (ns to months [5]). While the hole trapping mechanism has already been investigated using a multi-state defect model employing nonradiative multi-phonon (NMP) theory [6], a similar consistent description of electron trapping causing positive BTI (PBTI) is still missing. Besides the well investigated oxygen vacancy (OV), which has charge transition levels (CTLs) that are outside the energy window to interact with holes from the Si substrate [7], interest in hydrogen related defects, which are prevalent in Si/SiO$_2$ systems, has increased. H is deliberately introduced during device fabrication to passivate dangling bonds in the oxide and its interface to the substrate [8], however it can also create new defect types in the oxide such as the hydrogen bridge (HB) or the hydroxyl-E' center (H-E') [9], [10]. These defects have CTLs near the band edges of Si and SiC substrates [4], [11], [12], which act as reservoirs for holes (valence band maximum, VBM) and electrons (conduction band minimum, CBM). Nevertheless, the reported vicinity of the CTLs to the relevant band edges contradicts the experimentally observed weakened PBTI in Si MOSFETs when compared to negative BTI induced by hole trapping [13]. While the impact of the HB is likely weak, as these defects are assumed to only form at preexisting OVs

978-1-6654-8498-5/22 $31.00 © 2022 IEEE

which only occur in low concentrations ($\sim 10^{17}\,\text{cm}^{-3}$) [14], the concentration of the H-E' precursor (strained Si-O-Si sites) is estimated at around $5 \cdot 10^{19}\,\text{cm}^{-3}$ [15]. Hence, the H-E' is expected to be more relevant for device reliability and is thus the primary focus of this work.

Here we present a statistical *ab initio* analysis based on density functional theory (DFT) calculations of H-E' defects in an amorphous SiO$_2$ structure to analyze their potential role in PBTI. Similar to the already investigated hole trapping mechanism [4], we introduce a multi-state defect model for the H-E' for electron trapping as schematically shown in Fig. 1 (top), including newly discovered defect configurations in the negative charge state (state 3 in Fig. 1) which are significantly lower in energy compared to previously analyzed configurations. The different H-E' instances are analyzed by comparing their total energies and by calculating transition barriers between them. We show that in addition to the already reported 0/- trap level for the $1 \leftrightarrow 3'$ transition [11], the minimum configurations introduce $1 \leftrightarrow 3$ CTLs far below the CBM of Si and SiC substrates, which might give an explanation for the reduced PBTI effect observed in Si MOSFETs.

II. Hydroxyl-E' Center

In this section, the numerous configurations of the hydroxyl-E' center in the neutral and negative charge state are analyzed in terms of their atomic configurations, the localization of the electron at the defect site, total energy differences and energy barriers between them. We use the DFT calculation setup that was already successfully employed in previous investigations and is presented in detail in [11]. Transition barriers were calculated with the climbing image nudged elastic band algorithm (CI-NEB) [16] with seven intermediate images each, employing a PBE functional to reduce the computational costs.

A. Atomic Configurations

A hydroxyl-E' center defect is formed when a H atom attaches to an O of a strained Si-O bond, thereby breaking this bond and forming a hydroxyl group ($0 \rightarrow 1$ in Fig. 1). To create such a defect, a H was placed in the vicinity of an O before relaxing the structure within our DFT setup. The obtained structure was subsequently geometry optimized in the negative charge state. This procedure was repeated for all O sites in the structure to create 144 distinct H-E' defects in the neutral and negative charge state.

The resulting configurations are shown in Fig. 2 (a-f) with their highest occupied molecular orbital (HOMO) localized at the defect site at isovalues of $\pm 0.05\,\text{e}/\text{Å}^3$. In the neutral charge state, 117 out of the 144 initial configurations converged to H-E' defects (state 1 in Fig. 1). For the remaining O-sites, the H either became interstitial, the H moved to another already calculated O position in the oxide or formed a metastable [SiO$_4$/H]^0 center [17]. The neutral H-E' configuration is shown in Fig. 2 (a) with its HOMO localized at the Si dangling bond. In the negative charge state, several metastable defects, which all correspond to state $3'$ in Fig. 1 were found. Three of them are shown in Fig. 2 (b-d). The metastable

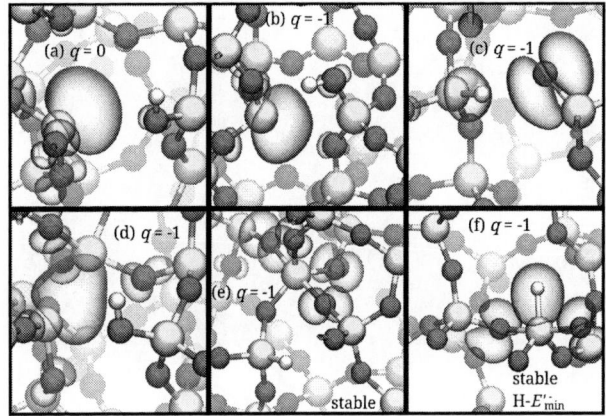

Fig. 2. Hydroxyl-E' configurations in neutral (a) and negative (b-f) charge state with the HOMO localized at the defect site. Configurations (b-d) are metastable and correspond to state $3'$ in Fig. 1 while the stable configurations (e-f) correspond to state 3.

configuration formed after trapping an electron depends on the atomic environment of the defect, with the HOMO either localizing at the Si dangling bond (b), hybridized between O and H (c) or at an oxygen vacancy-like site (d). The most stable states are shown in Fig. 2 (e-f) and correspond to state 3 in Fig. 1. For both configurations, the H diffuses away from the O and attaches to the Si atom opposite the hydroxyl group. Either the O of the hydroxyl group attaches to a different Si in the oxide making this Si fivefold coordinated with five O (e) or the O restores the initially broken Si-O bond, also making it fivefold coordinated with four O and one H (f). The latter will be denoted as the minimum configuration H-E'^{-}_{min} in the following. Furthermore, puckered configurations, where the Si on the opposite side of the hydroxyl group moves through the plane of the three adjacent O as schematically shown in Fig. 1 (bottom), were identified for neutral and metastable negatively charged H-E', and will be considered in the energy and trap level analysis.

B. Total Energies

To compare the stability of different negatively charged configurations, the H atom of a relaxed metastable state was deliberately moved to an adjacent Si position and subsequently geometry optimized to purposely create a H-E'^{-}_{min}. Furthermore, additional puckered configurations were created by pushing the Si facing the hydroxyl group through the plane of its adjacent three O atoms, followed again by a geometry optimization. In Fig. 3, energies of metastable states (b-d) are compared to H-E'^{-}_{min} at the same defect site as well as the energies of unpuckered and puckered configurations in the neutral and negative charge state. For all analyzed defects, H-E'^{-}_{min} is at least 0.6 eV lower compared to previously analyzed H-E' configurations at the same defect site [11]. Puckered configurations can be higher or lower in energy compared to unpuckered in both charge states, depending on the atomic environment of the defect site.

978-1-6654-8498-5/22 $31.00 © 2022 IEEE

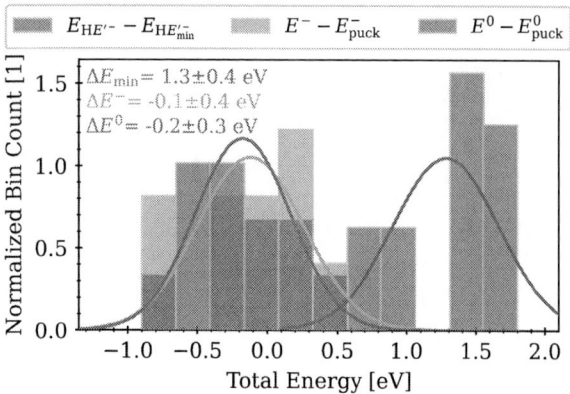

Fig. 3. Total energy differences between metastable (Fig. 2 (b-d)) and stable (Fig. 2 (e-f)) H-E' configurations in the negative charge state (blue) and total energy differences between unpuckered and puckered H-E' in negative (turquoise) and neutral (red) charge state with the parameters of the fitted normal distributions given in the plot.

C. Energy Barriers

Thermal energy barriers between negatively charged metastable and H-E'^-_{min} configurations ($3' \rightarrow 3$ in Fig. 1) were calculated with CI-NEB and are shown in Fig. 4 with the respective starting configuration according to Fig. 2 and the respective H position of the converged H-E'^-_{min} given in the legend. The total energy of the metastable state is used

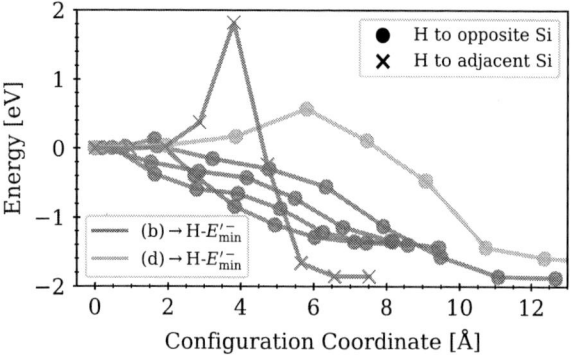

Fig. 4. Transition barriers between metastable and stable negatively charged hydroxyl-E' configurations of six different defects calculated with CI-NEB. Barriers correspond to the $3' \rightarrow 3$ transition in the state diagram in Fig. 1. The total energies of the metastable states are used as energy reference for each transition. The color of the barrier denotes the respective metastable configuration, (b) or (d), of Fig. 2 while the marker denotes the position of the Si where the H migrated to.

as an arbitrary zero for all defects. According to the NEB analysis, H moving to the Si opposite the hydroxyl group is clearly the preferred transition from state (b) with vanishing energy barriers, while the barrier for an attachment of H to the adjacent Si is rather enormous for the calculated site. The transition barrier for a trajectory from the OV like state (d) to H-E'^-_{min} is also relatively small with 0.57 eV. Thus we conclude that a H-E' defect will either instantaneously or by overcoming a small energy barrier transform to a minimum configuration H-E'^-_{min} after trapping an electron.

III. CHARGE TRANSITION LEVEL

The charge transition level (CTL) of a defect can be calculated by comparing its formation energy

$$E^q_{\text{Form}} = E^q_{\text{tot}} - E^{\text{bulk}}_{\text{tot}} - \sum_i \mu_i n_i + q E_{\text{F}} + E^{\text{corr}} \quad (1)$$

in two charge states. It corresponds to the energy of the Fermi level when E^q_{Form} in two charge states is equal. Here, E^q_{tot} is the total energy of the defect structure in charge state q, $E^{\text{bulk}}_{\text{tot}}$ is the total energy of the pristine bulk, $\mu_i n_i$ is the chemical energy needed to remove or add atoms to the bulk from type i, $E_{\text{F}} = E_{\text{VBM}} + \epsilon_{\text{F}}$ is the Fermi level given with respect to the highest occupied Kohn-Sham orbital of the bulk system, which is used as an approximation for the valence band edge of the oxide [18]. In order to correct for spurious interactions of charged defects in periodic cells, the scheme by Makov and Payne was employed, leading to an correction term $E^{\text{corr}} = 0.35$ when evaluated for our structures. By applying an electric field to the gate of a MOSFET, the CTL can be shifted relative to the substrate band edges due to the band bending near the interface and the electric field in the oxide. When the CTL of a defect coincides with the CBM of the substrate, the probability of the defect to exchange electrons with the reservoir and hence the power of a RTN signal is maximal. It is the only defect parameter than can be directly measured and is thus an important feature to identify a certain defect by linking experiments to theoretical predictions.

The calculated 0/- CTLs for the transitions $1 \leftrightarrow 3'$ and $1 \leftrightarrow 3$ corresponding to the states depicted in Fig. 1 are shown in Fig. 5 (top) with the parameters of a fitted normal distribution given in the plot. Puckered to puckered transitions are included in the $1 \leftrightarrow 3'$ CTLs as they are distributed at similar energies, quite contrary to the hole trapping mechanism [19]. The band gap of a Si substrate is depicted as a gray area with band alignments from [20] and the band gap of Si (1.38 eV) calculated with our DFT setup, the VBM and the CBM of a SiC substrate are shown as red and blue vertical lines respectively with band alignments from [21]. The fact that a certain H-E' can have two different CTLs is visualized in a correlation plot in Fig. 5 (bottom), where each data point corresponds to one single H-E' defect. While the $1 \leftrightarrow 3'$ CTLs are distributed near the CBM of both Si and SiC, which is in good agreement for defect parameters extracted from stress/recovery experiments in Si/SiC devices [7], [12], the $1 \leftrightarrow 3$ CTLs lie significantly lower and are thus in a rather inaccessible energy range for electron capture. A H-E' can thus easily become charged according to the $1 \leftrightarrow 3'$ level and subsequently quickly transition to a stable H-E'^-_{min} configuration as shown in Fig. 4. The large energetic distance of the inherent $1 \leftrightarrow 3$ level reduces the affinity of the defect to exchange electrons with the CBM of the substrate, given that the level is still accessible at all, and leads to much longer charge emission constants as typically seen in anomalous RTN [3]. This mechanism might explain the mitigated PBTI observed in Si devices [13] while the pronounced PBTI in SiC devices is expected to have a different origin, e.g. carbon defects like C-dimers in the oxide [22].

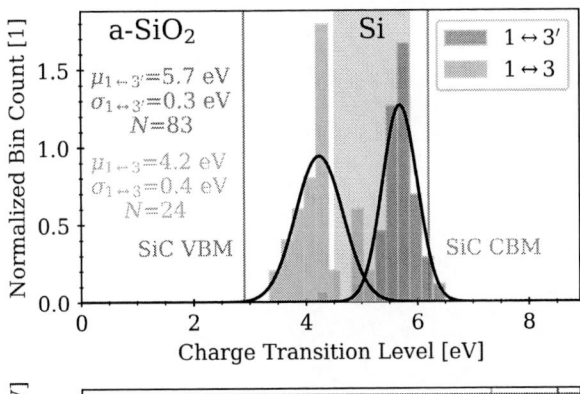

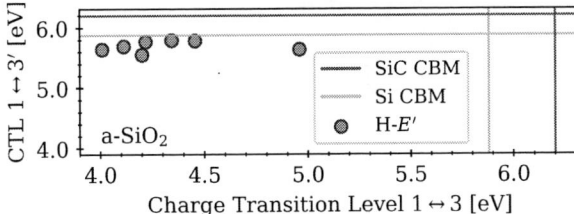

Fig. 5. **Top:** Thermodynamic charge transition levels of hydroxyl-E' centers relative to the a-SiO$_2$ VBM corresponding to the $1 \leftrightarrow 3'$ and $1 \leftrightarrow 3$ transitions of the state diagram in Fig. 1. CTLs are given in the context of a-SiO$_2$/Si and a-SiO$_2$/SiC band diagrams with the Si band gap shown as a grey area and the band edges of SiC given as a red and a blue line respectively. Normal distributions were fitted to the data with the fitting parameters and the sample size given in the plot. **Bottom:** Correlation plot of $1 \leftrightarrow 3$ and $1 \leftrightarrow 3'$ CTLs of single H-E' defects in a-SiO$_2$ with the CBM of Si and SiC substrates given as blue and turquoise lines.

However, the extra electron can also annihilate holes from the Si VBM under NBTI conditions when released from the defect site. Furthermore, by releasing the electron, the H might diffuse away from the defect site to an interstitial position and subsequently form a defect at a different O position in the oxide, exhibiting distinct defect parameters and thus could be a possible mechanism for defect volatility [4].

IV. CONCLUSIONS

We employed DFT to calculate a large distribution of neutral and negatively charged hydroxyl-E' defects in an amorphous SiO$_2$ structure to analyze their impact on PBTI, identifying two negatively charged minimum states H-E'^-_{min} which are $0.6 - 1.8\,\mathrm{eV}$ lower in energy compared to previously discovered configurations, making them suitable candidates for a multi-state defect model for electron capturing processes. The metastable negatively charged configurations quickly transition to a stable state by overcoming a vanishing energy barrier as shown by our NEB calculations. H-E'^-_{min} configurations introduce charge transition levels far below the conduction band of both Si and SiC substrates, which mitigates electron exchange with the respective charge reservoirs and gives an explanation for the experimentally observed reduced PBTI effect in Si MOSFETs while the pronounced PBTI in SiC might have a different origin such as carbon defects in the oxide. Furthermore, we show that individual defects can have distinct trap levels causing different charge transition time constants as seen e.g. in anomalous random telegraph noise.

ACKNOWLEDGMENT

This project was funded by the European Union's Horizon 2020 research and innovation program under grant agreement No. 871813 MUNDFAB. Furthermore, the financial support by the Austrian Federal Ministry for Digital and Economic Affairs, the National Foundation for Research, Technology and Development and the support from the Vienna Scientific Cluster is gratefully acknowledged.

REFERENCES

[1] T. Grasser *et al.*, "The paradigm shift in understanding the bias temperature instability: From reaction–diffusion to switching oxide traps," *IEEE Trans. Electron. Devices*, vol. 58, no. 11, pp. 3652–3666, 2011.

[2] K. S. Ralls *et al.*, "Discrete resistance switching in submicrometer silicon inversion layers: Individual interface traps and low-frequency ($\frac{1}{f}$) noise," *Phys. Rev. Lett.*, vol. 52, pp. 228–231, Jan 1984.

[3] T. Grasser, "Stochastic charge trapping in oxides: From random telegraph noise to bias temperature instabilities," *Microelectron. Reliab.*, vol. 52, no. 1, pp. 39–70, 2012.

[4] Y. Wimmer *et al.*, "Role of hydrogen in volatile behavior of defects in SiO$_2$-based electronic devices," *Proc. R. Soc. A*, vol. 472, no. 2190, p. 20160009, 2016.

[5] T. Grasser *et al.*, "Time-dependent defect spectroscopy for characterization of border traps in metal-oxide-semiconductor transistors," *Phys. Rev. B*, vol. 82, p. 245318, Dec 2010.

[6] F. Schanovsky *et al.*, "A multi scale modeling approach to non-radiative multi phonon transitions at oxide defects in MOS structures," *J. Comput. Electron.*, vol. 11, 09 2012.

[7] D. Waldhoer *et al.*, "Toward automated defect extraction from bias temperature instability measurements," *IEEE Trans Electron Devices*, vol. 68, no. 8, pp. 4057–4063, 2021.

[8] C. Kaneta *et al.*, "Defect states due to silicon dangling bonds at the Si(100)/SiO$_2$ interface and the passivation by hydrogen atoms," *MRS Proceedings*, vol. 592, p. 39, 1999.

[9] L. Skuja *et al.*, "An increased F$_2$-laser damage in 'wet' silica glass due to atomic hydrogen-related E'-center," *J. Non-Cryst. Solids*, vol. 352, no. 23, pp. 2297–2302, 2006.

[10] A.-M. El-Sayed *et al.*, "Hydrogen-induced rupture of strained Si-O bonds in amorphous silicon dioxide," *Phys. Rev. Lett.*, vol. 114, p. 115503, Mar 2015.

[11] C. Wilhelmer *et al.*, "Statistical ab initio analysis of electron trapping oxide defects in the Si/SiO$_2$ network," in *Proceedings of (ESSDERC)*, pp. 243–246, 2021.

[12] C. Schleich *et al.*, "Physical modeling of charge trapping in 4H-SiC DMOSFET technologies," *IEEE Trans Electron Devices*, vol. 68, no. 8, pp. 4016–4021, 2021.

[13] G. Rzepa *et al.*, "Comphy — A compact-physics framework for unified modeling of BTI," *Microelectron. Reliab.*, vol. 85, pp. 49–65, 2018.

[14] S. Dannefaer *et al.*, "Vacancy-type defects in crystalline and amorphous SiO$_2$," *J. Appl. Phys.*, vol. 74, no. 2, pp. 884–890, 1993.

[15] A.-M. El-Sayed *et al.*, "Identification of intrinsic electron trapping sites in bulk amorphous silica from ab initio calculations," *Microelectron. Eng.*, vol. 109, pp. 68–71, 2013.

[16] G. Henkelman *et al.*, "A climbing image nudged elastic band method for finding saddle points and minimum energy paths," *J. Chem. Phys.*, vol. 113, no. 22, pp. 9901–9904, 2000.

[17] A.-M. El-Sayed *et al.*, "Theoretical models of hydrogen-induced defects in amorphous silicon dioxide," *Phys. Rev. B*, vol. 92, p. 014107, Jul 2015.

[18] C. V. de Walle *et al.*, "First-principles calculations for defects and impurities: Applications to III-nitrides," *J. Appl. Phys.*, vol. 95, no. 8, pp. 3851–3879, 2004.

[19] W. Goes *et al.*, "Identification of oxide defects in semiconductor devices: A systematic approach linking DFT to rate equations and experimental evidence," *Microelectron. Reliab.*, vol. 87, pp. 286–320, 2018.

[20] E. Bersch *et al.*, "Complete band offset characterization of the HfO$_2$/SiO$_2$/Si stack using charge corrected x-ray photoelectron spectroscopy," *J. Appl. Phys.*, vol. 107, no. 4, p. 043702, 2010.

[21] V. V. Afanas'ev *et al.*, "Band offsets and electronic structure of SiC/SiO$_2$ interfaces," *J. Appl. Phys.*, vol. 79, no. 6, pp. 3108–3114, 1996.

[22] P. Deák *et al.*, "The mechanism of defect creation and passivation at the SiC/SiO$_2$ interface," *J. Phys. D: Appl. Phys.*, vol. 40, no. 20, pp. 6242–6253, 2007.

Silicon-Impurity Defects in Calcium Fluoride: A First Principles Study

Dominic Waldhoer[†] ⓘ, Bibhas Manna[†] ⓘ, Al-Moatassem B. El-Sayed[†] ⓘ,
Theresia Knobloch[†] ⓘ, Yury Illarionov[†*] ⓘ, Tibor Grasser[†] ⓘ

[†]Institute for Microelectronics, TU Wien, 1040 Vienna, Austria
[*]Ioffe Institute, Polytechnicheskaya 26, 194021 St-Petersburg, Russia

Abstract—**With the emergence of novel nanoelectronic devices based on 2D materials, the need for suitable dielectrics has become apparent in recent years. One particular promising candidate for future device dielectrics is calcium fluoride (CaF$_2$) due to its inert surface and the possibility to form a quasi van der Waals (vdW) interface with 2D semiconductors like molybdenum disulfide (MoS$_2$). In this work we explore silicon-impurity defects in CaF$_2$, which is typically grown on Si(111), and their potential relevance for device reliability using density functional theory. Defect parameters describing the charge trapping behavior of these defects within a nonradiative multiphonon theory are provided to be used in future TCAD simulations. We show that Si-impurity defects can act as amphoteric charge traps and hence might play a role in the observed hysteresis and bias temperature instability in Si/CaF$_2$/MoS$_2$ gate stacks.**

Index Terms—**2D materials, calcium fluoride (CaF$_2$), defects, reliability, nonradiative mulitphonon (NMP) theory, bias temperature instability (BTI), hysteresis**

I. INTRODUCTION

Being able to form high quality insulator/semiconductor interfaces is essential for creating operational nanoelectronic devices, metal-oxide-semiconductor field-effect transistors (MOSFETs) in particular. While conventional Si-based devices have settled on using a thin layer of native oxide SiO$_2$ followed by high-k materials like HfO$_2$ as gate dielectric, finding suitable dielectric materials for 2D devices is challenging [1]. First of all, chemical bonding with the dielectric layer disrupts the electronic structure of the delicate 2D channel material, leading to a significant reduction of carrier mobility compared to its theoretical maximum [2]. Furthermore, the usually employed amorphous 3D oxides, e.g. SiO$_2$ and HfO$_2$, have large defect concentrations at their surfaces [3]. Those defects are in close proximity to the device channel, since the 2D layer is directly exfoliated/transferred on a prepared dielectric surface for frequently manufactured back-gated prototype devices. This leads to a large electrostatic impact and enhanced bias temperature instability (BTI) or $I_D(V_G)$ hysteresis.

For these reasons, layered insulators which can form quasi van der Waals (vdW) interfaces with the channel have been considered [4]. The most prominent candidate within this

This project was in part funded by the European Union's Horizon 2020 research and innovation program under grant agreement No. 871813, within the framework of the project Modeling Unconventional Nanoscaled Device-FABrication (MUNDFAB). We also wish to acknowledge the support from the Vienna Scientific Cluster.

Corresponding authors: Dominic Waldhoer, Tibor Grasser (waldhoer|grasser@iue.tuwien.ac.at).

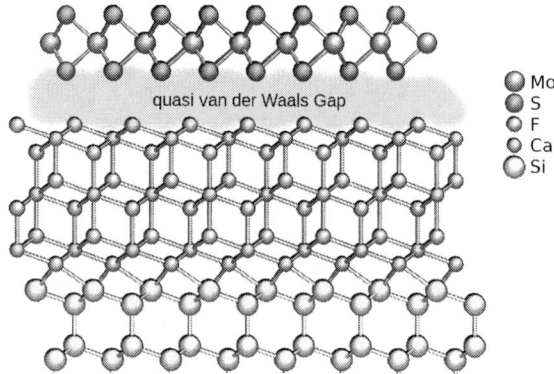

Fig. 1: Si(111)/CaF$_2$/MoS$_2$ gate stack envisioned in [6]. Due to its good lattice match, CaF$_2$ can be grown on Si(111) in thin layers. It provides a virtually defect-free quasi van der Waals interface to the 2D material and hence greatly improves the reliability of 2D devices.

category is hexagonal boron-nitride (hBN), which however is not scalable due to its rather small bandgap of $E_G = 5.95\,eV$ and a low permittivity, leading to excessive gate leakage currents [5]. We have recently suggested the use of the ionic crystal calcium fluoride (CaF$_2$) as a gate dielectric for 2D materials [6]. Its large bandgap of $E_G = 12.1\,eV$ and its fluorine-terminated inert (111) surface make it appear like an almost ideal insulator. Moreover, thin layers of CaF$_2$ have been deposited on Si(111) [7] but also free standing slabs of CaF$_2$ are dynamically stable according to theoretical studies [8]. Indeed, initial reports show that employing CaF$_2$ as a substrate leads to significantly reduced $I_D(V_G)$ hysteresis and BTI [6]. This improvement is attributed to the clean quasi vdW interface between MoS$_2$ and CaF$_2$ as schematically illustrated in Fig. 1. All these findings suggest that CaF$_2$ might be used as a reliable gate insulator in (2D) nanoelectronics in the near future.

Even though, the overall defect density is expected to be smaller in CaF$_2$ than in amorphous oxides, the defect levels and their prevalence will still determine the electrical stability of devices based on CaF$_2$ as a gate insulator. What is more, in order to include CaF$_2$ in TCAD simulation workflows, in particular when focusing on device reliability, not only do the basic material properties like permittivity, bandgap etc. have to be known, but also information about different defects in the material. In this work, we employ density functional theory (DFT) to determine important defect parameters like the

978-1-6654-8498-5/22 $31.00 © 2022 IEEE

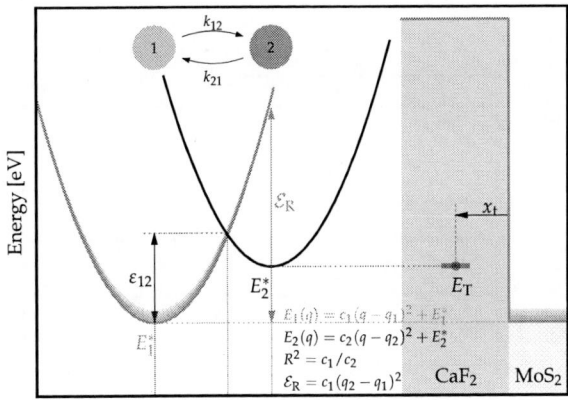

Fig. 2: Description of charge transfer within the NMP model. The differently charged states are treated as one-dimensional harmonic oscillators. The potential energy curves are parametrized by the relaxation energy $\mathcal{E}_R$, the curvature ratio R and the trap level E_T.

TABLE I: Comparison of CaF_2 bulk properties calculated by our DFT setup against experimental data.

		PBE0 $\alpha = 0.39$	exp.	ref.
Lattice Constant a	[Å]	5.454	5.463	[18]
Bulk Modulus K	[GPa]	77.29	81.67	[19]
Indirect Bandgap E_G	[eV]	11.90	11.8	[15]
Static Permittivity ε_0	[1]	6.43	6.81 - 6.84	[20], [21]
Optical Permittivity ε_∞	[1]	1.85	2.05	[20]

thermodynamic trap levels, relaxation and formation energies. These are important input parameters for physical charge trapping models based on nonradiative multiphonon theory, which allow the prediction of degradation over the course of the device lifespan.

Here, we focus our efforts on Si-impurity defects, which are relevant since many 2D device prototypes are fabricated as back-gated devices with a silicon wafer acting both as substrate for the dielectric and as a back-gate for the 2D layer. Furthermore, the stable (7×7) reconstructed Si(111) surface has four excess Si atoms per unit cell [9], which either enter the gas phase as SiF_4 during CaF_2 deposition or are incorporated in CaF_2. We show that Si-impurity defects like the Si interstitial (Si_i) or the Si substituting for Ca (Si_{Ca}) can act as amphoteric traps and might contribute to hysteresis and BTI in gate-stacks similar to Fig. 1. To the best of our knowledge and given the novelty of this material in 2D nanoelectronics, this is the first report on potential electrically active defects in CaF_2.

II. NMP MODEL FOR CHARGE TRAPPING

Charge transfer reactions between a defect in the dielectric and the (2D) semiconductor are governed by nonradiative multiphonon (NMP) transitions [10]. Within the NMP model the two interacting charge states, named 1 and 2, are treated as coupled quantum mechanical harmonic oscillators. In order to obtain a fast and compact model for describing charge transitions within a TCAD simulation, one typically employs several approximations. For instance, the vibrational motion is restricted to one effective mode, resulting in a so-called reaction coordinate. In this approximation, the two defect states are described by two parabolic potential energy curves (PEC) as illustrated in Fig. 2. Furthermore, in the high-temperature limit, the transition rate is solely determined by the classical barrier ε_{12} obtained at the crossing point of these energy curves. This is typically referred to as the classical

approximation. In order to avoid costly numerical integrals over the valence and conduction bands of the semiconductor [11], it is further assumed that the charge carriers are located exclusively at the band edges [12].

Using these approximations, the resulting transition rate can be expressed analytically, e.g. in the case of electron traps it is given by

$$k_{12} = n v_{\mathrm{th}} \sigma \vartheta \exp\left(-\varepsilon_{12}/k_B T\right) \quad (1)$$

where the prefactor contains the carrier concentration n, the thermal velocity v_{th}, the capture cross section σ and a WKB factor considering the tunneling amplitude between defect and semiconductor. In such a description the defect physics, i.e. essentially the barrier ε_{12} as a function of gate bias, is determined by only a few parameters: the trap level E_T giving the relative alignment between the states; the relaxation energy $\mathcal{E}_R$, which is a measure for the curvature of the PEC; the curvature ratio R between the two states; and the position of the defect x_t. Except for x_t, all these parameters can be obtained theoretically from DFT for a given defect candidate.

III. DFT METHODOLOGY

All calculations were performed on a $3 \times 3 \times 3$ cubic cell of CaF_2 using the Gaussian Plane Wave (GPW) method with the double-ζ GTH basis set as implemented in the CP2K code [13]. We used the PBE0_TC_LRC hybrid functional with the PBEsol [14] parametrization and a mixing factor $\alpha = 0.39$ for the Hartree-Fock (HF) exchange. This value was determined by calibrating the single-particle bandgap within our calculation to the known *indirect* bandgap $E_G = 11.8\,\mathrm{eV}$ [15]. Since the mixing factor is inversely related to the optical permittivity ε_∞ and thus material dependent [16], there is some theoretical justification for tuning the mixing factor to match certain experimental results. However, since this is still an empirical procedure, we independently verified our simulation setup by calculating various well-known properties of bulk CaF_2 and comparison to experiments as summarized in Tab. I. As shown, our calculation provides good agreement for lattice constants, static and optical permittivities and is thus considered well suited to describe this material. The permittivities were evaluated by the change in polarization after applying a finite electric field using the Berry phase formalism [17].

The lattice vectors of the initially pristine bulk cell were relaxed down to a residual pressure of $0.05\,\mathrm{GPa}$. Initial defect configurations were created manually within the structure followed by a geometry optimization with fixed lattice vectors

978-1-6654-8498-5/22 $31.00 © 2022 IEEE

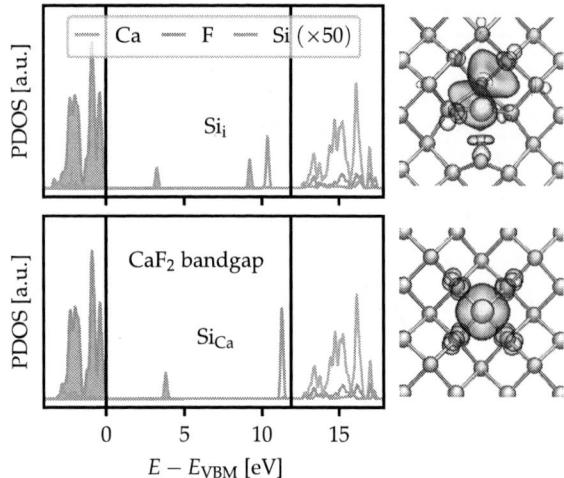

Fig. 3: Projected density of states (PDOS) and localized highest occupied molecular orbitals of the Si_i (**top**) and Si_{Ca} (**bottom**) defect. Both defects introduce localized states in the bandgap as can be seen from the DOS projected onto the Si impurity. Note that the Si PDOS is scaled by a factor of 50 for illustrative purposes.

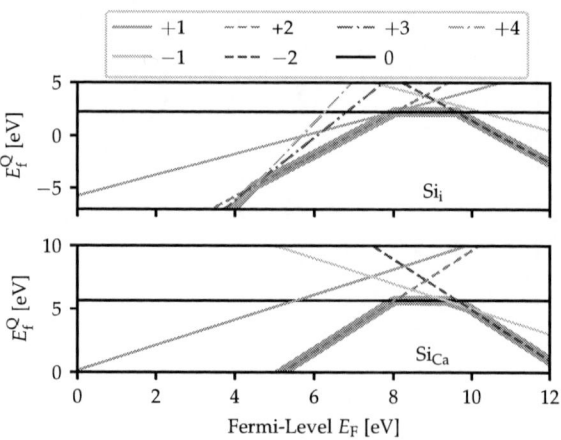

Fig. 4: Formation energies for the Si_i (**top**) and Si_{Ca} (**bottom**) defect as a function of the Fermi level E_F for different charge states. The lowest energy state at a given E_F is referred to as *stable* and highlighted by the thick gray line. Note that there are *unstable* states, which are never lowest in energy, e.g. the $+1$ state for both defects. E_F is taken w.r.t. the valence band maximum (VBM) of CaF_2.

down to a residual force of $25\,\mathrm{meV/\mathring{A}}$. In order to identify charged defect states, this was repeated for different total charges in the cell with the additional requirement that the excess charge has to localize at the defect site in order to be considered trapped. The formation energy of a charged defect is calculated by [22]

$$E_f^Q = E_{tot}^Q - E_{tot}^{bulk} - \sum_i \mu_i n_i + Q E_F + E_{corr}^Q \quad (2)$$

with E_{tot}^Q and E_{tot}^{bulk} being the total energies of the defective system in charge state Q and the pristine bulk respectively. $\mu_i n_i$ is the chemical energy to add/remove n_i atoms of species i, E_F is the Fermi level in the charge reservoir and E_{corr}^Q denotes the Makov-Payne correction term for charged defects [23].

IV. RESULTS

In this section, the obtained results for the Si_i and Si_{Ca} are presented followed by a discussion about their implications for 2D MOSFETs on CaF_2.

As shown in the atomistic models provided in Fig. 3 (right), the substitutional Si atom (bottom) is incorporated into the host crystal with small lattice distortions. On the other hand, the Si at an interstitial site (top) binds to the surrounding F atoms and displaces a nearby Ca in the process, leading to quite strong local distortions of the host lattice. As can be seen from the density of states projected onto the Si impurity in Fig. 3, both defects introduce localized electronic states within the CaF_2 bandgap. In particular, the Si_i defect exhibits two filled and one empty state, compared to only one filled and one empty state for the Si_{Ca}. Therefore, the Si_{Ca} potentially can carry the net charges $\{-2, -1, 0, +1, +2\}$, whereas Si_i additionally could also have a $+3$ and $+4$ state as well. The defects were

relaxed in all their possible respective charge states in order to determine whether or not a particular state actually exists, i.e. the excess charge stays localized at the impurity site after relaxation. For the defects studied in this work, we found that this is fulfilled for all the aforementioned charge states, hence they all have to be considered for subsequent studies on charge transitions.

Following the relaxation, we examined the thermodynamic stability of the differently charged states by calculating their respective formation energy E_f^Q as a function of the Fermi level according to Eq. 2. The resulting stability ranges are presented in Fig. 4. As can be seen, the Si_i and Si_{Ca} defect can be thermodynamically stable in their $\{-2, 0, +2, +4\}$ and $\{-2, -1, 0, +2\}$ states respectively, hence both are amphoteric defects which can act as hole *and* electron traps depending on the applied bias.

The crossovers between two stability regions mark the so-called thermodynamic charge transition levels (CTL), which are synonymous with the trap level E_T used within the NMP model. The CTLs together with all other obtained NMP model parameters for each individual charge transition are summarized in Tab. II. In our analysis we were primarily focused on transitions during which only one electron or hole is transferred. The reason is that transitions including multiple charges, e.g. $0/+2$, typically are associated with larger barriers [24], which is also reflected in the considerably higher relaxation energies compared to single-charge transitions. Furthermore, such many-particle processes naturally have a small capture cross section. Therefore, even though a direct transition would be thermodynamically optimal, this process is generally still more likely to occur via multiple steps, e.g. $0 \Rightarrow +1 \Rightarrow +2$ instead of $0 \Rightarrow +2$.

The impact of charge trapping on the device reliability is mostly governed by the energetic alignment between the

TABLE II: Theoretical Defect Transitions and Parameters predicted with DFT. E_T is given w.r.t. the CaF$_2$ valence band edge. $\mathcal{E}_R$ always refers to the state with the more positive charge, the relaxation energy of the other state is then given by $\mathcal{E}_R/R^2$.

Defect	Transition	E_f^0 [eV]	E_T [eV]	$\mathcal{E}_R$ [eV]	R [1]
Si$_i$		2.25			
	$0/+1$		7.99	1.16	0.77
	$+1/+2$		8.05	3.03	0.99
	$+2/+3$		4.39	1.94	0.95
	$+3/+4$		4.57	3.70	1.04
	$0/-1$		10.24	3.54	1.41
	$-1/-2$		8.95	0.73	0.50
Si$_{Ca}$		5.67			
	$0/+1$		5.53	2.23	0.90
	$+1/+2$		10.50	0.92	2.60
	$0/-1$		9.35	2.12	0.92
	$-1/-2$		9.92	1.67	0.91

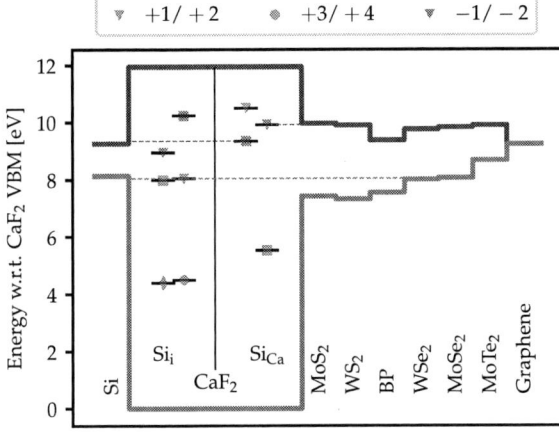

Fig. 5: Trap levels of the Si$_i$ and Si$_{Ca}$ defects within the bandgap of CaF$_2$ together with the band edges of silicon and commonly used 2D semiconductors. Alignments between band edges and active defect states are indicated with a dashed line. The band offsets were taken from [4].

trap level of the defect and the carrier reservoirs provided by the band edges of the semiconductor. Fig. 5 shows the calculated CTLs for both defects considered in this work together with the band edges of silicon and commonly used 2D semiconductors such as MoS$_2$. As can be seen, for a Si/CaF$_2$/MoS$_2$ stack as depicted in Fig. 1, the $0/+1$ level of the Si$_i$ is aligned with the valence band of the Si back-gate, while for the the Si$_{Ca}$ the $-1/-2$ and $0/-1$ level are aligned with the conduction band of MoS$_2$ and Si back-gate respectively. Hence charge trapping at those defects could provide an explanation for the comparatively low, but still existing hysteresis observed in such stacks. Based on the band diagram in Fig. 5 we would predict more severe effects of charge trapping for stacks including WSe$_2$ or MoSe$_2$ due to the additional alignment between the valence band of these semiconductors and the Si$_i$ $0/+1$ level. Note that the Si$_i$ $-1/-2$ level is also close to the conduction band of the back-gate. However, since the $0/-1$ level is higher up, the defect is initially in state 0 and going to state -2 requires two

electrons being transferred at once, which is very unlikely. Hence the $-1/-2$ level of this defect is considered inactive.

V. CONCLUSIONS

In this work we investigated Si-impurity defects in calcium fluoride (CaF$_2$) from first principles using density functional theory (DFT). With recently proposed Si/CaF$_2$ substrates for 2D transistors, such defects are expected to become relevant for improving the reliability of novel 2D devices. We studied the Si interstitial (Si$_i$) as well as the Si$_{Ca}$ substitutional defect, which to the best of our knowledge have not been mentioned in the literature so far, and showed that both can act as amphoteric charge traps in CaF$_2$. Furthermore, we provide defect parameters for a nonradiative multiphonon (NMP) charge trapping model, which can be used in reliability-aware TCAD simulations for this new material system and thus aiding the further improvement of 2D devices.

REFERENCES

[1] S. Das *et al.*, *Nature Electronics*, vol. 4, no. 11, pp. 786–799, 2021. DOI: 10.1038/s41928-021-00670-1.

[2] Z. Yu *et al.*, *Advanced Functional Materials*, vol. 27, no. 19, p. 1 604 093, 2017. DOI: 10.1002/adfm.201604093.

[3] C. Lee *et al.*, *Nanotechnology*, vol. 29, no. 33, p. 335 202, 2018. DOI: 10.1088/1361-6528/aac6b0.

[4] Y. Illarionov *et al.*, *Nature Communications*, vol. 11, p. 3385, 2020. DOI: 10.1038/s41467-020-16640-8.

[5] T. Knobloch *et al.*, *Nature Electronics*, vol. 4, no. 2, pp. 98–108, 2021. DOI: 10.1038/s41928-020-00529-x.

[6] Y. Illarionov *et al.*, *Nature Electronics*, vol. 2, pp. 230–235, 2019. DOI: 10.1038/s41928-019-0256-8.

[7] Y. Illarionov *et al.*, *Thin Solid Films*, vol. 545, pp. 580–583, 2013. DOI: 10.1016/j.tsf.2013.07.050.

[8] J. Weng *et al.*, *Journal of Physics and Chemistry of Solids*, vol. 148, p. 109 738, 2021, ISSN: 0022-3697. DOI: 10.1016/j.jpcs.2020.109738.

[9] M. A. Olmstead, *Thin Films: Heteroepitaxial Systems*, vol. 15, p. 211, 1999.

[10] T. Grasser, *Microelectronics Reliability*, vol. 52, no. 1, pp. 39–70, 2012, invited. DOI: 10.1016/j.microrel.2011.09.002.

[11] J. Michl *et al.*, *IEEE Transactions on Electron Devices*, vol. 68, no. 12, pp. 6365–6371, 2021. DOI: 10.1109/TED.2021.3116931.

[12] W. Gös *et al.*, *Microelectronics Reliability*, vol. 87, pp. 286–320, 2018. DOI: 10.1016/j.microrel.2017.12.021.

[13] T. D. Kühne *et al.*, *The Journal of Chemical Physics*, vol. 152, no. 19, p. 194 103, 2020. DOI: 10.1063/5.0007045.

[14] J. P. Perdew *et al.*, *Phys. Rev. Lett.*, vol. 100, p. 136 406, 13 2008. DOI: 10.1103/PhysRevLett.100.136406.

[15] G. W. Rubloff, *Phys. Rev. B*, vol. 5, pp. 662–684, 2 1972. DOI: 10.1103/PhysRevB.5.662.

[16] J. H. Skone *et al.*, *Phys. Rev. B*, vol. 89, p. 195 112, 19 2014. DOI: 10.1103/PhysRevB.89.195112.

[17] R. Resta *et al.*, in *Physics of Ferroelectrics: A Modern Perspective*. Berlin, Heidelberg: Springer Berlin Heidelberg, 2007, pp. 31–68. DOI: 10.1007/978-3-540-34591-6_2.

[18] M. Sugiyama *et al.*, *Microelectronics Journal*, vol. 27, no. 4, pp. 361–382, 1996, LDSD: Epitaxial Deposition Techniques and Materials Systems, ISSN: 0026-2692. DOI: https://doi.org/10.1016/0026-2692(95)00062-3.

[19] C. Wong *et al.*, *Journal of Physics and Chemistry of Solids*, vol. 28, no. 7, pp. 1225–1231, 1967, ISSN: 0022-3697. DOI: https://doi.org/10.1016/0022-3697(67)90065-0.

[20] R. P. Lowndes, *Journal of Physics C: Solid State Physics*, vol. 2, no. 9, pp. 1595–1605, 1969. DOI: 10.1088/0022-3719/2/9/309.

[21] J. Hartnett *et al.*, *IEEE Transactions on Ultrasonics, Ferroelectrics, and Frequency Control*, vol. 51, no. 4, pp. 380–386, 2004. DOI: 10.1109/TUFFC.2004.1295423.

[22] C. G. Van de Walle *et al.*, *Journal of Applied Physics*, vol. 95, no. 8, pp. 3851–3879, 2004. DOI: 10.1063/1.1682673.

[23] G. Makov *et al.*, *Phys. Rev. B*, vol. 51, pp. 4014–4022, 7 1995. DOI: 10.1103/PhysRevB.51.4014.

[24] Y. Wimmer, PhD thesis, TU Wien, 2017.

Impact of channel thickness scaling on the performance of GaN-on-Si RF HEMTs on highly C-doped GaN buffer

Alireza Alian[1], Raul Rodriguez[1], Sachin Yadav[1], Uthayasankaran Peralagu[1], Arturo Sibaja Hernandez[1], Vamsi Putcha[1], Ming Zhao[1], Rana ElKashlan[1,2], Bjorn Vermeersch[1], Hao Yu[1], Erik bury[1], Ahmad Khaled[1], Nadine Collaert[1], Bertrand Parvais[1,2]
[1]imec, Leuven, Belgium, Email: Alian@imec.be [2]VUB, Brussels, Belgium.

Abstract- **This work investigates scaling of the GaN channel thickness on top of a carbon-doped GaN buffer (cGaN) grown on 200mm Si substrates. Device performance tradeoffs are analyzed in terms of DC, RF, reliability and thermal behavior. A thinner channel improves *DIBL*, *I_off*, *V_th* roll-off and degrades *f_T*, *f_max*, *PAE*, *Pout*, charge trapping and thermal conductance characteristics. Transconductance was observed to increase under high saturation drain bias for thin and short channels indicating the lateral device scaling potential of cGaN based HEMTS.**
Keywords—GaN, HEMT, channel thickness, scaling

I. Introduction

GaN-based HEMT devices have been widely studied due to their promising RF, switching and power handling capabilities [1-4]. Scaling has pushed GaN HEMTs to higher operating frequencies [5]. However, as the lateral dimensions of the devices shrink, the short channel effects (like increased *DIBL*, degraded *I_off* and *V_th* roll off [6]) need to be addressed by proper vertical (barrier and channel thickness) scaling. This paper investigates scaling of the GaN channel thickness, t_{ch}, epitaxially grown on cGaN buffer and the resulting DC, RF, reliability and thermal performance trade-offs.

It has been widely acknowledged that besides scaling up the buffer thickness, doping with C is a very effective way to increase the resistance of an (Al)GaN buffer layer [7,8]. Therefore, using cGaN or cAlGaN as a back barrier can significantly suppress device off-state leakage and enhance the confinement of 2DEG [9,10]. However, this is often accompanied with an increase of the on-resistance (R_{on}) of the device [10] which is also impacted by the GaN channel thickness as will be shown in this work.

Fig 1 shows the cross-section TEM of the fabricated devices (fabrication flow in [2]) for t_{ch}=300nm. In the gate window, the 5nm Si₃N₄ cap is etched away, followed by the etching of the 15nm AlGaN barrier down to 8nm, a wet clean step and gate metal stack deposition to make the HEMT devices. Different channel thicknesses down to 0nm (= barrier directly deposited on the cGaN) were grown. AFM images of the different t_{ch} splits are shown in fig 2. For GaN channels of 20nm and thicker, the surface morphology remains similar, while higher defectivity is observed for t_{ch}=10nm, and the surface morphology strongly degrades without the GaN channel.

II. DC Characteristics

As shown in fig 3, when channel becomes thinner, the 2DEG sheet resistance (R_s) as well as the metal/2DEG contact resistance (R_c) increase. This can be understood from the Poisson-Schrodinger simulations in fig 4. Here, C_N acceptors pin the Fermi level of cGaN in the lower half of the band gap [11]. Thinner channels cause a larger vertical electric field in the GaN channel which further confines the 2DEG. This can

degrade the carrier mobility as the charge centroid is observed to shift towards the AlN/GaN interface. Furthermore, the 2DEG density decreases and thereby R_s of the 2DEG eventually increases as channel is thinned down. The degree of band bending due to cGaN depends on the C concentration in cGaN. When C concentration in cGaN reduces below the reference level of $2x10^{19}$ cm⁻³, the 2DEG charge density increases, especially for thinner channels. The measured 2DEG mobility drops as channel thins down (fig 5). In addition to the enhanced carrier confinement at the barrier interface, the mobility degradation can be caused by a possibly lower epitaxial quality GaN layer closer to cGaN.

As the channel is thinned down, a positive V_{th} shift, improved *DIBL* and I_{off} reduction are observed from the transfer characteristics of devices with a gate length (L_g) of 90nm (fig 6a). This is expected as the cGaN acts as a back barrier which improves the electrostatic control over the channel (fig 4a).

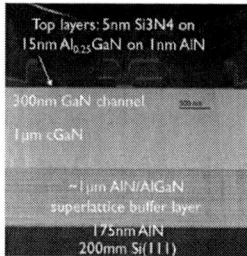

Fig 1- cross section HAADF-TEM of the devices epi stack.

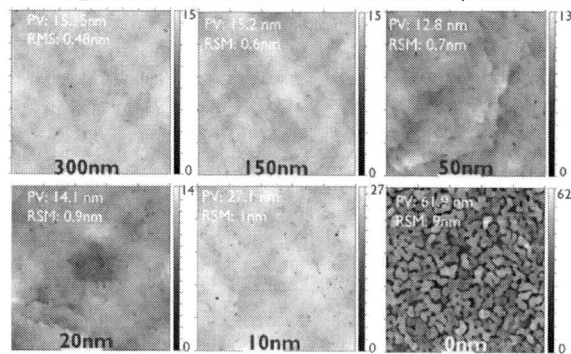

Fig 2- AFM image (10µm window) of different stacks with different t_{ch}

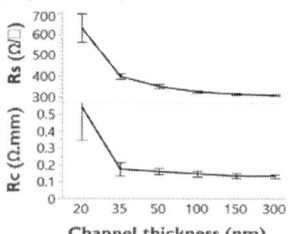

Fig 3- contact resistance (R_c) and 2DEG sheet resistance (R_s) vs. t_{ch}.

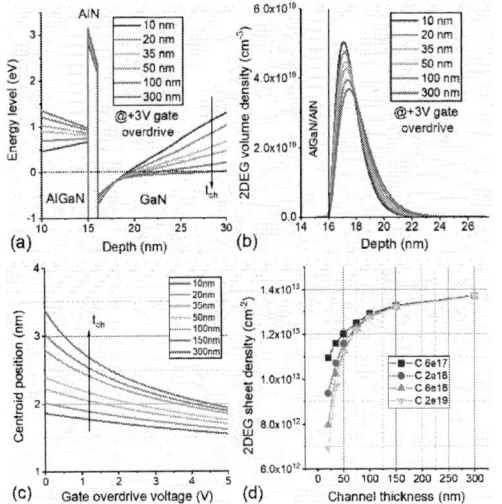

Fig 4- (a) 1D band diagram and (b) 2DEG distribution simulation for varied t_{ch} at 3V gate overdrive voltage. (c) 2DEG centroid distance to AlN/GaN interface. (d) 2DEG charge density.

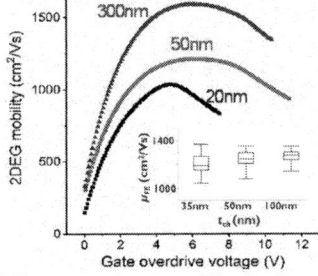

Fig 5- 2DEG mobility measured on a gated Van Der Paw structure with 5nm Si_3N_4 on 15nm/1nm AlGaN/AlN barrier resembling the access region of the HEMT. Inset shows the field effect mobility extracted at peak g_m.

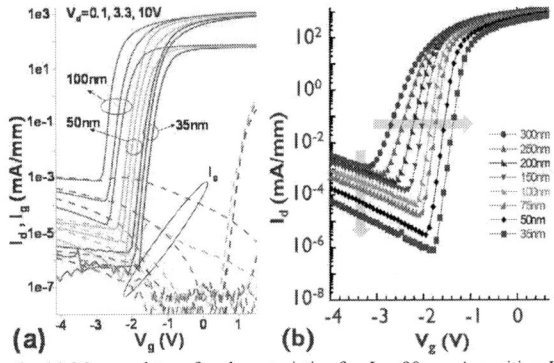

Fig 6 - (a) Measured transfer characteristics for L_g= 90nm. A positive V_{th} shift and I_{off} drop is observed for thinner t_{ch}. (b) simulated TCAD (Sentaurus Device) transfer curves (L_g=80nm) at V_d=3.3V showing similar V_{th} and I_{off} trend as observed experimentally for different channel thicknesses.

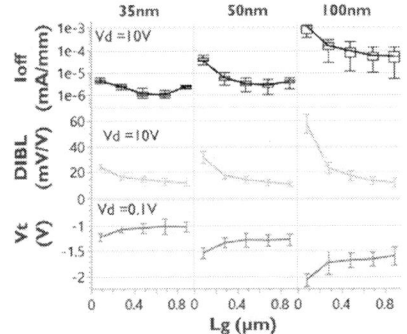

Fig 7- off state current (at 3V gate under drive), *DIBL* and threshold voltage roll-off improve as channel is thinned down.

Drift-diffusion TCAD simulations shown in fig 6b confirm the experimental results (positive V_{th} shift, I_{off} reduction and on-state performance degradation for the thinnest channels). These parameters are quantified (experimentally) in fig 7 for varying L_g and suggest improved short channel effects (SCE), expected from thinning down the channel.

The g_m shows a continuous drop at drain voltages (V_d) of 0.1V and 3.3V as channel is thinned down (fig 8). However, peak g_m is observed to increase significantly at t_{ch}=50nm compared to t_{ch}=100nm at V_d=10V. The peak g_m for t_{ch}=35nm is also observed to be higher than for t_{ch}=100nm at V_d=10V. The g_m is enhanced only for the shortest L_g at the highest measured V_d of 10V as channel is thinned down, while a degradation in g_m is observed for all other measured conditions as t_{ch} decreases. An electron velocity enhancement at V_d higher than the kink voltage is speculated to result in the enhanced g_m, as observed in the output characteristics. Corresponding output characteristics of L_g =90nm are shown in fig 9 which show a weaker kink response for the thinner channels.

III. Reliability aspects

Pulsed output characteristics (fig 9) show enhanced drive current as compared to DC measurements due to decreased self-heating effects. The kink is also significantly reduced compared to the DC output characteristics. Pulsed characteristics under gate lag, drain lag and high V_d stress Q-point bias conditions shown in fig 10 suggest stronger knee current collapse in case of t_{ch}= 35 nm and t_{ch}= 50 nm as compared to t_{ch}=100 nm. Fig 11 shows pulsed transfer

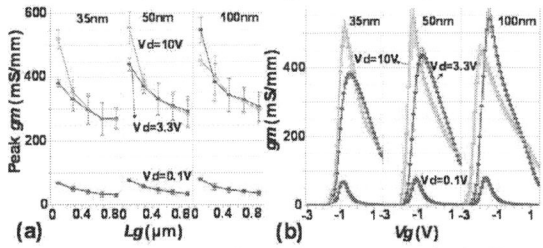

Fig 8- (a) peak g_m versus gate length. (b) g_m versus V_g for L_g=90nm. g_m decreases as channel is thinned down for V_d of 0.1V and 3.3V. At high V_d of 10V, g_m at 50nm and 35nm channel thicknesses is higher compared to the 100nm channel thickness.

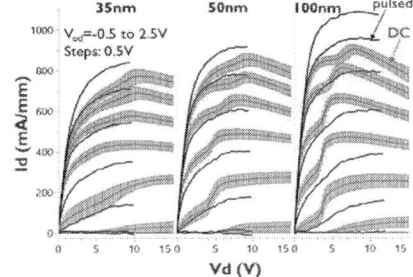

Fig 9- measured output characteristics for L_g =90nm (DC) and L_g =70nm (pulsed). Kink in the DC characteristics is weaker for thinner channels and is much less pronounced in the pulsed characteristics.

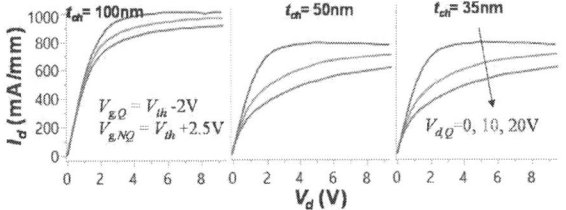

Fig 10- Pulsed output characteristics of device with L_g =70 nm under gate lag and drain lag Q-point bias conditions.

978-1-6654-8498-5/22 $31.00 © 2022 IEEE

characteristics of devices with L_g of 70 nm and varying channel thickness subjected to various Q-point biases. Increased dispersion of R_{on}, g_m and V_{th} is observed for thinner channels under off-state high V_d stress which can be attributed to increased interaction of C-defects with channel carriers. Therefore, the pronounced current collapse from off-state stress can impact device large-signal performance under deep class-AB biasing conditions for high V_d (>20 V).

Fig 12 shows the 2DEG sheet resistance dispersion measured with TLMs, after the buffer is subjected to different stress biases from -20V to -100V, following the methodology detailed in [12] (which characterises the gate-drain access region dispersion). A relative increase in 2DEG sheet resistance dispersion observed as channel thickness is decreased. The dispersion remains under 10% for the thinnest channel of 35nm. The increasing dispersion for thinner channels is attributed to the increased interaction between the 2DEG channel and the defects in cGaN [12].

The drain-source soft breakdown voltage measured for L_g of 90nm (fig 13a) suggest that the breakdown is not impacted by the channel thickness. Increasing the gate-drain distance (L_{gd}) significantly enhances the breakdown voltage and breakdown voltages as high as 450V are achieved for an L_{gd} of 5μm. Vertical buffer breakdown measurements as shown in fig 13b show no remarkable dependence on the channel thickness.

IV. Thermal aspects

Reducing GaN channel thickness worsens the self-heating (SH) inherently associated with RF power devices: both theoretical modelling (providing peak T at the active junction) and gate thermometry experiments (measuring average T of the gate electrode) show increases in thermal resistance (R_{th}) by 30%-50% (fig 14).

This somewhat counterintuitive effect arises because a thinner GaN layer leaves the heat less room to spread out laterally before it reaches the poorly thermally conductive AlGaN layer. This raises the R_{th} of the AlGaN due to the smaller flow path area. In addition, thinning the GaN reduces its effective thermal conductivity (TC) due to increased phonon boundary scattering. To account for thin film and interfacial filtering effects, we determine thickness-dependent in-plane and cross-plane GaN, AlGaN and AlN

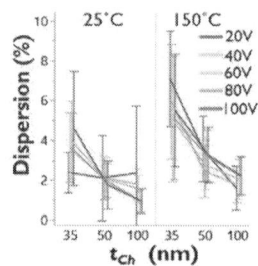

Fig 12- Relative TLM sheet resistance dispersion for different stress voltages between the TLM contacts and the backside of the wafer. stress duration was 10s and dispersion was measured after 5s relaxation.

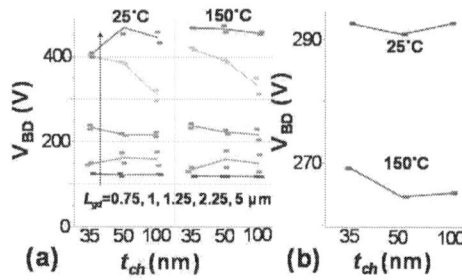

Fig 13- (a) Break down voltage defined at a drain-source current of 1mA/mm with the device biased at a gate voltage of -7V for varying L_{gd}. (b) vertical buffer breakdown defined at 1/10μA/mm² at 25/150°C.

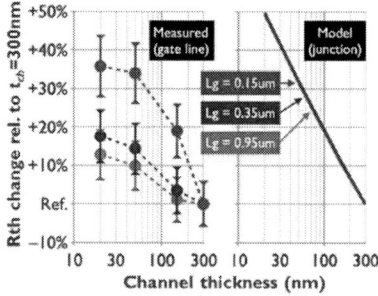

Fig 14-R_{th} increase for channel thinning observed through gate thermometry experiments (left) and junction temperature predictions (right). (structure is a 10μm wide single finger device)

TCs from Monte Carlo BTE simulations with first-principles phonon dispersions and scattering rates [13]. These TCs are then incorporated into a 2.5D thermal model [14] to assess SH for a variety of finger configurations. The impact of buffer doping is addressed by reducing the BTE-calibrated TC for undoped GaN by an additional factor 2, in line with studies on the effect of impurities in GaN films [15].

Our model with BTE-calibrated TCs reproduces S-parameter R_{th} measurements whereas the same model with nominal bulk TCs notably underestimates device SH (fig 15a), highlighting the importance of small-scale transport effects. Per the aforementioned heat spreading effect, R_{th} increases whenever the total GaN thickness (channel and/or buffer) is decreased (fig 15b). Device SH is a complex interplay of not only layer thicknesses and TCs but also device layout. Fig16a shows predicted R_{th} and temperature rise for a transistor with total effective width of 1 mm realised at different finger counts and pitch values. It is noted that the R_{th} peaks at the "squarest" layouts, i.e. those that minimise the aspect ratio and perimeter of the active region (fig 16b,c). Such configurations suffer most from finger cross-heating while benefiting least from lateral heat spreading along the sides so producing the highest R_{th}.

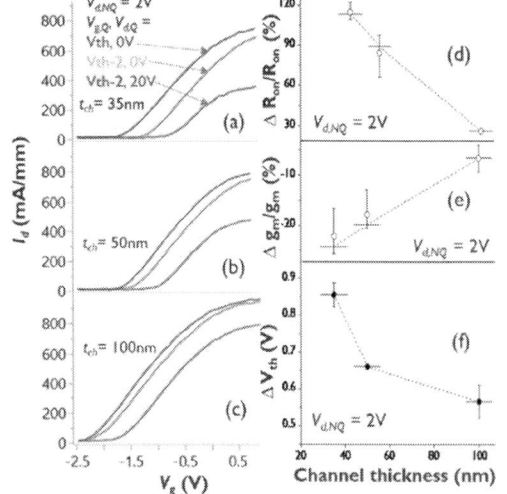

Fig 11- Pulsed I_d-V_g characteristics (L_g =70nm) at various Q-point biases. The dispersion of dynamic R_{on} (d), g_m (e), and V_{th} (f) is extracted using low-stress (V_{th},0V) and high-stress (V_{th} -2, 20V) Q-point conditions.

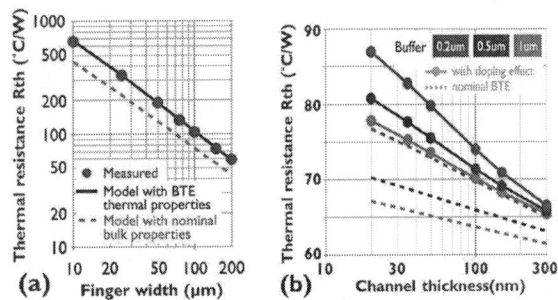

Fig 15- (a) Validation of our BTE-calibrated thermal model with S-parameter R_{th} measurements. (2 fingers at 60μm pitch, L_g=320nm, t_{ch}=300nm) – (b) Predicted thermal resistance (device self-heating) for various GaN channel and buffer thicknesses.

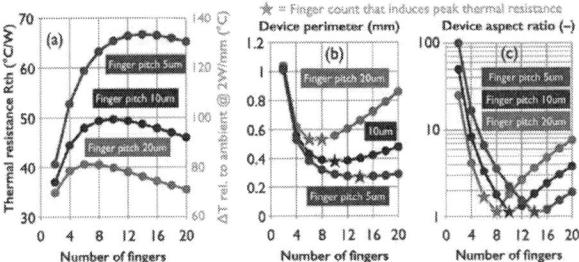

Fig 16 -(a) Predicted self-heating for devices at various layout configurations. Active region perimeter (b) and aspect ratio (c). (total device width =1mm, L_g=320nm, t_{ch}=300nm)

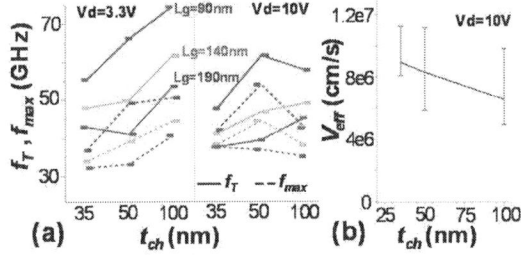

Fig 17- (a) f_T, f_{max} variations with channel thickness for varying L_g. (b) extracted effective electron velocity as a function of channel thickness.

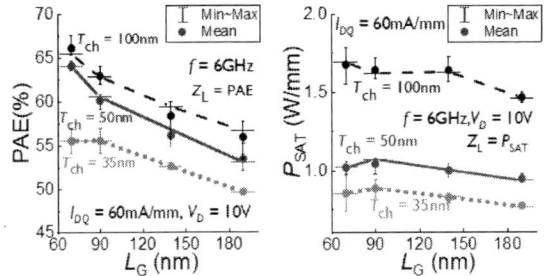

Fig 18- Passive load-pull characterisation at 6GHz .

V. RF characteristics

From the small signal measurement of f_T and f_{max} shown in fig. 17a, RF performance generally degrades for thinner t_{ch}. At V_d=10V, however, consistent with the g_m trend of fig 8, an enhanced performance is observed for t_{ch}= 50nm. Overall f_T and f_{max} are also limited by the source resistance (due to large gate-source spacing (L_{gs}) of 1.5μm) and field plate capacitances. Effective electron velocity, V_{eff}, extracted from the f_T data [16] shows that V_{eff} is increasing for thinner channels (fig 17b). The decreased charge density as shown in fig 4 can lead to increased carrier velocity in the access regions [17]. However, gate overdrive can compensate the charge difference in the gate region of the device. Passive load-pull characterization at 6 GHz was used to determine

HEMT large-signal performance metrics (*Pout* and *PAE*) under matched conditions (fig 18). *PAE* of >50% was achieved for devices with L_g in the range of ~70-150 nm at V_d of 10V. *PAE* for a device with L_g of 70 nm and t_{ch}= 100 nm is ~66% and *Pout* under power matching conditions is ~25 dBm (1.6 W/mm). *PAE* and *Pout* dropped for thinner channels which can be attributed to lower on-current, larger dispersion/current collapse and R_{th} as shown previously. Therefore, to enhance *Pout* and *PAE* simultaneously achieving SCE control using thinner channels, reduced parasitic resistance as well as engineering C-profile in cGaN for lower dispersion could be explored.

VI. Conclusions

Through extensive study of HEMT electrostatics, reliability and RF performance, we find that cGaN is a viable buffer option for sub-100 nm GaN HEMTs for mm-wave power amplifiers.

References

[1] H.W. Then et al, "GaN and Si Transistors on 300mm Si(111) Enabled by 3D Monolithic Heterogeneous Integration" VLSI symp., paper THL.2, 2020.

[2] U. Peralagu et al, "CMOS-compatible GaN-based devices on 200mm-Si for RF applications: Integration and Performance" IEDM, pp. 17.2.1-17.2.4, 2019.

[3] B. Parvais et al, "GaN-on-Si mm-wave RF Devices Integrated in a 200mm CMOS Compatible 3-Level Cu BEOL" , IEDM, pp. 8.1.1-8.1.4, 2020.

[4] K. Shinohara et al, "Self-aligned-gate GaN-HEMTs with heavily-doped n+-GaN ohmic contacts to 2DEG" IEDM, pp. 617-620, 2012.

[5] K. Shinohara et al, "Scaling of GaN HEMTs and Schottky Diodes for Submillimeter-Wave MMIC Applications" IEEE TED, vol. 60, no. 10, pp. 2982-2996, 2013.

[6] Y. Awano, M. Kosugi, K. Kosemura, T. Mimura, M. Abe "Short-channel effects in subquarter-micrometer-gate HEMTs: simulation and experiment" IEEE TED, vol. 36, no. 10, pp. 2260-2266, 1989.

[7] N. Ikeda et al, "GaN Power Transistors on Si Substrates for Switching Applications" Proc. IEEE 98, pp. 1151, 2010.

[8] I. B. Rowena, S. Lawrence Selvaraj and T. Egawa "Buffer Thickness Contribution to Suppress Vertical Leakage Current With High Breakdown Field (2.3 MV/cm) for GaN on Si" IEEE EDL. 32, pp. 1534, 2011.

[9] M. Zhao et al, "MOCVD growth of DH-HEMT buffers with low-temperature AlN interlayer on 200 mm Si (111) substrate for breakdown voltage enhancement" Phys. Status Solidi Curr. Top. Solid State Phys., vol. 13, no. 5-6, pp. 311-316, 2016.

[10] E. Bahat-Treidel et al, "AlGaN/GaN/GaN:C Back-Barrier HFETs With Breakdown Voltage of Over 1 kV and Low RON×A" IEEE TED, vol. 57, no. 11, pp. 3050-3058, 2010.

[11] J. L. Lyons, A. Janotti, and C. G. Van de Walle "Effects of carbon on the electrical and optical properties of InN, GaN, and AlN" Phys. Rev. B 89, pp. 035204-1-035204-8, 2014.

[12] V. Putcha et al, "On the impact of buffer and GaN-channel thickness on current dispersion for GaN-on-Si RF/mmWave devices" IEEE IRPS, pp. 1-8, 2021.

[13] J. Carrete et al, "almaBTE : A solver of the space–time dependent Boltzmann transport equation for phonons in structured materials" Computer Physics Communications 220, pp. 351-362, 2017.

[14] D. Maillet, S. André, J. C. Batsale, A. Degiovanni, C. Moyne "Thermal Quadrupoles: Solving the Heat Equation through Integral Transforms" ISBN 978-0-471-98320-0, Wiley, 2000.

[15] J. Zou, D. Kotchetkov, A. A. Balandin, D. I. Florescu, F. H. Pollak "Thermal conductivity of GaN films: Effects of impurities and dislocations" JAP 92, pp. 2534, 2002.

[16] M. Akita, S. Kishimoto, K. Maezawa, T. Mizutani "Evaluation of effective electron velocity in AlGaN/GaN HEMTs" Elec. Letters 36(20), pp. 1736-1737, 2000.

[17] S. Bajaj et al, "Density-dependent electron transport and precise modeling of GaN high electron mobility transistors" APL 107, pp. 153504, 2015.

Gap in pagination due to withheld paper.

Pages 388-391

A novel approach to analyze the reliability of GaN power HEMTs operating in a DC-DC Buck converter

Giuseppe Capasso*, Mauro Zanuccoli, Andrea Natale Tallarico and Claudio Fiegna

ARCES and DEI, University of Bologna, Cesena Campus, Italy

*giuseppe.capasso4@unibo.it

Abstract— **In this paper, we present a novel testbed based on a DC-DC synchronous Buck power converter, allowing the reliability analysis of GaN HEMTs with p-type gate. In particular, it is possible to monitor the drift of the device parameters to highlight the main degradation mechanisms affecting GaN transistors in power electronic applications. Stress is applied when HEMTs work within a practical 48/12V DC-DC converter operating at 1 MHz switching frequency and 4A output current. The reported analysis has been carried out under two different conditions, namely soft and hard stress, inducing a relatively low and high junction temperature, respectively. Results show that the high-side transistor of a DC-DC Buck converter is more prone to degradation, due to a larger threshold voltage and on-resistance drift. Moreover, based on the results of a validation analysis of the proposed characterization approach, the gate stack appears as the weaker transistor region causing device failure in the case of hard stress. Finally, a completely recoverable and a permanent V_{TH} and R_{ON} drift is observed in the case of soft and hard stress, respectively.**

Keywords—*GaN power HEMTs, reliability analysis, R_{ON} degradation, V_{TH} drift*

I. INTRODUCTION

Nowadays, several applications of power electronic, e.g. automotive industry, require more and more compact, efficient and high-power density DC-DC converters [1]. Gallium Nitride (GaN) technology, thanks to its low on-state resistance (R_{ON}), high switching speed, high breakdown voltage and comparable cost with respect to Silicon competitor, is gaining a primary role. However, to ensure a widespread adoption of GaN, a high level of reliability and robustness must be guaranteed. To date, GaN based high electron mobility transistors (HEMTs) are affected by degradation mechanisms which limit their reliability. For instance, these devices can be affected by current collapse increasing the dynamic R_{ON} or by threshold voltage (V_{TH}) instability, reducing the gate overdrive voltage. As a matter of fact, in order to identify the physical mechanisms responsible for degradation and to improve the device lifetime, GaN HEMTs long-term reliability is a relevant topic at research level [2]-[4].

V_{TH} instability is reported and investigated by several works [5]-[7]. In [5], the important role of the characterization methods, which can significantly affect the V_{TH} measurement, is discussed, both for pulsed and DC stress. In [6], it has been shown that the V_{TH} transients during positive gate bias stress are highly dependent on the

electrical current distribution in the gate stack of p-GaN HEMT. A link between gate current and V_{TH} shift during long-term positive bias temperature instability (PBTI) tests has been also observed in [7]. In addition to PBTI, the application of an off-state drain bias can induce positive V_{TH} shift in p-gate HEMTs [8].

Considerable effort has been devoted to monitor the dynamic R_{ON} under switching stress conditions (typical of a power converter application) [9][10]. Several methodologies and testbeds have been proposed to replicate switching conditions occurring in practical circuits [11]-[13]. In [11], the device response has been evaluated directly on-wafer, stressing the transistor through an off-state drain voltage sweep from 0 to 200V with 0.5 voltage step. Another technique consists of designing custom boards able to emulate current ringing, or voltage overshoot due to the inductor, and to measure R_{ON} value [12]. In [13], a non-invasive drain current sensing and drain-source voltage sampling block was added to a real circuit to continuously monitor R_{ON}. Many other papers study R_{ON} degradation under specific conditions, such as off-state voltage, drain current, junction temperature or on/off state time [14]-[16].

However, understanding the long-term reliability needs a more realistic context, involving realistic current and voltage transients at the device terminals; as a matter of fact, in a hard-switching power converter, due to relatively high switching frequency, voltage ringing affects sensitive nodes such as the transistors' gate and source terminals, possibly leading to significant power dissipation, hence, device heating, and electrical stress. In addition, a deep investigation about the degradation mechanisms requires a full I-V characterization of the device under test. In fact, the ΔR_{ON} simply measured as the variation of the ratio of drain voltage to drain current at a given bias, provides limited information as it is affected by different factors such as V_{TH} or transconductance (g_m) shift, which in turn can be triggered by different physical phenomena [17].

To allow the I-V characterization of GaN HEMTs operating in practical power circuits, a custom measurement test board able to carry out an in-situ characterization is proposed in this work. This board allows to straightforwardly acquire the input, transfer and output I-V characteristics of GaN transistors operating in a DC-DC synchronous Buck converter.

In the frame of a validation analysis, the devices behaviour under soft and hard switching stress is monitored,

978-1-6654-8498-5/22 $31.00 © 2022 IEEE

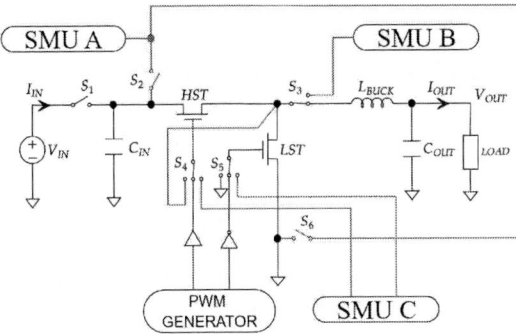

Fig. 1 - Simplified schematic of the setup adopted for the experimental in-situ characterization of the GaN HEMTs. The black wires and symbols denote the standard DC-DC synchronous Buck topology. The red portion of the network allows for the GaN I-V Characterization. Sj (j = 1..6) denotes the relays used for the hardware reconfiguration of the test circuit

highlighting the impact of the buck converter operation on the V_{TH}, R_{ON} and gate leakage of commercial GaN HEMTs.

II. SETUP DESCRIPTION

A simplified sketch of the novel setup adopted in this study is reported in Fig. 1. The circuit includes a standard type of synchronous Buck converter (in black) and the additional network (in red) to characterize the devices with a set of source measurement units (SMUs). In a synchronous Buck converter, the high-side transistor (HST) transfers energy to the output load, while the low-side transistor (LST) ensures continuity of the inductor current. The two transistors are subjected to different electrical and thermal stresses. For this reason, they both are characterized.

Stress is applied to transistors during power converter operation. After stress periods of pre-fixed duration, converter operation is stopped to allow transistors characterization by means of SMUs. Stress and characterization are successively repeated adopting a standard measure-stress-measure technique (Fig. 2). A hardware reconfiguration block based on electromechanical and solid-state relays enables the switching from Buck converter to characterization mode and viceversa. Such block as well as the characterization instruments are controlled by a virtual instrument. In addition, converter key-parameters, such as output voltage, input power and efficiency are continuously monitored during the stress. To estimate the junction temperature of the devices, a thermistor is mounted between the two GaN HEMTs. The performance deterioration of the converter circuit is linked to the device parameters degradation.

The monitoring of fast trapping and detrapping phenomena (<1s) is not possible because of the relatively long time required by the single device characterization (about 60s considering the soft shutdown of the converter and the subsequent device characterization). Therefore, the setup is aimed at studying the long-term reliability operating in a real-life application (DC-DC converter).

Commercial enhancement-mode p-type gate GaN-HEMTs featuring maximum drain-to-source voltage of 80V and maximum DC drain current of 6.8A are considered in this testbed validation.

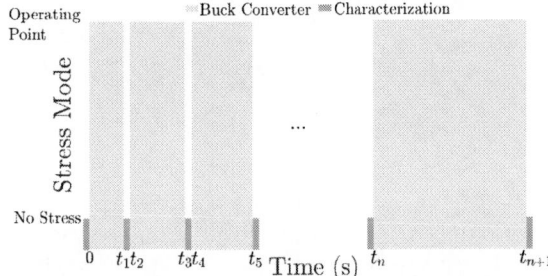

Fig. 2 – Illustration of the novel setup operation based on the measure-stress-measure technique. The stress phase is performed in the frame of power converter, whereas the measure one is carried out by SMUs.

A heat sink is required to avoid thermal stress to the transistors due to the significant junction to ambient thermal resistance of the small package.

III. RESULTS AND DISCUSSION

Experimental tests are performed by setting $V_{IN} = 48V$, $V_{OUT} = 12V$, $I_{OUT} = 4A$ (Fig.1) and switching frequency $f_S = 1$ MHz. This type of DC-DC converter plays a relevant role in electric and/or hybrid vehicles, because of its simplicity and high efficiency [18]. The stress is carried out under two different thermal conditions, namely:

- soft stress, where the heat sink is mounted to favour heat removal from the transistors;
- hard stress, where the transistors are exposed directly to the ambient to deliberately let the junction temperature of GaN transistors to increase.

The stress effect is monitored by measuring V_{TH}, R_{ON} and I_G from full I-V curves.

V_{TH} is extracted by constant-current method to avoid possible contributions coming from g_m degradation. V_{TH} is herewith defined as the value of V_{GS} corresponding to drain current $I_D = 3mA$. Thanks to the full I_D-V_{GS} characteristics, ΔR_{ON} is extracted at the same gate overdrive voltage ($V_{GS} - V_{TH}$) to evaluate the actual degradation, hence to avoid to include the contribution coming from ΔV_{TH} as in the case of online R_{ON} monitoring technique [13]. Finally, I_G-V_{GS} characteristic is measured.

Fig. 3 shows the V_{TH} drift during soft (top) and hard (bottom) stress for both HST and LST. First, it can be noted that, during soft stress (normal operation), the two transistors exhibit a different ΔV_{TH} dynamic: a non-monotonic time dependence for HST and monotonic increasing one for the LST. Such different behaviour is related to the asymmetric stress condition applied to the two GaN HEMTs in DC-DC converters. In particular, from numerical simulations of printed-circuit-board, power dissipation is estimated to be 1.6W and 0.5W for the HST and LST, respectively. Such difference induces a different junction temperature which is indirectly calculated (starting from the calculated power dissipation on transistors and thanks to the knowledge of thermal resistances provided by device and heat sink manufacturers) to be 60°C and 40°C for HST and LST, respectively. In Fig. 4, it is possible to note that the HST gate leakage, in the case of soft stress, starts to increase when the related V_{TH} (Fig. 3, upper tab) starts to decrease (1000 s). This is not observed in the case of LST. Concerning the ΔV_{TH} and I_G dynamics of the HST, similar behaviour is reported also in [7] under PBTI stress at device level, relating

978-1-6654-8498-5/22 $31.00 © 2022 IEEE   393

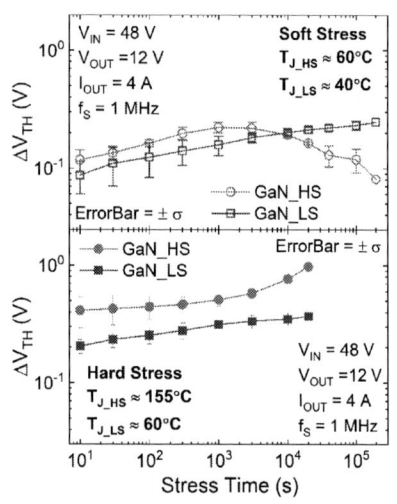

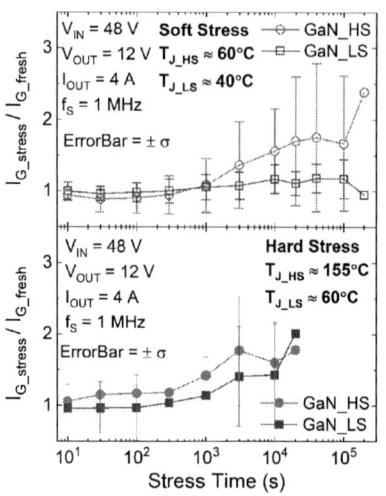

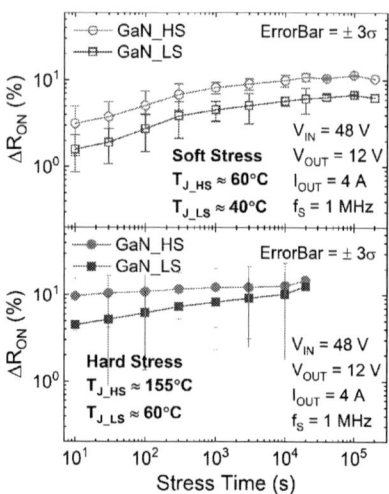

Fig. 3 – Threshold voltage degradation in the case of soft (top) and hard (bottom) stress; σ represents the standard deviation of the data.

Fig. 4 – Gate leakage degradation in the case of soft (top) and hard (bottom) stress; σ represents the standard deviation of the data.

Fig. 5 – On-state resistance degradation in the case of soft (top) and hard (bottom) stress; σ represents the standard deviation of the data.

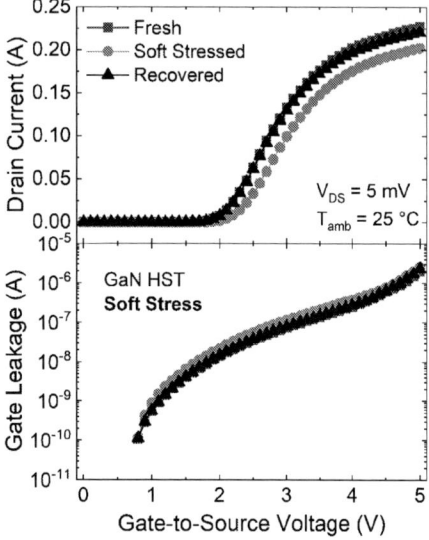

Fig. 6 – Transfer (top) and gate leakage (bottom) characteristics in the case of soft stress test.

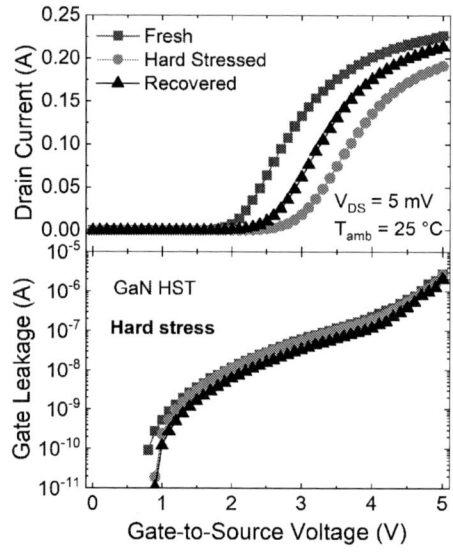

Fig. 7 – Transfer (top) and gate leakage (bottom) characteristics in the case of hard stress test.

the ΔV_{TH} reduction, occurring at relatively long stress-time, to the gate leakage increase caused by the combined effect of positive gate bias and relatively high temperature.

By removing the heat sink (hard stress condition) the junction temperature of the HST exceeds the maximum junction temperature operating rating (150 °C). In this case a larger V_{TH} drift (up to 1 V) is observed, leading to device failure after ∼ 25 ks. Also in this case, it is worth noting the larger degradation of the HST, which is the first one to fail.

Concerning the ΔRON (Fig. 5), the soft stress (top) induces the same monotonous increase in both transistors up to around 10 ks, after which a saturation effect is observed. Also in this case, the HST exhibits a larger ΔRON, highlighting the extent of the level of criticality of HST in terms of reliability. In the case of hard stress (Fig. 5 bottom), ΔRON exhibits a different trend with respect to the

soft stress test. In particular, the HST shows a 10% shift after only 10s, whereas the ΔR_{ON} of LST is characterized by a lower starting value and a higher slope, reaching approximately comparable drift at 20 ks, which is of the same magnitude of the hard stress case (HST).

Finally, a last characterization has been performed after 14 hours of recovery, showing that, parameters drifts (V_{TH} and R_{ON}) observed after soft stress are completely recoverable (Fig. 6), suggesting trapping mechanisms in pre-existing defects. A different picture is observed after the hard stress condition, where a permanent degradation persists after 90 hours, ascribed to the creation of new deep defects (Fig. 7).

Overall, from the reported testbed validation analysis, it is possible to recognize that a limited dispersion of the results obtained from experiments carried on different nominally identical devices allows to identify repeatable degradation trends, such as: i) HST is more prone to degradation because of the larger stress intrinsically induced by the DC-DC Buck converter; ii) a significant V_{TH} shift, indicating gate degradation, occurs. It is markedly larger in the case of hard stress; iii) the noticeably differences in ΔV_{TH} and ΔR_{ON} after soft and hard stress, suggest that the premature device failure may be ascribed to gate breakdown.

IV. CONCLUSIONS

A DC-DC power converter aimed at experimentally characterizing in-situ the reliability of GaN HEMTs has been implemented. The test circuit allows to stress power transistors under realistic electrical and thermal conditions and to monitor different signatures (drift of V_{TH}, R_{ON}, I_G, etc.) of their degradation, providing useful information for both GaN circuit designers and GaN technology manufacturers. The results of a validation analysis highlight that the developed testbed allows to identify the main degradation trends in the frame of realistic transistor operation. The obtained results show that: i) high-side transistor is more prone to degradation compared to low-side one; ii) the two transistors show a different ΔV_{TH} dynamic, suggesting that different degradation mechanisms are involved; iii) although the significant ΔR_{ON}, the gate stack seems to be the device region responsible for device failure under hard stress condition, supported by large ΔV_{TH} and gate leakage drift.

ACKNOWLEDGMENT

Part of this work is funded by iRel40, a European co-funded innovation project that has been granted by the ECSEL Joint Undertaking (JU) under grant agreement No 876659. The funding of the project comes from the Horizon 2020 research programme and participating countries. National funding is provided by Germany, including the Free States of Saxony and Thuringia, Austria, Belgium, Finland, France, Italy, the Netherlands, Slovakia, Spain, Sweden, and Turkey.

REFERENCES

[1] M. de Rooij and Q. Laidebeur, "Meeting the Power and Magnetic Design Challenges of Ultra-Thin, High-Power Density 48 V DC-DC Converters for Ultra-Thin Computing Applications," in IEEE Power Electronics Magazine, vol. 8, no. 3, pp. 30-36, Sept. 2021, doi: 10.1109/MPEL.2021.3099518.

[2] J S. Li, S. Yang, S. Han and K. Sheng, "Investigation of Temperature-Dependent Dynamic RON of GaN HEMT with Hybrid-Drain under Hard and Soft Switching," 2020 32nd International Symposium on Power Semiconductor Devices and ICs (ISPSD), 2020, pp. 306-309, doi: 10.1109/ISPSD46842.2020.9170048.

[3] A. N. Tallarico, S. Stoffels, N. Posthuma, S. Decoutere, E. Sangiorgi and C. Fiegna, "Threshold Voltage Instability in GaN HEMTs With p-Type Gate: Mg Doping Compensation," in IEEE Electron Device Letters, vol. 40, no. 4, pp. 518-521, April 2019, doi: 10.1109/LED.2019.2897911.

[4] N. Modolo et al., "Cumulative Hot-Electron Trapping in GaN-Based Power HEMTs Observed by an Ultra-Fast (10V/ns) on-Wafer Methodology," in IEEE Journal of Emerging and Selected Topics in Power Electronics, doi: 10.1109/JESTPE.2021.3077127.

[5] K. Murukesan, L. Efthymiou and F. Udrea, "Gate stress induced threshold voltage instability and its significance for reliable threshold voltage measurement in p-GaN HEMT," 2019 IEEE 7th Workshop on Wide Bandgap Power Devices and Applications (WiPDA), 2019, pp. 177-180, doi: 10.1109/WiPDA46397.2019.8998859.

[6] A. Stockman and P. Moens, "ON-State Gate Stress Induced Threshold Voltage Instabilities in p-GaN Gate AlGaN/GaN HEMTs," 2020 IEEE International Integrated Reliability Workshop (IIRW), 2020, pp. 1-4, doi: 10.1109/IIRW49815.2020.9312869.

[7] A. N. Tallarico et al., "Gate Reliability of p-GaN HEMT With Gate Metal Retraction," in IEEE Transactions on Electron Devices, vol. 66, no. 11, pp. 4829-4835, Nov. 2019, doi: 10.1109/TED.2019.2938598.

[8] L. Efthymiou, K. Murukesan, G. Longobardi, F. Udrea, A. Shibib and K. Terrill, "Understanding the Threshold Voltage Instability During OFF-State Stress in p-GaN HEMTs," in IEEE Electron Device Letters, vol. 40, no. 8, pp. 1253-1256, Aug. 2019, doi: 10.1109/LED.2019.2925776.

[9] N. Modolo et al., "Cumulative Hot-Electron Trapping in GaN-Based Power HEMTs Observed by an Ultra-Fast (10V/ns) on-Wafer Methodology," in IEEE Journal of Emerging and Selected Topics in Power Electronics, doi: 10.1109/JESTPE.2021.3077127.

[10] R. Li, X. Wu, S. Yang and K. Sheng, "Dynamic on-state resistance test and evaluation of GaN power devices under hard- and soft-switching conditions by double and multiple pulses", IEEE Trans. Power Electron., vol. 34, no. 2, pp. 1044-1053, Feb. 2019.

[11] B. Lu, T. Palacios, D. Risbud, S. Bahl and D. I. Anderson, "Extraction of dynamic on-resistance in GaN transistors: Under soft-and hard-switching conditions", Proc. Compound Semicond. Integr. Circuit Symp., pp. 1-4, 2011.

[12] J. -M. Zhang et al., "Bias- and temperature-assisted trapping/de-trapping of ron degradation in D-mode AlGaN/GaN MIS-HEMTs on a Si substrate," 2017 IEEE 24th International Symposium on the Physical and Failure Analysis of Integrated Circuits (IPFA), 2017, pp. 1-4, doi: 10.1109/IPFA.2017.8060181.

[13] M. Pizzotti, M. Crescentini, A. N. Tallarico and A. Romani, "An Integrated DC/DC Converter with Online Monitoring of Hot-Carrier Degradation," 2019 26th IEEE International Conference on Electronics, Circuits and Systems (ICECS), 2019, pp. 562-565, doi: 10.1109/ICECS46596.2019.8964721.

[14] Li, Yuan & Zhao, Yuanfu & Huang, Alex & Zhang, Liqi & Lei, Yang & Yu, Ruiyang & Ma, Qingxuan & Huang, Qingyun & Sen, Soumik & Jia, Yunpeng & He, Yunlong. (2019). Evaluation and Analysis of Temperature-dependent Dynamic $R_{DS,ON}$ of GaN Power Devices Considering High-Frequency Operation. IEEE Journal of Emerging and Selected Topics in Power Electronics. PP. 1-1. 10.1109/JESTPE.2019.2947575.

[15] K. Li, A. Videt, N. Idir, P. L. Evans and C. M. Johnson, "Accurate Measurement of Dynamic on-State Resistances of GaN Devices Under Reverse and Forward Conduction in High Frequency Power Converter," in IEEE Transactions on Power Electronics, vol. 35, no. 9, pp. 9650-9660, Sept. 2020, doi: 10.1109/TPEL.2019.2961604.

[16] T. Foulkes, T. Modeer and R. C. N. Pilawa-Podgurski, "Quantifying Dynamic On-State Resistance of GaN HEMTs for Power Converter Design via a Survey of Low and High Voltage Devices," in IEEE Journal of Emerging and Selected Topics in Power Electronics, vol. 9, no. 4, pp. 4036-4049, Aug. 2021, doi: 10.1109/JESTPE.2020.3024930.

[17] Nakayama, T. & Mannen, T. & Nakajima, Akira & Isobe, Takanori. (2020). Gate threshold voltage instability and on-resistance degradation under reverse current conduction stress on E-mode GaN-HEMTs. Microelectronics Reliability. 114. 113840. 10.1016/j.microrel.2020.113840.

[18] X. Huang, F. C. Lee, Q. Li, and W. Du, "High-Frequency High-Efficiency GaN-Based Interleaved CRM Bidirectional Buck/Boost Converter with Inverse Coupled Inductor," IEEE Transactions on Power Electronics, vol. 31, pp. 4343-4352, 2016.

A novel depletion mode p-GaN island HEMT and its use in a monolithically integrated start-up circuit

Loizos Efthymiou, Martin Arnold, Giorgia Longobardi, Florin Udrea
Cambridge GaN Devices Ltd.
Cambridge, United Kingdom
loizos.efthymiou@camgandevices.com

Abstract—**In this study we present the concept and features of a novel depletion mode p-GaN island HEMT. The HEMT presented was designed and fabricated on a state-of-the-art GaN-on-Si heterojunction process. The gate of the novel transistor comprises a row of p-GaN islands, displaced in the third dimension. Depending on the gate-source bias the device may operate in three distinctive regions identified by significant changes in its on-state resistance. The edges of the three distinctive regions are defined by the two threshold voltages of the device. The first threshold voltage is negative and signifies the first turn-on of the device. By adjusting the separation of the p-GaN islands the first threshold voltage can be controlled by layout design rather than only process design. The p-GaN island HEMT in this study was designed and fabricated as both a low voltage device (30V rating) and high voltage device (650V rating). Moreover, the device was demonstrated experimentally, as a key component in a start-up circuit, monolithically integrated within a fully functional power IC for 650V applications.**

Keywords—*HEMT, AlGaN/GaN, depletion mode, GaN IC, TCAD, start-up, normally-on, pGaN.*

I. INTRODUCTION

Gallium Nitride (GaN) is considered one of the most exciting materials for use in the field of power devices. The wide band gap of the material (Eg=3.39eV) results in high critical electric field (Ec=3.3MV/cm) which can lead to the design of devices with a shorter drift region, and therefore lower on-state resistance compared to a silicon-based device with the same breakdown voltage. The use of an Aluminium Gallium Nitride (AlGaN)/GaN heterostructure also allows the formation of a two-dimensional electron gas (2DEG) at the hetero-interface of high carrier density (e.g. 1×10^{13} cm^{-2} [1]) where carriers can reach very high mobility values (μ=2000cm^2/(Vs)) [2]. These properties allow the development of High Electron Mobility Transistors (HEMTs) with very competitive performance parameters.

The 2DEG which inherently exists at the AlGaN/GaN hetero-interface creates a challenge when attempting the design of enhancement mode rather than depletion mode devices. Nonetheless, several methods have been proposed which can lead to enhancement mode devices. Due to the relative maturity and controllability in the epitaxial growth of p-GaN layers, p-GaN/AlGaN/GaN HEMTs are currently the leading structure for commercialization. Fig. 1 shows schematically a cross section of the active area of a p-GaN HEMT. A thin cap GaN layer is typically added to form the gate with a Magnesium (Mg) p-type doping density greater than 1×10^{19}cm^{-3} [3] [4]. A typical p-GaN gate device has a threshold voltage of around 1.5V [4] and in operation has a maximum allowable gate-source bias of around 7V if a Schottky contact is used on the p-GaN [5].

While an enhancement mode device is used in most power electronic applications as the main power switch, there are numerous functions in a GaN power integrated circuit (IC) where a depletion mode device can be of use. Most commonly, a depletion mode AlGaN/GaN transistor is made by placing a Schottky metal contact, which acts as the gate terminal, directly on the AlGaN layer (see Fig. 2) excluding the p-GaN cap layer which exists in the enhancement mode devices.

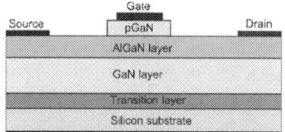 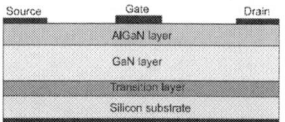

Figure 1: Enhancement mode HEMT with p-GaN gate

Figure 2: Depletion mode HEMT with Schottky gate

In depletion mode Schottky gate devices the threshold voltage of the device is dependent on process parameters such as, but not limited to, the AlGaN layer thickness, aluminium mole fraction and choice of gate metal. Therefore to adjust the threshold voltage to a level which is most suitable for a specific function, would require a change in the epitaxial growth and/or the gate metal processing, which can be time consuming and expensive. The ability to reliably adjust the device threshold voltage through layout modifications would be significantly less time consuming and more cost efficient in comparison. Furthermore, Schottky gate depletion mode devices in the on-state, have a limit on the maximum positive gate-source bias voltage which can be applied before the main on-state conduction path changes from drain-source to gate-source. This maximum bias voltage depends on the Schottky barrier height present at the gate contact and does not exceed 2V [6].

The patented depletion mode HEMT device [7] presented in this study utilizes p-GaN islands to create a novel device where the device threshold voltage is controllable by layout design and where allowable positive gate-source bias can significantly exceed what was previously available in a Schottky gate depletion mode HEMT. In section II of this study, the design and characterisation of the p-GaN island HEMT are presented. In section III, a TCAD model of the device was developed to support the experimental findings and strengthen the understanding of the operation of the device. In section IV, an attractive application for the p-GaN island HEMT in a GaN power IC start-up circuit is demonstrated experimentally.

II. RESULTS AND DISCUSSION

A three-dimensional drawing of the depletion mode p-island HEMT is shown in Fig. 3. This device was designed in a state-of-the-art 6-inch AlGaN/GaN process featuring an advanced Magnesium-based gate feature. The HEMT was designed as an interdigitated finger layout device with a gate width of 3mm. The device in Fig. 3 is designed as a low voltage device (30V rating) with a short drift region and no field plate design. Device dimensions are Lgs=0.8um, Lg=2um, Lgd=2um. In the third dimension, the gate of the device illustrated in Fig. 3 consists of a row of p-GaN islands rather than a continuous stripe of p-GaN, as is the case in an enhancement mode p-GaN HEMT. Fig. 4 illustrates a cross-section of the device along the gate finger shown in Fig. 3. A gate metal contact is made on each individual p-GaN island, and the contacts on the p-GaN islands are connected using a track metal layer. The p-GaN islands are separated by a dimension Wgap=0.4um and have a width Wisl=2.6um as shown. Through adjustments in these dimensions, the threshold voltage and resistance of the device under different gate-bias conditions may be controlled.

The measured transfer characteristic of a p-GaN island HEMT is illustrated in Fig. 5. According to the gate-source bias (V_{gs}) the device may operate in three distinctive regions (off-state, first conductive region, second conductive region) identified by significant changes in the resistance of the device. The edges of the three distinctive regions are defined by the two threshold voltages of the device shown in Fig. 5.

The first threshold voltage (Vth1) is negative and can be advantageously controlled by adjusting the separation of the p-GaN islands which form the gate. Vth1 signifies the first turn-on of the device where the device goes from the off-state region to the first conductive region of operation as V_{gs} bias increases (i.e. becomes less negative). In the device illustrated, the Wgap dimension is 0.4um. Increasing the separation between the pGaN islands leads to the first threshold voltage becoming more negative and vice-versa.

heterojunction under the p-GaN islands. As Vgs decreases the depletion region expands both vertically (y-dimension) and laterally (x-dimension) depleting the 2DEG channel in the gaps between the p-GaN islands. This is similar to a JFET effect. As the gate-source voltage is increased above the first threshold voltage, but remaining below the second threshold voltage, the formation of the 2DEG channel spreads from the middle of the pitch between adjacent pGaN islands towards the edges of the pGaN gate islands as the depletion region retracts. The current continues to increase as the on-state resistance decreases linearly. The resistance in the first conductive region can be controlled by the spacing (Wgap) and width (Wisl) of the p-GaN islands.

The second threshold voltage (Vth2) of the p-GaN island device is defined by the threshold voltage of the p-GaN gate as in an enhancement mode HEMT. It is therefore dependent on the process and is approximately around 1.5V [4]. As the device moves from the first conductive region to the second conductive region the resistance of the device decreases exponentially. In this region of operation, the gate-source bias applied modulates the 2DEG under the p-GaN islands causing the 2DEG carrier concentration to significantly increase as the gate-source bias is increased above the second threshold voltage. As such, in the second conductive region the current between the drain and source terminals flows along the entirety of the gate width of the interdigitated finger layout.

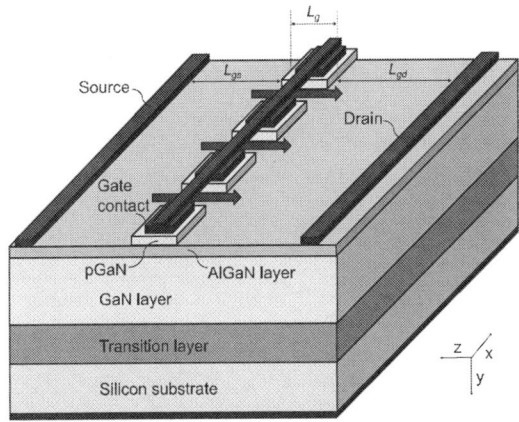

Figure 3: Depletion mode p-GaN island HEMT illustration

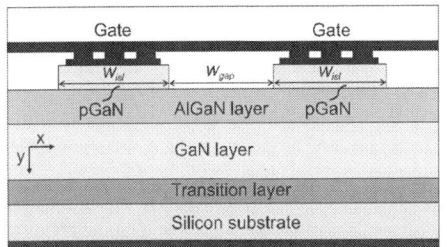

Figure 4: Cross section of p-GaN island HEMT along gate width

Moving from the off-state region to the first conductive region of operation, the current between the drain and source terminal starts flowing via the 2DEG channel present at the AlGaN/GaN interface where p-GaN islands are not present adjacently above the 2DEG. The bias of the gate contact on the p-GaN islands (relative to the source) modulates this conductive path between the drain and source terminal by adjusting the depletion region of the reverse biased

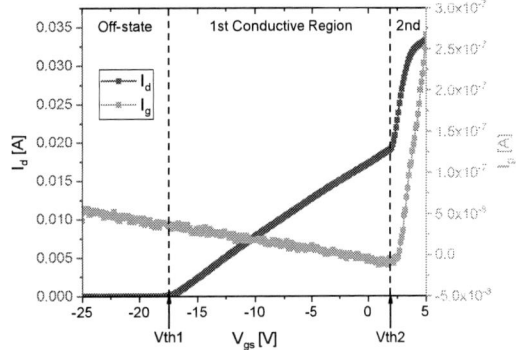

Figure 5: Transfer characteristic and gate leakage at V_{ds}=0.1V

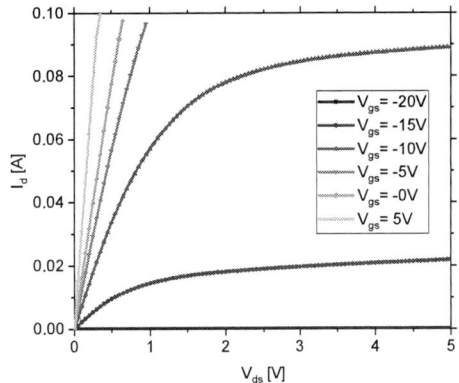

Figure 6: Output characteristic

The output characteristic of the p-GaN island device which further illustrates the three distinctive regions of operation is shown in Fig. 6. As described above, the device illustrated in Fig. 3 is designed as low voltage device. However, the p-GaN island HEMT concept can also be used in a high voltage device. Such a device was fabricated and used in the exemplar start-up application described in section IV.

III. TCAD MODELLING

A TCAD model using the Synopsys Sentaurus tool was developed to better understand the operation of the p-GaN island HEMT. The aim was to observe the carrier concentration in the 2DEG between two adjacent p-GaN islands as the gate-source bias is varied. The Vgs at which the 2DEG is fully depleted for a given p-GaN island separation distance (Wgap) can indicate the first threshold voltage (Vth1) of the device. Furthermore, the model can reveal which process related parameters can have a significant effect on Vth1.

A three-dimensional (3D) model was built (see Fig. 7). The model comprises a source terminal and two p-GaN islands. The drain terminal was not included in the model to limit the number of nodes and enable fast and robust simulations. Furthermore, the TCAD model ignores the complexity of the Si substrate, transition layer and the associated interfaces in these layers. The model comprises a thin AlN layer connected to a thin Si substrate to provide a back barrier. The AlN back barrier layer is connected to the C-doped GaN buffer layer. This simplification is acceptable in this case as simulations with a large vertical bias are not conducted. Fixed charges were included in the TCAD simulation deck according to Ambacher et al [1] to consider the piezo-polarisation effect observed in GaN devices. A p-type doping of $5x10^{16}$ cm^{-3} was included in the GaN buffer layer to take into account the carbon doping [8]. Finally, a thin cap GaN layer was added to form the p-GaN islands with a Magnesium (Mg) p-type doping density of $2x10^{19}$ cm^{-3}. Mg doping is included as acceptors at an energy level of 170meV above the valence band [9]. Mg out-diffusion into the AlGaN barrier and the GaN channel is simulated as well [4]. A work function of 4.6eV was used for the Schottky gate contact on the p-GaN islands [9]. The separate gate contacts illustrated on the two p-GaN islands are defined as the same electrode.

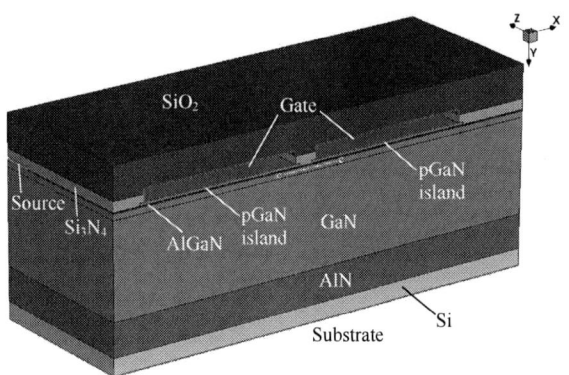

Figure 7: 3D TCAD model

Gate-source voltage, Vgs was swept from 0V to -20V in a transient simulation with the substrate electrode connected to source potential. Fig. 8 illustrates the electron carrier density across cutline cc in Fig. 7. As described in section II, a more negative Vgs leads to a depletion of the 2DEG from the edges of the p-GaN islands to the midpoint of the pitch of the two islands. The 2DEG appears depleted at Vgs=-18V. Wgap=0.4um for the TCAD results shown in Fig. 8 as is the case for the experimental device in section II. The electron carrier concentration at the mid-point of the p-GaN island separation was plotted against Vgs for various Wgap values as seen in Fig. 9. The simulation result confirms the influence of Wgap on Vth1 as expected.

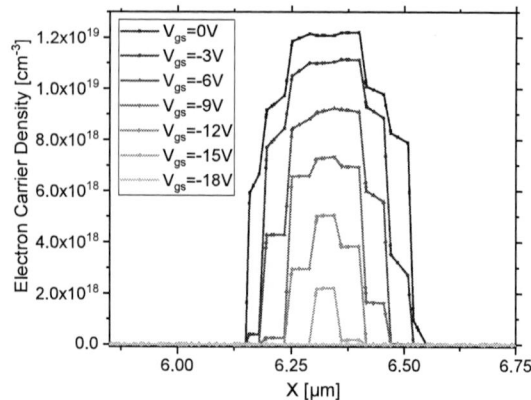

Figure 8: Electron carrier density across the pGaN island gap (W_{gap} = 0.4μm) as Vgs is varied.

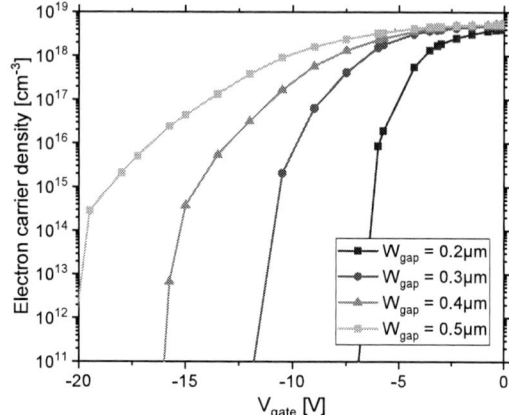

Figure 9: Electron carrier density as a function of Vgs at the mid-point of the p-GaN islands pitch

IV. START-UP CIRCUIT USING 650V P-GAN ISLAND HEMT

In this section an example of using the novel p-GaN island HEMT to provide a monolithically integrated start-up circuit is presented. Fig. 10 illustrates a common standard flyback converter which utilises a depletion mode device to charge the de-coupling capacitance of the PWM control IC. The depletion mode device powers up the IC, operating as a current source with the device being in saturation mode. Resistance, R, can be used to adjust the current level. As the de-coupling capacitor, C becomes increasingly charged, raising the voltage bias on the source of the depletion mode transistor, the gate-source voltage becomes increasingly negative. The depletion mode device gradually reaches a steady state as the source voltage approaches the device threshold voltage. The great advantage of the p-GaN island device is that the designer has a full control on the threshold voltage via layout design. The threshold voltage not only sets up the start-up rail voltage, but also determines the saturation current and hence the maximum allowable current during start-up. The p-GaN island HEMT presented in this study was used as the depletion mode device to provide this function and was monolithically integrated with the main enhancement mode power HEMT. The monolithically integrated circuit of a 650V p-GaN island HEMT and a p-GaN enhancement mode HEMT is shown in the schematic of Fig. 11. The p-GaN island HEMT and enhancement mode HEMT designs have identical drift regions and field plate designs. A current source circuit was integrated in series with the source of the p-GaN island HEMT to control the maximum current drawn from the high

978-1-6654-8498-5/22 $31.00 © 2022 IEEE

voltage rail if necessary. The current source comprises a Schottky gate depletion mode HEMT and a 2DEG resistor. The GaN IC was fabricated and packaged in an 8x8 DFN with the external pins illustrated in Fig. 11.

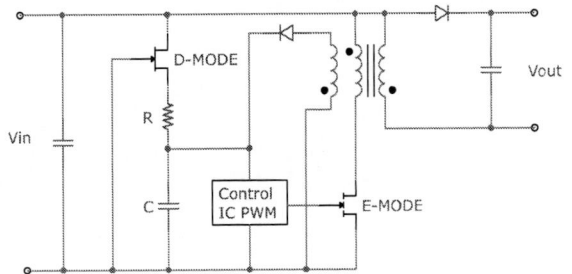

Figure 10: Flyback converter with depletion mode start-up

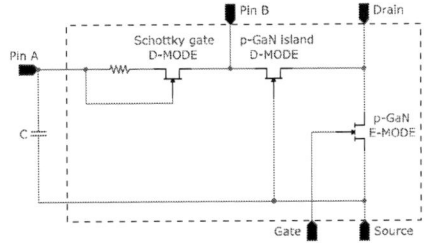

Figure 11: GaN power IC with start-up circuit monolithically integrated.

To test the functionality of the start-up circuit, 400V was applied (using an external power supply) to the drain terminal of the power IC and the different pins were probed using an oscilloscope. A 4.7uF capacitor was connected externally to terminal A. The waveforms of the Drain terminal, Pin A terminal and Pin B terminal are illustrated in Fig. 12. The current flow into the capacitor from Pin A is plotted in Fig. 13. As the drain voltage is applied to the GaN IC the voltage of Pin B rises abruptly. The voltage of Pin A which is connected to the 4.7uF capacitor and is being charged through the integrated current source rises more gradually controlled by the maximum current permitted by the current source (see Fig. 13). After a time, in the order of seconds, the voltage of Pin A reaches a steady state close to the Vth1 of the p-GaN island transistor as expected. Five devices were tested, and the steady state voltage reached is shown in Table 1. As mentioned in section II and illustrated in section III, the first threshold of the p-GaN island HEMT, and therefore the steady voltage reached by the start-up circuit, can be controlled by adjusting the spacing between the p-GaN islands. This provides an easy way to optimise the design at layout level.

V. CONCLUSION

We developed and demonstrated experimentally a novel depletion mode p-GaN HEMT with a gate which consists of a row of p-GaN islands with controlled pitch between them. It was shown that depending on the gate-source bias, the device may operate in three distinctive regions identified by significant changes in its on-resistance. By adjusting the separation of the p-GaN islands the first threshold voltage of the device can be controlled by layout design as verified by TCAD simulations. An enhancement mode power HEMT was monolithically integrated with a start-up circuit using a 650V p-GaN island HEMT. The feasibility of the design was demonstrated.

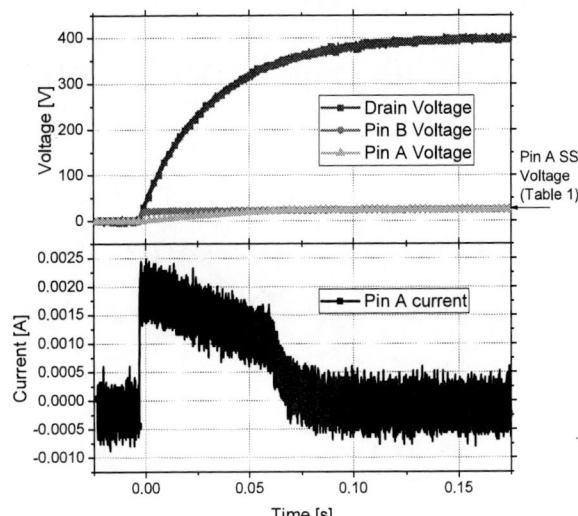

Figure 12 (top): Voltage waveforms when drain voltage is applied at 0s.
Figure 13 (bottom): Pin A current when drain voltage is applied at 0s.

	DUT1	DUT2	DUT3	DUT4	DUT5
Pin A SS voltage [V]	27.8	26.2	27.8	28.4	28.6

Table 1: Steady state voltage reached at Pin A when drain voltage is applied to the GaN power IC.

REFERENCES

[1] O. Ambacher et al., "Two-dimensional electron gases induced by spontaneous and piezoelectric polarization charges in N- and Ga-face AlGaN/GaN heterostructures," *Journal of Applied Physics,* vol. 85, no. 6, p. 3222, 1999.

[2] U. K. Mishra et al., "GaN-Based RF Power Devices and Amplifiers," *Proceedings of the IEEE,* vol. 96, no. 2, pp. 287-305, 2008.

[3] L. Efthymiou et al., "On the physical operation and optimization of the p-GaN gate in normally-off GaN HEMT devices," *Applied Physics Letters,* vol. 110, no. 12, p. 123502, 2017.

[4] B. Bakeroot et al., "Analytical Model for the Threshold Voltage of p-(Al)GaN High-Electron-Mobility Transistors," *IEEE Transactions on Electron Devices,* vol. 65, no. 1, pp. 79-86, 2018.

[5] J. He et al., "Frequency- and Temperature-Dependent Gate Reliability of Schottky-Type p-GaN Gate HEMTs," *IEEE Transactions on Electron Devices,* vol. 66, no. 8, pp. 3453-3458, 2019.

[6] A. K. Visvkarma et al., "Comparative study of Au and Ni/Au gated AlGaN/GaN high electron mobility transistors," *AIP Advances,* vol. 9, no. 12, p. 125231, 2019.

[7] F. Udrea et al., "A power semiconductor device with a double gate structure". UK Patent GB2564482B, 14 July 2017.

[8] G. Longobardi et al., "Suppression technique of vertical leakage current in GaN-on-Si power transistors," *Japanese Journal of Applied Physics,* vol. 58, SC, SCCD12, 2019.

[9] I. Hwang et al., "p-GaN Gate HEMTs With Tungsten Gate Metal for High Threshold Voltage and Low Gate Current," *IEEE Electron Device Letters,* vol. 34, no. 2, pp. 202-204, 2013.

Novel Normally-Off AlGaN/GaN-on-Si MIS-HEMT Exploiting Nanoholes Gate Structure

Mamta Pradhan*, Matthias Moser[†],
Mohammed Alomari[‡] and Joachim. N. Burghartz[§]
Neue Halbleiter Devices
Institut für Mikroelektronik Stuttgart (IMS CHIPS)
Stuttgart, Germany
{*pradhan, [†]moser, [‡]alomari, [§]burghartz}@ims-chips.de

Ingmar Kallfass
Institute of Robust Power Semiconductor Systems (ILH)
Universität Stuttgart
Stuttgart, Germany
ingmar.kallfass@ilh.uni-stuttgart.de

Abstract—A novel normally-off AlGaN/GaN staggered nanoholes MIS-HEMT concept is presented. Dielectrically isolated and metallized nanohole structures are arranged in staggered architecture to develop e-mode devices in AlGaN/GaN. Devices with two different dielectrics (SiO_2 and Al_2O_3) are compared. The e-mode HEMT with the SiO_2 gate dielectric features a threshold voltage of +0.5 V, a maximum transconductance of 166.7 mS/mm at V_{ds} = 5 V and R_{ON} of 5.86 $\Omega \cdot$mm in linear region at V_{gs} = 3 V.

Index Terms—e-mode devices, AlGaN/GaN nanohole MIS-HEMTs, gate recessed AlGaN/GaN devices, parallel nanoholes, staggered nanoholes

I. INTRODUCTION

Gallium nitride (GaN) High-Electron-Mobility-Transistor (HEMTs) are promising candidates for next generation power applications. Excellent material properties of GaN, such as wide bandgap, high breakdown field, high carrier density, and high saturation electron velocity makes it highly considerable for power and microwave applications [1]. GaN-on-Silicon substrate allows for development of commercial GaN devices on large diameter wafers [2]. The formation of 2-dimensional electron gas (2-DEG) channel at the AlGaN/GaN interface leads to a normally-on or depletion mode (d-mode) device. Since enhancement mode (e-mode) devices are required in power switching application to ensure fail-safe operation, additional steps are needed in order to deplete the highly mobile 2-DEG channel in off-state [3]. Various concepts such as additional p-GaN layer between barrier layer and gate contact [4], fluorine-based plasma treatment [5], partial or full barrier recess under the gate [6] [7] have been demonstrated. In comparison, tri-gate structures [8] and tri-gate with gate recess [9] provide better gate control and enhanced breakdown voltage.

This paper provides experimental evidence for a novel e-mode AlGaN/GaN Metal-Insulator-Semiconductor High-Electron-Mobility-Transistor (MIS-HEMT) [10]. This device features up to three rows of dielectrically isolated and metallized nanoholes (NH) underneath the gate contact, which lead to a lateral depletion of the 2-DEG in off-state. This e-mode MIS-HEMT may, thus, have a slightly higher on-resistance compared to its d-mode counterpart,

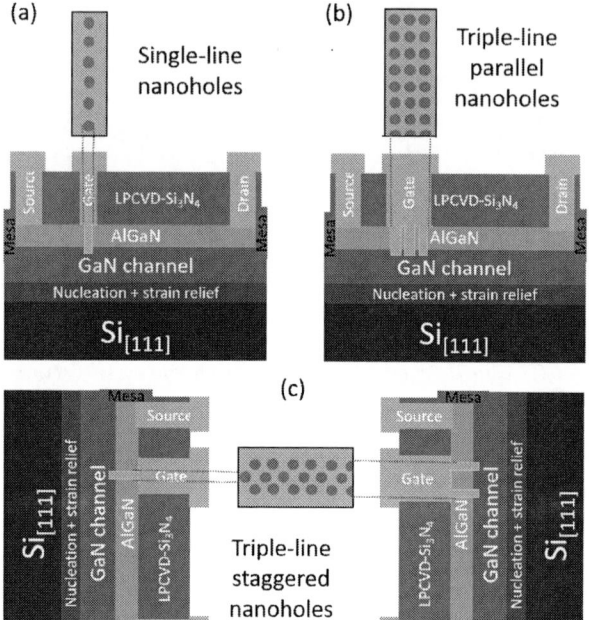

Fig. 1. Side view of the devices with top view of the gate foot (a) single-line nanoholes (b) triple-line parallel nanoholes and (c) triple-line staggered nanoholes.

i.e. an otherwise identical structure without any nanoholes underneath the gate. Therefore, the device design involves a trade-off of threshold voltage and on-resistance. Nanoholes can be arranged in a single row or in multiple rows, and such rows can be in parallel or staggered configurations (Fig. 1). In this work, three NH arrangements are compared. The impacts of different arrangements on the devices' characteristics are investigated in terms of threshold voltage (V_{th}), on-state resistance (R_{ON}), maximum transconductance ($g_{m.max}$) and off-state breakdown voltage. In addition, two different dielectrics, SiO_2 and Al_2O_3, are compared for MIS gate insulation.

978-1-6654-8498-5/22 $31.00 © 2022 IEEE

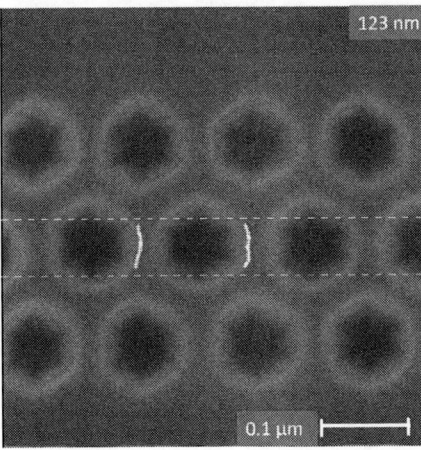

Fig. 2. SEM image of the top view of triple-line staggered nanoholes.

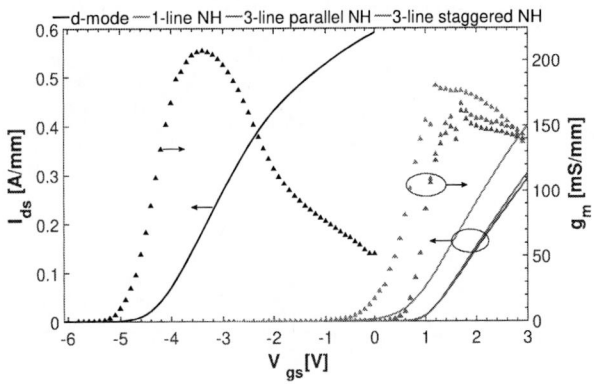

Fig. 3. Transfer characteristics at $V_{ds} = 5$ V for d-mode, single-line nanoholes, triple-line parallel and staggered nanoholes devices with SiO_2 gate dielectric.

II. DEVICE DESIGN AND PROCESSING

A. Device Design

The novel normally-off HEMT exploits a gate recess technology to deplete the 2-DEG laterally from the side walls of the nanoholes. Since the nanoholes of the gate electrode are designed to cut through the 2-DEG layer, it modifies the potential-state, and a depletion region is formed [10]. Different designs are compared experimentally: Single-line nanoholes (1-line NH) (Fig. 1a), triple-line parallel nanoholes (3-line parallel NH) (Fig. 1b) and triple-line staggered nanoholes (3-line staggered NH) (Fig. 1c). The staggered nanoholes configuration is adopted to additionally pinch-off the 2-DEG channel under the gate through obliquely spaced nanoholes. Based on initial Technology Computer Aided Design (TCAD) simulations, the nanoholes are defined with hole diameter of 100 nm and the lateral spacing between the holes is defined to be 25 nm. All the devices are designed with total device width of 50 μm, gate-source distance (L_{gs}) of 1.5 μm, gate-drain distance (L_{gd}) of 5 μm and varying gate foot lengths (L_g). The L_g is defined as 0.4 μm, 0.2 μm, 0.45 μm, and 0.42 μm for d-mode, 1-line NH, 3-line parallel NH, and 3-line staggered NH, respectively.

B. Device Processing

The devices are fabricated using CMOS-compatible GaN HEMT technology of IMS CHIPS™ [11]. They are processed on 150 mm Enkris™ GaN-on-HR-Si substrate, conceptually intended for RF applications, but featured the optimum wafer-thickness to accommodate for the e-beam lithography system's given wafer carrier. The epitaxial stack from top to bottom, consists of 2 nm GaN cap, 16 nm $Al_{0.25}Ga_{0.75}N$ barrier layer, 500 nm unintentional doped GaN and buffer layer, and 675 μm of High-Resistivity (HR) silicon. The wafer has a 2-DEG density of more than 8×10^{12} cm^{-2}. Processing starts with the definition and etching of 50 nm deep nanoholes in the

AlGaN/GaN layer using e-beam lithography. The Si_3N_4 layer is used as hardmask to etch the nanoholes. Fig. 2 shows the scanning electron microscope (SEM) image of the triple-line staggered nanoholes device (compare to Fig. 1c) after 50 nm of AlGaN/GaN etching. Next, the mesa isolation is defined using low damage chlorine based inductively coupled plasma (ICP) reactive ion etching (RIE). This is followed by low-pressure chemical vapor deposition (LPCVD), to passivate the surface with 130 nm thick Si_3N_4. The AlGaN barrier is partially recessed using atomic layer etching (ALE) for ohmic contacts. A Ti/Al metal stack is deposited and annealed at 450 °C for 10 minutes in an industrial batch-process furnace. Transmission line measurements (TLM) are done to extract the ohmic contact properties. The contact resistance and the sheet resistance is measured as 1 Ω·mm and 285 Ω/□, respectively. Then, the Si_3N_4 in the gate foot region is removed via etching, defining the gate lengths of 0.4 μm, 0.2 μm, 0.45 μm, and 0.42 μm for d-mode, 1-line NH, 3-line parallel NH, and 3-line staggered NH, respectively. The wafer is divided into quarters for applying different gate dielectric by atomic layer deposition (ALD) to be compared on the same wafer. The two quarters used in this work are separately deposited with 10 nm of SiO_2 and 10 nm of Al_2O_3, respectively. This is followed by the evaporation and structuring of Ni/Ti/Al gate metal.

III. RESULTS AND DISCUSSION

The fabricated devices are measured for the transfer and output characteristics using two Source Measure Units (SMUs) of Keithley 4200A-SCS parametric analyzer. The ALD gate dielectrics are characterized with Capacitance-Voltage (C-V) measurements to determine the relative permittivities of the insulators, which are extracted for SiO_2 and Al_2O_3 to be 4.12 and 7.10, respectively. The DC characteristics of the devices are normalized with respect to total device width of 50 μm. Fig. 3 shows the transfer characteristics of the d-mode planar MIS-HEMT device with SiO_2 gate dielectric in comparison to those of the nanoholes structures. The threshold voltage is extracted from the first derivative of the transfer

978-1-6654-8498-5/22 $31.00 © 2022 IEEE

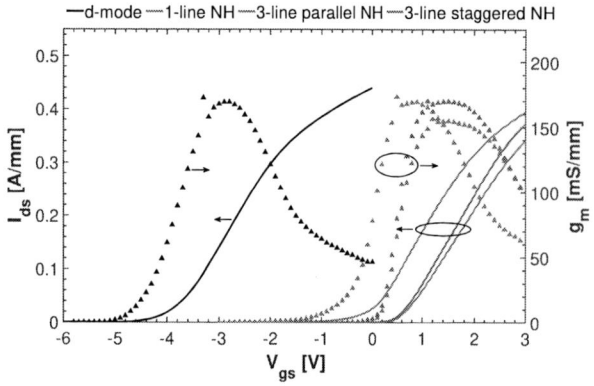

Fig. 4. Transfer characteristics at $V_{ds} = 5$ V for d-mode, single-line nanoholes, triple-line parallel and staggered nanoholes with Al_2O_3 gate dielectric.

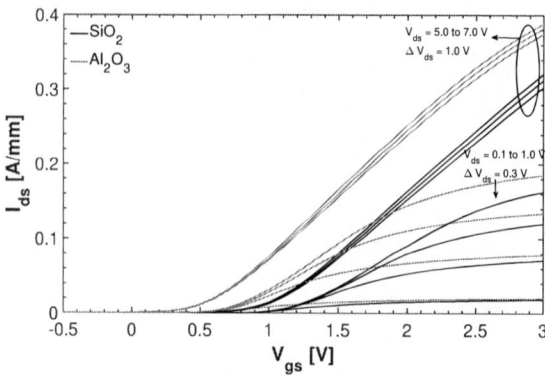

Fig. 5. Transfer characteristics of triple-line staggered nanoholes devices with gate dielectric SiO_2 and Al_2O_3.

characteristics and has a value of -5.5 V (Fig. 3) for the d-mode control device. For the 1-line NH device, the transfer curve shifts considerably towards more positive voltage, but is still shy of an e-mode device, given that the threshold voltage is -0.6 V. This might be due to insufficient depletion from the NH side walls [9]. With the 1-line NH structure, maximum depletion from the side walls is limited by the curved surface of the NH. The maximum transconductance for this device is 180.0 mS/mm. The transfer characteristics of the 3-line parallel NH, and 3-line staggered NH devices are quite similar, featuring a V_{th} of +0.5 V and $g_{m.max}$ of 161.0 mS/mm and 166.7 mS/mm, respectively. With three lines of nanoholes, the e-mode behavior is achieved due to two scenarios, (1) non-availability of the high mobility 2-DEG channel under the nanoholes and (2) depletion of the 2-DEG channel from NH side walls with gate modulation [12]. With three lines of NH, the side wall depletion appears to be more effective due to availability of depleting fields from nearby NH from all the lateral sides instead of two-sided lateral depletion in case of 1-line NH structure. Also, with the 3-line staggered NH structure, the depletion should likely be more effective, due to oblique arrangement of nanoholes, but the very small spacing of 25 nm makes both triple-line NH structures' characteristics very comparable. Relatively similar characteristics are observed for the Al_2O_3 dielectric MIS-HEMT (Fig 4), but only barely e-mode characteristics are achieved, leading to a threshold voltage of +0.1 V at best. Fig. 5 shows the transfer characteristics of staggered NH devices with SiO_2 and Al_2O_3 gate dielectrics both in linear and in saturation regimes. The threshold voltage of this device for SiO_2 and Al_2O_3 is observed to be +0.5 V and +0.1 V, respectively. The maximum transconductance at $V_{ds} = 5$ V is observed to be 166.7 mS/mm and 169.5 mS/mm, respectively. Also, the output characteristics of the 3-line staggered NH devices are presented in Fig. 6. At $V_{gs} = 3$ V, the R_{ON} for these normally-off devices is extracted to be 5.86 $\Omega \cdot$mm for SiO_2 and 5.17 $\Omega \cdot$mm for Al_2O_3. The degradation of maximum transconductance and R_{ON} as compared to the d-mode MIS-

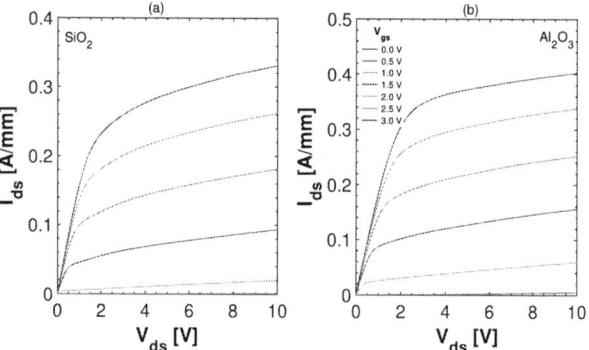

Fig. 6. Output characteristics of triple-line staggered nanoholes devices with gate dielectric (a) SiO_2 and (b) Al_2O_3.

HEMT is due to the degradation of the 2-DEG channel under the gate, caused by the NH [13]. Also, it is important to note that for NH, the effective width for the 2-DEG channel is physically reduced to the space between two consecutive NH, thus reducing the total 2-DEG high mobility conduction width to 10 μm. Therefore, in this experimental study, the effective width normalized current (not shown) for nanoholes devices is five times higher than the total width normalized current.

The two different dielectrics (SiO_2 and Al_2O_3) are further compared in Table I for all the devices in terms of V_{th}, R_{ON} and $g_{m.max}$ at $V_{ds} = 5$ V. For the d-mode device, R_{ON} is extracted at $V_{gs} = 0$ V, whereas for all the recessed devices, R_{ON} is extracted at $V_{gs} = +3$ V. In case of SiO_2 gate dielectric devices, the R_{ON} and $g_{m.max}$ degrade from d-mode to triple-line e-mode devices by an approximate value of 1.7 $\Omega \cdot$mm and 40 mS/mm, respectively. As mentioned before, this is due to degradation of 2-DEG in the recessed region [13]. For Al_2O_3 gate dielectric devices, the $g_{m.max}$ is comparable for all the device structures, except for triple-line parallel nanoholes device, where a slight dip is observed, so that R_{ON} degrades by 18.5 %. The SiO_2 is superior to Al_2O_3 in terms of higher threshold voltage for staggered NH devices without any significant degradation in

978-1-6654-8498-5/22 $31.00 © 2022 IEEE

TABLE I
EXTRACTED DEVICE PARAMETERS

Parameters	d-mode	1-line NH	3-line parallel NH	3-line staggered NH
SiO_2				
V_{th} [V]	-5.5	-0.6	+0.5	+0.5
$g_{m.max}$ [mS/mm]	207.0	180.0	161.0	166.7
R_{ON} [Ω.mm]	4.14	4.70	5.90	5.86
Al_2O_3				
V_{th} [V]	-5.0	-1.5	+0.1	+0.1
$g_{m.max}$ [mS/mm]	169.1	168.8	154.7	169.5
R_{ON} [Ω.mm]	4.54	4.52	5.38	5.17
Extracted values criteria : $g_{m.max}$ @ V_{ds} = 5 V $R_{ON_{d-mode}}$ @ V_{gs} = 0 V and $R_{ON_{NH}}$ @ V_{gs} = 3 V				

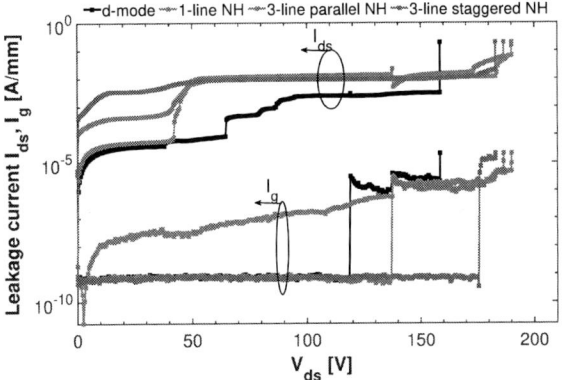

Fig. 7. Three terminal breakdown measurement in off-state for d-mode @ V_{gs} = -6 V, single-line nanoholes @ V_{gs} = -3 V, triple-line parallel and staggered nanoholes devices @ V_{gs} = 0 V with SiO_2 gate dielectric.

$g_{m.max}$ and R_{ON}.

The three terminal breakdown characteristic is measured for the SiO_2 gate dielectric devices as presented in Fig. 7. The e-mode devices such as the triple-line parallel and staggered NH devices are measured at V_{gs} = 0 V, whereas d-mode and single-line NH devices are measured in off-state at V_{gs} of -6 V and -3 V, respectively. The voltages at which the gate leakage is more than 1 μA/mm for d-mode, 1-line NH, 3-line parallel, and staggered NH devices are 118.5 V, 137.5 V, 137 V and 175.5 V respectively. Whereas, for a small gate-drain distance of 5 μm, the hard breakdown voltages of the devices are measured to be 158.0 V, 190.0 V, 186.5 V and 183.0 V respectively. The triple-line staggered NH device succeeds fairly well in both of the aforementioned breakdown voltages.

IV. CONCLUSION

A novel normally-off AlGaN/GaN MIS-HEMT using CMOS-compatible GaN HEMT technology of IMS CHIPS™ is demonstrated. Two gate dielectrics SiO_2 and Al_2O_3 are compared in terms of threshold voltage, on-state resistance

and maximum transconductance. Multiple design variations are investigated leading to the formation of an e-mode device. With gate dielectric SiO_2, the triple-line staggered nanoholes device features a threshold voltage of +0.5 V, a maximum transconductance of 166.7 mS/mm at V_{ds} = 5 V, R_{ON} of 5.86 Ω·mm in linear region at V_{gs} = 3 V, and lateral electric breakdown strength of 36.6 MV/m. Such nanohole e-mode MIS-HEMTs can easily be combined with d-mode devices to allow for GaN-only mixed-single circuit integration.

ACKNOWLEDGMENT

The authors would like to thank Dr. M. Alshahed, now with Robert Bosch GmbH, for his valuable inputs.

REFERENCES

[1] A. Fletcher et al., "A Survey of GaN HEMT for RF and High Power Applications," *Superlattices Microstruct.*, vol. 109, pp. 519–537, 2017.

[2] K. J. Chen, O. Häberlen, A. Lidow, C. l. Tsai, T. Ueda, Y. Uemoto, and Y. Wu, "GaN-on-Si Power Technology: Devices and Applications," *IEEE Transactions on Electron Devices*, vol. 64, no. 3, pp. 779–795, 2017.

[3] S. Liu, Y. Cai, G. Gu, J. Wang, C. Zeng, W. Shi, Z. Feng, H. Qin, Z. Cheng, K. J. Chen, and B. Zhang, "Enhancement-Mode Operation of Nanochannel Array (NCA) AlGaN/GaN HEMTs," *IEEE Electron Device Letters*, vol. 33, no. 3, pp. 354–356, 2012.

[4] Y. Zhou, Y. Zhong, H. Gao, S. Dai, J. He, M. Feng, Y. Zhao, Q. Sun, A. Dingsun, and H. Yang, "p-GaN Gate Enhancement-Mode HEMT Through a High Tolerance Self-Terminated Etching Process," *IEEE Journal of the Electron Devices Society*, vol. 5, no. 5, pp. 340–346, 2017.

[5] S. Jia, Y. Cai, D. Wang, B. Zhang, K. Lau, and K. Chen, "Enhancement-mode AlGaN/GaN HEMTs on silicon substrate," *IEEE Transactions on Electron Devices*, vol. 53, no. 6, pp. 1474–1477, 2006.

[6] D. Marcon, Y. N. Saripalli, and S. Decoutere, "200mm GaN-on-Si epitaxy and e-mode device technology," in *2015 IEEE International Electron Devices Meeting (IEDM)*, 2015, pp. 16.2.1–16.2.4.

[7] J. He, M. Hua, Z. Zhang, and K. J. Chen, "Performance and V_{TH} Stability in E-Mode GaN Fully Recessed MIS-FETs and Partially Recessed MIS-HEMTs With LPCVD-SiN$_x$/PECVD-SiN$_x$ Gate Dielectric Stack," *IEEE Transactions on Electron Devices*, vol. 65, no. 8, pp. 3185–3191, 2018.

[8] B. Lu, E. Matioli, and T. Palacios, "Tri-Gate Normally-Off GaN Power MISFET," *IEEE Electron Device Letters*, vol. 33, no. 3, pp. 360–362, 2012.

[9] M. Zhu, J. Ma, L. Nela, C. Erine, and E. Matioli, "High-Voltage Normally-off Recessed Tri-Gate GaN Power MOSFETs With Low on-Resistance," *IEEE Electron Device Letters*, vol. 40, no. 8, pp. 1289–1292, 2019.

[10] J. N. Burghartz, M. Alomari, and M. Alshahed, "Semiconductor element having an enhancement-type transistor structure," European Patent Office EP3707753, Mar. 2022.

[11] M. Pradhan, M. Alomari, M. Moser, D. Fahle, H. Hahn, M. Heuken, and J. N. Burghartz, "Physical Modeling of Charge Trapping Effects in GaN/Si Devices and Incorporation in the ASM-HEMT Model," *IEEE Journal of the Electron Devices Society*, vol. 9, pp. 748–755, 2021.

[12] Y.-P. Huang, C.-C. Huang, C.-S. Lee, and W.-C. Hsu, "High-Performance Normally-OFF AlGaN/GaN Fin-MISHEMT on Silicon With Low Work Function Metal-Source Contact Ledge," *IEEE Transactions on Electron Devices*, vol. 67, no. 12, pp. 5434–5440, 2020.

[13] C.-H. Wu, J.-Y. Chen, P.-C. Han, M.-W. Lee, K.-S. Yang, H.-C. Wang, P.-C. Chang, Q. H. Luc, Y.-C. Lin, C.-F. Dee, A. A. Hamzah, and E. Y. Chang, "Normally-Off Tri-Gate GaN MIS-HEMTs with 0.76 m·cm² Specific On-Resistance for Power Device Applications," *IEEE Transactions on Electron Devices*, vol. 66, no. 8, pp. 3441–3446, 2019.

ESSDERC 2022 AUTHOR INDEX

A

Abdulazhanov, Sukhrob 253
Adamopoulos, Christos 69
Adragna, Claudio 17
Affanni, Antonio 340
Agnew, Megan 185, 281
Ahn, Hong Keun 89
Ahsan, Ishtiaq 97
Albuquerque, Edgar 169
Alcotte, Reynald 261
Alian, Alireza 384
Aliane, Abdelkader 285
Alinezhad Chamazcoti, Saeideh 241
Alomari, Mohammed 400
Amir, Walid 324
Amrouch, Hussam 336
Anders, Jens 177, 217
Ando, Takashi 293
Andrieu, François 117, 121, 225, 233, 237
Angizi, Shaahin 145
Antolini, Alessio 101
Arcamone, Julien 233
Arnal, Victor M. 217
Arnold, Martin 396
Arrigo, Domenico 17
Aussenac, François 117, 225

B

Babaie, Masoud 45
Bae, Jin-Hee 113, 364
Bakeroot, Benoit 245
Baldo, Matteo 229
Balestra, Francis 257
Barajas, Carlos Chavez 189
Barbot, Justine 117
Bardon, Marie Garcia 241
Barik, Gourab 201
Baryshnikova, Marina 261
Baschirotto, Andrea 289
Bauer, Peter 312

Baumgartner, Oskar 265
Becker, Joachim 193
Bédécarrats, Thomas 37
Bedjaoui, Messaoud 117
Bendra, Mario 348
Benini, Luca 97
Bernard, Mathieu 225, 237
Bernardi, Thomas 344
Bernier, Nicolas 237
Bersano, Fabio 41
Bertrand, Benoit 37
Beugin, Virginie 233
Bhuwalka, Krishna K. 265
Bianchi, Raul-Andres 281
Blampey, Benjamin 332
Blond, Jérémy 285
Boccardi, Guillaume 261
Bol, David 149, 352
Boletti, Alessio 169
Bonizzoni, Edoardo 213
Bossuet, Alice 332
Bourgeois, Guillaume 121, 225, 233, 237
Braakman, Floris 41
Bragaglia, Valeria 297
Brändli, Matthias 49
Brebels, Steven 53
Brenneis, Andreas 65
Brew, Kevin 97
Brivio, Stefano 109
Brugger, Jürgen 153
Bu, Weihai 360
Buca, Dan 157, 364
Buchbinder, Sidney 69
Bugalho, Ricardo 169
Buj, Christel 185
Burghartz, Joachim 177, 400
Bury, Erik 384
Byun, Dong-Wook 249, 304

C

Caglar, Alican 53
Cagli, Carlo 121
Calmon, Francis 185
Campagna, Riccardo 12
Canegallo, Roberto Antonio 101
Cantatore, Eugenio 173
Capasso, Giuseppe 392
Capellini, Giovanni 157
Carabasse, Catherine 117
Cardoso Paz, Bruna 37
Carissimi, Marcella 101
Casse, Gianluigi 189
Castellani, Niccolo 225, 233, 237
Ceric, Hajdin 301
Cha, Eunjung 257
Chakraborty, Surajit 324
Chan, Mansun 308
Chan, Victor 97
Chanrion, Emmanuel 37
Charbon, Edoardo 269
Chatterjee, Baibhab 201
Chatterjee, Urmimala 245
Chen, Cheng 221
Chen, Yihan 308
Chen, Yong-Tai 73
Cherubini, Giovanni 97
Cherupally, Sai Kiran 81, 145
Chini, Alessandro 372
Choi, Sam 97
Cingu, Deepthi 245
Cioni, Marcello 372
Clerc, Raphaël 277
Coignus, Jean 117
Collaert, Nadine 261, 384
Concepción Diaz, Omar 364
Corinto, Fernando 109
Cosnier, Thibault 245
Coutier, Caroline 185
Covi, Erika 129
Craninckx, Jan 53
Cueto, Olga 121, 233
Cvitkovich, Lukas 376
Cyrille, Marie-Claire 121, 225, 233, 237

D

D'Amico, Antonio 269
D'Andragora, Alessio 189
Davy, Nil 320
De Camaret, Clement 233
De Franceschi, Silvano 37
de Milleri, Niccolò 289
De Palma, Franco 41
Decoutere, Stefaan 245
Dehaene, Wim 61
Dehos, Cedric 332
Delhaye, Thibault 352
Deng, Marina 320
Despois, Dominique 233
Di Biase, Ignacio 189
Di Marco, Mauro 109
Ding, Feilong 308
Ding, Yu-Chun 73
Divay, Alexis 332
Driussi, Francesco 340
Drost, Andreas 125
Droste, Dirk 12
Dünkel, Stefan 356
Duong, Q.T. 129, 340
Dussopt, Laurent 285

E

Eckmann, Bruno 181
Efthymiou, Loizos 396
Egger, Urs 97
El-Homsy, Victor 37
Elkashlan, Rana 384
El-Sayed, Al-Moatasem Bellah 376, 380
Ender, Johannes 348
Enz, Christian 269
Escudero, Manuel 109
Esseni, David 340, 344

F

Falcone, Donato 297
Fan, Deliang 145
Fantini, Andrea 241
Fargas Cabanillas, J.M. 69
Fiegna, Claudio 392
Fiorentini, Simone 348

Fontanini, Riccardo 340
Forti, Mauro 109
Fournol, Adrien 285
Francese, Pier Andrea 49
Franchi Scarselli, Eleonora 101
Frank, Alexander 177
Frank, Martin M. 293
Frauenrath, Marvin 364
Freye, Florian 141
Fujii, Hiroki 328
Furnémont, Arnaud 241

G

Gaddemane, Gautam 265
Galdi, Ivano 12
Garrione, Julien 225, 237
Gasparini, Leonardo 181
Gastaldi, Carlotta 153
Gaurav Kumar, K. 201
Geens, Karen 245
Gemmeke, Tobias 141
Geng, Kexing 360
Gerlach, Gerald 253
Ghibaudo, Gerard 257
Ghisu, Davide Ugo 213
Gießelmann, Timo 12
Gilhotra, Yatin 197
Glorieux, Olivier 117
Gnudi, Antonio 101
Goes, Wolfgang 348
Goh, Youngin 105
Golanski, Dominique 185, 281
Gondro, Elmar 312
Gouget, Gilles 281
Grasser, Tibor 376, 380
Grenmyr, Andreas 113
Grenouillet, Laurent 117
Gruijć, Miloš 65
Grützmacher, Detlev 113, 364
Gu, Jie 85
Guido, Roberto 297
Gunaputi Sreenivasulu, Aishwarya 45
Guo, Guoping 388
Guo, Shiyu 85
Gupta, Mohit 241
Gutheit, Tim 9
Gys, Benjamin 53

H

Halter, Mattia 297
Ham, Bumsub 89
Ham, Donhee 217
Han, Hung-Chi 269
Han, Yi 113, 364
Harpe, Pieter 173
Hartmann, Jean-Michel 364
Heim, David 49
Helleboid, Remi 185, 277
Hellings, Geert 265
Hernandez, Arturo Sibaja 384
Herrmann, Ingo 65
Hersche, Michael 97
Hinton, Henry 217
Hirtzlin, Tifenn 121
Horst, Folkert 297
Hu, Sirui 388
Huang, Chao-Tsung 73
Huang, Qianqian 360
Huang, Ru 308, 360
Huang, Zheng-Hong 245
Hung, Lee-Chi 273
Hurtado, Luis 57
Huynh, Dang Khoa 253
Hwang, Junghyeon 105

I

Ielmini, Daniele 125, 229
Illarionov, Yury 380
Innocenti, Giacomo 109
Ionescu, Adrian M. 41, 153

J

Jadot, Baptiste 37
Jang, Iksu 205
Jayant, Krishna 201
Jazaeri, Farzan 269
Jech, Markus 376
Jeon, Dongsuk 137
Jeon, Sanghun 105
Jeong, Dawon 328
Jeong, Sangsu 137
Jeong, Yeongseok 105
Jia, Tianyu 85

Jiang, Hongwu 93
Joo, Sunghwan 89
Ju, Yuhao 85
Juge, André 281
Jung, Seong-Ook 89, 105
Jung, Sung Jun 77
Junk, Yannik 364
Jutte, Peter 177

K

Kaiser, Md. Abdullah-Al 165
Kallfass, Ingmar 400
Kamaei, Sadegh 153
Kammerer, Jean-Baptiste 281
Kämpfe, Thomas 253
Kang, Jin 360
Kar, Gouri Sankar 241
Karner, Markus 273
Karunaratne, Geethan 97
Kaya, Hacile 285
Khaled, Ahmad 384
Kiene, Gerd 45
Kim, Byungsub 205
Kim, Chanho 205
Kim, Dae Sin 328
Kim, Dohyung 89
Kim, Minki 105
Kim, Min-Yeong 249, 304
Kim, Myoungsoo 328
Kim, Soosung 77
Kim, Tae-Woo 324
Kim, Woojin 241
Klemt, Bernhard 37
Kneip, Adrian 149
Knobloch, Theresia 380
Ko, Dong Han 105
Ko, Jaehyun 205
Kodandarama, Komal Vondkar 261
Koo, Sang-Mo 249, 304
Kootte, Bart 177
Kossel, Marcel 49
Kramnik, Danielius 69
Kretzschmar, Claudia 269
Krottenthaler, Peter 356
Krüger, Daniel 217
Kumar, Prem 69

Kunert, Bernardette 261
Kwon, Sangwoo 77
Kwon, Uihui 328

L

La Porta, Antonio 297
Lacerda de Orio, Roberto 301, 348
Laguna, Camille 225
Lallement, Christophe 281
Lama, Giusy 237
Lancaster, Suzanne 129, 340
Langenegger, Jovin 97
Langer, Robert 261
Lanius, Christian 141
Laurin, Luca 229
Le Brun, Grégoire 352
Le Gallo, Manuel 97
Le, Quang Huy 253
Lee, Geon-Hee 304
Lee, Hee-Jae 304
Lee, Hyunseung 77
Lee, Jae W. 77
Lee, Junghyup 89
Lee, Ki-Beom 89
Lee, Sooeun 205
Lee, Sumin 89
Lee, Young Kyu 105
Lefebvre, Martin 149
Le-Friec, Yannick 233
Lehmann, Steffen 269
Lehninger, David 253
Lei, Ka-Meng 217
Li, Hanyue 173
Li, Ning 97
Li, Wantong 93
Li, Xi 308
Li, Yiqing 360
Li, Yongfu 221
Lico, Andrea 101
Lim, Sehee 105
Lin, Chun-Yeh 73
Lin, Kai-Pin 73
Ling, Bill 165
Liu, Xia 153
Liu, Yaxin 221
Liu, Ying 360

Liu, Zhenhong 221
Liu, Zixuan 85
Lizzit, Daniel 340, 344
Longobardi, Giorgia 396
Lou, Jie 141
Luo, Chao 388
Luo, Jin 360

M

Ma, Zhouchen 221
Malcovati, Piero 213
Mamdy, Bastien 185, 281
Maneux, Cristell 320
Manna, Bibhas 380
Mannaert, Geert 261
Manouvrier, Jean-Robert 281
Marano, Vincenzo 17
Martin, Simon 225, 233
Martineau, Baudouin 332
Massari, Nicola 65, 189
Massarotto, Marco 340
Masud, Mohammad Ayaz 57
Matagne, Philippe 265
Mazzanti, Andrea 213
Meilhan, Jérôme 285
Melde, Thomas 356
Meli, Valentina 121, 225, 233
Meng, Jian 81
Menolfi, Christian 49
Meunier, Tristan 37
Mika, Nicki 356
Mikolajick, Thomas 129, 340, 356
Militaru, Liviu 225
Milozzi, Alessandro 125
Min, Seongsik 328
Mismer, Colin 320
Moldovan, Clara 153
Mols, Y. 261
Moon, Changjae 205
Moon, Soo-Young 249
Morandini, Yvan 332
Moreno García, Manuel 181
Morf, Thomas 49
Mortemousque, Pierre-André 37
Moser, Matthias 400
Motta, Alessandro 229

Mueller, Peter 49
Mukherjee, Chhandak 320
Myny, Kris 61

N

Narayanan, Vijay 97
Navarro, Gabriele 225, 233, 237
Neuner, Thomas 125
Ni, Kai 336
Nicholson, Isobel 185
Niebojewski, Heimanu 37
Niegemann, David 37
Niknejad, Tahereh 169
Nodjiadjim, Virginie 320
Nolot, Emmanuel 225
Novaresi, Lara 213

O

Ocola, Leonidas E. 293
Oehme, Michael 157, 161
Offrein, Bert Jan 297
Ok, Injo 97
Oliveira, Luis 169
Onaran, Güçlü 289
Oppliger, Fabian 41
Oropallo, Maria Vittoria 49
Ortmanns, Maurits 193
Oshinubi, Dayo 65
Ottaviani, Stefano 213
Otto, Michael 356
Overwater, Ramon W.J. 45

P

Park, Jeongwoo 137
Park, Jihoon 205
Parmesan, Luca 65
Parvais, Bertrand 53, 261, 384
Pasotti, Marco 101
Pavan, Paolo 316, 368
Pelgrims, Jonas 61
Pellegrini, Sara 185, 281
Pelzers, Kevin 173
Peralagu, Uthayasankaran 384
Perenzoni, Matteo 65, 181, 189
Perilli, Luca 101

Perruchot, François 37
Perumkunnil, Manu 241
Petroni, Elisa 229
Pettingell, John 189
Piazza, Gianluca 57
Pip, Alex 352
Pirson, Thibault 352
Poggio, Martino 41
Pollmann, Eric H. 197
Popoff, Youri 297
Popović, M.A. 69
Potocnik, Anton 53
Pozzi, Rachela 17
Pradhan, Mamta 400
Prathapan, Mridula 49
Prawoto, Clarissa Cyrilla 308
Puglisi, Francesco Maria 316, 368
Pulicelli, Fulvio 17
Pulvirenti, Francesco 17
Putcha, Vamsi 384

Q

Qi, Liang 221
Quaglia, Fabio 213
Quenette, Vincent 281

R

Radu, Ionut 41
Rae, Bruce 281
Rahimi, Abbas 97
Ramesh, Anirudh 69
Raskin, Jean-Pierre 352
Read, James 93
Redaelli, Andrea 229
Reich, Stefan 193
Reiser, Daniel 125
Ren, Ye 360
Renaudier, Jérémie 320
Rideau, Denis 185, 277, 281
Rieger, Stefan B. 193
Riel, Heike 25
Riet, Muriel 320
Rink, Sven 281
Rochat, Névine 225
Rochus, Veronique 209
Rodriguez, Raul 384

Rouchon, Denis 225
Roy, David 185
Ruffino, Andrea 49
Rzepa, Gerhard 265

S

Saeidi, Ali 153
Saikia, Jyotishman 81
Saint-Pierre, Dorian 277
Santermans, Sybren 265
Saulnier, Nicole 97
Savoia, Alessandro Stuart 213
Sawan, Mohamad 221
Scarlino, Pasquale 41
Schäfer, Sören 157
Schleipen, Jean 177
Schulze, Jörg 157, 161
Schüttler, Martin 193
Schwarz, Daniel 157, 161
Schweitz, Michael A. 249, 304
Sebastian, Abu 97
Sebastiano, Fabio 45
Segatto, Mattia 340
Seidel, Lukas 157, 161
Selberherr, Siegfried 301, 348
Selijak, Andrej 189
Sen, Shreyas 201
Seo, Jae-Sun 81, 145
Seo, Seong Hoon 77
Serra Di Santa Maria, Francesco 257
Shen, Yuting 173
Shepard, Kenneth L. 197
Shin, Chan-Soo 324
Shin, Hunbeom 105
Shin, Ju-Won 324
Shin, Ki-Yong 324
Shin, Myeong-Cheol 249
Sicre, Mathieu 185
Sideris, Constantine 165
Silva, Jose 169
Silvestre, Claire 97
Slesazeck, Stefan 129, 340
Song, Peilin 293
Song, Yi-Qiao 217
Song, Zhitang 308
Souifi, Abdelkader 225

Sousa, Marilyne 297
Sousa, Veronique 121
Spieth, Christian 161
Spiga, Sabina 109
Sporer, Markus 193
Sridharan, Amitesh 145
Stadtmann, Tim 141
Stanojević, Zlatan 273
Stecconi, Tommaso 297
Stefanov, André 181
Stellari, Franco 293
Stojanović, Vladimir 69
Strohm, Thomas 65
Suh, Han-Sok 81
Sun, Haiding 388
Sun, Jui-Hung 165
Sun, Nan 133
Sun, Yue 388
Sverdlov, Viktor 348
Symanzik, Horst 12
Syshchyk, Olga 245

T

Tabruyn, Dries 209
Tallarico, Andrea Natale 392
Tang, Yiqiao 217
Taylor, Alan 189
Taylor, Jon 189
Terenzi, Marco 213
Tesi, Alberto 109
Theodorou, Christoforos 257
Thiney, Vivien 37
Thomann, Simon 336
Tonietto, Davide 1
Tontini, Alessandro 65
Tornow, Marc 125
Torres, Mattia Luigi 101
Trabelsi, Ahmed 121
Triozon, Francois 117
Tsai, Chen-Ming 273
Turconi, Lorenzo 229

U

Udrea, Florin 396
Uhring, Wilfried 281
Ulm, Markus 12
Urdampilleta, Matias 37

V

Vais, Abhitosh 261
Van de Put, Maarten 265
van der Struijk, Mariska 173
Van Helleputte, Nick 209
Van Winckel, Steven 53
Varela, João 169
Vaxelaire, Nicolas 117
Vecchi, Sara 368
Veliadis, Victor 31
Venkataramanaiah, Shreyas Kolala 81
Verbauwhede, Ingrid 65
Verecken, Julien 149
Vermeersch, Bjorn 261, 384
Vianello, Elisa 121
Vinet, Maud 37
Vohra, Anurag 245
Vudumula, Pavan 245

W

Waldhoer, Dominic 376, 380
Waltl, Michael 376
Wambacq, Piet 53, 261
Wang, Guoxing 221
Wang, Huan-Ching 73
Wang, Imbert 69
Wang, Kaifeng 360
Wang, Li-Wei 73
Wang, Runsheng 308
Wang, Zhixuan 360
Wanitzek, Maurice 161
Wei, Renjie 360
Weisbuch, Fancois 356
Wellekens, Dirk 245
Wen, Mingjie 388
Weng, Chi-Wen 73
Wiesbauer, Andreas 289
Wilhelmer, Christoph 376
Willemen, Joost 312
Wong, Oi-Ying 209
Wu, Ernest Y. 293
Wu, Hao 265
Wu, Tian-Li 245
Wu, Yan 221
Wu, Yongqin 360

X

Xi, Fengben 113
Xiang, Zikun 388
Xiao, Shulan 201
Xue, Qiwen 388

Y

Yadav, Sachin 261, 384
Yang, Lei 388
Yang, Libo 360
Yang, Mengxuan 360
Yang, Xiangxing 133
Ye, Le 360
Yeo, Injune 81
Yin, Heyu 197
Yoo, Jaehyun 328
Yu, Hao 384
Yu, Shimeng 93

Z

Zagni, Nicolò 372
Zanotti, Tommaso 316
Zanuccoli, Mauro 392
Zarghami, Majid 181
Zarkos, Panagiotis 69
Zeng, Bolun 388
Zhang, Aoyang 217
Zhang, Fan 145
Zhang, Fangxing 360
Zhang, Haochen 388
Zhang, Jiayuan 113
Zhang, Lining 308, 360
Zhang, Yichi 81
Zhang, Yuanke 388
Zhang, Zhiru 81
Zhao, Jian 221
Zhao, Ming 384
Zhao, Qing-Tai 113, 364
Zhao, Zhixing 269
Zheng, Kai 360
Zhou, Linfeng 221
Zota, Cezar 49
Zota, Cezar B. 257

IEEE
445 Hoes Lane
Piscataway, NJ 08854-4141

ISBN 978-1-6654-8498-5